Applied
FIBRE SCIENCE

Applied
FIBRE SCIENCE

Volume 2

Edited by

F. HAPPEY

School of Postgraduate Studies in Textile Technology,
University of Bradford, Bradford, Yorkshire, England

1979

ACADEMIC PRESS

LONDON   NEW YORK   SAN FRANCISCO

A Subsidiary of Harcourt Brace Jovanovich, Publishers

ACADEMIC PRESS INC. (LONDON) LTD.
24/28 Oval Road.
London NW1

United States Edition published by
ACADEMIC PRESS INC.
111 Fifth Avenue
New York, New York 10003

Library of Congress Catalog Number 77 75367
ISBN: 0 12 323702 5

Printed in Great Britain by
J. W. ARROWSMITH LTD, BRISTOL BS3 2NT

Contributors

ALGIE, J. E., C.S.I.R.O., Division of Food Research, North Ryde, N.S.W. Australia.

ASQUITH, R. S. Department of Industrial Chemistry, Queen's University of Belfast, Belfast, Northern Ireland.

FRITZSCHE, PETER, Institut fur Polymerenchemie der Akademie der Wissenschaften dder D.D.R., Teltow-Seehof, G.D.R.

JENKINS, A. D., School of Molecular Sciences, University of Sussex, Brighton, Sussex, England.

KUDLÁČEK, L. Department of Fibrous Materials, University of Chemical Technology, Pardubice, Czechoslovakia.

LORIMER, J. W. Department of Chemistry, University of Western Ontario, Ontario, Canada.

MARTIN, J. RONALD, Textile Research Institute, Princeton, New Jersey, U.S.A.

MILLER, BERNARD, Textile Research Institute, Princeton, New Jersey, U.S.A.

OTTERBURN, M. S., Department of Industrial Chemistry, Queen's University of Belfast, Belfast, Northern Ireland.

PETERS, R. H., University of Manchester Institute of Science and Technology, Manchester, England.

PHILIPP, BURKART, Institut fur Polymerenchemie der Akademie der Wissenschaften der D.D.R., Teltow-Seehof, G.D.R.

REBENFELD, LUDWIG, Textile Research Institute, Princeton, New Jersey, U.S.A.

ROWLAND, STANLEY P., Southern Regional Research Center, Agricultural Research Service, U.S. Department of Agriculture, New Orleans, Louisiana, U.S.A.

STILL, R. H., University of Manchester Institute of Science and Technology, Manchester, England.

TRELOAR, L. R. G., University of Manchester Institute of Science and Technology, Manchester, England.

WOOD, EDWARD J., Department of Biochemistry, University of Leeds, Hyde Terrace, Leeds, England.

Foreword

Fibers, as the prime and essential basic raw materials for such important industries as paper, textiles, ropes, nets and cords and as reinforcing elements for many composites, have always been preferred objects for descriptive and analytical scientific efforts. Already at the beginning of this century we find an extensive literature on wood, cotton, wool, hair, silk and bast fibers. In many cases the chemical composition of the fiberforming substance—polysaccharide or polypeptide—was approximately known and as a result of the microscopy of those days there also existed elaborate and sometimes almost artistic descriptions of the amazing morphology of such natural miracles as a cotton or a wool hair.

In between there was a vacuum; many essential questions had no answer.

How are the basic units—glucose, pentose, amino acid—connected with each other? What are the structural elements between 5 and 5,000 ÅU? Are there any? If so, how do they build up to the visible morphological elements: the fibrils, membranes, scales and ligaments? What provides the high modulus and high tensile strength together with considerable extensibility and some rapid and some delayed recovery? What is the mechanism of the delicate moisture metabolism and the remarkable resistance against the action of organic solvents and of chemical agents? How can fibers possess the high softening ranges they have and still be tough and flexible at very low temperatures—at that time some −80°C? It took about two decades—from 1900 to 1920—until the basic facts for the answers to these questions were established, mainly with the aid of new experimental techniques which will all be duly described in these volumes and almost two more decades—from 1920 to 1940—until the systematic application of these principles led to one of the most spectacular breakthroughs of contemporary chemistry—the science and technology of man-made fibers.

Most fundamental aspects were learned from nature but superimposed on them was the endless variability of polymer synthesis and the controlled response of macromolecules to rheological processing. As a result there emerged the eminently useful concept of a "fiber former" being a linear polymer of controlled molecular weight and molecular weight distribution; having the capacity to "crystalize" under simple rheological conditions; softening and melting within a convenient temperature range—190° to 280°C—or being soluble in an available but not too common solvent such as DMF, DMAcTA, etc.; exhibiting permanent resistance against normal solvents including dry cleaning agents and, if possible, accepting various dyestuffs under convenient conditions and being not too susceptible to static charges. In spite of these numerous and partly conflicting demands more than a hundred "fiber formers" were prepared and tested over the last two or three decades and covered an enormous range and variety of properties and property combinations. But with added restrictions of economics, ecology, comfort and safety only a few of the major families of polymers have emerged as successful contenders. They are polyesters, polyamides, polyacrylics, polypropylene and certain composite polyvinyls.

At present, however, these ought to be considered just to be the *first generation* of man-made fibers which has succeeded in developing a considerable acceptance appeal and a correspondingly large market but which all need *substantial improvements* technically and economically to establish a long lasting favorable relationship between producers and consumers.

Let us, therefore, briefly attempt to speculate on the *second generation*; how they will be classified, market oriented and produced.

Up to date it was common practice to classify man-made fibers according to their chemical composition such as polyamides, polyesters, etc. It would seem, however, to be more progressive to consider them according to their end use and speak of: comfort fibers; safety fibers; and structural fibers.

Comfort fibers are those which we *wear* and which come in direct, or almost direct *contact with our body*: underwear, shirts, skirts, pants, dresses, suits, coats, etc. Most existing fibers—natural and synthetic—are useful for these but none of them is perfect, not even for a single individual use. To be satisfactory comfort fibers, which are usually in the range between 1.5 and 5.0 denier, should have moduli in the range between 30 and 80 g/den, tensile strength of between 3 and 6 g/den, and elongation to break from 20 to 40% with elastic recoveries around 80% from a 3–5% strain. They also should withstand temperatures of up to 180°C without softening or shrinking, should be resistant to dry cleaning agents and should be able to be repeatedly laundered without damage. But on top of all of this they should be voluminous, should have a soft and pleasant touch and a moisture metabolism which renders them antistatic, antisoiling and antiodorant, and they should be readily dyeable and printable under standard conditions. Finally, and very importantly, the fiber should be reasonably inexpensive. Right now it is not very probable that a novel polymer or family of polymers will provide a *new fiber former* which will be substantially superior to all existing ones from the point of view of manufacturing, applications and cost. Consequently most present efforts are directed towards incorporate *improvements* of existing compositions either by processing devices such as appropriate cross-sections, multicomponent filament spinning or by chemical modification of the base polymer. Important progress has already been made but even the most advanced synthetics of the polyester (Kodell) and Polyamide (Cantrece, Qiana) type are still not yet satisfying all demands. Evidently, if one wants to modify and improve an existing fiber in certain desired directions one must know as much as possible about its molecular and supermolecular structure. It has become clear in the past that no single experimental method is capable of providing this information and it is, therefore, the purpose of this and subsequent Volumes to enumerate and describe in considerable detail all presently existing techniques. Each of them approaches the problem from a different angle and with the aid of a different fundamental phenomenon—scattering, absorption, diffusion, sedimentation, conduction, convection, etc. Altogether they represent much more than "Applied Fibre Science"; they are *fundamental polymer science* in the broadest sense and at the highest existing level. A galaxy of distinguished experts has been assembled to tackle this problem and a review of each single chapter confirms the prevalence of excellence.

Safety fibers are those which *surround* us, wherever we are—at home, at work, in school, on travel and during vacation. They are needed to make bedding, upholstery, carpets, curtains, wall covers, tents and other similar items.

These fibers should permit designing and styling, should be easy to clean and maintain, should be durable mechanically and chemically and, above all, should resist ignition or, if ignitable should be self-extinguishing or at least have a slow flame spread and a low heat release. Finally, if caught in a conflagration they should not produce smoke and should not give off toxic fumes. On top of all of that, they should be readily processible by existing textile techniques and the end product should be available at reasonable prices.

As in the case of the comfort fibers no *existing* fiber former is capable of meeting *all* the demands enumerated above but several of them are already satisfactory in certain respects and there is every reason to believe that modifications in composition and structure, together with the incorporation of appropriate ingredients, will provide a sound basis for the design and manufacture of superior products. Impressive progress has been made in the area by the introduction of new fiber formers belonging to the families of aromatic polyester, polyamides and related macromolecules which, however, are still too expensive to capture the larger markets of the safety fibres. Again, however, any rapid and successful progress is going to depend on a good and reliable knowledge concerning the molecular and supermolecular characteristics of the best existing representatives of this class, and the present work should be as useful here as for the design of better comfort fibers.

Structural fibers. Since cotton cord was used in automobile and bicycle tires some seventy years ago, fibers have served as *reinforcing elements* in thermoplastic and thermosetting systems. The prevailing demands for this application are high modulus and tensile strength, low extensibility and excellent bonding to the surrounding matrix; sometimes there have to be added high softening point and chemical resistance against such influences as hydrolysis and oxidation. To date, progress in this field has been achieved by chemical change of the fiber former—first from cellulose (cotton, rayon) to aliphatic polyamides (66-nylon), then to standard polyester (PET) and finally to aromatic polyamides (Kevlar). Several individual members of the aromatic compositions still have to be further explored and tested but it appears that after accomplishment of these studies the possibilities for chemical innovations are substantially enhanced. As a consequence, additional progress might have to depend on improvements of the supermolecular characteristics of the structural fibers which, again, falls in the domain of the methods which are described in the book.

These methods also should be particularly useful and important for the structural investigation of the most sophisticated reinforcing filaments namely the *carbon* or *graphite fibers* which display most unusual characteristics and are still in their infancy concerning the best precursors, the most efficient method of preparation and the exact knowledge of their solid state structure.

Polytechnic Institute of New York   H. MARK
Brooklyn, New York

Preface

In the specialist chapters of Volume I an emphasis was given to the development of particular physical techniques and their application in many specific polymer and fibre studies was expanded. Volume 2 continues this work, but an effort has been made to broaden the interdisciplinary fields of study to accent the advances obtained in the technology of polymers in general and of fibrous materials in particular.

In its early stages fibre science was confined to the materials available in the 1920s and 1930s, but the fundamental findings on these fibres provided a basis for the production of man-made fibres and their introduction into the textile industry. Volume 2 shows the further development of this process and also that with the introduction of fibrous model structures many of the results obtained could be applied in the refinement of the structural details of the original natural polymers used. This increased the technological knowledge which could be applied to fibre study, and many fibrous properties which had been matters of purely empirical knowledge became understandable. For the first time the craft textile industries became capable of explanation in scientific terms. Throughout this volume the various techniques are expanded in detail in the hope of emphasising the strong interdisciplinary nature of polymer and fibre science and the dependence of textile development on a scientific basis in recent years. Once again I would express my thanks to all my friends at home and abroad who have collaborated in the production of this volume and with whom I have shared in the development of fibre science over the years. Finally it is a pleasure to acknowledge my former colleagues in the University of Bradford whose cooperation I have enjoyed for over a quarter of a century.

F. HAPPEY
Emeritus Professor of Textile Industries.

University of Bradford
Bradford, England
and
Whirlow, Lon Refail
Llanfairp. g.
Isle of Anglesey.

Contents

Acknowledgements

My thanks are due to the Leverhulme Research Trust for the award of an Emeritus Fellowship to assist in the completion of this work. Also I should like to thank my wife for help with the manuscript and proofs. Acknowledgement is due to Mrs G. Marshall for secretarial assistance, proof-reading and indexing, and to Mrs J. Burdon for help with the index.

Dedication

To my parents, wife and family.

1. Analytical Ultracentrifuge Techniques[*]

EDWARD J. WOOD

Department of Biochemistry, University of Leeds, Hyde Terrace, Leeds, England

I. Introduction

An ultracentrifuge is an instrument in which a small sample of liquid can be subjected to very high angular velocities, and ultracentrifuges have been used as analytical and preparative tools in polymer science, and especially protein chemistry, for nearly half a century. The rate at which particles in a fluid move under the influence of a centrifugal field depends upon a number of factors including the force applied, the size, shape, and density of the particles, and the density and viscosity of the fluid. These factors are exploited in preparative instruments to effect the separation of particles of different sizes or densities. On the other hand, by observing the behaviour of particles in an analytical ultracentrifuge by the use of some suitable optical arrangement, one can obtain information on the sizes and shapes of the particles as well as on heterogeneity and particle size distributions. This chapter will be concerned chiefly with the analytical ultracentrifuge.

Ultracentrifuge techniques have been particularly useful to biological chemists because most soluble proteins represent a unique type of polymer in that they are usually highly monodisperse. Furthermore solutions of globular proteins in dilute aqueous buffers frequently show almost ideal behaviour. Thus the analytical ultracentrifuge has been used a great deal to determine molecular weights of proteins. One consequence of this is that the theory and methods for dealing with monodisperse solutes in simple solvent systems are highly developed. Much recent work has been concerned with the behaviour of interacting particles in the analytical ultracentrifuge and here again it is protein systems which have received

---

[*] *Glossary of Symbols: B*: second virial coefficient; c_0: initial concentration; $f(M)$: differential molecular weight distribution function; $g(s)$: differential distribution of sedimentation coefficient; M: molecular weight (of a homogeneous solute); M_{app}: apparent molecular weight; M_n: number-average molecular weight; M_w: weight average molecular weight; M_z: z-average molecular weight; r_a: radius of the meniscus; r_b: radius to bottom of liquid column; s: sedimentation coefficient; $\bar{v}$: partial specific volume; y: activity coefficient of solute; ρ: density of solution; ω: rotor speed (angular velocity, radians/s).

most attention because their ability to interact is one of their most important features.

Although biological molecules such as enzyme proteins show behaviour in dilute aqueous buffer solutions not far removed from the ideal, this is rarely true with most synthetic high polymers. It is important therefore that as far as possible experiments with such materials are performed under theta conditions (Flory, 1953; Williams and Yphantis, 1971; J. W. Williams, 1972). However the discovery of such conditions is often very time-consuming, and the conditions themselves frequently involve the use of high temperatures and "difficult" solvents, leading to technical problems not normally encountered by biological chemists working with proteins. Details of the methods for finding theta conditions are not within the scope of the present chapter but reference to two recent articles will illustrate the problems involved and how they may be solved. Nakazawa and Hermans (1971) studied a styrene-methylacrylate co-polymer by density gradient ultracentrifugation, but first had to find a theta solvent for the material. This was particularly time-consuming and they eventually used a mixture (by volume) of 30% methylethyl ketone, 13% 1,2-dibromo-1,1-difluoroethane, 29·4% isopropanol, and 27·6% dichlorooctafluorocyclohexane-1. They established by osmometry that the second virial coefficient was zero at 26·5°C (the Flory temperature) in this solvent. Similarly, with aqueous solvent systems, Cerny et al. (1973) describe the determination of theta conditions for samples of dextran and poly(vinyl pyrolidone), and eventually chose a mixture of 40·5% methanol and 59·5% water (by volume) at 30°C.

It should be mentioned that the theory of sedimentation behaviour has only recently attempted to deal with compressibility effects. Some solutes, and certain of the solvents and solvent mixtures that have been used to attain theta conditions, have relatively high compressibilities and it is necessary to take account of such effects either by minimizing them by appropriate experimental design or by making the appropriate corrections in the interpretation of results (Weiss and Yphantis, 1972).

In what follows, recent developments in analytical centrifugation have been considered under three main headings: instrumentation and techniques, sedimentation velocity, and sedimentation equilibrium. A brief section on density gradient centrifugation follows these. The field is large and in many instances limited space has decreed that rather than attempting to present an exhaustive survey, the author has had to choose a few examples to illustrate potentialities and developments in a particular area. Those wishing to follow the subject in greater detail will find an abundance of excellent books and reviews (Creeth and Pain, 1967; Chervenka, 1969; Coates, 1970; Bowen, 1971; Williams and Yphantis, 1971; J. W. Williams, 1972; Lloyd, 1974).

II. Instrumentation and Methodology

A. GENERAL

A number of analytical ultracentrifuges are on the market at the present time and details of their construction and performance have been given in several reviews (Bowen, 1966, 1971; Coates, 1970). It should be remembered that manufacturers are constantly redesigning and improving instruments, so that the manufacturer's literature is the best source of technical details of the performance of a particular machine.

In all instruments the samples, often of less than 1 ml volume, are held in small cells with quartz or sapphire windows. The cells are placed in an aluminium or titanium rotor along with appropriate counterbalances and the rotor is spun in a vacuum (Fig. 1 shows a typical cell, and Fig. 2 shows a typical alignment of rotor and optical system). The power to drive the rotor is usually supplied by an electric motor, sometimes via a gear box, through

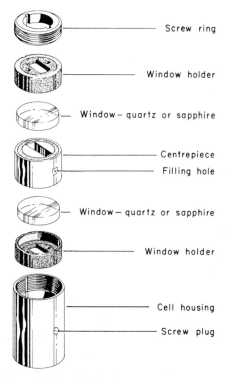

FIG. 1.   Basic parts of a standard analytical cell. The centrepiece shown holds approx. 0·7 ml of sample and the light path is 12 mm. Other centrepieces are available holding approx. 0·3 and 1·8 ml (4 mm and 30 mm light path, respectively). Figure by courtesy of Beckman-RIIC Ltd.

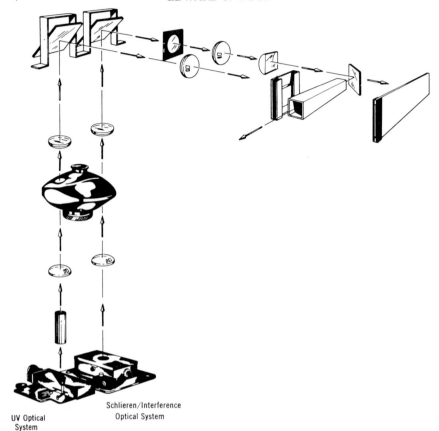

UV Optical
System

Schlieren/Interference
Optical System

FIG. 2. Arrangement of light sources, rotor, and optical systems in a Beckman model E analytical ultracentrifuge. Figure by courtesy of Beckman-RIIC Ltd.

a flexible drive to reduce vibrations. The evacuated chamber in which the rotor spins is armoured in case of accident, and contains refrigeration and heating coils for temperature regulation. Temperature measurement and control may be achieved by a number of methods such as by means of a thermistor set in the rotor which is connected to a bridge circuit, or by an infra-red sensor set near the rotor. Most instruments are capable of 60,000 rev/min (equivalent to about 300,000 g in the cell) and operate over the temperature range 0–30°C. This is adequate for most biochemical applications, but for applications in polymer science higher temperatures are frequently called for. Usually manufacturers offer a high temperature kit as an accessory to the basic instrument, allowing temperatures up to 130°C to be used. In fact, certain proteins have been studied at elevated temperatures and the technical problems of running an analytical ultracentrifuge at such temperatures have been discussed by Neet and

Putnam (1965). The maximum speed at which the centrifuge will run is rarely the limiting factor in experimental work. Rather, the features that make an instrument attractive are accurate speed control and stability, especially at low speed, accurate temperature control and measurement, and ease of operation in such things as the changing of optical systems in mid-run.

Apart from the rotor and its associated drive mechanism, the other major part of an analytical ultracentrifuge is the optical system including light sources and cameras by means of which the distribution of solutes under the influence of the centrifugal field is investigated. All the instruments on the market at the present time have some form of schlieren and Rayleigh interference optical systems, but an absorption optical system is usually also available. It is in this latter field that many recent developments have taken place. For details of the construction and alignment of the schlieren and interference optical systems reference may be made to Bowen (1971), Coates (1970), Haschemeyer and Haschemeyer (1973), and, especially, Lloyd (1974), as well as manufacturers instruction manuals.

It is clear that an analytical ultracentrifuge system involving high-performance materials, precision components, high quality optical systems, and considerable amounts of electronic circuitry, is a very expensive item of laboratory equipment. It is not surprising therefore that there has been a trend amongst manufacturers to try to produce cheaper, if perhaps slightly less precise and flexible, instruments. Designs available range from a simple attachment to an existing preparative instrument involving the addition of an optical system, to the redesign of what was basically a preparative machine to incorporate a new drive and optical system. Table I lists some of these newer designs. Several optical systems are usually available on such instruments, and while for a variety of reasons such instruments might not be expected to produce the highly accurate data expected of a full-scale machine, some of them appear to be surprisingly good, no doubt partly because they have the advantage of being fitted with modern electronic control and measuring circuitry. Potential purchasers should therefore consider carefully which machine would be most appropriate for their purposes in view of the great difference in price between these and full-scale instruments.

B. Use of Laser Light Sources

As investigators try to extract more and more information from analytical ultracentrifuge data it is inevitable that they should continue to try to improve the performance of the optical systems themselves. This may involve radical redesign of the system, but more often an improvement can

TABLE I

Some newer, smaller analytical ultracentrifuges

Manufacturer and name of instrument	Comments
Beckman-RIIC Ltd. Schlieren optics and UV scanner accessories attached to model L5 preparative ultracentrifuge	Simple, removable accessories, requires different rotor to normal analytical or preparative instrument. Schlieren pictures on Polaroid® film. Double beam UV Scanner output on X-Y recorder: multiplexer allows use of several cells in same rotor.
MSE Instruments Ltd., Centriscan®	Radical redesign of a preparative instrument. Triple-phase motor controlled by AC generator (no gearbox). Output: X-Y recorder, modified schlieren or scanning-absorption optics. Can also be used preparatively. Multiplexer.
Hungarian Optical Works, MOM 3170	Completely redesigned for compactness. Drive: timing belt and hydraulic shaft clutch. Pictures on 120-size film: schlieren and interference optics standard, UV-absorption optics optional. Can also be used preparatively.

be made by a small modification to an existing arrangement. One such improvement which is actively being pursued by a number of groups of workers is to replace the normal light source by a laser (R. C. Williams, 1972; Paul and Yphantis, 1972). The use of a laser light source has several advantages with the interferometric technique. For example the coherence of a laser beam is so much greater than that of a conventional light source (usually a high pressure mercury lamp) that the resolution of interference fringes and observable path differences is greatly improved. Furthermore the fringe blurring normally caused by having to have a finite slit length can be reduced because the narrow laser beam can be focussed to form an almost perfect point source, and in any case the intensity of light available is greater. This latter fact permits the use of slower fine-grain photographic emulsions for example. R. C. Williams (1972) has used a 3 mW helium-neon laser which emits at 632·8 nm, and whereas the observable concentration range with a conventional light source was about seven fringes, with the laser source it was about 40 fringes. Another advantage of a laser source is that it can be pulsed and this has been exploited by Paul and Yphantis (1972). A pulsed light source can be used to perform multiplexing, that is to "look at" any one of a number of cells in the same rotor. This can also be done electronically (see below). Perhaps more important is the fact that pulsing can be used to improve fringe definition. In the interference optical system one uses a double-sector cell and a system of double slits to generate interference fringes. The slits are stationary, and as the rotor rotates it will be seen that there will be three configurations in which light can pass through the cell channels and the slits—two asymmetric configurations in which one slit is aligned with the "opposite" channel, and one symmetrical alignment. Only the latter, of course, leads to the production of interference fringes, but light passing in asymmetrical configurations also reaches the photographic plate and blackens it, thus reducing fringe definition. By arranging that laser pulses coincide with the symmetrical configuration of slits and cell channels, fringe photographs were much improved. In addition there are other minor ways in which a laser source is better than a conventional source. These are problems connected with, for example, variation in effective slit width and precession which could be overcome with a conventional source by narrowing the slits. In practice this is not possible simply because of loss of light intensity, but with a laser source, these difficulties largely disappear.

C. PHOTOELECTRIC SCANNING SYSTEMS

The use of the light-absorbing properties of the solute under investigation in the ultracentrifuge cell is an attractive technique for following rates of movement of boundaries and for determining concentration distributions through the cell, and indeed the technique was used a great deal in early ultracentrifuges. As a technique it is of course not only highly specific but also highly sensitive if some means of recording the amount of light absorbed can be found. Originally photographic plates were used, but unfortunately the degree of blackening of photographic emulsion is not linearly, nor even simply, related to the amount of light striking it. Thus, to relate the degree of blackening to the concentration of material at a point in the cell was difficult. It was for example necessary to perform a number of calibration runs at low speed (so there was no sedimentation), with solutions of different strengths in the cell. Furthermore direct viewing of the cell was not usually possible and so the run had to be to a large extent blind until the photographic plates were developed. These difficulties provided the impetus for workers to attempt to devise a system in which light passing through the cell was measured by a photoelectric device. Several such systems have been developed and some of them are available on commercial instruments or can be added to existing instruments. A number of systems exist: some involving a slit being driven across the image of the cell, some involving a rotating mirror, and some involving a television camera.

The original system developed by Schachman and his associates (Schachman *et al.*, 1962; Schachman, 1963; Hanlon *et al.*, 1962; Lamers *et al.*, 1963) was a slit system. This system, somewhat improved and modified, is available on Beckman instruments, and has been described in some detail (Chervenka, 1971). Light from a monochromator passes through the cell and a photomultiplier tube, mounted behind an adjustable slit which scans across the composite image of the cell, is placed at the focal plane of what would normally be the camera lens of the optical system. At the end of each scan the slit returns automatically. With each revolution of the rotor, the image of the cell moves vertically across the slit and the light level through each sector of the double-sector cell is detected separately. The voltage pulses so generated are translated into a scan of optical density against position in the cell. A calibration circuit is included in the system which produces a plot of a series of steps proportional to a change of $0 \cdot 2$ optical density units over the range $0–1 \cdot 0$. Also included is an electronic differentiator circuit which gives a derivative trace (equivalent to the schlieren trace). An electronic multiplexer is available as a separate accessory enabling up to five cells in a single rotor to be scanned independently.

The scanning system used in MSE instruments (based on a design by Spragg *et al.*, 1956) is a unit which can be housed in the eyepiece and camera-mounting assembly, replacing the film used in conventional recording. A front-aluminized mirror is brought into the light path just in front of the camera. This reflects the image of the cell on to an adjustable slit in front of a photomultiplier. When a scan is required the mirror is tilted slowly by means of a stepping motor causing the image of the cell to move perpendicularly across the slit. The duration of the scan can be either 30 seconds or 2 minutes. Suitable electronic circuitry feeds a twin-channel pen recorder which simultaneously displays concentration and its differential against radial position in the cell. In addition a multiplexing unit permits the examination of any single sample channel, or any pair of sample channels, from any position in a two, four, or six hole rotor fitted with either single- or double-sector cells.

A novel system has been devised by Lloyd and Esnouf (1974) which uses a vidicon television camera instead of a photomultiplier tube as the detector. There are therefore no moving parts at all as the scanning is done with a beam of electrons, and the whole image is scanned 50 times per second. The camera tube does not have to respond in the very brief moment when the cell passes through the beam of light, but gathers light from all parts of the cell each time the rotor comes round and stores it ready for read out from time to time by the scanning beam of electrons. The analogue circuits which produce the extinction of the solution in the cell from the pulses derived from the television camera are slightly different to those in other scanning systems in that the pulses are not stored in a holding circuit which stores their peak amplitude, but in an integrator which stores the product of their amplitude and duration. Such a system has the advantage over peak detection that it is less affected by electronic noise in the system. Finally, with the television system the images of the cells can be continuously displayed on a monitor screen, but a chart record can also be obtained with a high speed chart recorder.

The main advantages of any scanner system are that by choice of wavelength one can in effect choose the component in a solution one wishes to look at. As the signal is electronic, it can be amplified, providing great sensitivity. Experiments with proteins have been performed at concentrations well below 50 $\mu g/ml$, and Schachman and Edelstein (1966) measured a sedimentation coefficient at a concentration of 3 $\mu g/ml$. (With schlieren optics the lower concentration level useful for the determination of sedimentation coefficients is about 0·5 mg/ml.) Not only are such concentrations close to infinite dilution, but also it is clear that extremely small amounts of material may be analysed. Finally the fact that the data are in electronic form means that the ultracentrifuge scanner may be directly interfaced with a computer. This can be illustrated by a consideration of the system devised by Crepeau *et al.* (1972, 1974). These workers have interposed an analogue–to–digital converter between an ultracentrifuge (a Beckman model E with scanner) and a computer (a Data General Nova system with a 20K memory). The input of data that the computer receives is therefore (a) *digital* voltage values relating to the light signals detected during the course of the scan, and (b) informational pulses from the multiplexer giving hole and cell sector information. In fact the system collects a point on each revolution of the rotor from each cell that is present. Figure 3 shows the arrangement in the form of a block diagram and also indicates the other equipment used with the system, including an X-Y plotter, a high-speed printer, a disc storage system, an oscilloscope, a teletype, and a cassette recorder. The versatility and capacity of the system is impressive. By using five six-channel Yphantis cells in a six-hole rotor, up to 15 solution–reference solvent pairs may be brought to equilibrium and analysed, the results then being available as an oscilloscope display, a print-out, or a plot, either in the form of optical density vs radius, or as log (optical density) vs radius squared. In an experiment, when equilibrium has been reached, 5000 transmittance points are collected for each sample–reference pair. The data are averaged in groups of 50 pairs of points and log (reference transmittance/sample transmittance) is calculated giving finally a set of 50 values of radius, optical density, and standard deviation.

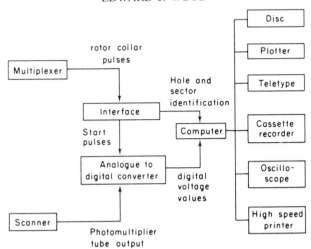

FIG. 3. Block diagram of arrangement of ultracentrifuge, computer, and peripheral equipment as used by Crepeau *et al.* (1974) for directly interfacing ultracentrifuge output with computer system. Reprinted with permission from *Biochemistry* **13**, 4860 (1974). Copyright by The American Chemical Society.

The advantages of such a system in terms of convenience or data acquisition and handling and analysis will be obvious to anyone who has ever measured an ultracentrifuge photograph and performed calculations with the data so obtained, and must surely point to the future trend in ultracentrifuge technique. Moreover the ability to examine up to 15 solutions simultaneously and to process the data rapidly to obtain M_w vs concentration has allowed interacting systems to be analysed with a precision previously unattainable (see p. 38).

D. MISCELLANEOUS DEVELOPMENTS

A number of articles and reviews have appeared on the subject of the alignment of the optical systems of analytical ultracentrifuges. In addition to the useful book by Lloyd (1974), there have been articles describing highly precise techniques for improving optical alignment (Richards *et al.*, 1972). These include consideration of modifications to the design and placement of the Rayleigh mask for use with the interference system. Other workers have been more concerned with improving the design and flexibility of ultracentrifuge cells. Ansevin *et al.* (1970) have described a new type of three-channel cell which has six compartments, and so may be used with the interference system. These compartments may be filled without disassembly of the cell. These workers analysed the optical aberrations often observed with conventional cells when operated at high speed, and they claim that when their new cells are operated and used in accord with their suggestion, they are substantially free of aberration even after long operation above 50,000 rev/min.

Dyson (1970) has described a simple test for optical alignment. A double exposure is made with and without the cylindrical lens for a spinning single-sector cell which has been assembled with interference windows and filled with a protein solution. In a correctly aligned system the fringes should be absolutely straight, and should be symmetrically placed between the images formed by the slits in the condensing lens mask.

Schumaker *et al.* (1970) have devised a system for pressurizing Beckman ultracentrifuge cells with nitrogen so that pressure-dependent reactions can be studied independently of the pressures generated by the centrifugal force. These latter may reach 100 to 500 atmospheres at the bottom of the cell (Kegeles *et al.*, 1967), and may have a profound effect when for example interacting systems are under investigation. Kegeles and his coworkers have studied pressure-dependent reactions by overlaying the liquid in the ultracentrifuge cell with a layer of oil. The effects of pressure are considered below, but it may be mentioned at this stage that at least one group of workers has investigated the effect of high pressures by means of a special light-scattering apparatus in which the liquid in the scattering cell can be subjected to very high pressures.

Finally Lewis and Barker (1971) provided a mathematical formulation for the effects of cell distortion and liquid height compression in the analytical ultracentrifuge. They found that the effects of rotor stretch were significant at high speeds (r can increase by 0·5–0·75% at 59780 rev./min) and suggested correction procedures. Previously Baghurst and Stanley (1970) had measured the stretch in a Beckman titanium (AnH) rotor at nine speeds between 2000 and 68,000 rev./min. Measurements at the interference counterbalance showed a radial stretch of about 0·4 mm at top speed when the change at the rotor reference was less than half this amount.

III. Sedimentation Velocity

A. Sedimentation Coefficients

If the centrifuge be run at such a speed that particles in solution sediment, one may define a sedimentation coefficient, s, as the rate of movement of the solute, in the radial direction, per unit field:

$$s = \frac{1}{\omega^2 r}\frac{dr}{dt} \tag{1}$$

where t is the time in seconds, ω is the radial velocity, and r is the radial distance.

It is convenient to use the quantity 10^{-13} seconds, known as the Svedberg unit, in expressing sedimentation coefficients. For a homogeneous solute, originally uniformly distributed through the cell, sedimentation results in the formation of a more or less diffuse boundary between a centrifugal region containing solute (the plateau region) and a centripetal region essentially devoid of solute (Fig. 4).

For a rapidly sedimenting homogeneous ideal solute the boundary formed is very nearly symmetrical and s may be calculated from the rate of movement of the peak maximum (with the schlieren optical system) or from the point in the boundary corresponding to the radius at which the concentration is half that in the plateau region (with the absorption or interference optical systems). This method may not be valid when a skew boundary is observed: this can result from a dependence of s on concentration, from non-ideality, or from heterogeneity. In fact the distribution is approximately Gaussian even when the concentration dependence of s is significant. The radial position of the position of maximum gradient is then essentially identical with the square root of the second moment (Schachman, 1959). If the skewness results from ordinary non-ideality of a single solute then the error is quite small and may usually be disregarded (Dishon et al., 1967). Where the boundary is skew because of the sedimentation of a heterogeneous solute then the approximation is not valid. With synthetic polymers where the distribution of sedimentation coefficients is practically continuous, this distribution may be obtained by examining the way in which boundary spreading occurs during sedimentation (see below).

The plot of $\ln r$ against t used for calculating the sedimentation coefficient of a single solute will not be linear in concentration dependent systems because radial dilution causes a progressive increase in s. Schachman (1959) has described ways of overcoming this by, for example, an extrapolation procedure. With single homogeneous solutes it is often adequate to calculate the mean slope over a finite time interval: the concentration relating to the resulting s is then that of the mean plateau region (Creeth and Pain, 1967). A knowledge of this concentration is of importance in the determination of the concentration dependence parameter, k_s, in

$$s = s^0(1 - k_s c) \qquad (2a)$$

or

$$s = s^0/(1 + k_s c) \qquad (2b)$$

The relation between k_s and the shape of molecules is described by Creeth and Knight (1965).

For water soluble solutes such as proteins it is normal to reduce sedimentation coefficients (determined in dilute salt solution) to what they

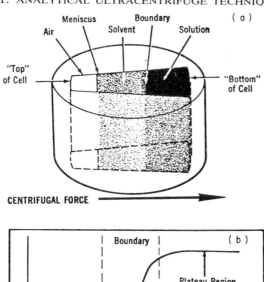

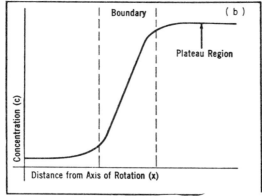

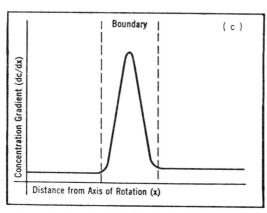

FIG. 4. Sedimentation velocity method. (a) Distribution of solute in the cell and plots of (b) concentration, and (c) concentration gradient, against distance from axis of rotation. (b) might be the record obtained using scanner optics, while (c) might be that obtained using schlieren optics. Figure by courtesy of Beckman-RIIC Ltd.

would be in water at 20°C, using appropriate factors involving the viscosity and density of the solvent. Sedimentation coefficients of proteins are therefore normally expressed as $s_{20,w}^{0}$ values, i.e. such values extrapolated to infinite dilution.

It will be noted that because of radial dilution (and, of course, because of diffusion), the peak height observed with the schlieren optical system decreases as the peak moves down the cell. If the concentration existing in the plateau region is c_p, it may be shown that

$$c_0 = c_p(r^2/r_a^2) \tag{3}$$

where c_0 is the initial concentration, r is the radial position of the boundary and r_a is the position of the meniscus. Fulfilment of this condition may be used to check that a system in fact contains only a single solute component (Schachman, 1959, 1963) and in the case of polydisperse solutes it is important that optical recoveries should be quoted. A diffusion coefficient may be calculated from the spreading of the boundary with time (see below).

It may be remarked that in individual experiments in the analytical ultracentrifuge, sedimentation coefficients may be measured with an accuracy of about ±1%. The technique of differential sedimentation is, in contrast, capable of measuring very small *differences* in sedimentation coefficient. Richards and Schachman (1957, 1959) showed that the difference between two sedimentation coefficients could be measured with great accuracy using the interference optical system. The optical system was used to subtract the concentration distribution curves for two solutions contained in a double-sector ultracentrifuge cell. Kirschner and Schachman (1971) developed the technique further and identified the systematic errors which originally held back the development and use of the method. These errors result from a number of causes and may be to a large extent overcome by the use of a new interference mask along with improved alignment procedures. In the region of large concentration gradients fringe bowing occurs and this may be eliminated by measurement of the zero order fringe selected by making an exposure with white instead of monochromatic light (Richards and Schachman, 1959).

It was later shown (Lamers *et al.*, 1963; Schachman, 1963) that the scanning absorption system could also be used for this differential measurement, and more recently the schlieren optical system has been used. Schumaker and Adams (1968), for example, used two cells, one with wedge windows, in the same rotor. They concluded that errors in measuring sedimentation coefficients separately were not errors in taking data from photographic plates but rather were mainly caused by actual variations in the position of the peak due to such things as fluctuation in rotor speed, rotor precession, optical misalignment, and thermal gradients within the rotor. Clearly, comparing the two solutions in two cells in the same rotor eliminates many of these sources of error. When they measured differences in sedimentation coefficient by means of the wedge window method, the results were so good that it was desirable to take into account the change due to the concentration dependence of the sedimentation coefficient as radial dilution took place. Schumaker (1968) went on to develop equations for interpreting changes in sedimentation coefficient in terms of frictional ratios. These equations are especially suitable for macromolecules which change their shape and volume when binding a low molecular weight ligand. This technique clearly has particular importance in the study of enzymes and proteins which bind small molecules. For example, Gerhardt and Schachman (1968) studied the enzyme aspartate transcarbamylase in the presence and absence of certain ligands and found a difference of 0·4 S in the sedimentation coefficient. Similarly Charlwood (1971) found that $s_{20,w}$ of the protein apoferritin increased by 1·8% when it bound two iron atoms per molecule. It is claimed that differences in s of 0·01 S can be measured readily to an accuracy of better than 5%. However when small differences in sedimentation coefficient can be measured, it is tempting to ascribe the small changes observed when proteins bind small molecules to conformational changes in the protein molecule causing a change in the frictional ratio. In fact, in many cases, such changes in sedimentation coefficient can be accounted for by increased molecular weight and a decrease in apparent specific volume. Thus great caution is necessary in interpreting results from such experiments.

B. MOLECULAR WEIGHTS FROM TRANSPORT MEASUREMENTS

The sedimentation velocity of a solute depends on the applied field, its buoyant mass and its mobility. The particle also has a diffusion velocity dependent upon the chemical-potential gradient and the mobility. Sedimentation and diffusion data may therefore be combined to give the molecular weight, and this is the basis of the well-known Svedberg equation:

$$\frac{s}{D} = \frac{M(1 - \bar{v}\rho)}{RT(1 + d \ln y/d \ln c)} \tag{4}$$

which, in the limiting case using the values of the sedimentation and diffusion coefficients at zero concentration, i.e. s^0 and D^0, becomes

$$M = \frac{s^0 RT}{D^0(1 - \bar{v}\rho^0)} \tag{5}$$

This has been extremely useful in protein chemistry where the solutes being investigated are often single homogeneous ones. As will be shown below, the diffusion coefficient, D may be evaluated in an independent experiment, but in principle the data are already present in the spreading of the boundary observed in the ultracentrifuge. Hence there have been a number of attempts to determine the parameter s/D from sedimentation velocity experiments (see, for example, McCallum and Spragg, 1972).

C. MEASUREMENT OF DIFFUSION COEFFICIENTS

Whereas sedimentation coefficients are relatively easy to determine, diffusion coefficients have been more difficult to determine with accuracy. Several solutions to this problem have been sought and have ranged from refining the method of calculating D from ultracentrifuge data (Baldwin, 1957; Van Holde, 1960; Kawahara, 1969) to the determination of D by a completely independent technique. Diffusion coefficients may be calculated in the ultracentrifuge by static methods or by dynamic methods. In the first technique, solvent is layered at low speed over solution using a double sector, capillary-type synthetic boundary cell, and a low centrifugal field is maintained to stabilize the boundary without causing significant sedimentation of solute molecules. The spreading of the boundary with time may be followed by schlieren optics and D calculated from the reduced height-area ratio (Gosting, 1956). However for a number of technical and theoretical reasons (discussed by Creeth and Pain, 1967) high precision cannot be expected of this method even for an homogeneous protein dissolved in buffer. Kawahara (1969) has tried to improve the precision of this method. He used a simple and approximate form of the Fujita equation (Fujita, 1962), and showed that the effect of the concentration dependence of s on the evaluation of D decreased with decreasing speed. Most of the experiments he performed were done at 12,500 rev/min in a synthetic-boundary double-sector cell. The second approach is the dynamic method in which the spreading of the moving boundary is followed under high gravitational fields. For example, Van Holde (1960) used a simplified form of Fujita's expression for the variation of the height–area ratio (H/A) with time:

$$\left(\frac{H}{A}\right)t^{\frac{1}{2}}(1 - \tfrac{1}{2}\omega^2 st)^{-1} = (4\pi D)^{-\frac{1}{2}} + \left(1 - \frac{2}{\pi}\right)\frac{r_a\omega^2 st^{\frac{1}{2}}}{4D}\left(\frac{k_s c_0}{1 - k_s c_0}\right)(1 - \tfrac{1}{2}\omega^2 st) + \dots$$

$$(6)$$

Nevertheless the values of D obtained by such procedures are not without error, and this has led workers to search for other methods for measuring D which do not involve the analytical ultracentrifuge. The technique of examining the spectral distribution of scattered laser light has been particularly useful for a number of large particles, such as large protein molecules, viruses, and polystyrene spheres, though in principle the method also applies to smaller molecules (Dubin et al., 1967; Foord et al., 1971). The spectrum of the scattered light may be measured either by means of a spectrum analyser (Dubin *et al.*, 1967) or by autocorrelation (Foord *et al.*, 1971) and the resulting values for D can be shown to have a high precision. The diffusion coefficients for certain small protein molecules have been measured, including those for lysozyme (Foord *et al.*, 1971), chymotrypsinogen and bovine serum albumin (Sellen, 1972), and attempts were made to determine the diffusion coefficients of two components in solution (Foord *et al.*, 1971). This may turn out to be a useful method for providing diffusion coefficient distributions. Another method potentially capable of yielding values for D and the distribution of D values is the method of gel filtration on a calibrated column. This actually yields values for the Stokes radius (Ackers, 1967).

A different approach has been investigated by McCallum and Spragg (1972). They attempted to evaluate the parameter s/D from boundary data taken from sedimentation velocity experiments in the analytical ultracentrifuge, using non-linear statistical methods to fit a mathematical model, and they later extended the method to analysing boundaries formed in difference experiments. A similar approach has been pursued theoretically and practically by McNeil and Bethune (1973) with encouraging results. These investigators used computer-generated concentration distributions to develop numerical techniques for the determination of sedimentation and diffusion coefficients from the integrated Lamm equation (Lamm, 1929). The methods they developed were then applied to experimental data obtained for two homogeneous proteins, ribonuclease A and chymotrypsinogen A. This type of approach has a number of advantages. Determination of s and D simultaneously under identical experimental conditions (i.e. in the same ultracentrifuge cell) allows the estimation of an apparent molecular weight by means of the Svedberg equation. In addition, for certain systems where the attainment of equilibrium is prevented by, for example, polymerization, aggregation, or autolysis, the method allows a rapid estimate of M_{app} to be made for all but the most rapidly reacting irreversible system.

It is clear that for monodisperse solutes we have a number of ways of evaluating s and D either independently or simultaneously, thus allowing calculation of molecular weights. It may be asked how much such methods and calculations may be applied to heterogeneous solutes. This problem was considered by Creeth and Pain (1967), as applied to solutes not giving evidence of gross heterogeneity. However it is probably more fruitful to consider what can be discovered about frankly polydisperse systems by a consideration of distributions of sedimentation coefficients and this is dealt with below.

Information may be obtained from sedimentation velocity experiments on molecular weight distributions and especially second virial coefficients, by combining data from sedimentation, osmotic pressure, and light-scattering studies. Meyerhof and Sütterlin (1973) used such procedures to investigate the behaviour of ethylcellulose samples in organic solvents. Each method had its own particular advantages and disadvantages with this rather difficult system. The intensity of scattered light was strongly influenced by dust and microgels, whereas on the other hand osmosis and sedimentation–diffusion measurements were unable to measure directly the radius of gyration.

D. DISTRIBUTION OF SEDIMENTATION COEFFICIENTS

When a non-uniform solute is exposed to a centrifugal field partial fractionation occurs and one may therefore measure the distribution of sedimentation coefficients in sedimentation velocity experiments. The sedimentation coefficient is related to the molecular weight by

$$s = KM^\alpha \tag{7}$$

in which the values of the constants K and α depend on the nature of the solvent, the conformation of the polymer molecules in it and the extent to which the latter is permeated by the solvent. At the Flory temperature, α should have the value $\frac{1}{2}$. Correction must be made for the effects of concentration, pressure and diffusion. Biological chemists have been able to study proteins in aqueous buffer solution where the effects of pressure are often, but not always (see Josephs and Harrington, 1967; Morimoto and Kegeles, 1971), negligible, whereas polymer chemists have had to use organic solvents. In this latter case, working at the Flory temperature largely eliminates the concentration-dependence effects but introduces pressure effects.

In order to determine the distribution of sedimentation coefficients it is usual to make the initial assumption that the extent of the boundary spreading depends only on the distribution of s with no contribution from diffusion, that s is independent of concentration for each component, and that solvent and solute are incompressible. (All of these are of course unjustified to some degree and the procedure adopted is to extrapolate them out as far as possible.) The distribution of sedimentation coefficients, $g(s)$, making the above assumptions, is given by

$$g(s) = \frac{1}{c_0} \frac{dc}{dr} \left(\frac{r}{r_a}\right)^2 (\omega^2 rt) \tag{8}$$

The effects of diffusion can be eliminated by taking regard of the fact that sedimentation is proportional to time, whereas diffusion is proportional to the square root of the time. Thus, apparent distribution curves are obtained at several times and extrapolated to infinite time. A plot against $1/t$ was shown to be approximately linear when $\omega^2 st \ll 1$ (Gosting, 1952; Kotaka and Donkai, 1968).

Concentration effects are more difficult to deal with since both s and the diffusion coefficient, D, vary not only with the concentration of a given solute but also with the concentration of other solutes present. The procedure is essentially to extrapolate to infinite dilution, where the effects disappear. The problems involved in such a procedure have been discussed by Williams (Williams and Yphantis, 1971; see also Williams and Saunders, 1954). Oncley (1969) has also described a method for the treatment of distribution functions obtained in sedimentation velocity experiments, involving adequate corrections for diffusion and for the dependence of the sedimentation coefficient on concentration. This treatment used modified forms of Gosting's equations for the transformation of the apparent distribution function $g^*(s)$ to $g(s)$. Oncley used the method to examine a β-lipoprotein from blood plasma which had a bivariate normal distribution. Experimental difficulties were considerable because the extensive heterogeneity caused the boundary to disappear quite quickly, making the extrapolation to $1/t = 0$ difficult.

Pressure effects are also difficult to deal with but successful correction for such effects have been based on the theoretical treatment of Fujita (1956). In this treatment the value of the sedimentation coefficient at 1 atmosphere, s^0, is related to s_{obs} by Eqn. (9) where m is a coefficient containing the dependence on pressure of the frictional coefficient and specific volume of the solute, and of the density of the solvent:

$$s^0 = s_{obs}/1 - m\left[\left(\frac{r}{r_a}\right)^2 - 1\right] \qquad (9)$$

The importance of the pressure correction may be seen in the results obtained for the distribution of sedimentation coefficients of a sample of polystyrene in cyclohexane (Billick, 1962). He found that for this polymer of low polydispersity ($M_w = 251,000$; $M_w/M_n = 1 \cdot 05$), correcting for pressure effects resulted in an appreciable shift of the sedimentation coefficient to higher values (the average sedimentation coefficient increased by about 8%), and in addition there was a slight broadening of the distribution (Fig. 5). The ratio M_w/M_n however did not change appreciably.

J. W. Williams (1972) gives a detailed discussion of how a molecular weight distribution, $f(M)$, may be obtained from the distribution of sedimentation coefficients, $g(s)$. In general the constants α and K in Eqn. (7) ($s = KM^\alpha$) are adjusted from a knowledge of other data. Such other data may involve for example osmotic pressure or sedimentation equilibrium measurements. One example chosen by J. W. Williams (1972) was a polystyrene-cyclohexane system at $34 \cdot 2°C$ (the Flory temperature) for which $\alpha = 0 \cdot 50$ and K $= 1 \cdot 47 \times 10^{-2}$ were selected in order to obtain a

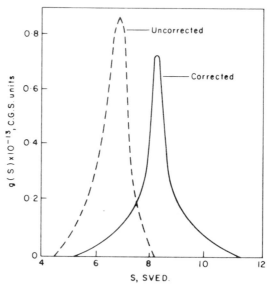

FIG. 5.   Differential distribution of sedimentation coefficients at $t = \infty$ and $c = 0$ in a sample of polystyrene in cyclohexane, showing the effect of correcting for the influence of pressure. Taken from Billick (1962) by permission of the copyright owner.

plot of $f(M)$ against M after correction for pressure dependence. Finally the general validity of the curve obtained was confirmed by performing numerical integration to obtain number and weight average molecular weights.

Tung (1969) has previously reported a method for the determination of distribution for long chain branching polymers which uses a 2-D distribution function which made the calculation cumbersome and tedious. More recently, it was found (Tung, 1971) that for a special class of branched chain molecules this long method could be avoided, and for linear polydisperse samples the M_{app} distributions obtained from gel permeation chromatography and from sedimentation velocity experiments were the same and were identical with the true molecular weight distribution.

Wales and Rehfeld (1962) determined the molecular weight distribution for polystyrene in cyclohexane at 34°C. Instead of calibrating the method with fractions of known molecular weight they combined boundary spreading data from sedimentation velocity experiments with data on the intrinsic viscosity. There was only a small dependence of sedimentation coefficient upon concentration in the chosen solvent at the chosen temperature, but corrections were applied for pressure dependence. The molecular weight distribution obtained agreed closely with that obtained by precipitation chromatography (Table II). Kotaka and Donkai (1968)

TABLE II

Collected molecular weight averages for polystyrene fractions in cyclohexane. (Taken from Wales and Rehfeld (1962) by permission of the copyright owner.)

Material	s_0 max Svedbergs	$[\eta]$	M_n		M_w			M_z		M_{sv}	$\dfrac{M_w}{M_n}$
			Ultracentrifuge	Precipitation chromatography	Ultracentrifuge	Light scattering	Precipitation chromatography	Ultracentrifuge	Precipitation chromatography		
4008-69 pooled fractions[a]	13·15	0·604[b]	633,000	—	666,000	637,000	—	700,000	—	629,000	1·05
4008-49-2 pooled fractions[a]	6·55	0·316[b]	152,000	—	165,000	150,000	—	177,000	—	160,000	1·08
Polystyrene No. 2	6·58	0·36	42,100	60,000	241,000	234,000	242,000	435,000	410,000	167,000	5·72

[a] Experiments performed in 12 mm cell, 59,780 rpm. [b] Calculated from viscosities in toluene.

also studied polystyrene in cyclohexane at 35°C (the Flory temperature), again taking advantage of the fact that under these conditions the dependence of sedimentation coefficient on concentration was minimal. They derived a semi-empirical method for deducing the distribution of molecular weights from a single sedimentation experiment, and corrected the observed boundary shape for the effects of diffusion and pressure dependence. The method was applied to several polystyrenes with narrow, broad, and very broad distributions. Again the results were compared with data obtained by other methods. In this case, gel permeation chromatography predicted a broader distribution of molecular weight than either the sedimentation transport method or elution chromatography.

Pita and Müller (1972, 1973) have used a non-optical approach to study polydisperse and paucidisperse systems. The sedimentation cells were either capillary tubes holding 4 μl solution or small cellulose tubes of 0·8 ml capacity used in a swing-out preparative rotor. One of the advantages of the method is that very small amounts (a few nanograms) of a biological material could be examined. Several methods of data acquisition were used such as successive aliquot extraction along the tube length, analysis of integral concentration in the centripetal segment of the cell, and centrifugation at different speeds. The materials examined were serum albumin (monodisperse), a proteoglycan subunit (paucidisperse), and a proteoglycan complex (pauci- and polydisperse).

Other workers have used both analytical and preparative ultra-centrifuges to distinguish two types of polymer in a mixture. Yamamoto *et al.* (1970) used the sedimentation velocity method to resolve two components of a test mixture of two samples of monodisperse poly (α-methylstyrene). The sedimentation method showed the two peaks in the molecular weight distribution a little better than other methods such as precipitation chromatography, elution and gel permeation chromatography. Indeed it was the only method which resolved a 50:50 mixture of polymers with molecular weights of 342,000 and 433,000. Chauvel and Daniel (1974) used both analytical and preparative ultracentrifugation to study A.3S copolymers obtained by polymerizing a mixture of styrene and acrylamide in the presence of a polybutadiene latex using an emulsion radical process. The resulting copolymer was a mixture of a graft copolymer and a random (linear) copolymer and these were separable by dispersing the A.3S powder in an appropriate solvent (MEK) in which the linear copolymer was soluble. When the centrifugal field was applied the graft copolymer sedimented more rapidly than the linear one in spite of its lower density. Analytical ultracentrifugation with schlieren optics was used to study the process and to choose suitable times and speeds, and the polymers were then actually separated in the preparative ultracentrifuge and their quantities determined. Ultracentrifugation was a sufficiently reproducible and efficient technique for separating the two components and enabled the effects of different polymerization conditions on the grafting parameters to be studied.

In summary, one sees that sedimentation coefficient distribution obtained from analytical ultracentrifuge experiments may be transformed into molecular weight distributions but that ancillary data are usually required and that considerable corrections have to be applied. It is not surprising therefore that much attention has focussed on the determination of molecular weight distributions directly in the ultracentrifuge by means of the technique of sedimentation equilibrium.

IV. Sedimentation Equilibrium

A. THEORY

The state of sedimentation equilibrium may be defined as the state in which equilibrium is attained between the transport of solute due to sedimentation and the reverse transport due to diffusion. The description of the concentration distribution at sedimentation equilibrium has been rigorously derived by thermodynamic analysis, starting with the condition for equilibrium in a field, and reference may be made to numerous articles and books for such a treatment (see Williams *et al.*, 1958; Fujita, 1962). Because of the great range of centrifugal fields available in the modern analytical ultracentrifuge, with rotor speeds from 800 to 60,000 rev/min or more, conditions may be found for the investigation of molecules with molecular weights ranging from several hundred to several million. There is probably no other method of providing molecular weight data over such a wide range. In addition to providing molecular weights, or molecular weight averages in the case of polydisperse materials, the method also provides data from which it is also possible to calculate molecular weight distributions and to investigate non-ideality. The method has the advantage that ancillary data are not usually required for the determination of molecular weights and molecular weight distribution (cf. sedimentation velocity).

Before the application of the centrifugal field, solute is distributed uniformly throughout the cell. Upon the application of the centrifugal field, transport of solute begins, and diffusion tends to nullify the concentration gradients so produced. For a single homogeneous solute in an incompressible solvent equilibrium is reached when it may be said that the total potential, chemical plus gravitational, is independent of radius. Under these conditions it can be shown (Williams *et al.*, 1958) that

$$M\omega^2 r(1 - \bar{v}\rho) = (\partial\mu/\partial c)_{P,T}(dc/dr) \qquad (10)$$

The chemical potential of the solute can be expressed by the use of an activity coefficient, y:

$$\mu = \mu^0 + RT \ln (yc) \tag{11}$$

Thus Eqn. (10) becomes

$$\frac{1}{rc}\frac{dc}{dr} = \frac{\omega^2 M(1 - \bar{v}\rho)}{RT[1 + c(\partial \ln y/\partial c)_{\mathrm{P,T}}]} \tag{12}$$

which is the fundamental expression used for all experiments involving sedimentation equilibrium. For a single solute at sufficiently low concentration the non-ideality term, $(\partial \ln y/\partial c)_{\mathrm{P,T}}$, may be considered to be essentially constant. A limiting relationship then applies:

$$\frac{1}{M_{\mathrm{app}}} = \frac{1}{M} + Bc \tag{13}$$

where the apparent molecular weight, M_{app}, is given by

$$M_{\mathrm{app}} = \frac{RT}{(1 - \bar{v}\rho)\omega^2}\frac{1}{rc}\frac{dc}{dr} \tag{14}$$

This equation describes the concentration gradient at any point in the cell. How it is used to determine molecular weight depends on the optical system used, and there are many excellent accounts of the different ways of handling the data (Creeth and Pain, 1967; Van Holde, 1967; Chervenka, 1969).

B. EXPERIMENTAL

The general technique for sedimentation equilibrium experiments is to centrifuge short columns of solution at constant speed and to take photographs, or scan, at intervals until the patterns change no more. The concentration ranges that may be investigated depend to a great extent upon what optical systems are available.

If we integrate from the meniscus, r_a, to some point in the cell, r, we obtain the integral form of Eqn. (14) which may be written:

$$\ln \frac{c_r}{c_a} = \frac{M_{\mathrm{app}}(1 - \bar{v}\rho)\omega^2}{2RT}(r^2 - r_a^2) \tag{15}$$

Clearly a plot of $\ln c$ vs r^2 can be used for the determination of M, but any method of evaluating M requires a knowledge of concentration, or some quantity proportional to it, at points in the liquid column. The absorption optical system gives concentration directly whereas the Rayleigh interference system gives the difference in concentration between any r and the meniscus. Two general ways will be outlined of how concentration distributions may be determined.

If the centrifuge is run at a sufficiently high speed one obtains the situation where the concentration at the meniscus is essentially zero, and experiments employing this technique are called variously "meniscus depletion", "high-speed", or "Yphantis" experiments. Using the interference optical system, increments of fringe shift along any fringe are thus directly proportional to the actual concentration at points down the liquid column (Yphantis, 1964; Chervenka, 1970). This method has the advantage that no other data are needed for the determination of M_{app} (and this includes c_0), and that with heterogeneous systems it gives, from the region nearest to the meniscus, the molecular weight of the *smallest* species present. This is of great interest to protein chemists studying subunit structure in proteins which have been caused to dissociate. On the other hand there are disadvantages in that for smaller molecules the high speeds required may lead to significant window distortion, and that only about one-third of the liquid column provides data of use. This latter consideration is important as the fact that a plot of $\ln c$ vs r^2 is linear is commonly taken as a criterion that a single, homogeneous ideal solute is present.

In contrast to this technique, the so-called low-speed experiment provides more data and is usually the method of choice for analysis of complex systems involving polydispersity and non-ideality. Rearranging Eqn. (14) and integrating a different way we obtain

$$\int_{r_a}^{r_b} \frac{dc}{dr}\, dr = \frac{M_{app}(1 - \bar{\nu}\rho)\omega^2}{RT} \int_{r_a}^{r_b} rc_r\, dr \tag{16}$$

It can be shown that in a sector-shaped cell, conservation of mass requires that

$$\int_{r_a}^{r_b} rc_r\, dr = c_0 \frac{(r_b^2 - r_a^2)}{2} \tag{17}$$

and hence

$$\frac{c_b - c_a}{c_0} = \frac{M_{app}(1 - \bar{\nu}\rho)\omega^2}{2RT}(r_b^2 - r_a^2) \tag{18}$$

Using the interference optical system a fringe count gives Δj_{eq}, which is proportional to the quantity $(c_b - c_a)$, and a quantity Δj_0, similarly in fringes, proportional to c_0 can be obtained by a separate synthetic boundary experiment or by differential refractometry (Richards and Schachman, 1959; Van Holde, 1967). Figure 7 shows the sort of result that may be obtained with a single homogeneous solute in dilute aqueous buffer solution using this technique and Fig. 6 shows the form of the data.

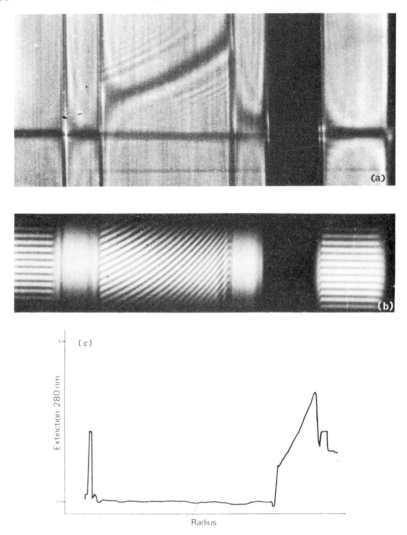

FIG. 6. Sedimentation equilibrium method (low speed). (a) and (b) show the form of the raw data from schlieren and Rayleigh interference systems, respectively. These two photographs relate to the determination of the molecular weight of the enzyme, superoxide dismutase (see Fig. 7). (c) is an experiment with the protein bovine serum albumin ($M = 67,000$) using absorption-scanner optics in the Centriscan ultracentrifuge (courtesy of MSE Instruments Ltd). In all the experiments a short column of liquid was used (approx. 0.1 ml).

Data on the equilibrium concentration distribution are also available from the use of the schlieren optical system. However, except perhaps for the determination of M_z values by the Lamm method (logarithmic schlieren plot, see Chervenka, 1969), the form of the data (differential

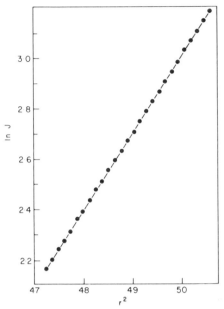

FIG. 7. Low speed sedimentation equilibrium experiment with a protein in dilute aqueous buffer solution. The protein was the enzyme superoxide dismutase, initial concentration about 4 mg/ml. Run performed at 24°C for 48 h and 12,590 rev/min using the interference optical system. That a straight line was obtained indicates that the material was homogeneous: molecular weight $(M_{w\,app}) = 32,800$. Taken from Wood *et al.* (1971) by permission of the copyright owner.

refractive index vs r), the lack of precision in measuring the photographic plates, and the comparatively high solute concentrations that have to be used, rule out the use of this optical system as a method of choice for observing concentration distributions in sedimentation equilibrium experiments. No doubt the trend in the future will be to the use of absorption scanning systems, particularly as their output may be arranged to be in a form directly handleable by computer systems (Crepeau *et al.*, 1974).

The time taken to reach equilibrium depends on a number of factors and varies from a few hours to several days. It depends on the length of the liquid column, the viscosity of the solution and the diffusion coefficient of the solute. Because of the time taken to reach equilibrium a number of ways have been suggested of either shortening the time or of performing several experiments simultaneously on the same instrument. Into this latter category fall the use of cells with several, usually three, independent pairs of channels in which three different concentrations of the same solute may be placed, as well as the use of multiplexing arrangements to "look-at" several different cells in the same rotor. The ways of shortening the time required for the attainment of equilibrium involve the use of short columns of liquid (3 mm or even 1·5 mm) and also the technique of overspeeding. In this latter technique the rotor is accelerated to a speed somewhat above that selected for the final attainment of equilibrium. After a number of hours at the higher speed, during which the solute undergoes sedimentation towards the bottom of the cell, the speed is lowered to the equilibrium speed. Howlett and Nichol (1972b) have recently re-examined the considerations that apply for the selection of the higher speed and for the time required at that higher speed, and they present their results in the form of a graph from which the speeds and times can be read. In an experiment with the enzyme lysozyme they used an overspeed of 28,000 rev/min for 1·2 h followed by an equal period of time at 20,000 rev/min. After 1·2 h of overspeeding, the concentration distribution very closely approximated that at equilibrium. An alternative approach has been to build up, by means of synthetic boundary cells, a concentration gradient approximating that at equilibrium and to let the centrifuge run for an hour or two to achieve the final equilibrium distribution.

It is clear such methods have great value in the study of very large particles (with low diffusion coefficients) by sedimentation equilibrium. For example Bancroft and Freifelder (1970) have determined, by sedimentation equilibrium, the molecular weight of a bacteriophage (T7) and its DNA to be $49·4 \times 10^6$ and $25·3 \times 10^6$ respectively. In the case of the bacteriophage, the ultracentrifuge was run at 800 rev/min for 120 h to attain equilibrium. This was in fact a high-speed or meniscus depletion experiment! It is noteworthy that the instrument, a Beckman model E with electronic speed-control, appeared to hold the speed at 800 ± 7 rev/min. In similar experiments Kado and Black (1968) determined the molecular weight of small virus. Equilibrium was attained in 92 h of operation at 2379 rev/min and the data yielded a molecular weight of $6·3 \times 10^6$.

Another way of shortening the length of the experiment is to make measurements as the system proceeds towards equilibrium and then to make an extrapolation to the equilibrium condition.

The Archibald method (Archibald, 1947) is a so-called approach to equilibrium method which enables the run time to be shortened, though "transient state method" would be a more appropriate name for the technique. Its basis is that at the upper and lower limits of the solution column one necessarily has zero flow, since no solute can pass through the air-solution meniscus or through the bottom of the cell. Though conditions at the meniscus and cell bottom may be far removed from equilibrium, a particular equilibrium relationship always applies at these two points in the liquid column, and an equation similar to the equilibrium equation may be used to determine the relation s/D at the air–liquid meniscus:

$$\frac{1}{r_a c_a}\left(\frac{dc}{dr}\right)_a = \omega^2 \frac{s}{D} \tag{19}$$

A similar equation can be written for data at the cell bottom, but because of the high pressures operating in this region of the liquid column the use of these data is not usually desirable. The molecular weight is given by

$$\frac{1}{r_a c_a}\left(\frac{dc}{dr}\right)_a = \frac{M_{app}(1 - \bar{v}\rho)\omega^2}{RT} \tag{20}$$

and thus both the gradient and the absolute concentration at the meniscus (or at the base of the cell) have to be known. This demands an extrapolation to the meniscus (or cell bottom) and this is not straightforward. It is also essential that the optical system used is correctly focussed in order to obtain good initial data. Data can be obtained using the schlieren optical system in a transient state experiment to obtain (dc/dr), and in a synthetic boundary experiment to determine the concentration. Alternatively, interference optics may be used to determine the concentration simultaneously with the use of schlieren optics to obtain the gradient (Richards and Schachman, 1959; Yphantis, 1960). Details of the somewhat tedious measurements and calculations are given by Schachman (1959), Chervenka (1969), and by Bowen (1971). The use of computers would clearly be advantageous in removing the tedium from the calculations. The precision of the method when used properly is quite good (see Creeth and Pain, 1967), and extrapolation to zero time gives M_w^0. This extrapolation should be against $\sqrt{t}$.

If the material under investigation is polydisperse then relative depletion of the heavier components occurs progressively at the air–liquid meniscus leading to a decrease in $M_{w\,app}$ values with time. This is a reasonably sensitive test for heterogeneity in fact, but as in the case of normal sedimentation equilibrium, heterogeneity and non-ideality produce opposing effects and may mask one another. One advantage of the use of the method with heterogeneous solutes is that M_w^0 can be found by extrapolation to zero time and so problems resulting from redistribution of the solute molecules can be avoided. Problems do, however, arise with this extrapolation and the slope of the plot of $M_{w\,app}$ against $\sqrt{t}$ for polydisperse systems can be markedly non-linear (Creeth and Pain, 1967). Another potential advantage of the method is that chemically reacting systems may be studied, and the results that may be expected in this case depend upon whether the reaction is slow compared with the period of observation or whether it is rapidly reversible. Pandit and Rao (1974) used the method to study the self-association of the enzyme, papain, but had difficulty in distinguishing a dimerization equilibrium from a model involving indefinite self-association. Creeth and Pain (1967) have discussed the relative merits of the method as applied to proteins, and they list a number of "difficult" systems that have been studied. These include a submaxillary mucoprotein ($M \sim 10^6$, polydisperse) and fibrinogen ($M \sim 4 \times 10^5$, asymmetric molecule) both native and in $5M$ guanidinium chloride (Gottschalk and McKenzie, 1961; Johnson and Mihalyi, 1965). More recently Paetkau (1967) has attempted to improve data analysis by this method and has proposed a general theory for data extrapolation. He used a computer to test the influences of various parameters and compared them with experimental data for the protein β-lactoglobulin B. The molecular weights obtained by this method were almost unaffected by the practical difficulty encountered in the Archibald method, that of measuring the true meniscus position accurately.

Baurain et al. (1973) have used an approach to equilibrium method using a photo-electric scanner. The data they use consist of sets of values of ΔC, the difference in solute concentration between the bottom of the cell and the meniscus. These are used to extrapolate, with the help of a computer, an equilibrium value, ΔC_{eq}, which permits the estimation of the molecular weight. The durations of the experiments with a range of proteins of molecular weights in the range 14,000 to 37,000 were all of less than one hour, and the molecular weights obtained agreed well with the literature values. They also consider the effects of polydispersity on the results obtained by this method. Finally, using methods analogous to those used for the detection of very small differences in sedimentation coefficient by means of the differential sedimentation experiment (p. 14), Schachman's group has made a detailed consideration of difference sedimentation equilibrium as a method of analysing very small changes in molecular weight (Springer et al., 1974; Springer and Schachman, 1974). In this technique for obtaining differences in effective molecular weight ($\Delta\sigma$), $\Delta c/\bar{c}$ is plotted against r^2. Δc, the difference in concentration between the two sectors of a double sector cell is measured with interference optics while $\bar{c}$, the average concentration, can be determined by integration of the schlieren pattern. In initial computer simulation experiments it appeared that changes in effective molecular weight as small as 1% ought to be detectable. In the second paper of the series (Springer and Schachman, 1974) the technique was tested by measuring known changes in effective molecular weights of 1–10% produced by the addition of various quantities of D_2O to one of a pair of sample solutions. The technique was successful, but was restricted with respect to concentration range by difficulties in measuring $\bar{c}$. A special three-compartment cell and corresponding Rayleigh mask was therefore developed in order to permit the interferometric determination of $\bar{c}$ as well as of Δc. The advantages and disadvantages of this latter development have not yet been fully assessed. However, Teller (footnote to Springer and Schachman, 1974) has suggested an alternative way of measuring both $\bar{c}$ and Δc interferometrically by means of a six-channel Yphantis cell with the use of a transform which permits concentrations distribution of the solute molecules to be transformed from one co-ordinate system to another.

C. HETEROGENEITY AND NON-IDEALITY

The slope of a plot of ln c vs r^2 gives the molecular weight. For a mixture of homologous macromolecules, assuming $\bar{v}$ is independent of molecular weight and is equal to the (weight-average) $\bar{v}$ values of the mixture, at any radius point in the cell average molecular weights are given by

$$M_w = \frac{RT}{\omega^2(1 - \bar{v}\rho)} \frac{1}{rc} \frac{dc}{dr} \tag{14}$$

and

$$M_z = \frac{RT}{\omega^2(1 - \bar{v}\rho)r} \left[\frac{d^2c}{dr^2} \middle/ \frac{dc}{dr} - \frac{1}{r^2} \right] \tag{21}$$

In fact the number average molecular weight at any radius is also available, provided the rotor is run at a speed high enough to deplete the solute at the meniscus, so that the concentration near the meniscus becomes infinitely small compared with the initial concentration

$$M_n \sim c \middle/ \int_a^r rc \, dr \tag{22}$$

These averages will vary with radius because of the fractionating effect of the centrifugal field and will vary with rotor speed at a given radius. These variations may be exploited in order to obtain information on the molecular weight distribution of the solute.

If a solute behaves non-ideally then the molecular weights obtained at finite concentrations will be apparent molecular weights, and the general form of the fundamental equation is Eqn. (12).

$$\frac{1}{rc} \frac{dc}{dr} = \frac{\omega^2 M(1 - \bar{v}\rho)}{RT[1 + c(\partial \ln y / \partial c)_{P,T}]} \tag{12}$$

In the limiting relationship

$$\frac{1}{M_{app}} = \frac{1}{M} + Bc \tag{13}$$

B is a constant to be determined in the course of the experiment. It is often called the second virial coefficient and has *twice* the value of the analogous coefficient derived through the variation of osmotic pressure with concentration, but the same value as the light-scattering virial coefficient (Creeth and Pain, 1967). If a polydisperse solute behaves non-ideally, then the situation is more complex and has been dealt with in some detail by J. W. Williams (1972). The complexity is probably the main reason for the sedimentation equilibrium method not gaining the general acceptance in high polymer chemistry that it has received in protein physical chemistry. J. W. Williams (1972) comments on this and gives examples to show that the method can indeed provide information about molecular weight averages and distributions for high organic polymers, and that measurements do not have to be made at the Flory temperature to be interpretable. It is clear that this field is developing slowly because the interpretation of the data from sedimentation equilibrium experiments with polydisperse non-ideal solution is considerably more involved than the treatment of light-scattering and osmotic pressure information. Nevertheless Utiyama *et al.* (1969) claimed that the weight average molecular weight of the dissolved polymer could be obtained with an experimental error of $\pm 2\%$ irrespective of the degree of polydispersity. The error involved in the determination of the second virial coefficient is greater, but in favourable cases may not be more than 5%. The question of polydispersity is further dealt with below.

Non-ideal behaviour can result from the presence of electric charge on the macromolecule being examined. Charge has the effect of lowering the apparent molecular weight, and the largest effect ("primary charge effect") is dependent on the concentration of macromolecules. It can usually be eliminated, at least in the case of protein in aqueous solution by using dilute solution and adding electrolytes of ionic strength 0·1 or greater. Even at zero concentration there is a residual effect ("secondary charge effect") but the correction involved for this is rarely more than a few percent in most cases (see Van Holde, 1967). For proteins near to their isoelectric point the correction will be negligible.

If macromolecular components can undergo association or dissociation reactions the analysis of sedimentation data becomes further complicated but in favourable cases much can be learned about equilibria (Nielsen and Beckwith, 1971).

D. The Investigation of Polydispersity

It has already been indicated that the concentration distribution at equilibrium can give information on heterogeneity and can yield molecular weight distributions. The main problems are how to extract the information from the raw data, and in particular, how to deal with the effects of non-ideality.

Biochemists working with proteins in dilute aqueous (buffer) solutions have long used sedimentation equilibrium as a method not only for providing highly accurate molecular weight data, but also as a check that the material is homogeneous. In the simplest terms, treatment of the data (Tanford, 1961) yields straight lines for ideal homogeneous solution, upward-curved plots for heterogeneous solutions and downward-curved plots for non-ideal solutes. Unfortunately non-ideality tends to mask curvature due to heterogeneity. Munk and Cox (1972) have recently re-examined this problem with special reference to protein subunits in very concentrated $(6M)$ guanidinium chloride containing thiols. The use of such solvents is a common technique for dissociating multisubunit proteins but under these conditions proteins do not behave ideally, and the virial coefficients of certain proteins are at best an order of magnitude higher than those of native proteins in normal solvents (Castellino and Barker, 1968). Munk and Cox show, using simulated data, that conventional treatments neglecting non-ideality and heterogeneity may yield erroneous molecular weights particularly with the low speed technique. They suggest a simple procedure which allows the true molecular weight, and an estimate of the virial coefficient, to be extracted from the data. The basic plot of $\ln c$ vs r^2 is not very sensitive as the non-ideality effect is not visually striking. Similarly in plots of $1/r \cdot dc/dr$ vs c, the effects of non-ideality completely masked those of heterogeneity. In contrast, plots of $1/M_{app}$ vs c produced much more difference in slopes for homogeneous and heterogeneous solutes, with and without the effects of non-ideality. This test was also used successfully with several proteins dissolved in $6M$ guanidinium chloride. Munk and Cox (1972) also draw attention to errors resulting from such factors as the evaporation of water from small volumes of these highly concentrated solutions during handling.

J. W. Williams (1972) gives examples of several ways in which heterogeneity and non-ideality have been dealt with for synthetic high polymers in organic solvent systems. For instance Fujita et al. (1960) determined M_w, M_z and B for polystyrene in cyclohexane at concentrations of up to 0.6 g$/100$ ml. They did experiments at temperatures on either side of, but near to, the Flory temperature and showed that to a first approximation plots of $1/M_{w\,app}$ against c_0, the initial concentration, extrapolated to the same point but that the plots were not as linear as might have been desired (Fig. 8). More recently Williams and his co-workers (Osterhoudt and Williams, 1965; Albright and Williams, 1967) developed methods of obtaining weight-average molecular weights and second virial coefficients of non-ideal polydisperse systems by performing a series of equilibrium experiments at different rotor speeds. This technique was used with considerable success with the system polystyrene-toluene ($\bar{M}_w = 267,000$) by Albright and Williams (1967). An alternative method, also involving the

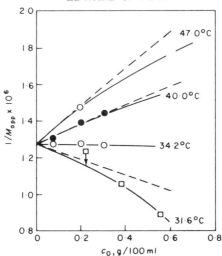

FIG. 8. Measurement of M_w and second virial coefficient of polystyrene in cyclohexane at temperatures near the Flory temperature. Dashed lines show result expected from simple non-ideality. Reprinted with permission from Fujita *et al.*, *J. Amer. Chem. Soc.* **82**, 379. Copyright by the American Chemical Society.

use of several rotor speeds, has been described by Scholte (1968) in which points in the solution column other than the end points are used and this method was applied to the polyethylene-biphenyl system at its theta temperature (123·2°C in this case). Scholte claimed that the overall precision of the method was about ±5% in $\bar{M}_w$, about ±10% in $\bar{M}_n$ and $\bar{M}_z$, and ±25% in $\bar{M}_{z+1}$. At the present time this method does not appear to have been adapted to deal with non-ideal solutions. Both these methods involve experiments at different rotor speeds and require extrapolation of the data to zero rotor speed, which has the advantage that effects due to compressibility of the solvent are eliminated.

Compared with molecular weight averages, the determination of molecular weight distribution is a more formidable task and the reader is referred to the articles by Williams (Williams and Yphantis, 1971) and by Adams *et al.* (1974) for detailed expositions on this topic. Wan and Adams (1974) used sedimentation equilibrium techniques to determine the molecular weight distribution of a sample of dextran, and their approach may be taken as an illustration of the potentialities and also the problems invoved in such determinations. They studied the dextran in aqueous solution at 25°C and used osmotic pressure measurements to obtain $\bar{M}_n$, and sedimentation equilibrium experiments at different speeds in the manner of Albright and Williams (1967) to evaluate $\bar{M}_w$ and B_{LS}, the light scattering second virial coefficient. For example values for $\bar{M}_{w\,app}$ were obtained from

$$\Delta j/j_0 = A M_{w\,app} \tag{23}$$

when j_0 is the original concentration in fringes, Δj is the difference in fringes across the solution column, and $A = (1 - \bar{v}\rho)\omega^2(r_b^2 - r_a^2)/2RT$. The limiting slope of a plot of $\Delta j/j_0$ against A gave the quantity M_w^0 app, the value of $M_{w\,app}$ at zero speed. B_{LS} was then obtained from Eqn. 13. The next step was to correct for non-ideal behaviour and this was done assuming the non-ideality term was approximately the same as B_{LS}. This gave ideal values of c and $dc/d(r^2)$ needed for the determination of molecular weight distributions. Wan and Adams investigated two different methods for determining this distribution. The basis of Donnelly's method (1966) is that the concentration distribution at sedimentation equilibrium is in the form of a Laplace transform, while Scholte's method involves varying the rotor speed and the use of data from several different radii in the solution column at each of the speeds. Scholte (1968, 1970) originally used the method for the study of polystyrene in cyclohexane at 35°C, and the method was restricted to ideal systems. Wan and Adams (1974) compared the distributions obtained by the two methods with the distribution obtained by gel chromatography. They considered that the application of the Donnelly method gave the best results. In any event it seems that one can obtain molecular weight distributions from sedimentation equilibrium experiments on non-ideal solutions and the procedure used in this example is a general one. The contribution of the separation of polymeric components by the centrifugal field to the non-ideal behaviour has been ignored. It is also clear that such results as have been obtained so far have been obtained only by processing the data by computer and several workers have been giving consideration to the computing techniques involved. For example Wiff and Gehatia (1973) considered that the problem of inferring the molecular weight distribution of a polymer in solution from sedimentation equilibrium data was mathematically ill-posed, and they formulated a technique of regularization applicable to the sedimentation equilibrium equation which casts this into a well-posed problem.

At a simpler level Gibbons *et al.* (1973) have developed a new method for assessing the molecular weight distributions of polydisperse materials from sedimentation equilibrium experiments under meniscus depletion conditions. In this approach the quantity $c^{-\frac{1}{3}}$ is plotted against ξ, the reduced radius. This plot is linear when the molecular weight distribution approximates to the most probable, i.e. when $M_n : M_w : M_z : M_{1+z} \ldots$ is as $1:2:3:4$. The amount of and direction of curvature of the plots of $c^{-\frac{1}{3}}$ against ξ gives information on whether the distribution is narrow or broader than the most probable one. For this context the "most-probable" distribution of molecular weights is a Schulz distribution with $h = 1$ (Schulz, 1964) and this is of especial significance as it is the distribution resulting from random break-up of an infinitely large molecule or from macromolecular synthesis by random chain lengthening. Gibbons *et al.*

(1973) investigated two glycoproteins from pig intestine and two dextrans by this procedure. Whereas the dextran (which gave markedly curved plots of $\ln c$ against c/r^2) gave linear plots of $c^{-\frac{1}{3}}$ against ξ, an epithelial glycoprotein gave curved plots of $c^{-\frac{1}{3}}$ against ξ which indicates that these materials have a wider distribution of molecular weights than that expected from random synthesis or breakdown.

E. AGGREGATION

Many substances in solution undergo self-association reactions (Adams, 1967a, b):

$$nP_1 \rightleftarrows P_n \quad \text{where } n = 2, 3 \ldots \tag{24}$$

or

$$nP_1 \rightleftarrows qP_2 + mP_3 + \ldots \tag{25}$$

and such systems may be studied by ultracentrifugation. Protein chemists have, for example, found the sedimentation equilibrium method particularly useful, but in fact many substances are known to undergo self-association, including soaps and detergents, chlorophylls, organic dyes, and synthetic polymers. Often the interactions are extremely complex and may additionally involve small ions or molecules. The sedimentation velocity method has frequently been useful in studying the latter types of interactions. The literature on interacting systems is considerable and much of it deals with protein systems in dilute aqueous environments. A brief survey of the field is given below, and further information on this topic as applied to proteins may be found in the articles by Nichol et al. (1964) and by Adams (1967a, b). Though both sedimentation velocity and approach to equilibrium methods can be used to investigate an aggregating system, the sedimentation equilibrium method is particularly applicable. Not only does it rest on rigorous thermodynamic foundations, but also a very wide concentration range may be studied.

The investigation of an aggregating system by sedimentation equilibrium can be divided into three sections: (a) determinations of concentration as a function of radius for the system at sedimentation equilibrium; (b) calculation of apparent molecular weight as a function of concentration; and (c) interpretation of the data in terms of aggregation. From data on c as a function of r, apparent molecular weights can be calculated in a number of ways but most commonly the equation

$$M_{w\,app} = \frac{2RT}{\omega^2(1 - \bar{v}\rho)} \frac{d \ln c}{dr^2} \tag{26}$$

is used. Often the plot of $\ln c$ vs r^2 has only slight curvature and determination of this curvature puts severe demands on the accuracy of the data.

The final plot of $M_{w\,app(c)}$ vs c obtained is curved upwards, but this may result not from self-association, but from heterogeneity. To distinguish between these two it is necessary to perform other equilibrium experiments at different initial concentrations. For an associating system all sets of values will lie on one curve whereas for a heterogeneous system different initial concentrations will yield separate curves.

It is in the interpretation of the data that most problems arise. Adams (1967a, b) has described in detail how analysis may be achieved. In outline one starts by writing down the type of association one thinks is present. If n species are present, one sets down $n + 1$ equations, one for each associating species plus one for the non-ideal behaviour. One then eliminates terms starting with that for the highest associating species, until one equation in two unknowns, c_1 and BM_1 is obtained. Using equation

$$c_1 = \alpha \exp\left(-BM_1c\right) \qquad (27)$$

where α is the apparent concentration of monomer, see Adams, 1967a), one then obtains an equation for BM_1 which is solved by making successive approximations of the unknown BM_1. With the proper value for BM_1 one works backwards successively to obtain values of c, equilibrium constants and molecular weight ratios ($M_1/M_{n(c)}$ and $M_1/M_{w(c)}$). Finally it is necessary to test for the type of association one has proposed by comparison of the various quantities calculated and measured. Adams (1967a) gives a description of how this is done step-by-step, for a proposed monomer–dimer–trimer system and the reader is referred to this review and to other of Adams' papers (Adams, 1967b; Adams and Lewis, 1968) for further details of this analysis. It is also shown how the analysis can fairly simply be extended to cope with indefinite self-associations (Adams and Lewis, 1968). More recently Chun and Kim (1970) introduced simplified graphical procedures for the evaluation of the mode of association from weight-average molecular weights as a function of concentration. These workers were critical of the techniques of Roark and Yphantis (1969) which called for a double differentiation, thus introducing a high probability of error especially in the region near zero concentration. Instead they eliminated the $\bar{M}_z$ value and replaced it by the more workable weight fraction monomer of Adams and Williams (1964). In a later paper Yphantis and Roark (1972) considered the effects of non-ideality in associating systems in great detail and used reciprocal weight or number average molecular weights to generate new averages, the "ideal moments", that were independent of the second virial coefficients. These ideal moments were useful for characterizing non-ideal self-associating systems.

In addition to non-ideality, another complication of studying associating systems is the effect of pressure, and in several cases protein systems have been shown to be sensitive in this way (Josephs and Harrington, 1967; Kegeles *et al.*, 1969).

Probably the biggest impetus to the study of associating systems has come from the use of computers in handling the data. It is of course a great help if the ultracentrifuge output is in the form of computer-readable data, but even if it is not it seems likely that progress would have been slow without computers to cope with the calculations. Chun *et al.* (1968) and Chun and Fried (1967), for example, showed how equilibrium constants could be compiled using stepwise polynomial regression analysis. They re-investigated the polymerization of insulin (dimer, trimer, tetramer, hexamer, and possibly higher polymers) taking into account the effect of the unit charge of the monomeric species.

The use of what is probably the ultimate in ultracentrifuge technology at the present time, that is an ultracentrifuge directly interfaces with a computer, to examine interacting protein systems, was the analysis of haemoglobin performed by Crepeau *et al.* (1974). The equipment used by these investigators has already been described (p. 10). The oxygen-carrying protein haemoglobin is one of the most well studied proteins, and is normally thought of as existing as a tetramer. It was particularly suitable for study with this system because of the great amount of information on its dissociation already available, and also because it is known to be a system which does not involve infinite self-association. There has been controversy as to whether the dimer or the tetramer is the functional unit, and whether monomers exist under certain experimental conditions.

The data Crepeau *et al.* obtained (optical density vs position in the cell) were analysed by the computer to fit an equation of the form

$$\text{Optical Density} = \text{baseline} + A \exp(\alpha M_i r^2) + B \exp(2 + M_i r^2) \tag{28}$$

where $\alpha = 2RT\omega^2/(1 - \bar{v}\rho)$ and M_i is the molecular weight of monomer in a monomer–dimer equilibrium, or of dimer in dimer–tetramer equilibrium. After calculating the constants A and B, dissociation constants can be calculated from:

$$K_{4,2} = [\text{dimer}]^2/[\text{tetramer}] \tag{29}$$

where $[\text{dimer}] = A/1000\epsilon$ and $[\text{tetramer}] = B/1000\epsilon$, ϵ being the haem millimolar extinction coefficient. The relationship $K_{4,2} = A^2/1000\epsilon B$ was therefore evaluated. A programme in Fortran 4 was used for the least squares fitting procedure, based on a general multiple linear regression fit to a function which is linear in its coefficients. The programme also provided three statistical parameters: the coefficient of the determination, the chi-squared parameter, and the F-test. From data obtained with the carbonmonoxyderivative of haemoglobin at pH 7 and 20°C, a value for $K_{4,2}$ of $1\cdot09 \times 10^{-6}$ (s.d. $= 5\cdot5 \times 10^{-8}$)M was calculated. Examination of the statistical parameters showed that the data could be described in terms of two exponentials corresponding to dimer and tetramer.

Howlett and Nichol (1972a, b) have recently presented methods for computer simulation of sedimentation equilibrium distributions obtained with interacting systems involving a polymerizing acceptor and a series of acceptor–ligand complexes. This treatment permitted consideration of both non-ideality effects and of apparent volume changes. The advantage of simulated distributions is that they may be computed for various possible models and compared directly with the experimental result without recourse to differentiation procedures needed to evaluate average molecular weights.

Brief mention must also be made of how the transport or sedimentation velocity method may be used to study interacting systems. Gilbert (1955) initiated the interpretation of sedimentation patterns produced by reversibly associating macromolecules by considering the puzzling data presented for the sedimentation behaviour of the enzyme α-chymotrypsin at pH 7·9 and low ionic strength. He concluded that a monomer–dimer system would yield a single schlieren peak, whereas hexamer formation would lead to the appearance of two peaks. In these early studies the effects of diffusion had to be neglected, but since that time it has been possible to consider not only diffusional effects, but also the effects of the binding of small molecules or ions (Cann, 1970), and the influence of kinetically controlled interactions on the schlieren patterns (Cann and Oates, 1973; Cann and Kegeles, 1974). Cann and Oates computed sedimentation patterns for several kinetically controlled macromolecular interactions: irreversible isomerization, dimerization, dissociation into identical subunits, and irreversible and reversible dissociation of a complex, and Cann and Kegeles computed theoretical sedimentation patterns for reversible, kinetically controlled macromolecular dimerization reactions, both nonmediated and ligand-mediated. It is interesting to note from the latter work that the patterns obtained for half-times of reaction as long as 20 to 60 sec. are essentially the same as those for instantaneous re-equilibration during differential transport of monomer and dimer. For half-times of dimer dissociation of less than about 200 sec. the reaction boundary is only resolved into two peaks if dimerization is ligand-mediated.

Mention has already been made of the possibility of pressure-dependence of associating systems and such systems have been investigated by the sedimentation velocity method. Indeed, the association-dissociation behaviour of lobster haemocyanin originally investigated by Morimoto and Kegeles (1971) and interpreted by Cann and Kegeles (1974) is such a pressure-dependent system (see Harrington and Kegeles, 1974). Recently Johnson et al. (1973) have considered instabilities in the pressure-dependent sedimentation of monomer–polymer systems. For such systems it is predicted that the occurrence of negative gradients in the total concentration of solute should lead, in the absence of a stabilizing density gradient, to convective disturbances. Johnson *et al.* (1973) present accurate numerical solutions to the Lamm equation to illustrate this phenomenon.

Finally an interesting technique for the enumeration of species in a self-associating protein system may be mentioned. Godschalk (1971) described a method for analysing self-associating systems from the observation of boundary spreading in the ultracentrifuge recorded by means of the absorption optical system. The concentration profiles obtained at intervals after the formation of the boundary were subjected to an integral transformation, and the resulting transforms were investigated by matrix rank analysis, thus yielding the number of interacting species in the system. This method has long been used as a method of spectral analysis.

V. Sedimentation in a Density Gradient

Both velocity and equilibrium sedimentation may be performed in density gradients. In the former instance the gradient serves to stabilize sedimenting zones and in the latter case zones of macromolecules form in regions of isodensity. The biological sciences have greatly benefited from the application of such methods to the study and the separation of subcellular particles and of nucleic acids. The methods appear to have found less application in polymer science so far. Nevertheless a brief indication of the principles and potentialities will be given.

A. RATE ZONAL METHOD

Velocity or zone sedimentation may be performed in an artificially preformed gradient of sucrose, NaCl, or other suitable substance usually in a swing-out preparative ultracentrifuge rotor. Macromolecules or particles initially applied as a zone centripetally, move through the gradient and would eventually settle at the bottom of the tube if centrifugation continued indefinitely. The purpose of the sucrose gradient here is to stabilise the sedimenting zone(s) and to prevent convection. At some suitable time the rotor is very gently brought to a halt and fractions are carefully drawn from the tubes for analysis. The method has chiefly been used as a separatory technique, but of course sedimentation coefficients may be calculated and these may be used to identify particular fractions. Components in the initial zone sediment roughly in the order of their sedimentation coefficients. However it is difficult to relate such coefficients to those ($s_{20,w}^0$) normally obtained by analytical ultracentrifugation in dilute buffer solutions. Often standard, identifiable "markers" of known sedimentation coefficient are included in a mixture to be separated and are used for internal calibration.

Schumaker and Rosenbloom (1965) have presented a theoretical treatment for the determination of sedimentation coefficients from observations of the migrations of zones. On the basis of the assumption that mass transport is caused by sedimentation and diffusion alone, these authors showed that the velocity of the centre of gravity of the migrating zone was equal to the average sedimentation velocity of the individual particles in a centrifuge cell of uniform cross-section area in a centrifugal field. They also considered in detail the effects of concentration dependence of both sedimentation and diffusion coefficients, and the effects of superimposed viscosity and density gradient. Both single-component and polydisperse systems were considered. Belli (1973) has discussed the standardization of the calculation of sedimentation coefficients of macromolecules from band sedimentation techniques. When a sample of macromolecules suspended in a solvent is layered on to a denser bulk solution, the molecules do not sediment in this solution alone, but sediment in a mixture of the bulk solution and sample solvent. Belli showed that in the case of band sedimentation of ribosomes under typical experimental conditions, standardization made in the usual way leads to the overestimation of the sedimentation coefficient. If the diffusion effect of sample solvent into the bulk gradient solution is taken into account then a correct coefficient results. A method for calculating the standard coefficient by means of a computer is proposed by Belli.

It is possible to obtain information about diffusion coefficients from the spreading of the zones with time (Vinograd *et al.*, 1963). Halsall and Schumaker (1972) used this technique to determine the diffusion coefficient of the DNA from the virus M13. They used low rotor speeds and 24–72 h runs with a zonal rotor, and observed zone spreading. Using their value for the diffusion coefficient (0.95×10^{-7} cm^2 s^{-1}) together with a sedimentation coefficient obtained from band sedimentation experiments in the analytical ultracentrifuge, they calculated a molecular weight of 1.7×10^{6} for this DNA.

It has been customary when using sucrose gradients to regard the density distribution as unchanging during the period of centrifugation. This assumption is justified up to rotor speed of about 40,000 rev/min, but not at the higher speeds now available. McEwen (1967) has computed sucrose (and also NaCl) distributions at sedimentation equilibrium in swing-out rotors at a number of speeds up to 65,000 rev/min, for various initial gradients. Using the tables provided, the solution properties are computed and a graphic method is then used for the rapid determination of density distribution.

In general, results from zonal and band sedimentation experiments agree with those from boundary experiments and thus the method is validated

(Creeth and Pain, 1967). It may also be extended for use in the analytical, as opposed to the preparative, ultracentrifuge by (a) generating a "diffusion gradient" during the run instead of preforming it (see below), and (b) using special band-forming centrepieces in the ultracentrifuge cells to apply the initial zone when the rotor is spinning. Advantages of the method include sensitivity to heterogeneity and economy of material (Vinograd et al., 1963; Vinograd and Bruner, 1966): since very low concentrations are used, long extrapolations to infinite dilution may be avoided. Disadvantages include uncertainty in the parameters relating to the initial distribution.

B. Equilibrium Density Gradient Centrifugation

If it is arranged for a given macromolecule sedimenting in a density gradient that the factor $(1 - \bar{v}\rho)$ becomes equal to zero at some point, then the macromolecule comes to rest at that point. Meselson et al. (1957) and Meselson and Stahl (1958) were the first to establish this as one of the most important tools of molecular biology. DNA in 7·7 M caesium chloride was centrifuged in an analytical ultracentrifuge at 31,410 rev/min. At the beginning of the experiment the solution was homogeneous, containing CsCl and DNA throughout; after 5 h the CsCl had formed an equilibrium distribution of concentration and therefore of density. After 30 h the DNA had collected into a zone at the position of isodensity. Under these equilibrium conditions, the width of the more or less Gaussian profile of the band depends on the molecular weight and density homogeneity of the macromolecular species. Thus if two types of DNA are present, two bands form. The sensitivity of the method is very great and this was exploited in what has turned out to be one of the classical experiments of molecular biology. Bacterial DNA was prepared containing one or other of the isotopes of nitrogen, ^{14}N or ^{15}N, or both, one type in each strand of the double helix. All three types of DNA, i.e. ^{14}N-, ^{15}N-, and ^{14}N,^{15}N- were resolvable on a CsCl gradient. The density differences involved in this experiment were of the order of 0·007 g ml^{-1}, and by the use of even more shallow gradients differences as low as 0·001 g ml^{-1} may be resolved.

The theory of the technique has been discussed in detail elsewhere (Vinograd and Hearst, 1962; Vinograd, 1963). Two articles by Szybalski (1968a, b) are particularly useful as an introduction to the practical aspects of the technique.

In the ideal case the band of material formed at equilibrium has a Gaussian distribution and its width is inversely proportional to the square root of the molecular weight of the macromolecule. The partial specific volume is affected by solvation and pressure effects and of course the solvent environment changes throughout the gradient. The molecular weight, assuming a knowledge of $\bar{v}_0$, the partial specific volume under the experimental conditions, is given by

$$M_0 = \frac{RT}{\sigma^2 \bar{v}_0 (d\rho/dr)\omega^2 r_0}$$

(30)

where σ is the standard deviation of the distribution of solute in the zone, $d\rho/dr$ is the effective density gradient, and r_0 is the radial distance to the centre of the band. The method has been extensively exploited as a technique for determining molecular weights of nucleic acids (see Szybalski, 1968a). For a consideration of the problems involved in such determinations the reader is referred to the review by Creeth and Pain (1967).

Originally gradients of CsCl were used, but more recently the use of Cs_2SO_4 has allowed the formation of steeper gradients, allowing RNA to be investigated. Ludlum and Warner (1965) have determined the basic thermodynamic data necessary for calculating gradients in Cs_2SO_4. They determined density, refractive index, and osmotic and activity coefficient data for Cs_2SO_4 over the range 0–3·6 M. Mixed $CsCl$–Cs_2SO_4 gradients have also been used for the investigation of the equilibrium sedimentation of RNA (Szybalski, 1968a). Hahn (1972) examined DNA and RNA samples in Cs_2SO_4 gradients in the presence of aprotic polar solvents. Here Cs_2SO_4 was used rather than CsCl because of its better solubility, and the solvents used were water with ethylene glycol, formamide, and dimethyl sulphoxide.

References to the use of equilibrium density gradient centrifugation in polymer science are more sparse. A theoretical basis for the study of synthetic polymers has been presented by Hermans and Ende (1963). It should be possible to use the method to measure molecular weight distributions of mixtures of homologous polymers as well as composition distributions of copolymers whose components vary in density. However there appear to be many practical difficulties (Williams and Yphantis, 1971), and the difficulties of finding a theta solvent have already been mentioned (Nakazawa and Hermans, 1971). These workers tried to investigate the compositional distribution in styrene-methacrylate copolymers, but their conclusion was that data from density gradient centrifugation alone were insufficient and the final interpretation of the data was somewhat ambiguous. Cowie and Toporowski (1969) used density gradient ultracentrifugation to investigate the various tactic forms of poly–α-methylstyrene. They employed a benzene-chloroform gradient and used comparative rather than absolute measurements to determine apparent specific volumes. They found a difference of $0\cdot0094$ ml g^{-1} in apparent specific volumes between atactic and a highly syndiotactic poly-α-methyl styrene.

VI. Conclusions

It is seen that the analytical ultracentrifuge is a tool of considerable versatility for the study of high polymers and macromolecules. The range of molecular sizes that may be investigated is large, and information may be obtained on heterogeneity, polydispersity, non-ideal behaviour, and association–dissociation reactions. Probably the sedimentation equilibrium method has the most to offer in all these areas, particularly if data from the use of other techniques such as osmotic pressure measurement and light-scattering are not available. Nevertheless the sedimentation velocity method is often very useful as a rapid way of checking purity, estimating the amounts of a number of non-interacting components in a solution, and especially of investigating the interaction of small ions and molecules with associating or dissociating macromolecules, and of studying pressure effects. The gradient methods have found particular applications in molecular biology, but less so in the field of polymer science so far. Perhaps the innovation most likely to speed up the development of ultracentrifuge applications is the technique of using a photoelectric scanner with the absorption optical system, particularly because the output from such an arrangement is eminently suitable for direct input to a computer.

References

Ackers, G. K. (1967). *J. Biol. Chem.* **242**, 3237–8.

Adams, E. T. (1967a). *Fractions, No. 3.* Beckman Instruments Inc., Palo Alto, Calif., U.S.A.

Adams, E. T. (1967b). *Biochemistry* **6**, 1864–1871.

Adams, E. T. and Lewis, M. S. (1968). *Biochemistry* **7**, 1044–1053.

Adams, E. T. and Williams, J. W. (1964). *J. Amer. Chem. Soc.* **86**, 3454–3461.

Adams, E. T., Wan, P. J., Soucek, D. A. and Barlow, G. H. (1974). *In*: "Recent Trends in the Determination of Molecular Weights", ACS Advances in Chemistry Series.

Albright, D. A. and Williams, J. W. (1967). *J. Phys. Chem.* **71**, 2780–2786.

Ansevin, A. T., Roark, D. E. and Yphantis, D. A. (1970). *Analyt. Biochem.* **34**, 237–261.

Archibald, W. J. (1947). *J. Phys. Coll. Chem.* **51**, 1204.

Baghurst, P. A. and Stanley, P. E. (1970). *Analyt. Biochem.* **33**, 168–173.

Baldwin, R. L. (1957). *Biochem. J.* **65**, 503–512.

Bancroft, F. C. and Freifelder, D. (1970). *J. Molec. Biol.* **54**, 537–546.

Baurain, R. M., Moreux, J. C. and Lamy, F. (1973). *Biochim. Biophys. Acta* **293**, 18–29.

Belli, M. (1973). *Biopolymers* **12**, 1853–1864.

Billick, I. H. (1962). *J. Polymer Sci.* **62**, 167–177.

Bowen, T. J. (1966). *Biochem. Soc. Symp.* **26**, 1–24.

Bowen, T. J. (1971). "An Introduction to Ultracentrifugation". Wiley-Interscience Ltd., London.

Cann, J. R. (1970). "Interacting Macromolecules". Academic Press, New York and London.

Cann, J. R. and Kegeles, G. (1974). *Biochemistry* **13**, 1868–1874.

Cann, J. R. and Oates, D. C. (1973). *Biochemistry* **12**, 1112–1119.

Castellino, F. J. and Barker, R. (1968). *Biochemistry* **7**, 2207–2217.

Cerny, L. E., McTiernan, J. and Stasiw, D. M. (1973). *J. Polymer Sci. Symp.* **42**, 1455–1465.

Charlwood, C. A. (1971). *Biochem. J.* **125**, 1019–1026.

Chauvel, B. and Daniel, J. C. (1974). *Polymer Preprints* **15**, 329–333.

Chervenka, C. H. (1969). "A Manual of Methods for the Analytical Ultracentrifuge". Spinco Division of Beckman Instruments, Inc., Palo Alto, Calif., U.S.A.

Chervenka, C. H. (1970). *Analyt. Biochem.* **34**, 24–29.

Chervenka, C. H. (1971). *Fractions, No. 1.* Beckman Instruments, Inc., Palo Alto, Calif., U.S.A.

Chun, P. W. and Fried, M. (1967). *Biochemistry* **6**, 3094–3098.

Chun, P. W. and Kim, S. J. (1970). *Biochemistry* **9**, 1957–1961.

Chun, P. W., Fried, M. and Yee, K. S. (1968). *J. Theoret. Biol.* **19**, 147–158.

Coates, J. H. (1970). *In*: "Physical Principles and Techniques of Protein Chemistry (Part B)" (Leach, S. J., ed.), pp. 2–98. Academic Press, New York and London.

Cowie, J. M. G. and Toporowski, P. M. (1969). *Eur. Polymer J.* **5**, 493.

Creeth, J. M. and Knight, C. G. (1965). *Biochim. Biophys. Acta* **102**, 549–558.

Creeth, J. M. and Pain, R. H. (1967). *Progr. in Biophys. and Mol. Biol.* **17**, 219–287.

Crepeau, R. H., Edelstein, S. J., and Rehmar, M. J. (1972). *Analyt. Biochem.* **50**, 213–233.

Crepeau, R. H., Hensley, C. P. and Edelstein, S. J. (1974). *Biochemistry* **13**, 4860–4865.

Dishon, M., Weiss, G. H. and Yphantis, D. A. (1967). *Biopolymers* **5**, 697–713.

Donnelly, T. H. (1966). *J. Phys. Chem.* **70**, 1862–1871.

Dubin, S. B., Lunacek, J. H. and Benedek, G. B. (1967). *Proc. Natl. Acad. Sci. U.S.A.* **57**, 1164–1171.

Dyson, R. D. (1970). *Analyt. Biochem.* **33**, 193–199.

Flory, P. J. (1953). *In*: "Principles of Polymer Chemistry", p. 546. Cornell Univ. Press, Ithaca, U.S.A.

Foord, R., Jakeman, E., Oliver, C. J., Pike, E. R., Blagrove, R. J., Wood, E. J. and Peacocke, A. R. (1970). *Nature, Lond.* **227**, 242–245.

Fujita, H. (1956). *J. Amer. Chem. Soc.* **78**, 3598–3604.

Fujita, H. (1962). "The Mathematical Theory of Sedimentation Analysis", Academic Press, New York and London.

Fujita, H., Linklater, A. M. and Williams, J. W. (1960). *J. Amer. Chem. Soc.* **82**, 379–386.

Gerhardt, J. C. and Schachman, H. K. (1968). *Biochemistry* **7**, 538–552.

Gibbons, R. A., Dixon, S. N. and Pocock, D. H. (1973). *Biochem. J.* **135**, 649–655.

Gilbert, G. A. (1955). *Discuss. Faraday Soc.* **20**, 68.

Godschalk, W. (1971). *Biochemistry* **10**, 3284–3289.

46   EDWARD J. WOOD

Gosting, L. J. (1952). *J. Amer. Chem. Soc.* **74**, 1548–1552.

Gosting, L. J. (1956). *Adv. Prot. Chem.* **11**, 429–554.

Gottschalk, A. and McKenzie, H. A. (1961). *Biochim. Biophys. Acta* **54**, 226–235.

Hahn, C. W. (1972). *J. Polymer Sci. (C)* **39**, 293–304.

Halsall, H. B. and Schumaker, V. N. (1972). *Biochemistry* **11**, 4692–4695.

Hanlon, S., Lamers, K., Lauterback, G., Johnson, R. and Schachman, H. K. (1962). *Arch. Biochem. Biophys.* **99**, 157–174.

Harrington, W. F. and Kegeles, G. (1974). *Methods in Enzymol.* **27**, 306–345.

Haschemeyer, R. H. and Haschemeyer, A. E. V. (1973). "Proteins—a Guide to Study by Physical and Chemical Methods", pp. 140–157 and 181–195. Wiley and Son, New York.

Hermans, J. J. and Ende, H. A. (1963). *J. Polymer Sci. (C)* **Symp 1**, 161–177 and 179–185.

Howlett, G. J. and Nichol, L. W. (1972a). *J. Biol. Chem.* **247**, 5681–5685.

Howlett, G. J. and Nichol, L. W. (1972b). *J. Phys. Chem.* **76**, 2740–2743.

Johnson, P. and Mihalyi, E. (1965). *Biochim. Biophys. Acta* **102**, 467–475.

Johnson, M., Yphantis, D. A. and Weiss, G. H. (1973). *Biopolymers* **12**, 2477–2490.

Josephs, R. and Harrington, W. F. (1967). *Proc. Natl. Acad. Sci. U.S.A.* **58**, 1587–1594.

Kado, C. I. and Black, D. R. (1968). *Virology* **36**, 137–139.

Kawahara, K. (1969). *Biochemistry* **8**, 2551–2557.

Kegeles, G., Rhodes, L. and Bethune, J. L. (1967). *Proc. Natl. Acad. Sci. U.S.A.* **58**, 45–51.

Kegeles, G., Kaplan, S. and Rhodes, L. (1969). *Ann. N.Y. Acad. Sci.* **164**, 183–191.

Kirschner, M. W. and Schachman, H. K. (1971). *Biochemistry* **10**, 1900–1918 and 1919–1926.

Kotaka, T. and Donkai, N. (1968). *J. Polymer Sci. A-2*, **6**, 1457–1479.

Lamers, K., Putney, F., Steinberg, I. Z. and Schachman, H. K. (1963). *Arch. Biochem. Biophys.* **103**, 379–400.

Lamm, O. (1929). *Z. Physik. Chem.* **A143**, 247.

Lewis, J. A. and Barker, N. F. (1971). *J. Phys. Chem.* **75**, 2507–2515.

Lloyd, P. H. (1974). "Optical Methods in Ultracentrifugation, Electrophoresis and Diffusion". Clarendon Press, Oxford.

Lloyd, P. H. and Esnouf, M. P. (1974). *Analyt. Biochem.* **60**, 25–44.

Ludlum, D. B. and Warner, R. C. (1965). *J. Biol. Chem.* **240**, 2961–2965.

McCallum, M. A. and Spragg, S. P. (1972). *Biochem. J.* **128**, 380–402.

McEwen, C. R. (1967). *Analyt. Biochem.* **19**, 23–39.

McNeil, B. J. and Bethune, J. L. (1973). *Biochemistry* **12**, 3244–3253 and 3254–3259.

Meselson, M. and Stahl, F. W. (1958). *Proc. Natl. Acad. Sci. U.S.A.* **74**, 671–682.

Meselson, M., Stahl, F. W. and Vinograd, J. (1957). *Proc. Natl. Acad. Sci. U.S.A.* **43**, 581–588.

Meyerhof, G. and Sutterlin, N. (1973). *J. Polymer Sci. Symp.* **42**, 943–949.

Morimoto, K. and Kegeles, G. (1971). *Arch. Biochem. Biophys.* **142**, 247–257.

Munk, P. and Cox, D. J. (1972). *Biochemistry* **11**, 687–697.

Nakazawa, A. and Hermans, J. J. (1971). *J. Polymer Sci. A-2*, **9**, 1871–1885.

Neet, K. E. and Putnam, F. W. (1965). *J. Biol. Chem.* **240**, 2883–2887.

Neilsen, H. C. and Beckwith, A. C. (1971). *Agric. Food Chem.* **19**, 665–668.

Nichol, L. W., Bethune, J. L., Kegeles, G. and Hess, E. L. (1964). *In*: "The Proteins" (Neurath, H. ed.), Vol. 2, pp. 305–403. Academic Press, New York and London.

Oncley, J. L. (1969). *Biopolymers* **7**, 119–132.

Osterhoudt, H. W. and Williams, J. W. (1965). *J. Phys. Chem.* **69**, 1050–1056.

Paetkau, V. H. (1967). *Biochemistry* **6**, 2767–2774.

Pandit, M. W. and Rao, M. S. N. (1974). *Biochim. Biophys. Acta* **371**, 211–218.

Paul, C. H. and Yphantis, D. A. (1972). *Analyt. Biochem.* **48**, 588–604 and 605–612.

Pita, J. C. and Müller, F. J. (1972). *Analyt. Biochem.* **47**, 395–407.

Pita, J. C. and Müller, F. J. (1973). *Biochemistry* **12**, 2656–2665.

Richards, E. G. and Schachman, H. K. (1957). *J. Amer. Chem. Soc.* **79**, 5324–5325.

Richards, E. G. and Schachman, H. K. (1959). *J. Phys. Chem.* **63**, 1578–1591.

Richards, E. G., Bell-Clark, J., Kirschner, M., Rosenthal, A. and Schachman, H. K. (1972). *Analyt. Biochem.* **46**, 295.

Roark, D. E. and Yphantis, D. A. (1969). *Ann. N.Y. Acad. Sci.* **164**, 245–278.

Schachman, H. K. (1959). "Ultracentrifugation in Biochemistry", Academic Press, N.Y.

Schachman, H. K. (1963). *Biochemistry* **2**, 887–905.

Schachman, H. K. and Edelstein, S. J. (1966). *Biochemistry* **5**, 2681–2705.

Schachman, H. K., Gropper, L., Hanlon, S. and Putney, F. (1962). *Arch. Biochem. Biophys.* **99**, 175–190.

Scholte, T. G. (1968). *J. Polymer Sci. A-2*, **6**, 111–127.

Scholte, T. G. (1970). *Eur. Polymer J.* **5**, 51.

Schulz, G. V. (1944). *Z. Phys. Chem.* **A193**, 168.

Schumaker, V. N. (1968). *Biochemistry* **7**, 3427–3431.

Schumaker, V. N. and Adams, P. (1968). *Biochemistry* **7**, 3422–3427.

Schumaker, V. N. and Rosenbloom, J. (1965). *Biochemistry* **4**, 1005–1011.

Schumaker, V. N., Wlodawer, A., Courtney, J. T. and Decker, K. M. (1970). *Analyt. Biochem.* **34**, 359–365.

Sellen, D. B. (1973). *J. Polymer Sci. Symp.* **42**, 1205–1208.

Spragg, S. P., Travers, S. and Saxton, T. (1965). *Analyt. Biochem.* **12**, 259–270.

Springer, M. S. and Schachman, H. K. (1974). *Biochemistry* **13**, 3726–3733.

Springer, M. S., Kirschner, M. W. and Schachman, H. K. (1974). *Biochemistry* **13**, 3718–3725.

Szybalski, W. (1968a). *Fractions, No. 1.* Beckman Instruments Inc., Palo Alto, Calif., U.S.A.

Szybalski, W. (1968b). *Methods in Enzymology* **12**, 330–360.

Tanford, C. (1961). "Physical Chemistry of Macromolecules", John Wiley & Sons, New York.

Tung, L. H. (1969). *J. Polymer Sci. A-2*, **7**, 47–55.

Tung, L. H. (1971). *J. Polymer Sci. A-2*, **9**, 759–762.

Utiyama, H., Tagata, N. and Kurata, M. (1969). *J. Phys. Chem.* **73**, 1448–1454.

Van Holde, K. E. (1960). *J. Phys. Chem.* **64**, 1582–1584.

Van Holde, K. E. (1967). *Fractions, No. 1*, Beckman Instruments Inc., Palo Alto, Calif., U.S.A.

Vinograd, J. (1963). *Methods in Enzymol.* **6**, 854–870.

Vinograd, J. and Bruner, R. (1966). *Biopolymers* **4**, 131–156 and 157–170.

Vinograd, J. and Hearst, J. E. (1962). *Fortschr. Chem. org. Natstoff.* **20**, 372–422.

Vinograd, J., Bruner, R., Kent, R. and Weigle, J. (1963). *Proc. Natl. Acad. Sci. U.S.A.* **49**, 902–910.

Wales, M. and Rehfeld, S. J. (1962). *J. Polymer Sci.* **62**, 179–196.

Wan, P. J. and Adams, E. T. (1974). *Polymer Preprints* **15**, 509–514.

Weiss, G. H. and Yphantis, D. A. (1972). *J. Polymer Sci.* **A10**, 339–344.

Wiff, D. R. and Gehatia, M. (1973). *J. Polymer Sci. Symp.* **43**, 219–234.

Williams, J. W. (1972). "Ultracentrifugation of Macromolecules", Academic Press, New York and London.

Williams, J. W. and Saunders, W. M. (1954). *J. Phys. Chem.* **58**, 854–859.

Williams, J. W., van Holde, K. E., Baldwin, R. L. and Fujita, H. (1958). *Chem. Rev.* **58**, 715–806.

Williams, R. C. (1972). *Analyt. Biochem.* **48**, 164–171.

Williams, R. C. and Yphantis, D. A. (1971). *In*:"Encyclopedia of Polymer Chemistry (Techniques)" (Bikales, N. M., Ed.), Vol. 14, pp. 97–116. Interscience Publishers, New York.

Wood, E. J., Dalgleish, D. G. and Bannister, W. H. (1971). *Eur. J. Biochem.* **18**, 187–193.

Yamamoto, A., Noda, I. and Nagasawa, M. (1970). *Polymer J.* **1**, 304.

Yphantis, D. A. (1960). *Ann. N.Y. Acad. Sci.* **88**, 586–601.

Yphantis, D. A. (1964). *Biochemistry* **3**, 297–317.

Yphantis, D. A. and Roark, D. E. (1972). *Biochemistry* **11**, 2925–2934.

NOTE ADDED IN PROOF: A compendium of articles on the ultracentrifuge has appeared, since this article was written, as Volume 5 of "Biophysical Chemistry" entitled "Fifty Years of the Ultracentrifuge", proceedings of a conference in Bethesda, Maryland, in February, 1975 (edited by Lewis, M. S. and Weiss, G. H.). North Holland Publishing Co., Amsterdam, New York and Oxford, 1976.

2. Viscometry

L. KUDLÁČEK

*Department of Fibrous Materials, University of Chemical
Technology, Pardubice, Czechoslovakia*

1. Introduction

Much valuable information on polymer characterization and technological
processing can be obtained from measuring polymer solutions and melts
viscosity. Measurement techniques thereof are not exacting as far as
equipment is concerned, their accuracy being relatively very good.

Flow properties of solutions and melts of linear polymers depend upon a
number of parameters. Their relationships towards viscosity were most
carefully treated from the theoretical point of view for very diluted solutions; here they are mostly used for determination of the size and shape of
the macromolecule. Analogous relationships are also available for concentrated solutions and for melts, though no such accuracy has so far been
obtained.

Viscosity alone has been defined as a ratio of shear stress τ, between
parallel layers of liquid, and velocity gradient γ perpendicularly to the flow
direction:

$$\eta = \frac{\tau}{\gamma} \qquad \begin{array}{l} \mathrm{kg\,m^{-1}\,s^{-1}} \\ \mathrm{N\,s\,m^{-2}} = \mathrm{Pa\,s} \end{array}$$

$$\tau = \frac{f}{A} \qquad \begin{array}{l} \mathrm{kg\,m^{-1}\,s^{-2}} \\ \mathrm{N\,m^{-2}} = \mathrm{Pa} \end{array} \qquad (1)$$

$$\gamma = \frac{dv}{dy} \qquad \mathrm{s^{-1}}$$

f = force $(\mathrm{kg\,m\,s^{-2}}) = (\mathrm{N})$, A = square $(\mathrm{m^2})$, v = velocity $(\mathrm{m\,s^{-1}})$, y =
perpendicular direction to the direction of the acting force (m).

The above-given definition holds for dynamic viscosity. The unit of
viscosity, poise P, equals therefore $0 \cdot 1\ \mathrm{N\,s\,m^{-2}}$, formerly 1 dyne s $\mathrm{cm^{-2}}$.
Kinematic viscosity, as follows, is then represented by the ratio of dynamic

viscosity of the liquid and its density, i.e. $\nu = \eta . \rho^{-1} (m^2 s^{-1})$. Its unit is the stoke $St = 1 \, cm^2 \, s^{-1}$.

For Newtonian liquids, the ratio between shear stress and velocity gradient is constant. Polymer solutions, however, mostly behave in a non-Newtonian manner even at a small concentration, being dependent upon the velocity gradient.

II. Diluted Polymer Solutions

A. DEPENDENCE OF VISCOSITY UPON CONCENTRATION

Viscosity increase due to dissolved polymer in very diluted solutions (< 1% weight) is expressed by a dimensionless ratio, relative viscosity

$$\eta_r = \frac{\eta_s}{\eta_0} = \frac{t_s}{t_0} \tag{2}$$

where s = solution, 0 = solvent, or by specific viscosity

$$\eta_{sp} = \frac{\eta_s - \eta_0}{\eta_0} = \eta_r - 1 \approx \frac{t_s - t_0}{t_0} \tag{3}$$

For capillary viscometers and low polymer concentrations ($\rho_s \approx \rho_0$) viscosity values are currently replaced by flow times t.

Viscosity ratio dependence upon concentration is usually given in the form of a series

$$\eta = \eta_0 . (1 + [\eta] . c + k_1 [\eta]^2 . c^2 + \ldots) \tag{4}$$

$$\eta_r = 1 + [\eta] . c + k_1 . [\eta]^2 . c^2 + \ldots \tag{5}$$

$$\frac{\eta_{sp}}{c} = [\eta] + k_1 . [\eta]^2 . c \tag{6}$$

Equation (6) is called Huggins' equation (Huggins, 1942); its members with the second and higher powers of concentration are supposed to be negligible. The ratio η_{sp}/c is called viscosity number (reduced viscosity); it has the reciprocal dimension of concentration ($cm^3 \, g^{-1}$ or $dm^3 \, g^{-1}$). Also the logarithmic viscosity number (inherent viscosity) $\ln \eta_r / c$ can be expressed in the form of a series whose two initial members form the Kraemer's equation (Kraemer, 1938)

$$\frac{\ln \eta_r}{c} = [\eta] - k_2 . [\eta]^2 . c \tag{7}$$

In many polymer–solvent systems

$$k_1 + k_2 = 0·5 \tag{8}$$

holds for constants k_1, k_2.

The other widely used equations expressing dependence of η upon concentration are the Schulz–Blaschke (Schulz and Blaschke, 1941; Eqn. 9) and Martin (Martin, 1942; Eqn. 10) equations.

$$\frac{\eta_{sp}}{c} = [\eta] + k_3 . [\eta] . \eta_{sp} \tag{9}$$

The value of k_3 is 0.3–0.4 and is almost independent of the molecular weight. It is recommended that the concentration of the solution should be such in case η_{sp} should exceed the value 0.3–0.5. The upper interval limit at which the negligibility of the quadratic member is still considered to be admissible is for the remaining equations at $\eta_{sp} \leqslant 0.7$–0.8.

The least concentration dependence shows the Martin's equation whose applicability is stated to reach up to 5%.

$$\log \frac{\eta_{sp}}{c} = \log [\eta] + k_4 . [\eta] . c \tag{10}$$

$$k_4 = \frac{k_1}{2.303}$$

In addition to these equations others have also been published (e.g. Maron and Reznik, 1969; Elliot et al., 1970).

By means of Eqns. (6) to (10) a limiting viscosity number $[\eta]$ (intrinsic viscosity) is determined. From the given equations it follows that

$$[\eta] = \lim \left(\frac{\eta_{sp}}{c} \right) \qquad c \to 0 \tag{11}$$

$$[\eta] = \lim \left(\frac{\ln \eta_r}{c} \right) \qquad c \to 0 \tag{12}$$

$$\log [\eta] = \lim \log \frac{\eta_{sp}}{c} \qquad c \to 0 \tag{13}$$

The plots of these dependences represent lines with an intercept equal to $[\eta]$ on the ordinate axis, upon extrapolation to $c = 0$; Fig. 1 gives a schematic representation. The slopes of lines in Fig. 1 have the value $k_1 . [\eta]^2$, $k_2 . [\eta]^2$, respectively. Plotting of reciprocal values c/η_{sp}, $c/\ln \eta_r$ and their arithmetic mean $c/2 . (c/\eta_{sp}) + (c/\ln \eta_r)$ against concentration makes the extrapolation more accurate (Heller, 1954). The intercept taken up on the ordinate axis equals then to $1/[\eta]$ (Fig. 2). The Schulz–Blaschke equation (9) is extrapolated by plotting η_{sp}/c against η_{sp}. It is often made use of for directly calculating $[\eta]$.

$$[\eta] = \frac{\eta_{sp}/c}{1 + k_3 . \eta_{sp}} \tag{14}$$

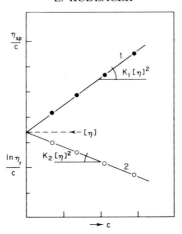

FIG. 1. Graphic extrapolation of η_{sp}/c to $c = 0$ (1), $\ln \eta_r/c$ to $c = 0$ (2).

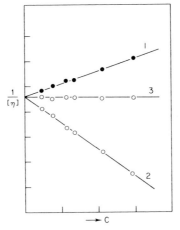

FIG. 2. Graphic extrapolation of $c/\ln \eta_r$ to $c = 0$ (1), c/η_{sp} to $c = 0$ (2), $c/2 . (1/\eta_{sp} + 1/\ln \eta_r)$ to $c = 0$ (3).

At very low concentration deviations sometimes make themselves felt due to adsorption of macromolecules on the capillary walls of the viscometer, thus reducing the polymer concentration in the measured solution as well as the effective diameter of the capillary; it is therefore not recommended to work with solutions whose $\eta_{sp} < 0 \cdot 1$. The intrinsic viscosity can be calculated by means of a computer programme with a better confidence, eliminating errors inherent to graphic extrapolation (Hofreiter *et al.*, 1973).

The constants of Eqns. (6) to (10) remain almost unchanged for a given polymer–solvent system at a given temperature. They are, however, subject to change depending upon molecular mass and velocity gradient. The changes result from hydrodynamics, a thermodynamic interaction taking place between macromolecules, or of their aggregation, respectively. Constant values are tabulated together with their range of validity (Meyerhoff, 1961; Kurata and Stockmayer, 1963; Brandrup and Immergut, 1966; Lipatov *et al.*, 1971).

B. VISCOSITY–VELOCITY GRADIENT DEPENDENCE

In a polymer solution the velocity gradient brings about the orientation and deformation of macromolecules. The orientation and elongation of the polymer coil in the flow direction causes viscosity increase to be less than would correspond to Newtonian liquid. A flow curve of polymer solution has, as a rule, two regions in which the dependence of shear stress upon velocity gradient γ is of linear character (Fig. 3). This occurs at a very low value of γ (as a rule a very short, Newtonian region $-\eta_0$) and at a very high value of γ (Newtonian region $-\eta_\infty$). Viscometry of diluted polymer solutions operates in the first region (Fig. 3).

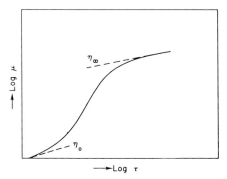

FIG. 3.   Schematic representation of shear stress–velocity gradient dependence.

In this region it is possible to express the influence of the rate gradient upon a limiting viscosity number in the form of a linear relationship which is in agreement with many experimental data.

$$[\eta]_\gamma = [\eta]_0 . (1 - A . \gamma) \tag{15}$$

Exact theoretical treatment of this influence (Scheraga, 1955) involves also the contrary influence exercized by heat (Brownian) movement expressed by rotational diffusion coefficient D_r. The two competing factors involved in the parameter are

$$\beta = \frac{\gamma}{D_r} = \frac{\gamma[\eta]_0\eta_0 M}{RT} \tag{16}$$

where η_0 = solvent viscosity. The dependence which has been theoretically derived is then of quadratic nature

$$\frac{[\eta]_\gamma}{[\eta]_0} = 1 - B \cdot \beta^2 \tag{17}$$

This relationship is fulfilled, above all, by polymers having a stiff chain (Peterlin, 1968). Influence of rate gradient upon viscosity decrease is the more pronounced, the greater is the hydrodynamic volume of the macromolecule, i.e. the greater are the anisotropy and stiffness of the chain, the better is the solvent. The viscosity decrease is, from a distinct value of molecular weight, independent of molecular weight; it is, however, steeper for lower molecular weights. A more pronounced dependence upon rate gradient is shown by polyelectrolyte solutions because of coil expansion due to electrostatic repulsion of ionized parts of the chain.

At high rate gradients, $[\eta]_\gamma/[\eta]_0$ becomes gradually independent of concentration (Merill, 1959; Lohmander and Svensson, 1963). As there is, during shear deformation, dissipation of mechanical energy into heat energy, the temperature of the liquid rises, which results in viscosity decrease. The energy W thus released and temperature increase are proportional to the square of rate gradient

$$W/t = \tau \cdot \gamma = \gamma^2 \cdot \eta \tag{18}$$

To suppress this effect a satisfactory cooling system must be provided for in viscosity measurements.

Extrapolation to the zero value γ is not easy from the experimental point of view as it is difficult to carry out measurements at very low rate gradients when a linear relationship between $[\eta]_0$ and γ is guaranteed. In current measurements, as a rule, only extrapolation to $c = 0$ is carried out at a given rate gradient. It is conventionally established by, for example, the length of the liquid column being 10% higher than the length of the capillary (Schulz and Cantow, 1954). Also the capillary diameter has been determined to suit a given range of kinematic viscosity η/ρ. For example, for η/ρ within the range $3-7 \cdot 10^{-3}$ a capillary diameter of 0.03 cm is recommended.

C. SINGLE-POINT MEASUREMENT OF $[\eta]$

A number of equations have been suggested for quick determination of the limiting viscosity number from single-point measurement of viscosity at one concentration. The use of these equations has been derived under the supposition of constancy of k-values in Eqns. (6) to (10) and the validity of the $k_1 + k_2 = 0.5$ relationship. This condition, however, at the same time restricts their applicability.

The joined Eqns. (6) and (7) create, for example, the relationship (Solomon and Ciuta, 1962)

$$[\eta] = \frac{1}{c} . [2(\eta_{sp} - \ln \eta_r)]^{\frac{1}{2}} \tag{19}$$

Further equations have been derived from other authors equations (Solomon and Gotesman, 1967, 1969; Palit and Kar, 1967; Berlin, 1968; Elliot et al., 1970; Varma and Sengupta, 1971). At the same time, a series of critical papers was published confirming that the above-mentioned equations are valid within a narrow range of polymer concentrations and for good solvents. Inaccuracies arising from neglecting these conditions are considerable (Pechoč, 1964; Budtov, 1967; Schroff, 1968; Ackerman and McGill, 1972).

D. MOLECULAR WEIGHT DEPENCENCE OF $[\eta]$

The equation which gives the possibility of obtaining an average value of molecular weight from viscosity measurements is linked up with the names of a number of scientists (Staudinger, Mark, Kuhn, Houwink, Sakurada):

$$[\eta] = K . M^a \tag{20}$$

In bilogarithmic coordinates this dependence should be of linear character. In some polymers and at low values of M a deviation occurs which can be expressed by an additive member (Imai, 1969).

$$[\eta] + b_\eta = K . M^a \tag{21}$$

Parameters a, K of Eqn. (20) depend upon the polymer solvent system. The value of exponent a varies according to the shape and rigidity of macromolecules. Theoretically, a can reach values ranging from 0.5 up to 2 (coil–rod). Gaussian molecular coil of unperturbed dimensions (impermeable coil, no excluded volume), i.e. under θ-conditions, has $a = 0.5$. For real flexible macromolecules (vinyl polymers, polyamides, polyesters, polyethers, etc.) a varies from 0.5 to 0.85. It is higher in good solvents, which corresponds to expansion of molecular coil.

For more rigid molecules, e.g. cellulose and its derivatives, or rigid chains of aromatic and heterocyclic polymers (Burke, 1973) a is about 1 (freely permeable coil). The theoretical value a for rigid rod-like particles equals 2. Near values for rigid spiral-like molecules of synthetic α-amino acids have been found in some solvents. Branched molecules, on the other hand, may have an exponent $a < 0.5$.

The value of molecular weight, obtained from Eqn. (20), is the so-called viscosity average value $\bar{M}_\eta$ which can be defined as follows:

$$\bar{M}_\eta = \left(\frac{\sum\limits_i n_i \cdot M_i^{a+1}}{\sum\limits_i n_i \cdot M_i} \right)^{\frac{1}{a}} \tag{22}$$

For $a = 1$, $\bar{M}_\eta$ is equal to the weight average $\bar{M}_w$. This kind of polymer–solvent system is quite rare in practice. For $a < 1$, $\bar{M}_\eta$ lies between number $\bar{M}_n$ and weight $\bar{M}_w$ average, nearing more or less the $\bar{M}_w$ value. For $a = -1$, $\bar{M}_\eta$ is identical with the number average $\bar{M}_n$.

Experimental determination of the constant K and the exponent a is best reached for a fractionated polymer. In fractions of different molecular weights the $\bar{M}_w$ is determined by means of any method for determination of the weight average (light scattering, ultracentrifuge).

From the plot of log $\bar{M}_w$ against log $[\eta]$ for these fraction values K, a can be obtained. The accuracy of this correlation is considerably influenced by the polydispersity of samples, particularly if the method leading to number average $\bar{M}_n$ is used for calibration. This procedure is, however, applicable only for monodisperse samples of polymer.

To make K, a values more accurate we correct the procedure so as to eliminate the influence of polydispersity (Elias et al., 1973). If the type of the distribution function in the samples is known, their $[\eta]$, $\bar{M}_w$, and $\bar{M}_n$ is determined. By the above-mentioned procedure provisional values K, a are obtained. For normal logarithmic distribution $\bar{M}_\eta$ is calculated from the relationship (Chiang, 1959)

$$\frac{\bar{M}_\eta}{\bar{M}_w} = \left(\frac{\bar{M}_w}{\bar{M}_n} \right)^{a-1/2} \tag{23}$$

For Schulz–Flory distribution the following relationship holds in the region $0 < a < 2$ (Schulz et al., 1959) approximately

$$\bar{M}_\eta = 0 \cdot 5(\bar{M}_w + \bar{M}_n) + 0 \cdot 5(\bar{M}_w - \bar{M}_n) \cdot a \tag{24}$$

The value log $\bar{M}\eta$, calculated in this way, is plotted again against log $[\eta]$ and the whole procedure is repeated until K, a values stop changing. With the growing distribution width, K rises while the exponent a falls. At normal logarithmic distribution $\bar{M}\eta$ and $\bar{M}_w$ differences can be considerable; at Schulz–Flory distribution they are of less importance. The K, a values for a number of polymers, including the fibre-forming ones, are given in literature together with the methods which were used for calibrating the Eqn. (20) (Meyerhoff, 1961; Kurata and Stockmayer, 1963; Brandrup and Immergut, 1966; Lipatov et al., 1971).

E. DETERMINATION OF THE COIL SIZE

The Flory–Fox theory enables practical advantages of viscometry to be made use of for determination of the molecular coil size (Fox and Flory, 1949, 1951). The theoretical relationships existing between $[\eta]$ and the size of the macromolecule in the solution originate from Kirkwood and Riseman's (1948) calculations. Under θ-conditions, according to Flory and Fox

$$[\eta]_\theta = 6^{\frac{3}{2}} . \phi_0 . \left(\frac{\bar{S}_0^2}{M}\right)^{\frac{3}{2}} . M^{\frac{1}{2}} = K_\theta . M^{\frac{1}{2}} \qquad (25)$$

where M = molecular weight, ϕ_0 = Flory's constant, whose most frequent quoted theoretical value is $2 \cdot 87 . 10^{23}$, if $[\eta]$ is given in ml g^{-1}, the best acceptable experimental value being cited as $2 \cdot 5 . 10^{23}$ (Yamakawa, 1970, 1971). S_0 = radius of gyration in unperturbed state.

The coil size in other than θ-solvent can be expressed by means of expansion factor α. In conjunction with Eqn. (25) the following expression is obtained for a good solvent:

$$[\eta] = 6^{\frac{3}{2}} . \phi . \left(\frac{\bar{S}_0^2}{M}\right)^{\frac{3}{2}} . M^{\frac{1}{2}} . \alpha_\eta^3 \qquad (26)$$

$$\alpha_\eta^3 = \frac{[\eta]}{[\eta]_\theta} \sim M^{a_\eta - 0\cdot 5} \qquad (27)$$

The constancy of ϕ value holds only under θ-conditions, i.e. the macromolecular coil has spherical symmetry and no draining effect. In improving solvents, however, ϕ is falling. To express this dependence the following relationship has been suggested (Pticyn and Eizner, 1958, 1959):

$$\phi = \phi_0 . (1 - 2\cdot 63\epsilon + 2\cdot 86\epsilon^2) \qquad (28)$$

where

$$\epsilon = \frac{2a_\eta - 1}{3}$$

Equation (25) can easily be used to determine the unperturbed $\bar{S}_0^2/M$ if θ-solvent is available. If not, then the most frequently employed method is that of Stockmayer and Fixman (Stockmayer and Fixman, 1963), extrapolating $[\eta]/M^{\frac{1}{2}}$ against $M^{\frac{1}{2}}$ to the zero value of M. Viscosity of fraction series having a known weight average of molecular weight is measured. According to the equation

$$\frac{[\eta]}{M^{\frac{1}{2}}} = K_\theta + 0\cdot 51 . \phi_0 . \chi . M^{\frac{1}{2}} \qquad (29)$$

where χ = interaction parameter. The theoretical validity of this equation should be limited to the environs of θ-temperature. Experiments, however, confirm a broader validity according to which K_θ equals the intercept on the ordinate axis at the given extrapolation. To express more accurately the relationship $([\eta]/M^{\frac{1}{2}}) - M^{\frac{1}{2}}$ further equations have been suggested with a modified first member and exponent at M (Kurata and Stockmayer, 1963; Hearst, 1964; Ullman, 1964; Bohdanecký, 1964, 1966, 1970). Extrapolation methods lose accuracy with the increasing coil expansion. Extrapolation is not suitable for polymers having a rigid chain ($a > 0.85$) because conditions for which the said relationship has been derived do not hold for them, i.e. independence of ϕ_0 and $(\bar{S}_0^2/M)^{\frac{3}{2}}$ of molecular weight.

In some cases θ-conditions can be realized in mixed solvents (Elias, 1961; Beachell and Peterson, 1967). The K_θ value, obtained in this way, is however considerably influenced by the character and ratio of solvents in the mixture.

For unperturbed dimensions of rigid macromolecules and small values of the expansion factor ($\alpha < 1.2$) the relationship is used existing between viscosity, expansion factor and second virial coefficient A_2 (Orofino and Flory, 1957)

$$\frac{A_2 . M}{[\eta]} = 165 . \log [1 + 4.49(\alpha_\eta^2 - 1)] \tag{30}$$

This expansion allows expansion factor α_η^2 to be calculated from measuring viscosity of a given polymer fraction. The accuracy is, however, limited by the accuracy of determination of the second virial coefficient of the osmotic pressure A_2. Equation (30) is suitable for stiff molecules.

F. VISCOSITY–TEMPERATURE DEPENDENCE

The viscosity of diluted solutions is influenced basically by the change of the solvent viscosity with temperature. In general, for a given temperature interval, in which the stability of activation energy of flow E is valid, viscosity–temperature dependence is expressed by the Andrade equation

$$\eta = K . e^{E/RT} \tag{31}$$

The limiting viscosity number depends upon temperature according to the character of the macromolecule and its interaction with the solvent. This follows from the derivation of Eqn. (25) with respect to temperature

$$\frac{d \ln [\eta]}{dT} = \frac{d \ln \phi}{dT} + \frac{3}{2} . \frac{d \ln \bar{S}_0^2/M}{dT} + \frac{d \ln \alpha^3}{dT} \tag{32}$$

The first member expressing the changes of intramolecular hydrodynamic interaction is thought to be negligible. Of importance are, however, the two remaining members. Flexible chains are more influenced by the expansion coefficient. The latter is, as a rule, small in stiff chains; of more consequence then is the temperature dependence of the characteristic ratio $\bar{S}_0^2/M$.

The temperature coefficient of $[\eta]$ also depends upon the quality of the solvent employed. It makes itself particularly felt in a bad solvent wherein, due to temperature increase, coil expansion takes place. In a good solvent the temperature coefficient is usually small. The influence of temperature is greatest in the neighbourhood of θ-temperature and drops with the growing distance from that temperature. In the given region the expansion coefficient α^3 is a linear function of the expression $c \cdot [1 - (\theta/T)] \cdot M^{\frac{1}{2}}$, c being a parameter independent of temperature (Berry, 1967).

G. INFLUENCE OF THE CHAIN STRUCTURE UPON $[\eta]$

The hydrodynamic size of the macromolecule in the solution is further influenced by structural changes modifying its geometry, flexibility and interaction with the solvent, e.g. the case of branching and copolymerization.

It is only very seldom that branched structures appear among fibreforming polymers. The branching reduces the gyration radius of the macromolecule due to changes in segment distribution taking place around the centre of gravity (Zimm and Stockmayer, 1949). The ratio of gyration radii $g < 1$:

$$g = \frac{\bar{S}_{0,b}^2}{\bar{S}_{0,1}^2} \qquad (33)$$

where $\bar{S}_{0,b}^2$ = mean-square unperturbed gyration radius of a branched polymer, $\bar{S}_{0,1}^2$ = mean-square unperturbed gyration radius of a linear polymer with the same molecular weight.

Also the ratio of limiting viscosity numbers g' is less than 1. The theoretical relationship existing between g and g' is of the shape (Zimm, 1956)

$$g' = g^{\frac{3}{2}}$$

$$\frac{[\eta]_{0,b}}{[\eta]_{0,1}} = \frac{(\bar{S}_{0,b}^2)^{\frac{3}{2}}}{(\bar{S}_{0,1}^2)^{\frac{3}{2}}} \qquad (34)$$

It is generally true that g and g' decrease with increasing degree of branching. Experimental results indicate, however, that g' is in fact smaller than $g^{\frac{3}{2}}$. The g' ratio is rather proportional to $g^{\frac{1}{2}}$ (Zimm and Kilb, 1959; Orofino, 1961; Shultz, 1965; Nagasubramanian et al., 1969). The branching reduces also the exponent a in equation (20) and, as the branching degree depends upon the length of the chain, the logarithmic dependence $[\eta] - M$ is not of linear character. The value of the g' parameter can also be made use of for the relative estimate of branching. For a branched polymer, $[\eta]$ and M are determined experimentally, while $[\eta]$ of a linear polymer, having the same molecular weight, is calculated.

Viscosity behaviour of polymer solutions is influenced in a most complex manner by copolymerization, the former depending largely upon the monomer type and the chain structure. A good example of different behaviour is offered by block and statistical copolymers.

H. Viscosity of Polyelectrolyte Solutions

The presence of the charge in the chain causes the coil to expand, the expansion being dependent upon the size and distribution of the charge, ionic strength of the solution and upon the presence of low-molecular electrolytes. Linear polyelectrolytes can be exemplified by polyacrylic acid, carboxymethylcellulose, poly-4-vinylpyridine, chains carrying sulfo-groups etc. Polyelectrolytic behaviour is exhibited also by polyamides dissolved in formic acid.

Determination of limiting viscosity number in polyelectrolyte solutions by extrapolation to the zero concentration can be exhibited schematically following three forms depending upon the ionic strength of the solution (Fig. 4). In solutions of weak electrolytes (1) there is, at a high degree of dilution, a rapid increase of η_{sp}/c. Counterions travel from the polyion whose charges repulse one another, which results in a rapid expansion of the coil. The curve 2 corresponds to an addition of a certain amount of foreign ions. The expansion occurs in this case up to a maximum. Then even at further dilution, ion-ic strength does not change. This is due to the

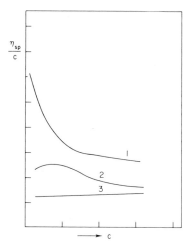

Fig. 4. Determination of $[\eta]$ in polyelectrolyte solutions. (1) Weak ionic strength; (2) medium ionic strength; (3) high ionic strength.

presence of foreign ions, viscosity decreasing as in an uncharged particle. In case 3, foreign ions predominate from the beginning, the ionic strength does not change and the $(\eta_{sp}/c) - c$ relationship is linear.

Influence of repulsive forces upon coil expansion (curve 1) and, therefore, also upon viscosity can be described by the relationship derived by Fuoss (Fuoss, 1948):

$$\frac{\eta_{sp}}{c} = \frac{A}{1 + B \cdot \sqrt{c}} \tag{35}$$

where A, B = constants. According to the given equation the value $1/[\eta]$ can be obtained as intercept from the plot c/η_{sp} against $\sqrt{c}$.

Measurement is carried out also under conditions when ionic strength of the solution does not change by dilution, i.e. in an isoionic way. Polyelectrolytic effects can be suppressed by adding a sufficient quantity of neutral salt into the solution (Noda et al., 1970).

Expansion of polyion chains is also the reason why there is a strong dependence of viscosity of these solutions upon the rate gradient.

III. Concentrated Solutions and Polymer Melts

If the volume fraction φ_2 of polymer is greater than $0 \cdot 1$, the solution shows a typical dependence of Newtonian viscosity ($\gamma \to 0$) upon molecular weight and temperature. A detailed survey of these and related problems can be found in Berry and Fox (1968), Semjonow (1968), Simha (1971) and Tager (1974).

Taking numerous experimental results into consideration, the following empirical equation has been suggested to express dependence of iso-thermal viscosity η_T upon the chain length:

$$\eta_T = K_T \cdot Z^a \tag{36}$$

where Z = number of atoms or their groups in the chain, K = constant.

For exponent a it is true

$$a = 3 \cdot 4 \quad \text{for} \quad Z \geqslant Z_c$$

$$2 \cdot 5 \geqslant a \geqslant 1 \quad \text{for} \quad Z \leqslant Z_c$$

Instability of a for values Z lower than critical Z_c follows from the dependence of the friction coefficient ξ upon molecular weight. Viscosity can be defined as the product of structural factor $F(Z)$ which is mainly dependent upon the length of the macromolecule and the friction coefficient $\xi(\rho)$ which depends upon temperature or density.

$$\eta = F(Z) \cdot \xi(\rho) \tag{37}$$

In converting isothermal data η_T to constant friction coefficient η_ξ at $Z < Z_c$, the exponent $a = 1$ for a whole series of polymers. Z_c values for various polymers lie within the interval of about 300–800.

A sudden change in the linear dependence $\eta - Z$ at Z_c was theoretically accounted for by Beuche (1952, 1956). The author starts from the theory of the motion of an isolated chain in the medium which exercises only the friction resistance expressed by ξ. In concentrated solutions and melts the mentioned medium is represented by neighbouring segments. For $Z < Z_c$ the melt viscosity equals

$$\eta = \left[\frac{N_0}{6} \cdot \frac{\bar{S}_0^2}{M} \cdot \frac{Z}{v_2} \right] \cdot \xi \tag{38}$$

The expression in square brackets corresponds to the structural factor F in equation (37) and expresses the dependence of F upon the coil size. In the case of concentrated solutions v_2/φ_2 is introduced instead of specific volume v_2.

In the region $Z \geqslant Z_c$ Bueche presupposes energy loss due to mutual intertwining and entanglements of chains. Chains perform a snaking motion during dislocation, exerting a draw, at the same time, upon neighbouring chains. Calculation of this effect was included into the structural factor by Bueche and, in this way, a relationship was obtained according to which η is proportionate to $Z^{3\cdot5}$. This is in good agreement with the empirical equation (36).

Fox and Allen (1964) introduced parameters X and X_c into Eqn. (36).

$$X = \frac{Z}{v_2} \cdot \frac{\bar{S}_0^2}{M} \tag{39}$$

$$X_c = \frac{Z_c}{v_2} \cdot \frac{\bar{S}_0^2}{M} \tag{40}$$

If this relationship is employed to represent the dependence $\log \eta - \log X$ of various flexible polymers (Berry, 1968) approximately the same value is obtained for X_c

$$10^{17} X_c = 400$$

The slope of the line in the region $X < X_c$ is then equal to 1, the region $X > X_c$ being equal to 3·4 (Fig. 5).

Also for polymer solutions it is approximately true

$$X_c = \frac{Z_c \cdot \varphi_2}{v_2} \cdot \frac{\bar{S}_0^2}{M} = \text{const.} \tag{41}$$

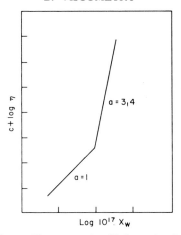

FIG. 5. Dependence of $\log \eta$ upon $\log X_w$ for melts of various polymers

$$\log X_w = \frac{\bar{S}_0^2}{M} \cdot \frac{Z_w}{v_2}$$

where Z_w = weight average of a number of atoms or their groups in the chain.

For polymers with chains longer than Z_c the viscosity is then well expressed by the equation with parameter X

$$\eta = \frac{N_0}{6} \cdot \left(\frac{X}{X_c}\right)^{2 \cdot 4} \cdot X \cdot \xi \tag{42}$$

Polymers having a stiff chain can deviate considerably from this equation.

The temperature dependence of the friction coefficient is used to express influence of temperature upon viscosity. Influence of temperature upon the structural factor $F(Z)$ is negligible. The dependence is, as a rule, expressed in the form of the Vogel equation

$$\ln \xi = \ln \xi_0 + \frac{1}{\alpha(T - T_0)} \tag{43}$$

where ξ_0, α and T_0 are constants for a given temperature interval and a given chain length. This equation is equivalent to the WLF equation for viscosity

$$\log a_T = \log \frac{\eta_0(T)}{\eta_0(T_0)} = \frac{C_1(T - T_0)}{C_2 + T - T_0} \tag{44}$$

C_1 and C_2, being empirical constants, have the value $-8 \cdot 86$, or $101 \cdot 6$ respectively. The validity of this equation is limited to the temperature interval $(T - T_0) < 50$ grad.

As T_0 is usually 30–50 grad above T_g, the interval of the WLF equation validity for Newtonian viscosity of amorphous polymers reaches up to 70–100 grad above T_g. For T_g in the role of reference temperature, values of the constants C_1 and C_2 are $-17·44$, or $51·6$ respectively.

At temperatures high above $T_g(T_g + 200)$ Eqn. (31) is employed, according to which ln η is proportionate to $1/T$. Activation energy of the viscous flow being dependent upon temperature, this equation is not valid in a wide temperature range. So far no relation has been suggested for the whole temperature range from T_g to account for the temperature influence upon viscosity.

Effect of the solvent quality upon viscosity of concentrated solutions ($\varphi_2 > 0·2$) is small. This is in connection with the fact that, in these solutions and melts, the size of the macromolecule is very much near unperturbed dimensions.

IV. Viscosity Measurement

The most widely used method for measuring viscosity consists in determining the flow-rate of a liquid in the capillary viscometers. For this purpose, a viscometer with a hanging level (Ubbelohde type) is, as a rule, made use of; it can be modified for measurements at constant pressure, at elevated temperatures, for diluting the measured solution etc. (Quadrat et al., 1966; Nakajima et al., 1966; Grammain and Libeyre, 1970; Chan, 1971; Gruber and Sezen, 1974).

For Newtonian liquids the effluxed volume is given by the Hagen–Poisseuille equation

$$V = \frac{\pi . R^4 \Delta P . t}{8L . \eta_N} \tag{45}$$

from which an absolute viscosity value can be calculated

$$\eta_N = \frac{\pi . R^4 . \Delta P . t}{8LV} = K . t \tag{46}$$

if flow-rate is measured and if the viscometer dimensions are known. R = capillary radius, ΔP = difference of pressures between both ends of the capillary, L = capillary length, V = effluxed volume, t = time.

Shear stress in Newtonian liquid changes linearly from the centre to the wall of the capillary. The rate gradient is given by the relationship

$$\gamma = \frac{\Delta P . R}{2L . \eta_N} = \frac{4V}{\pi . R^3 t} = \frac{4Q}{\pi . R^3} \tag{47}$$

where $Q = (V/t)$, efflux rate.

Polymer solutions are non-Newtonian often at small concentrations, and therefore the above-mentioned relationships give only apparent values η and γ. Real viscosity is calculated using a correction (Rabinowitsch); its value can be determined from measurements of flow-times of the solvent and solution at varying ΔP. Relative viscosities are thus obtained with satisfying approximation

$$t_{r,\Delta P} = \left(\frac{t}{t_0}\right)_{\Delta P}$$

The ratio

$$\frac{d \ln t_{r,\Delta P}}{d \ln \Delta P}$$

is used to calculate corrected viscosities

$$(\eta)_R = t_{r,\Delta P} \cdot \left(1 - \frac{1}{4} \cdot \frac{d \ln t_{r,\Delta P}}{d \ln \Delta P}\right)^{-1} \tag{48}$$

For current measurements serving for extrapolation of viscosity values to zero shear stress mostly apparent values are used.

Further corrections follow from energy losses occurring at the inlet and outlet of the liquid into and from the capillary (end effect) and also follow from that part of energy, supplied by ΔP pressure, which is converted into kinetic flow energy. The equation for η is, with this regard, corrected to the form

$$\eta = \frac{\pi \cdot R^4 \cdot \Delta P \cdot t}{8LV(1 + nR/L)} - \frac{m \cdot \rho \cdot V}{8\pi L(1 + nR/L) \cdot t} \tag{49}$$

Values of coefficients m, n are not universally constant (Schurz and Pippan, 1963). When nearing unity equation (49) is, as a rule, modified into the form

$$\eta = B \cdot \rho \cdot t - C \cdot \frac{\rho}{t} \tag{50}$$

Parameters B, C are determined for a given viscometer by measuring the flow-time of several liquids of known η and γ.

As

$$\frac{\eta}{\rho \cdot t} = B - C \cdot \frac{1}{t^2} \tag{51}$$

B, C are obtained from the plot in which $\eta/\rho \cdot t$ against $1/t^2$ is extrapolated to $1/t^2 = 0$.

For exact measurements, carried out in the capillary viscometer, constant temperature should be maintained ($\pm 0, 01°C$) and flow-time should be measured with an accuracy of $0 \cdot 1$ s. The next condition is perfect purity of the solution and a reproducible vertical fastening of the viscometer.

Rotational viscometers are less accurate than capillary ones. Their main advantages, however, are a high rate of measuring, the possibility of measuring concentrated solutions and changing rate gradient, and also the possibility of carrying out long-term measurements at constant shear stress.

Tangential flow in these apparatuses takes place in a slot between two co-axial cylinders or between a plate and a cone with apex angle nearing 180°.

In the Couette viscometer with co-axial cylinders the outer cylinder is motor-driven. If liquid is present in the slot reaching up to the height h, it begins to move simultaneously and, by friction, transfers the motion to the inner cylinder. This is deflected by the angle δ which is directly proportional to the torsion momentum J.

$$\delta = C \cdot J \qquad (52)$$

C is the constant of the torsion-string upon which the cylinder is hung. In another apparatus the inner cylinder is represented by a rotor with a metallic core which is rotated by an outside magnet. The rotor floats in the liquid to be measured contained in the outer cylinder acting as stator. The friction resistance of the liquid causes the rotor to lag as compared with magnet rotation (Berry, 1967).

Measurement of the relative viscosity in a Couette viscometer is carried out by comparing the solution and the solvent

$$\eta_r = \frac{P - P_m}{P_0 - P_m} \qquad (53)$$

P = time of 1 rotation in solvent (0), solution or of driven cylinder (m). Shear stress at the walls of cylinders (radii R_1, R_2) is defined as force J/R acting on the surface $2\pi Rh$.

$$\tau = \frac{J}{2\pi R^2 h} \qquad (54)$$

With small $R_2 - R_1$ mean value τ is used. The rate gradient at angular rate Ω in the distance R is,

$$\gamma = \frac{2\Omega}{R^2} \cdot (R_1 \cdot R_2)^2 (R_2^2 - R_1^2) \qquad (55)$$

For viscosity of Newtonian liquids according to the Margules equation

$$\eta_N = \frac{J(R_2^2 - R_1^2)}{4\pi R_1^2 \cdot R_2^2 \cdot h \cdot \Omega} \qquad (56)$$

For non-Newtonian liquids the value of apparent viscosity must be corrected

$$\frac{\eta}{\eta_N} = (1 - \Delta_s)^{-1}$$

$$\Delta_s = \frac{s^2 - 1}{6s^2} \cdot \frac{d \log \eta_N}{d \log \tau} \left[3 + 2 \ln s - \ln s \cdot \frac{d \log \eta_N}{d \log \tau_1} \right]$$

$$s = \frac{R_2}{R_1} \tag{57}$$

where τ_1 = shear stress at the wall of the inner cylinder. The correction will be obtained by graphic derivation of $\log \eta$ to $\log \tau_1$ dependence.

The cone-plate viscometer has the advantage that, at the angle of the slot $\theta < 3°$, the rate gradient and, consequently, also the shear stress may be considered constant in the overall volume of the liquid, just the same as in the ideal viscometer. Then

$$\eta = \frac{3 \theta J}{2 \pi R^3 \Omega} \tag{58}$$

will be obtained for both Newtonian and non-Newtonian viscosity.

References

Ackerman, L. and McGill, W. J. (1972). *J. Polymer Sci. A* **10**, 3689–3690.

Beachell, H. E. and Peterson, J. C. (1967). Paper presented at the 153 National Meeting of American Chemical Society, Miami Beach, Florida.

Berlin, A. A. (1968). *Vysokomol. sojed.* **10B**, 21–23.

Berry, G. C. (1967). *J. Chem. Phys.* **46**, 1338–1352.

Berry, G. C. and Fox, T. G. (1968). *Adv. Polymer Sci.* **5**, 261–357.

Bohdanecký, M. (1964). *Faserforsch. Textilechn.* **15**, 605–608.

Bohdanecký, M. (1966). *Collection Czech. Chem. Commun.* **31**, 3979–3984.

Bohdanecký, M. (1970). *Collection Czech. Chem. Commun.* **35**, 1972–1990.

Brandrup, J. and Immergut, E. H. (eds.) (1966). "Polymer Handbook", IV, pp. 1–72, Interscience Publishers, New York.

Budtov, V. P. 1967). *Vysokomol. sojed.* **9A**, 2765–2766.

Bueche, F. (1952). *J. Chem. Phys.* **20**, 1959–1964.

Bueche, F. (1956). *J. Chem. Phys.* **25**, 599–600.

Burke, J. J. (1973). *In:* "High-modulus wholly aromatic fibres" (W. B. Black and J. Preston, eds.), pp. 187–200, M. Dekker, New York.

Chan, F. S. (1971). *J. Appl. Polymer Sci.* **15**, 1703–1708.

Chiang, R. (1959). *J. Polymer Sci.* **36**, 91–103.

Elias, H. G. (1961). *Makromol. Chem.* **50**, 1–19.

Elias, H. G., Bareiss, R. and Watterson, J. G. (1973). *Adv. Polymer Sci.* **11**, 111–204.

Elliot, J. H., Horowitz, K. H. and Hoodcock, T. (1970). *J. Appl. Polymer Sci.* **14**, 2947–2963.

Fox, T. G., Jr. and Flory, P. J. (1964). *J. Phys. Chem.* **41**, 344–352.

Fox, T. G., Jr. and Flory, P. J. (1949). *J. Phys. Chem.* **53**, 197.

Fox, T. G., Jr. and Flory, P. J. (1951). *J. Amer. Chem. Soc.* **73**, 1909–1915.

Fuoss, R. M. (1948). *J. Polymer Sci.* **3**, 603.

Gramain, P. and Libeyre, R. (1970). *J. Appl. Polymer Sci.* **14**, 383–391.

Gruber, E. and Sezen, M. C. (1974). *Angew. Makromol. Chem.* **36**, 57–65.

Hearst, J. E. (1964). *J. Chem. Phys.* **40**, 1506–1509.

Heller, W. (1954). *J. Colloid Sci.* **9**, 547–573.

Hofreiter, B. T., Ernst, J. O. and Williams, W. L. (1973). *J. Appl. Polymer Sci.* **5**, 1449–1454.

Huggins, M. L. (1942). *J. Amer. Chem. Soc.* **64**, 2716–2718.

Imai, S. (1969). *J. Chem. Phys.* **51**, 1732–1741.

Kirkwood, J. G. and Riseman, J. (1948). *J. Chem. Phys.* **16**, 565–573.

Kraemer, E. O. (1938). *Ind. Eng. Chem.* **30**, 1200–1203.

Krozer, S. (1974). *Makromol. Chem.* **175**, 1905–1915.

Kurata, M. and Stockmayer, W. H. (1963). *Adv. Polymer Sci.* **3**, 196–312.

Lipatov, Ju. S., Nesterov, A. E., Gricenko, T. M. and Veselovskij, R. A. (1971). "Spravochnik pro chimii polimerov", pp. 370–388, Naukova dumka, Kiev.

Lohmander, U., Svensson, A. (1963). *Makromol. Chem.* **65**, 202–223.

Maron, S. H. and Reznik, R. B. (1969). *J. Polymer Sci.* **A2**, 7, 309–324.

Martin, A. F. (1942). Paper presented at ACS meeting, Memphis.

Merill, E. W. (1959). *J. Polymer Sci.* **38**, 539–543.

Meyerhoff, G. (1961). *Fortschr. Hochpolym.-Forsch.* **3**, 59–105.

Nagasubramanian, K., Sanito, O. and Graessley, W. W. (1969). *J. Polymer Sci. A2*, **7**, 1955–1964.

Nakajima, A., Hamoda, F. and Hayashi, S. (1966). *J. Polymer Sci.* **C15**, 285–294.

Noda, I., Tsuge, T. and Nagasawa, M. (1970). *J. Phys. Chem.* **74**, 710–719.

Orofino, T. A. (1961). *Polymer* **2**, 305–314.

Orofino, T. A. and Flory, P. J. (1957). *J. Chem. Phys.* **26**, 1067–1076.

Palit, S. R., Kar, J. (1967). *J. Polymer Sci. A1*, **5**, 2629–2636.

Pechoč, V. (1964). *J. Appl. Polymer Sci.* **8**, 1281–1286.

Peterlin, A. (1968). "Non-Newtonian viscosity and the macromolecule". *In:* "Advances in Macromolecular Chemistry", Vol. 1, pp. 225–281 (W. M. Pasika, ed.), Academic Press, New York and London.

Pticyn, O. B., Eizner, J. J. (1958). *Zh. fiz. chim.* **32**, 2464–2466.

Pticyn, O. B., Eizner, J. J. (1959). *Zh. techn. fiz.* **29**, 1117–1134.

Quadrat, O., Bohdanecký, M. and Munk, P. (1966). *Chem. listy* **60**, 825–828.

Scheraga, H. A. (1955). *J. Chem. Phys.* **23**, 1526–1532.

Schroff, R. N. (1968). *J. Appl. Polymer Sci.* **12**, 2741–2742.

Schulz, G. V. and Blaschke, F. (1941). *J. Prakt. Chem.* **158**, 130.

Schulz, G. V. and Cantow, H. J. (1954). *Makromol. Chem.* **13**, 71–75.

Schulz, G. V., Henrici-Olivé, G. and Olivé, S. (1959). *Z. Physikal. Chem. Neue Folge* **19**, 125–141.

Schurz, J. and Pippan, H. (1963). *Monatsh. Chem.* **94**, 859–889.

Schultz, A. R. (1965). *J. Polymer Sci. A*, **3**, 4211–4225.

Semjonow, V. (1968). *Adv. Polymer Sci.* **5**, 387–450.

Simha, R. (1971). *J. macromol. Sci. B, Phys.* **5**, 425–428.

Solomon, O. F. and Ciuta, J. Z. (1962). *J. Appl. Polymer Sci.* **6**, 683–686.

Solomon, O. F. and Gotesman, B. S. (1967). *Makromol. Chem.* **104**, 177–184.

Solomon, O. F. and Gotesman, B. S. (1969). *Makromol. Chem.* **127**, 153–164.

Stockmayer, W. H. and Fixman, M. (1963). *J. Polymer Sci. C.* **1**, 137–141.

Tager, A. A. (1974). *Rheologica Acta* **13**, 831–840.

Ullman, R. (1964). *J. Chem. Phys.* **40**, 2193–2201.

Varma, T. D. and Sengupta, M. (1971). *J. Appl. Polymer Sci.* **15**, 1599–1605.

Yamakawa, H. (1970). *J. Chem. Phys.* **53**, 436–443.

Yamakawa, H. (1971). Lecture presented at "IXth Macromolecular Micro-symposium, ČSAV Praha", pp. 179–199 (B. Sedláček, ed.), Butterworths, London.

Zimm, B. H. (1956). *J. Chem. Phys.* **24**, 269–278.

Zimm, B. H. and Kilb, R. W. (1959). *J. Polymer Sci.* **37**, 19–42.

Zimm, B. H. and Stockmayer, W. H. (1949). *J. Chem. Phys.* **17**, 1301–1314.

3. Osmosis

J. W. LORIMER

*Department of Chemistry, University of Western Ontario,
London, Ontario, Canada*

I. Introduction

The phenomenon of osmosis is, in general terms, the flow of a fluid (liquid or gas) across a boundary that separates two different phases. In most textbooks, osmotic flow is considered to arise only when a difference in concentration exists across a membrane. The term osmosis is used to describe other types of flow across interfaces, many of which, however, do not involve differences in concentrations.

In Fig. 1, a typical osmosis experiment that involves concentration differences is shown in schematic form. Two compartments, I and II, at constant temperature, are separated by a membrane M that is permeable to a solvent and not to the solute. Compartment I contains pure solvent,

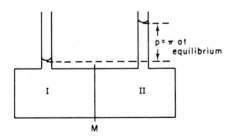

FIG. 1. Schematic osmosis experiment.

while compartment II contains a solution. As soon as the system is set up, solvent begins to pass spontaneously from compartment I to compartment II; this is the phenomenon of osmosis, or osmotic flow. With the experimental arrangement shown, the solvent level in compartment I decreases with time, while the solution level in compartment II increases with time. A difference in hydrostatic pressure builds up across the two compartments, and opposes the flow of liquid. Eventually the hydrostatic pressure reaches a maximum value, the *osmotic pressure*, where the tendency for osmotic flow is just balanced by the tendency for reverse flow. In other words, a state of osmotic equilibrium is set up. If an external pressure greater than the osmotic pressure is applied to compartment II, a flow of solvent will take place from II to I. This flow is called *reverse osmosis*.

The fundamental "driving force" or *thermodynamic affinity* in osmosis is the difference across the membrane of the chemical potential μ_i of the species that permeates the membrane. In accordance with the Second Law of thermodynamics spontaneous osmotic flow occurs from a region of higher chemical potential to a region of lower chemical potential of the permeating species.

In the above example, the difference in chemical potential arises from a difference in concentration between the two phases. The condition for thermodynamic equilibrium is then (cf. Guggenheim, 1959)

$$d\mu_0^{I}(T, p) = d\mu_0^{II}(T, p + \pi)$$

where μ_0^{I}, μ_0^{II} are the chemical potentials of the permeating species in compartments I and II, respectively, and π is the osmotic pressure. Since the pressure p is the same on each side (the excess pressure in II is π), then we have

$$0 = V_0 \, d\pi + (d\mu_0)_{T,p} \tag{1}$$

or, using the thermodynamic Gibbs–Duhem equation at constant T and p,

$$d\pi = -(d\mu_0)_{T,p}/V_0 = c_s(d\mu_s)_{T,p}/c_0 V_0 \tag{2}$$

In these equations, V_0, V_s are the partial molar volumes of solvent and solute, and c_0, c_s are the volume concentrations of solvent and solute, respectively. The subscripts T and p indicate constant temperature and pressure. For the experiment described above, the partial molar volume of the solvent is very nearly the molar volume V_0^{*} of the pure solvent in dilute solutions, and

$$(d\mu_0)_{T,p} = RTd \ln a_0 \tag{3}$$

where R is the gas constant, and a_0 is the activity of the solvent in the solution. Thus, if the dependence of the molar volume of the solvent on pressure is negligible,

$$\pi = -RT \ln a_0/V_0^{*} \tag{4}$$

Equation (4) shows that the osmotic pressure is essentially a measure of the activity of the solvent in the solution. In dilute solutions, where activity coefficients may be neglected, Eqn. (2) gives the classical van't Hoff equation

$$\pi = RTc_s = RTw_s/M_s \tag{5}$$

where w_s, M_s are the mass concentration and molar mass of the solute. This equation forms the basis of obtaining molecular weights from measurement of osmotic pressures across a membrane that is truly semipermeable; i.e. permeable only to the solvent. The method is discussed in detail in many books on experimental techniques and on polymer chemistry—for example Flory (1953), Wagner and Moore (1959)—and an extensive collection of osmotic pressures of solutions of macromolecules has been made by Meyerhoff (1960). The use of Eqn. (4) in the measurement of activities of low molecular weight solutes is discussed by Robinson and Stokes (1959).

Osmotic phenomena are not even restricted to systems containing a membrane that separates two solution phases. One phase may be a swollen, cross-linked macromolecular network, or gel (Hermans, 1953). In this case, solvent (phase I) can enter the gel (phase II), again if the chemical potential of the solvent is lower in phase II. The gel network swells, and the chemical potential of the solvent in the gel increases as the network becomes more and more dilute and as the contractile force of the swelling network increases. At equilibrium, there is an equilibrium pressure called the equilibrium *swelling pressure*. The swelling pressure has its origin in the contractile force of the swollen network. No new thermodynamic principles are involved, since the swelling pressure is simply the term used for osmotic pressure in this case (Hermans, 1953).

Other osmotic phenomena can be observed easily. If the two phases I and II are at different temperatures, the phenomenon of *thermo-osmosis* can occur, even with pure water on each side of the membrane. Passage of electric current through many membranes causes *electro-osmosis*. The general "driving force", or thermodynamic affinity, for osmotic flow is, in fact, a difference in chemical potential across the membrane arising from differences in concentration, pressure, temperature or electric field, and possibly from differences in other variables as well. The number of osmotic phenomena is therefore very large, and includes all those phenomena that are classified as *membrane transport phenomena.*

There is an immense quantity of publications pertaining to membrane transport phenomena, much of which has been classified and tabulated in the monograph by Lakshminarayanaiah (1969) and, by the same author, in the Specialist Periodical Reports on Electrochemistry (1972, 1974) published by the Chemical Society.

In fibre science, osmotic phenomena are important in two situations: in membrane materials composed all or in part of fibres where the whole range of membrane transport phenomena may be encountered, and in single fibres, where permeation and swelling are connected with many practical aspects such as absorption of liquids and vapours, wettability and dyeing. The aim of this chapter is to provide workers in the general area of fibre science with a comprehensive guide to the voluminous literature on transport in membranes. The basis of this guide is the thermodynamics of irreversible processes, first applied in a sufficiently comprehensive manner to membrane transport by Staverman (1952). Over the past two decades, irreversible thermodynamic theories have proceeded along several different but interconnected paths. Partial accounts of the approaches are available, especially Lakshminarayanaiah (1969); Katchalsky and Curran (1965); Caplan and Mickulecky (1966); our purpose here is a critical catalogue and comparison. The necessity of a guide to these theories is indicated by claims (Bresler and Wendt, 1969a, b) that irreversible thermodynamics gives an incorrect result when applied to reverse osmosis. These claims have gained a certain degree of credence (Lakshminarayanaiah, 1972; Kesting, 1971), even though they are based on a misinterpretation of the basic equations.

To set the theories in proper perspective, a classification of membranes based on structure and mechanism of transport is described first. The irreversible thermodynamic theory is presented next in two main parts. In the first part, the membrane and its adjoining phases are considered as a discontinuous system. This permits the classification of osmotic phenomena on a rational basis, and also makes clear the type of information and the parameters that are useful in describing the transport properties of membranes. In the second part, the membrane is treated as a continuous system, which enables the transition to observable parameters on the one hand, and to molecular theories on the other hand to be given a sound basis in general theoretical concepts. The construction of molecular theories of transport is discussed briefly, and some recent comprehensive work in two areas, reverse osmosis and transport in ion-exchange membranes, illustrates the use of both thermodynamic and extra-thermodynamic ideas to problems in membrane transport.

II. The Classification of Membranes

A membrane may be defined (Lakshminarayanaiah, 1969) as a phase which separates two other phases, prevents bulk mixing of these two phases, but permits passage of energy and of one or more species between the phases.

There are many ways of classifying membranes (Hwang and Kammermayer, 1975): by the chemical nature of membranes, by their structure, by their intended applications or by the mechanism of their action. For our purposes, the classification used by Meares et al. (1967) on the basis of structure seems to be most useful from the standpoint of theories of transport. The following discussion uses their terms, but includes aspects of other classification schemes (Lakshminarayanaiah, 1969; Kesting, 1971;

Hwang and Kammermayer, 1975). Biological membranes of the lipid bilayer type are specifically excluded here, as are models of biological membranes such as bimolecular leaflets. Their structure and mechanism deserve separate classification (Lakshminarayanaiah, 1969).

Macroporous membranes have relatively large pores (i.e., greater than about 5 nm in diameter), where "pore" refers to a pathway for a permeating species. Ordinary filter papers and felts belong in this category, as do membranes made by etching radiation tracks in thin polymer sheets (available under the trade name "Nuclepore"). Other common types are cellulose that has been swollen in alcohol–water mixtures, then solvent-exchanged with alcohol, and "hollow fibre" membranes. These types and others are described in detail by Kesting (1971).

Transport in macroporous membranes is mainly by convective flow; that is, there is little or no discrimination among permeating species of different molecular types. Such membranes find wide use in separation of macromolecules, colloids and larger particles from fluids, and as membranes for measuring the molecular weights of macromolecules by osmometry. Many of the materials that make up membranes of this type possess small concentrations of ionic groups (e.g., cellulose) or can absorb ions. These charges can give rise to electro-osmosis and other effects (see below).

Microporous membranes are polymeric membranes in which the pore diameters are comparable to the width of the polymer chains. Typical examples are cellulose acetate and nitrocellulose membranes prepared by casting from a solvent, followed by complete removal of the solvent.

Porous phase inversion membranes, such as the Sourirajan–Loeb cellulose acetate membranes for reverse osmosis membranes (Sourirajan, 1970), may be included as well. These membranes are made by casting from a solvent, but removal of solvent is controlled so that the system separates into two coexisting phases. Further removal of solvent causes formation of a gel, with cross-links (see below) supplied by intermolecular forces between the polymer chains. For details of these and other types of dense membranes, see Kesting (1971).

The micropores result from irregularities in the packing of the more-or-less randomly-coiled chain molecules. Bulky chains (as in cellulose) or side-groups result in larger pores. Most of the materials used for membranes of this type have glass transition temperatures well above room temperature. Below the glass transition temperature, thermal random motions of the polymer chains are restricted, so that the concept of pores is a useful one. The properties of these membranes can be modified by annealing, swelling or shrinking in suitable liquids, or mechanical stretching.

Transport in these membranes occurs by both convection (as in macro-porous membranes) and diffusion in the denser regions surrounding the pores. There are strong interactions between the permeating molecules and the polymer matrix.

Solvent-type membranes consist of elastomers or flexible polymers above their glass transition temperatures. The rapid penetration of many organic solvents through silicone rubbers is typical of membranes of this type (Suwandi and Stern, 1973). The mobile polymer chains preclude the exis-tence of any permanent pore-like structures. Substances of low molecular weights can dissolve in these membranes, because of a considerable decrease in free energy on mixing, until an equilibrium state is reached. The main driving force for the process of dissolving is a decrease in the entropy of mixing, while the equilibrium state is determined by the net effects of entropy, heat of mixing, the extent of cross-linking in the polymer and polymer–solvent interactions in the solution. The permeating substance can be transported by diffusion, if its equilibrium concentrations at opposite faces of the membrane are different. As in liquid mixtures, the rate of diffusion is determined largely by the random thermal motions of the polymer chains and the solvent molecules.

Gases have relatively low solubilities in solvent-type membranes, and the rate of permeation depends on the solubility of the gas (which tends to increase with molecular size) and the diffusion coefficient (which tends to decrease with molecular size). Because of the low solubility, molecules of different types do not interfere greatly with one another, and practical separations based on differences among individual gases can be achieved.

Liquids or organic vapours usually have high solubilities in solvent-type membranes, i.e. they cause excessive swelling. These substances act as plasticizers; the thermal motions of the polymer chains are increased, and the chain separations become greater. Both these effects lead to an increase in the rate of diffusion.

Gel membranes consist of cross-linked polymers containing a swelling liquid. Membranes in which the solvent is incorporated during poly-merization are usually stronger than those in which the solvent is incorporated by inhibition and swelling, since the osmotic pressure due to the contractile forces of the chains is smaller in the former case.

Both the liquid and substances that are soluble in the liquid can be transported through the membrane by a combination of convection and diffusion. Interactions among solvent, solute and polymer molecules are all of importance, and depend on molecular nature, size and shape.

The network structure of the gel is determined by the kinetics of the polymerization reaction and polymer–liquid interactions, and appears to consist of regions where cross-linking is dense and interconnecting regions of less dense cross-linking. Variations in cross-linking result in an inhomogeneous distribution of polymer chain concentration and liquid concentration, since the regions of less dense cross-linking swell more than the regions of more dense cross-linking. There is limited thermal motion of the polymer chains, so that pore-like regions centred on the regions of less dense cross-linking can act as pathways for permeation. The distribution and extent of cross-links thus determines the permeation properties of the membrane.

Cross-linked polystyrene gels are examples of gel membranes or packings for columns that are used to fractionate polymer molecules in solution. The interstices in the gel are made to have the appropriate range of sizes to accommodate and retard smaller polymer molecules more effectively than larger ones, and thus effect separations.

Ion-exchange gel membranes are of the type discussed above, but in addition the polymer chains carry "fixed charges", ionic groups that have been introduced during polymerization or by chemical modification after polymerization. Water is the usual swelling liquid, and both ions and uncharged molecules can be transported in the liquid. The strong hydration-type interactions between both polymer ions and small ions lead to strong coupling effects between the flows of ions and of solvent. Strong interactions between the fixed polymer ions and the mobile ions affect both flows of ions and the selectivity of the membrane towards different ions.

Spectroscopic studies by Zündel (1969) have shown that counter ions and water are closely associated with the fixed polymer ions, while studies on the thermodynamics of ion-exchange equilibria have shown that, in a resin containing two kinds of counter ions, the preferred counter ions (those that interact most strongly with the polymer chains) occupy the regions of greatest density of cross-linking. Co-ions (with the same charge as the fixed ions) accumulate in regions of highest swelling. Reichenberg (1966), Marinsky (1966) and Goldring (1966) have reviewed the relation of structure to equilibrium properties of ion exchange resins.

The above brief sketch of the variety and complexity of membrane types suggests that any general approach to the description of transport in membranes will be very involved. Fortunately, such is not the case; the use of irreversible thermodynamics permits a convenient classification of transport phenomena, and as well indicates what parameters are relevant in comparing these phenomena among different types of membranes.

III. Irreversible Thermodynamics of Transport Process in Membranes

A. DISCONTINUOUS SYSTEMS

Figure 2 shows a more general osmotic system than that of Fig. 1. It consists of a membrane M separating two sub-systems I and II. The

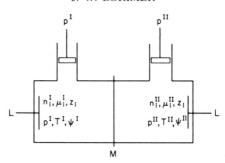

FIG. 2.   Generalized osmotic experiment.

material in each sub-system consists of thermodynamic components of amounts n_i, chemical potentials μ_i and charge numbers z_i, at pressure P, tempereature T and electric potential ψ. Superscripts I and II refer to the values of these quantities in the two sub-systems. The pressure difference $P^{II}-P^{I}$ is maintained by suitable pistons, the electric potential difference $\psi^{II}-\psi^{I}$ is maintained by electrodes L reversible to at least one charged species in the system, and the temperature difference $T^{II}-T^{I}$ is maintained by suitable thermostats. The properties of the sub-systems I and II are assumed to be uniform throughout. The size of the membrane phase is not of importance, providing we consider processes in or close to *steady states*, i.e. flows of matter and energy take place through the membrane, but the variables that define the properties of the sub-systems are independent of time. A system of this type is said to be *discontinuous*; the membrane is treated as a discontinuity between two uniform phases, and all irreversible processes take place in the membrane.

According to the Second Law of thermodynamics, entropy changes $\Delta_e S^{I} = q^{I}/T^{I}$, $\Delta_e S^{II} = q^{II}/T^{II}$ occur in the open sub-systems through interactions with the surroundings. As well, there is a change in entropy $\Delta_i S$ due to internal changes in the system. This change is always positive and arises from the transport of matter and energy through the membrane.

The rate of change of $\Delta_i S$ with time, or the *rate of production of entropy* due to irreversible processes is calculated from the thermodynamics of irreversible processes. This calculation has been done to various degrees of rigour. Highly rigorous treatments are available (de Groot, 1952; de Groot and Mazur, 1962; Haase, 1963, 1969; Hanley, 1969), along with good accounts that concentrate on applications (Lakshminarayanaiah, 1969; Katchalsky and Curran, 1965; Tyrrell, 1961). Glansdorff and Prigogine (1971) have provided a succinct summary of the basic principles of irreversible thermodynamics and have stressed the importance of the assumption of *local equilibrium* in the thermodynamic discussion of non-equilibrium states. Landsberg (1961, 1972) calls this assumption the "basic trick" of thermodynamics: a system which may have gradients of temperature, pressure and composition may be considered to be composed of elements to each of which equilibrium thermodynamics applies. For small departures from a stable equilibrium state, this assumption enables the Gibbs equation of equilibrium thermodynamics to be used (where U is the energy):

$$dU = T\,dS - p\,dV + \sum_i \mu_i\,dn_i \qquad (6)$$

The exact meaning of "small' departures is difficult to define, and can be decided on only by experiment or by calculations from statistical mechanics of transport processes. The general conclusion that has been reached from many studies on irreversible processes is that such phenomena as osmotic flow, diffusion heat conduction and laminar hydrodynamic flow are described adequately by this assumption, but that such phenomena as chemical reactions (in most cases), shock waves and turbulent flow are too "far" from equilibrium (cf. Miller, 1969).

The result of the calculation of the rate of production of entropy, σ, or the *dissipation function* $\Phi' = T\sigma$ due to irreversible processes is (see, e.g., Katchalsky and Curran, 1965)

$$\Phi' = -\sum_i j_i(d\tilde{\mu}_i)_T - j_q'\,dT/T \qquad (7)$$

Here, T is the temperature,

$$(d\tilde{\mu}_i)_T = (d\mu_i)_{T,p} + V_i\,dp + z_iF\,d\psi \qquad (8)$$

is the difference in electrochemical potential of species i at constant temperature between phases I and II, $(d\mu_i)_{T,p}$ is the corresponding difference at constant T and pressure p, V_i and z_i are the partial molar volume and charge number of species i, and F is the Faraday constant. A positive flow

$$j_i = -dn_i^{\mathrm{I}}/dt \qquad (9)$$

is the rate of decrease of the amount n_i of species i in sub-system I due to transport of i through the membrane. This choice of sign results in a positive flow towards a lower concentration; some authors use the opposite convention. The *reduced heat flow* is

$$j_q' = j_q - \sum_i (H - H_i)j_i \qquad (10)$$

where j_q is the heat lost per unit time by sub-system I, H is the molar enthalpy of the sub-system and H_i is the partial molar entropy of species i. This definition results in a flow of heat from a higher to a lower temperature; again, some authors use the opposite convention.

The significance of the heat flows has been discussed by Tyrrell (1961). The flow j_q is the total flow of heat, while j_q' is the "pure" heat flow; that is, the total heat flow less the flow of enthalpy that accompanies the flow of matter, or, equivalently, the heat transferred between the two sub-systems due to transport of matter at constant temperature.

Two features of Eqn. (7) that differ from customary notation should be noted. We have written the dissipation function in terms of infinitesimal deviations of the state variables from their equilibrium values, rather than as finite but "small" deviations (Lorimer et al., 1956; Michaeli and Kedem, 1961; Foley et al., 1974). In doing so, we take into account the possibility that the fluxes may be functions of the state variables of the two sub-systems, but may still be linear in differences in these variables. This point will be discussed further below.

We now wish to express the dissipation function, Eqn. (7), in terms of observable quantities of flow. These quantities are the total electric current

$$I = F \sum_i z_i j_i \tag{11}$$

the total volume flow j_V, and the flows of the thermodynamic components. We will discuss the flow j_0 of a solvent component 0 and the flow j_s of a solute component s, where the solute is a salt $A_{\nu_1}B_{\nu_2}$, one mole of which ionizes completely in the solvent to give ν_1 mole of cations of charge number z_1 and ν_2 mole of anions of charge number z_2. There will also be a new heat flux j_q''. Clearly, electrical neutrality requires that

$$\nu_1 z_1 + \nu_2 z_2 = 0 \tag{12}$$

For the salt, the chemical potential is

$$\mu_s = \nu_1 \mu_1 + \nu_2 \mu_2 \tag{13}$$

and the partial molar volume is

$$V_s = \nu_1 V_1 + \nu_2 V_2 \tag{14}$$

The electric current enters and leaves the system via electrodes that are reversible to the anions. This is for convenience only; a change to cation-reversible electrodes will not alter the general conclusions reached below. If the difference in electrical potential between the two wires that connect the electrodes to the external circuit is dE, then the difference in electrochemical potential of the anions in the two sub-systems is

$$d\tilde{\mu}_2 = z_2 F \, dE = d\mu_2 + z_2 F \, d\psi \tag{15}$$

Here, the dependence of $d\mu_2$ on T and p must be taken into account by the appropriate thermodynamic relations. The salt flow through the membrane is

$$j_s = j_1/\nu_1$$

$$= j_2/\nu_2 - I/Fz_2\nu_2 = j_2'/\nu_2 \tag{16}$$

Substitution of Eqns. (11), (15) and (16) into (7) and use of (8), (12), (13) and (14) gives

$$\Phi' = -j_q'' dT/T - j_s(d\mu_s)_{T,p}$$

$$-j_0(d\mu_0)_{T,p} - (j_s V_s + j_0 V_0) dp - I dE \tag{17}$$

Equation (17) contains the new heat flow j_q'' which has been discussed recently (Lorimer, 1976). Since relatively few investigations on non-isothermal transport processes have been published (see Lorimer, 1976, for a review), no account will be taken of them except for the purpose of classification.

The quantity

$$j_V = j_1 V_1 + j_2' V_2 + j_0 V_0$$

$$= j_s V_s + j_0 V_0 \tag{18}$$

is the volume flow, excluding any volume changes associated with the solid electrode phases. Weinstein and Caplan (1973; see also Kedem, 1973) have discussed Eqn. (17) for isothermal phenomena. However, they made several errors which are corrected in this discussion. Substitution of Eqn. (2) in (17) gives, after some manipulation,

$$\Phi = -j_s d\pi/c_s - j_V d(p - \pi) - I dE - j_q'' dT/T \tag{19}$$

where the relation $c_s V_s + c_0 V_0 = 1$ has been used. This is the dissipation function written in terms of observable flows of matter, heat and electricity. The significance of j_s, j_V and j_q'' should be kept clearly in mind; in many papers, the relation between flows of individual species and observable net flows is not set out clearly. It should also be noted that dE is not the potential difference dE' between two isobaric, isothermal wires connected to the electrodes if there is a difference in temperature or pressure across the membrane, since dE' then includes thermocouple effects due to gradients of temperature in the wires (Agar, 1963), or strain effects due to gradients of pressure. Pressure effects are usually unimportant. A system of the type considered here in which a temperature difference exists corresponds to a Class I thermocell for systems that do not contain a membrane. The theory of the electromotive force of such cells will not be pursued here. For a thorough discussion of cells containing electrolytes, see Agar (1963), and for non-isothermal membrane cells, see Hills et al. (1957, 1961).

B. The Phenomenological Equations

The essential point of the restriction of the thermodynamic state variables to "small" deviations from their equilibrium values is that the flows of matter and heat may be expressed as *linear* functions of these deviations. Mathematically, this procedure is simply the construction of a Taylor expansion of the flows as functions of the deviations of the state variables from their equilibrium values. For small deviations, only linear terms need be retained. A number of studies (Rastogi *et al.*, 1969) purport to demonstrate that higher-order terms in the Taylor expansion can and should be retained. These studies are misleading, since they imply (but do not state) that the Gibbs equation, Eqn. (6), remains valid even when non-linear terms in the expansion are used. This implication is contrary to the results of detailed statistical mechanical calculations (Weiss, 1969; Cohen, 1969). They also confuse linearity in the deviations and linearity in the state variables themselves. A simple example of the two types of linearity is found in the conduction of heat. The historical law of Fourier

$$j_q = -\lambda \nabla T \tag{20}$$

where λ is the thermal conductivity and ∇T is the gradient of temperature in a medium, follows from the *linear* laws of irreversible thermodynamics; it is a *linear* relation between j_q and ∇T. The thermal conductivity itself can be, and usually is, a function of temperature. The differential equation that describes the distribution of *temperature* in the medium will then be a non-linear differential equation:

$$\partial T/\partial t = \nabla \cdot (D(T)\nabla T) \tag{21}$$

where D is the temperature-dependent thermal diffusivity of the material. This non-linearity has its origin in the molecular properties of the material, and not in the relation between flow and affinity. Glansdorff and Prigogine (1971) refer to this as "extended linearity".

For the discontinuous system described above, the phenomenological equations are found from the dissipation function by expressing the flows as linear functions of the "forces" or affinities:

$$
\begin{bmatrix} j_s \\ j_V \\ I \\ j_q'' \end{bmatrix} = - \begin{bmatrix} L_D & L_{Dp} & L_{DE} & L_{DT} \\ L_{pD} & L_p & L_{pE} & L_{pT} \\ L_{ED} & L_{Ep} & L_E & L_{ET} \\ L_{TD} & L_{Tp} & L_{TE} & L_T \end{bmatrix} \begin{bmatrix} d\pi/c_s \\ d(p - \pi) \\ dE \\ dT/T \end{bmatrix} \tag{22}
$$

The equations are written here in matrix form in order to display the *phenomenological coefficients L_D*, etc. clearly. The individual flow equations can be written using the usual laws of matrix equality and matrix multiplication. For example, the electric current is

$$I = -L_{ED} \, d\pi/c_s - L_{Ep} \, d(p - \pi) - L_E \, dE - L_{ET} \, dT/T \tag{23}$$

The relation between the phenomenological equations and observable transport processes can be seen more easily if Eqn. (23) is solved for dE, and the resulting equation is substituted in Eqn. (22). The results are:

$$j_s = L_{DE}I/L_E - M_D \, d\pi/c_s - M_{Dp} \, d(p - \pi) - M_{DT} \, dT/T$$

$$j_V = L_{pE}I/L_E - M_{Dp} \, d\pi/c_s - M_p \, d(p - \pi) - M_{pT} \, dT/T \tag{24}$$

$$j_q'' = L_{TE}I/L_E - M_{DT} \, d\pi/c_s - M_{pT} \, d(p - \pi) - M_T \, dT/T$$

The M-coefficients in this equation are combinations of the L-coefficients in Eqn. (22):

$$M_D = L_D - L_{ED}^2/L_E$$

$$M_{Dp} = L_{Dp} - L_{DE}L_{Ep}/L_E = M_{pD}$$

$$M_{DT} = L_{DT} - L_{DE}L_{ET}/L_E = M_{TD}$$

$$M_p = L_p - L_{pE}^2/L_E \tag{25}$$

$$M_{pT} = L_{pT} - L_{pE}L_{ET}/L_E = M_{Tp}$$

$$M_T = L_T - L_{ET}^2/L_E$$

An important theorem also holds among the phenomenological coefficients. Onsager (1931a, b) showed that the matrix of phenomenological equations is symmetric:

$$L_{Dp} = L_{pD}, \qquad L_{DE} = L_{ED}, \qquad L_{DT} = L_{TD}$$

$$L_{pE} = L_{Ep}, \qquad L_{pT} = L_{Tp}, \qquad L_{ET} = L_{TE} \tag{26}$$

These relations are known as the *Onsager reciprocal relations*.

Experimentally, the validity of the Onsager principle is reasonably clear, in the relatively few cases where tests have been made. The evidence has been summarized by Miller (1960, 1969). Not all cases are unambiguous, a point that has been stressed by Truesdell (1969). However, the necessary precision in experimental data that is required to test the Onsager principle in even the simplest cases is difficult to achieve. At the present time, observed deviations from the Onsager principle would be ascribed to either both poor experimental precision, or deviations from equilibrium states that are too large.

The Onsager principle effects a great economy in the description of the system; in Eqn. (22) it means that only 10 distinct phenomenological coefficients appear instead of 16. Clearly, if the Onsager reciprocity theorem holds for the L-coefficients, it also holds for the M-coefficients.

The M-coefficients all have a similar structure: a *primary* contribution (Churaev and Derjaguin, 1966) which is present even if there are no electrical effects between the permeating species and the membrane, e.g. there are no external electric fields, no charged species or no electrical double layers. There is also a *secondary* contribution, due to the effect of an induced electrical potential. The sign of this potential can be assigned unambiguously for the main coefficients M_D, M_p, M_T and L_E. Since the rate of production of entropy (Eqn. 19) is necessarily positive, substitution of Eqn. (22) into Eqn. (19) gives

$$\Phi = L_D(d\pi/c_s)^2 + L_{Dp}\,d(p - \pi)\,d\pi/c_s + L_{DE}\,dE\,d\pi/cs$$
$$+ L_{DT}\,dT\,d\pi/Tc_s + 12 \text{ similar terms.} \tag{27}$$

This is a positive-definite quadratic form, and has the properties that the main diagonal coefficients L_D, L_p, L_E and L_T are all positive, while relations like

$$L_D L_E - L_{DE} L_{ED} \geq 0$$

also hold. It follows that the contributions involving the induced electric potentials, $-L_{ED}^2/L_E$, etc. in the coefficients M_D, M_p and M_T are always *negative*, and that M_D, M_p and M_T themselves are always positive. No conclusions concerning the signs of the cross-coefficients L_{DE}, L_{Dp}, L_{DT}, L_{pE}, L_{pT} and L_{ET} can be drawn, in general.

C. Classification of Osmotic Phenomena

Equations (24) have an immediate use as a convenient framework for the classification of osmotic phenomena. The following list does not exhaust all possibilities, but includes most of the phenomena that have been observed and named for membrane systems.

1. Phenomena Connected Directly with Flows

(i) $I = dp = dT = 0$; $d\pi \neq 0$; i.e., only a difference in solute concentration exists across the membrane.

Osmosis: j_0, flow of solvent.

Dialysis: j_s, flow of solute.

Diasolysis: selective flow of solute and solvent. This term is restricted usually to liquid mixtures, although in most practical membranes, osmosis and dialysis usually accompany one another.

A few specialized membranes are known that are impermeable to solutes with molecular weights as low as that of sucrose (Wagner and Moore, 1959).

(ii) $I = d\pi = dT = 0; dp \neq 0$.
Ultrafiltration: j_0, flow of solvent
(or reverse osmosis)
Pressure permeation: selective flow of components of a mixture.

(iii) $d\pi = dp = dT = 0; I \neq 0$.
Electro-osmosis: j_0, flow of solvent
Electro-dialysis, or electrical transport: j_s, flow of solute
Conductivity: I, flow of electric current.
Peltier effect: j_q'', flow of heat accompanying an electric current.

(iv) $I = d\pi = dp = 0; dT \neq 0$.
Thermo-osmosis: j_0, flow of solvent.
Thermal diffusion (Soret effect): j_s, flow of solute.

(v) $d\pi = dE = dT = 0; I \neq 0$.
Streaming current: flow of electric current per unit volume flow. Solute, volume and heat flows in a short-circuited cell ($dE = 0$) could be defined in this case.
Thermal conductivity: j_q'', flow of heat.

2. Phenomena Connected with Affinities

(i) $j_V = 0$: no volume flow.
  Osmotic pressure: $dp/d\pi$ $(I = dT = 0)$
  Electro-osmotic pressure: dp/I $(d\pi = dT = 0)$
  Thermo-osmotic pressure: dp/dT $(I = d\pi = 0)$

(ii) $I = 0$: no electrical current; cell potentials
  Diffusion potential: $dE/d\pi$ $(I = dp = dT = 0)$
  Streaming potential: dE/dp $(I = d\pi = dT = 0)$
  Thermal potential: dE/dT $(I = d\pi = dp = 0)$

The values of these quantities found from Eqn. (23) are the values of the *total* cell potential (excluding effects of T and p on the solid electrode phases; see above).

D. ALTERNATIVE FORMS OF PHENOMENOLOGICAL COEFFICIENTS

1. Constituent Flow Equations

The sections above are based on writing the dissipation function in terms of observable flows of solute, total volume, electric current and heat. Instead of the resulting phenomenological Eqns. (22), a set of $n + 1$ pheno-menological equations may be based directly on Eqn. (7). For isothermal systems,

$$j_i = - \sum_{j=1}^{n} L_{ij}(d\tilde{\mu}_j)_T \qquad i = 1, 2, \ldots n \qquad (28)$$

As before, the total electric current is given by Eqn. (11). If Eqn. (28) is substituted in (11) to eliminate $d\psi$, we find, after some manipulation,

$$j_i = t_i I/F - \sum_{j=1}^{n} M_{ij}(d\mu_j)_T \qquad (29)$$

In these equations

$$M_{ij} \equiv L_{ij} - t_i t_j L_E', \qquad (30)$$

and

$$L_E' \equiv \sum_{i,j} z_i z_j L_{ij} \qquad (31)$$

is analogous to the quantity E that appears in Eqn. (22); it can be expressed as

$$L_E' = (-I/F^2 \, d\psi) \text{ at } (d\mu_i)_{T,p} = dp = 0 \qquad (32)$$

Under these conditions, $dE = d\psi$, so that $L_E = L_E' F^2$. The current density is $i = I/A$ and the potential gradient is $d\psi/l$, where l is the thickness and A the area of cross-section of the membrane. It follows that

$$F^2 L_E' l/A = - (Il/A \, d\psi) = \varkappa \qquad (33)$$

where $\varkappa$ is the conductivity of the membrane.

The flow of species i divided by the total flow of electric charge is the *mass transport number* of species i:

$$t_i = j_i F/I = \sum_j z_j L_{ij}/L_E' \text{ at } (d\mu_i)_T = 0, \quad dT = 0 \qquad (34)$$

Note that uncharged species can have finite mass transport numbers. Only charged species have *charge transport numbers* $T_i = z_i t_i$.

Clearly,

$$\sum_{i=1}^{n} z_i t_i = \sum_{i=1}^{n} T_i = 1 \qquad (35)$$

The Onsager reciprocal relations lead to

$$\sum_i z_i M_{ij} = \sum_j z_j M_{ji} = 0, \quad \text{and} \quad M_{ij} = M_{ji}, \quad i = j \qquad (36)$$

When applied to a system containing the ions from a single salt $A_{\nu_1}B_{\nu_2}$ that ionizes completely in a solvent 0 to give ν_1 mole of cations of charge z_1 and ν_2 mole of anions of charge z_2, the Eqns. (29) can be expressed in the form of Eqn. (22). The correspondence among the coefficients is found to be, for constant temperature,

$$L_D = L_{11}/\nu_1^2$$

$$L_{Dp} = L_{pD} = L_{11}V_s/\nu_1^2 + L_{10}V_0/\nu_1$$

$$L_{DE} = L_{ED} = t_1 L_E' F/\nu_1$$

$$L_p = (V_s/\nu_1)(L_{11}V_s/\nu_1 + L_{10}V_0) + V_0(L_{10}V_s/\nu_1 + L_{00}V_0)$$

$$L_{pE} = L_{Ep} = (t_1 V_s/\nu_1 + t_0 V_0)FL_E'$$

$$L_E' = z_1^2 L_{11} + 2z_1 z_2 L_{12} + z_2^2 L_{22} = L_E/F^2$$

(37)

This approach gives a small amount of extra information about charged membranes in one limiting case. If there are no anions (co-ions) in the membrane, we may set $L_{22} = L_{12} = L_{20} = 0$. Then, $t_1 = 1$, $t_0 L_E' = L_{10}$, $L_E' = z_1^2 L_{11}$. There are then only three independent coefficients L_{11}, L_{10} and L_{00}. This case may be realized experimentally for a well-washed cation exchange membrane in contact with pure water (a "salt-free" membrane). The only mobile ions present are then the positive counter ions that neutralize negative fixed ions on the membrane matrix.

The L_{ij}-coefficients can be expressed in terms of the coefficients L_D, L_{Dp}, etc. The results are:

$$L_{11} = \nu_1^2 L_D$$

$$L_{12} = \nu_1 \nu_2 L_D + \nu_1 L_{DE}/Fz_2$$

$$L_{10} = (\nu_1/\nu_0)(L_{Dp} - V_s L_D)$$

$$L_{22} = \nu_2^2 L_D + L_E/z_2^2 F^2 - 2\nu_2 L_{DE}/Fz_2$$

(38)

$$L_{20} = -\nu_2 V_s L_D/V_0 + \nu_2 L_{Dp}/V_0 + L_{pE}/Fz_2 V_0$$
$$- L_{DE}V_s/z_2 V_0 L_E$$

$$L_{00} = L_p/V_0^2 - 2\nu_1 L_{Dp}/V_0 + \nu_1 V_s L_D/V_0.$$

These rather complicated relations arise because the L_{ij} coefficients and the coefficients in Eqn. (22) are based on dissipation functions of different forms.

2. Diffusive Flow, Reflection Coefficients and Permeability Coefficients

Other coefficients are also in use. The dissipation function, Eqn. (7), may be written (at constant temperature), where f_s is the activity coefficient of the salt,

$$\Phi = -(j_s - c_s j_0/c_0)(d\mu_s)_{T,p} - j_V\, dp - I\, dE$$

$$= -(j_s/c_s - j_0/c_0)RT\, dc_s\left(1 + \nu\frac{d\ln f_s}{d\ln c_s}\right) - j_V\, dp - I\, dE$$

$$= -j_D c_s(d\mu_s)_{T,p} - j_V\, dp - I\, dE \tag{39}$$

where $\nu = \nu_1 + \nu_2$ and a *diffusive flow*

$$j_D = (j_s/c_s - j_0/c_0)(1 + \nu c_s d \ln f_s/dc_s)$$

$$= (j_s/c_s - j_V)(1 + \nu c_s d \ln f_s/dc_s)/c_0 V_0 \tag{40}$$

may be defined that is analogous to diffusive flow in free solution. Note, however, that j_D, unlike j_s, is not measurable directly, but must be obtained from separate measurements of j_s and j_V. The phenomenological equations become:

$$\begin{bmatrix} j_D \\ j_V \\ I \end{bmatrix} = -\begin{bmatrix} L_D' & L_{Dp}' & L_{DE}' \\ L_{pD}' & L_p' & L_{pE}' \\ L_{DE}' & L_{pE}' & L_E' \end{bmatrix}\begin{bmatrix} c_s(d\mu_s)_{T,p} \\ dp \\ dE \end{bmatrix} \tag{41}$$

The coefficients in the two formulations are related as follows:

$$L_D'(c_0 V_0)^2 = L_D/c_s + L_p - 2L_{pD}/c_s$$

$$L_{Dp}'(c_0 V_0) = L_{Dp}/c_s - L_p = L_{pD}'(c_0 V_0)$$

$$L_{DE}'(c_0 V_0) = L_{DE}/c_s - L_{pE} = L_{DE}'(c_0 V_0) \tag{42}$$

$$L_p' = L_p, \qquad L_{pE}' = L_{pE}, \qquad L_E' = L_E$$

Thus, the Onsager relations are satisfied by both sets of coefficients.

Another quantity used extensively is the *reflection coefficient*, σ, defined for uncharged systems as:

$$\sigma \equiv (dp/d\pi)_{j_V = dE = 0}$$

$$= 1 - L_{pD}/L_p c_s$$

$$= -c_0 V_0 L_{pD}'/L_p' \tag{43}$$

In terms of the reflection coefficient,

$$j_V = L_p(\sigma\, d\pi - dp) \tag{44}$$

The significance of σ is clear: for a membrane that is permeable only to the solvent, zero volume flow will occur when $p = \pi$, i.e. $\sigma = 1$. For a membrane that is permeable only to the solute, $\sigma = 0$. The definition can be extended easily to electrolyte systems:

$$\sigma = (dp/d\pi)_{j_V = I = 0}$$

$$= 1 - M_{pD}/M_p c_s = -c_0 V_0 M'_{pD}/M'_p \tag{45}$$

by Eqn. (24). The coefficients M'_{pD} and M'_p are defined analogously to M_{pD} and M_p, Eqn. (24). Another coefficient, the *solute permeability*, ω, is sometimes used:

$$\omega = -(j_s/d\pi)_{j_V = I = 0}$$

$$= (M_D - M^2_{Dp}/M_p)/c_s \quad \text{(electrolyte systems)} \tag{46}$$

$$= (L_D - L^2_{Dp}/L_p)/c_s \quad \text{(non-electrolyte systems)}$$

In terms of the solute permeability, at zero electric current,

$$j_s = -\{\omega - \sigma M_p c_s(1-\sigma)\}\, d\pi - M_p c_s(1-\sigma)\, dp \tag{47}$$

$$j_V = M_p(\sigma\, d\pi - dp) \tag{48}$$

Other definitions of σ and ω are in use, namely:

$$\omega' = -(j_s/d\pi)_{j_0 = I = 0}; \qquad \sigma' = (dp/d\pi)_{j_0 = I = 0} \tag{49}$$

These are related to the other values by:

$$\omega' = \omega/(c_0 V_0 + c_s V_s \sigma)$$

$$\sigma' = \sigma + V_s \omega/M_p(c_0 V_0 + c_s V_s \sigma) \tag{50}$$

In dilute solutions, where $c_s V_s \ll c_0 V_0 \approx 1$,

$$\omega' \approx \omega, \qquad \sigma' \approx \sigma + V_s \omega/M_p. \tag{51}$$

3. Resistance Coefficients

The L-type and M-type coefficients used above have been called "conductance coefficients", since they have the same general form as electrical conductance. It is just as reasonable to write, instead of the linear relations (22),

$$\begin{bmatrix} d\pi/c_s \\ d(p-\pi) \\ dE \end{bmatrix} = - \begin{bmatrix} R_D & R_{Dp} & R_{DE} \\ R_{pD} & R_p & R_{pE} \\ R_{ED} & R_{Ep} & R_E \end{bmatrix} \begin{bmatrix} j_s \\ j_V \\ I \end{bmatrix} \tag{52}$$

i.e. the forces are written as linear functions of the flows, and the phenomenological coefficients are called "resistance coefficients". The conductance and resistance coefficients matrices are inverse to each other.

For $dE = I = 0$, the relations connecting them are:

$$R_D = L_p/\Delta, \qquad R_{Dp} = -L_{Dp}/\Delta = R_{pD}, \qquad R_p = L_D/\Delta$$

$$\Delta = L_D L_p - L_{pD}^2 \tag{53}$$

Note that, if $L_{pD} = 0$, $R_D = 1/L_D$, $R_p = 1/L_p$, confirming the resistive nature of the R-coefficients, compared to the conductive nature of the L-coefficients.

The whole development thus far could now be repeated in terms of R-coefficients. The R-coefficients are not widely used, but related coefficients are used in the discussion of continuous systems (see Eqn. 75 below).

E. THE MEMBRANE AS A CONTINUOUS SYSTEM

1. The Dissipation Function and Phenomenological Equations

In contrast to the treatment given above, where the membrane is treated as a "black box" through which flows of matter and energy can take place, an alternative approach may be taken in which thermodynamic state variables are defined for each volume element in the membrane. "Volume element" is used in the thermodynamic–hydrodynamic sense: small enough to be considered infinitesimal in a macroscopic experiment, but large enough to permit definition of thermodynamic state variables. The dissipation function per unit volume (see de Groot, 1952) is similar in this case to that given in Eqn. (7), but possesses several new features:

$$\Phi = -\sum \mathbf{j}_i \cdot \mathbf{x}_i - \mathbf{j}_q' \cdot \mathbf{x}_q + \text{viscous terms} \tag{54}$$

The flows j_i are now vectorial fluxes (flows per unit area) having both direction and magnitude, and measured relative to the velocity $\mathbf{v}$ of the local centre of mass:

$$\mathbf{j}_i = \rho_i(\mathbf{v}_i - \mathbf{v}) \tag{55}$$

where $\mathbf{v}_i$ is the local velocity of species i, and ρ_i is the mass of i per unit volume, or "mass concentration" of i. The reduced heat flux $\mathbf{j}_q'$ has the same physical significance as in a discontinuous system: it is the total heat flux less the enthalpy flux due to transport of matter. The conjugate forces are, per unit mass,

$$\mathbf{x}_i = -(\nabla \mu_i')_T + \mathbf{F}_i \tag{56}$$

and

$$\mathbf{x}_q = -\nabla T/T \tag{57}$$

For external electrical forces,

$$\mathbf{F}_i = -z_i F \nabla \psi \tag{58}$$

i.e. the forces are similar in form to those for the discontinuous system, but gradients replace differences.

In order to apply these equations to transport in membranes, we note first of all that if the concentrations ρ_i are taken as molar concentrations c_i and the chemical potentials μ_i as molar chemical potentials, then, where M_i is the molar mass of species i,

$$\rho_i(\nabla \mu_i')_T = (\rho_i/M_i)\{M_i(\nabla \mu_i')_T\} = c_i(\nabla \mu_i)_T \tag{59}$$

The reference velocity remains as the velocity of the centre of mass, however. We also wish to consider the membrane to be stationary relative to the laboratory, and define molar fluxes $\mathbf{j}_i^m$ relative to the stationary membrane:

$$\mathbf{j}_i^m = c_i(\mathbf{v}_i - \mathbf{v}_m) = c_i \mathbf{v}_i$$

$$= \mathbf{j}_i + c_i \mathbf{v} \tag{60}$$

In addition, the chemical potentials obey the Gibbs–Duhem equation

$$\sum_{i=1}^{n} c_i(\nabla \mu_i)_T = -\nabla p \tag{61}$$

where the right-hand side is the gradient of pressure in the system. Equations (55)–(60) give, when substituted in Eqn. (54),

$$\Phi = -\sum_{i=1}^{n-1} \mathbf{j}_i^m \cdot (\nabla \mu_i)_T - \mathbf{v} \cdot \sum_{i=1}^{n} \rho_i \mathbf{F}_i + \mathbf{v} \cdot \nabla p + \text{viscous terms} \tag{62}$$

The summation in the first term is over the $n - 1$ permeating species; species n is the membrane. In the absence of viscous flow, the second and third terms on the right-hand side of Eqn. (62) cancel under conditions of mechanical equilibrium. This condition means that all pressure pulses have been damped out, or that, in a system with no external forces, the pressure is uniform. Such a condition can be expected to be established rapidly in most types of flow. In membranes, however, viscous flows can retard the attainment of mechanical equilibrium. Mickulecky and Caplan (1966) showed that, if only steady-state isothermal processes are considered, the viscous terms are negligible if the fluxes are averaged over a thin section parallel to the face of the membrane. These average flows correspond, in fact, to the flows that are measured. The dissipation function becomes an average:

$$\langle \Phi \rangle = -\sum_{i=1}^{n-1} \langle \mathbf{j}_i^m \rangle \cdot (\nabla \tilde{\mu}_i)_T \tag{63}$$

If it is now assumed that there is equilibrium at the surfaces of the membranes (designated by single and double primes) then in the x-direction normal to the membrane,

$$\Phi' \equiv \int_{'}^{''} \langle \Phi \rangle \, dx = -\sum_{i=1}^{n-1} \langle j_i^m \rangle \int_{'}^{''} \left(\frac{d\tilde{\mu}_i}{dx} \right)_T dx$$

$$= -\sum_{i=1}^{n-1} \langle j_i^m \rangle (d\tilde{\mu}_i)_T \tag{64}$$

where $(d\tilde{\mu}_i)_T$ now refers to differences in the *external* solutions. The dissipation function for the discontinuous system has therefore been recovered, and further discussion of the dissipation function and the phenomenological equations proceeds as above.

Equation (63) can be used in another way. The average fluxes $\langle j_i^m \rangle$ can be replaced by average fluxes measured relative to the centre of mass:

$$\Phi = -\sum_{i=1}^{n-1} \{\langle \mathbf{j}_i \rangle + \langle c_i \mathbf{v} \rangle\} \cdot (\nabla \tilde{\mu}_i)_T \tag{65}$$

Phenomenological equations can now be written for the fluxes j_i:

$$\mathbf{j}_i = -\sum_k L_{ij} (\nabla \tilde{\mu}_k)_T \tag{66}$$

and

$$\langle \mathbf{j}_i^m \rangle = -\sum_k \langle L_{ij} \rangle (\nabla \tilde{\mu}_k)_T + c_i \langle \mathbf{v} \rangle \tag{67}$$

This equation implies that there is a convective, or bulk contribution $c_i \mathbf{v}$ to the flux of species i relative to the membrane, and separate contributions relative to the velocity of the centre of mass. In dilute solutions, the velocity of the centre of mass is essentially the same as the velocity of the solvent. The recognition of two separate types of contribution to the flux also implies that the membrane is macroporous, or microporous with pores large enough to permit convective motion, or gel-type membranes. Solution-type membranes are excluded.

Equations of the type (67) have been used extensively for descriptions of fluxes in the interior of the membrane, but the assumptions underlying their use were first made clear by Mickulecky and Caplan (1966). The phenomenological equations may be written by analogy with the constituent Eqns. (29) as:

$$\langle \mathbf{j}_i^m \rangle = t_i \mathbf{I}/F - \sum_{j=1}^{n-1} M_{ij} (\nabla \mu_j)_T + c_i \mathbf{v} \tag{68}$$

A particular model of the membrane must now be specified. One common type of model [which dates back to Helmholtz and to Bikerman (1933)] assumes that uniform pores exist whose walls are covered with a uniform density of electrical charge. In the case of ion-exchange membranes, these charges arise from the fixed ions on the polymer matrix. Two limiting cases may be distinguished. (a) For the *salt-free* case, the external salt concentration is very small compared to the concentration of fixed ions. If the fixed ions have negative charges, the average concentration of negative, mobile co-ions in the membrane is also small, so that the concentration of the positive counter ions is nearly equal to that of the fixed ions. In this case, no gradients of concentration can be set up in a homogeneous membrane. (b) For the *salt-flooded* case, the concentration of fixed charges on the polymer matrix is small compared to the external salt concentration. Both co-ions and counter ions invade the membrane and concentration gradients can arise. In either case, the velocity of the centre of mass is found by solution of the hydrodynamic equations of motion (Navier–Stokes equations) and the Poisson–Boltzmann equation, which describes the distribution of electrical potential in the electrical double layer that adjoins the walls of the pore, or even fills the pore. The general result is, where η is the velocity of the pore fluid (Mickulecky and Caplan, 1966)

$$\mathbf{v} = -(\alpha \nabla p + \beta \nabla \psi)/\eta \tag{69}$$

where α depends on the geometry of the pore, while β depends on the charge density and dielectric constant of the pore fluid as well as on the geometry. It is then possible to transform Eqns. (67) into a set of the form

$$\langle \mathbf{j}_i^m \rangle = -\sum_k L'_{ik}(\nabla \tilde{\mu}_k) \tag{70}$$

where the L'_{ik} coefficients are symmetric, and include contributions both from convection and motion relative to the solvent. It is then assumed that the contributions relative to the solvent can be described in terms of ionic mobilities and diffusion coefficients in free solution. There has been a considerable amount of theoretical work in this area, notably by Dresner (1963) for case a, and by Kobatake (1958) and Kobatake and Fujita (1964a, b, c) for case b. Errors in the Kobatake–Fujita work have been pointed out (Mickulecky and Caplan, 1966), and systematic application of the theories has yet to be made. In particular, the important intermediate case where the concentrations of co-ions, counter ions and fixed ions in the membrane are of comparable magnitudes has not been discussed quantitatively. Swelling of the membrane matrix (see Rice and Nagasawa, 1961; Marinsky, 1966) should also be taken into account, so that the problem is a formidable one for theoreticians.

The traditional Nernst–Planck equations that have been used widely in describing fluxes in a membrane are special cases of Eqn. (67). If all the cross-coefficients L_{ij} with $i \neq j$ are set equal to zero in the x-direction,

$$\mathbf{j}_i^m = -L_{ii} \, d\mu_i/dx - L_{ii} \, dp/dx - L_{ii}z_1 F \, d\psi/dx + c_i \mathbf{v} \tag{71}$$

A diffusion coefficient is defined:

$$D_i = L_{ii}RT/c_i \tag{72}$$

so that, for ideal chemical potentials,

$$\mathbf{j}_i^m = -D_i\,dc_i/dx - (c_iD_i/RT)\,dp/dx - (z_ic_iD_iF/RT)\,d\psi/dx + c_i\mathbf{v} \tag{73}$$

The approximations used in this equation imply that any cross effects involving the off-diagonal L_{ij} coefficients must arise from the convective term. The approximations are known to be unrealistic in free solution (Miller, 1969; Pikal, 1971), and clearly should not be used in discussing membrane phenomena, except as rough approximations. A careful experimental comparison of the Nernst–Planck and irreversible thermodynamic descriptions has been given by Gardner and Paterson (1972).

2. Resistance and Friction Coefficients

A different kind of phenomenological description of transport in membranes is also used. Resistance coefficients r_{ik} can be defined by the equations:

$$\nabla\tilde{\mu}_i = \sum r_{ik}c_k(\mathbf{v}_i - \mathbf{v}_k) \tag{74}$$

with $r_{ik} = r_{ki}$. As well, friction coefficients f_{ik} can be defined by

$$f_{ik} = c_k r_{ik} \tag{74a}$$

with $f_{ik}/c_k = f_{ki}/c_i$. Sometimes (e.g., Staverman, 1972) the r_{ik} are called "friction coefficients", but the definition used here is more consistent historically (Lorimer, 1978). Other definitions have also been used (Lamm, 1947, 1957; Klemm, 1953). Either resistance or friction coefficients are consistent with the dissipation function, Eqn. (54), for isothermal, non-viscous processes in general. Each term in the sum represents the average force exerted by molecules of component k on molecules of component i, where c_k is the concentration of k inside the membrane, and v_i, v_k are the velocities of species i and k.

Spiegler (1958) was the first to apply these ideas to membrane transport. Recently, Staverman (1972) has reviewed the relations between friction coefficients and the phenomenological coefficients in Eqns. (22) and (24). For uncharged systems consisting of a solute 1, a solvent 0 and a membrane m, there are four friction coefficients, defined by:

$$\nabla\mu_1 = r_{10}c_0(\mathbf{v}_1 - \mathbf{v}_0) + r_{1m}c_m(\mathbf{v}_1 - \mathbf{v}_m)$$

$$\nabla\mu_0 = r_{01}c_0(\mathbf{v}_0 - \mathbf{v}_1) + r_{0m}c_m(\mathbf{v}_0 - \mathbf{v}_m) \tag{75}$$

Flows $\mathbf{j}_1 = c_1\mathbf{v}_1$ and $j_0 = c_0\mathbf{v}_0$ are measured relative to the membrane, so that $\mathbf{v}_m = 0$. The ill-defined quantity c_m (the concentration of the membrane matrix) is incorporated into the combinations

$$f_{0m} = r_{0m}c_m$$
$$f_{1m} = r_{1m}c_m$$

(76)

so that

$$\nabla\mu_1 = (f_1m + r_{01}c_0)\mathbf{j}_1/c_1 - r_{10}\mathbf{j}_0$$
$$\nabla\mu_0 = (f_0m + r_{01}c_1)\mathbf{j}_0/c_0 - r_{10}\mathbf{j}_1$$

(77)

In terms of the coefficients in Eqn. (22), it is found that

$$r_{10}/V_0V_1 = (\sigma + x)(1 - \sigma)/L_1 - 1/L_p$$
$$f_{1m}/V_1 = (1 - \phi_m)\{1/L_p - (\sigma + x)(1 - \sigma - K)/L_1\}$$
$$f_{0m}/V_0 = (1 - \phi_m)\{1/L_p - (1 - \sigma)(1 - \sigma - K)/L_1\}$$

(78)

where

$$x = -V_0(d\mu_1)_T/V_1(d\mu_0)_T = c_0V_0/c_1V_1$$

(79)

ϕ_1, ϕ_0 and ϕ_m are volume fractions in the membrane:

$$\phi_0 + \phi_1 + \phi_m = 1$$

(80)

and

$$K = (1 + x)\phi_1/(1 - \phi_m)$$

(81)

is the partition coefficient for the solute, i.e. the ratio of the volume fraction of solute inside the membrane to that outside the membrane. Staverman has also expressed the coefficients L_1, L_p and σ in terms of the friction coefficients, and has worked out the more complicated relations when the solute is completely ionized.

It is often stated that the friction coefficients are independent of the choice of reference velocity, or more loosely, independent of the frame of reference. This statement has been based on the fact that the reference velocity does not occur in Eqn. (74). A rigorous proof, based on transformation properties of friction coefficients, has been given recently (Lorimer, 1978).

IV. Other Phenomenological Theories of Transport in Membranes

A set of phenomenological equations that is popular with chemical engineers, and combines features of the L-coefficients and friction coefficients is called the Stefan–Maxwell relations. Their application to membrane transport has been described by Scattergood and Lightfoot

(1968). The equations can be written down from Eqn. (75) by replacing $r_{ik}c_k$ by $-RTc_k/cD_{ik}$, where $D_{ik} = D_{ki}$. The quantity D_{ik} is called a Stefan–Maxwell diffusivity.

Theories based on kinetic rate processes have also been used to some extent, and are summarized by Lakshminarayanaiah (1969, sect. 3.22). A related but more modern approach is the use of stochastic descriptions of rate processes, especially for biological membranes where a continuum approach is probably unrealistic. Examples are the works of Heckmann *et al.* (1972) and of Hill and Kedem (1966).

V. Examples of the Use of Phenomenological Equations

A. REVERSE OSMOSIS

Spiegler and Kedem (1966) and Kedem (1972) have discussed the phenomenology of reverse osmosis or hyperfiltration. There is a distinction between the two terms (Kesting, 1971). Reverse osmosis is the flow of solvent from a concentrated to a dilute solution through a membrane under influence of an applied pressure that exceeds the osmotic pressure. The implication is made often that convective flow is the dominant process. Ultrafiltration is taken to imply that transport of solute is by diffusion, and is often used in referring to separation of relatively large solute molecules or particles at relatively low concentrations, where the osmotic pressure is very small. The phenomenological description of the two processes is the same, however.

For uncharged systems, Eqns. (47), (48) may be used. For steady-state fluxes j_s and j_V, the membrane may be divided into thin layers parallel to its faces and of width dx, so that the combination of (47) and (48) gives, since ω is inversely proportional to dx,

$$j_s = -P\,dc_s/dx + (1 - \sigma)c_s j_V \tag{82}$$

where the constant

$$P = -(\omega RT\,dx/c_0 V_0)(1 + c_s d \ln f_s/dc_s) \tag{83}$$

is the solute permeability and f_s is the activity coefficient of the solute. Integration of (82) with boundary conditions $c_s = c_s'$ at $x = 0$, $c_s = c_s''$ at $x = l$ gives, where l is the thickness of the membrane:

$$c_s'' = c_s' \exp\left\{\frac{(1 - \sigma)j_V}{P}\right\} + j_s/(1 - \sigma)j_V\left[1 - \exp\left\{\frac{(1 - \sigma)j_V}{P}\right\}\right] \tag{84}$$

For reverse osmosis, c_s' is kept fixed, and c_s'' is determined completely by the ratio of solute flow to volume flow:

$$c_s'' = j_s/j_V \tag{85}$$

Thus, Eqn. (82) may be rearranged to give:

$$R \equiv 1 - c_s''/c_s' = \sigma\left\{\frac{\exp\left\{(1 - \sigma)j_V/P\right\} - 1}{\exp\left\{(1 - \sigma)j_V/P\right\} - \sigma}\right\} \tag{86}$$

The quantity R is known as the *solute rejection*, while $s = 1/(1 - R)$ is the *desalination ratio*. Clearly, for perfect removal of solute by reverse osmosis or ultrafiltration, $R \to 1$, $s \to \infty$ and $\sigma = 1$, while for negligible removal of solute, $R \to 0$, $s \to 1$ and $\sigma \to 0$.

Bresler and Wendt (1969a, b) used an equation derived for gases from simple kinetic ideas by Hertz (1923) and used for solutions by Manegold and Solf (1932) that relates j_s to j_V:

$$j_s = j_V\left\{\frac{c_s'' \exp\left(-j_V l/D\right) - c_s'}{\exp\left(-j_V l/D\right) - 1}\right\} \tag{87}$$

which may be rearranged to

$$c_s'' = c_s' \exp\left(j_V l/D\right) + (j_s/j_V)\{1 - \exp\left(j_V l/D\right)\} \tag{88}$$

Clearly, this equation is simply Eqn. (84) with $\sigma = 0$ and $P = D/l$. Bresler and Wendt compared Eqn. (82) with Eqn. (87), and concluded that, since Eqn. (87) agreed with experimental results, Eqn. (82) was erroneous (except when $j_V l/D$ was small compared to unity, when (82) and (87) become identical). Clearly, it is the basis for their comparison, and not Eqn. (82), that is erroneous.

Kedem (1972) used Eqn. (86) as the basis for deciding what transport coefficients are useful in predicting the efficiency of the hyperfiltration process. Various possible mechanisms of salt rejection in reverse osmosis membranes are discussed in the same paper, using the flow equations of irreversible thermodynamics as a basis.

B. Flows in Ion-exchange Membranes

Recently, Meares and his co-workers (Foley *et al.*, 1974) have presented the first complete set of experimental phenomenological coefficients for an ion-exchange membrane. The isothermal coefficients in Eqn. (22) were measured for a Zeo–Karb phenolsulphonate resin membrane and aqueous solutions of $NaBr$, $CaBr$ and $SrBr_2$ at 25°C and over a wide range of salt concentrations. The six coefficients that can be measured under a given set of conditions were found as follows. At uniform concentration across the membrane, L_E was found from conductance measurements, L_{DE} from measurements of ion transport numbers combined with L_E, and L_{pE} from measurements of water transport numbers (electro-osmosis) combined with ion transport numbers and conductivity. (See Eqn. 37.) The quantities

$$(j_V c_s / d\pi)_{I=dp=0} = M_p c_s - M_{Dp}$$

$$(j_V / dp)_{I=0} = (M_p - M_{Dp}/s)(d\pi/dp)_{I=0} - M_p \qquad (89)$$

$$(c_s / dp)_{I=0} = (M_{Dp} - M_D/c_s)(d\pi/dp)_{I=0} - M_{Dp}$$

and $(d\pi/dp)_{I=0}$ were all measured, and provided sufficient information to calculate M_D, M_{Dp} and M_p. From these and values of L_E, L_{DE} and L_{pE}, the remaining coefficients L_D, L_{Dp} and L_p were calculated. Foley *et al.* (1974) refer to these coefficients as "differential conductance coefficients", following Michaeli and Kedem (1961). Our discussion above is concerned with such coefficients, since we consider that only differential coefficients have significance in the dissipation function.

As an example of these results, data for NaBr solutions are given in Fig. 3. The strong dependence of all the coefficients on concentration is apparent, and Meares remarks that any attempt to describe membrane transport processes by using flows that are linear in finite differences in concentration cannot be successful for the concentration differences used in practical separation processes or found in biological systems.

C. Convective Processes in Membrane Transport

The phenomenological equations that have been discussed above all assume that the concentrations in the solutions immediately adjoining the membrane surfaces are well known. In many experimental apparatus and industrial processes, this is not the case; diffusion films exist. The rate of net transport processes in the combination of membrane plus diffusion films may be controlled mainly by transport in the membrane or by transport in the diffusion films (Helfferich, 1962; Hwang and Kammermayer, 1975). Usually, processes in both regions are significant. Correction for processes in diffusion films is of great importance in experimental investigations of membrane transport. As an example, many conflicting reports of the dependence of electro-osmotic flow on electric current density have appeared. It has been found that current flow induces concentration changes, which in turn cause observed electro-osmotic flows to depend on current density, on time and on rate of stirring. If these effects are not allowed for, accurate values of electro-osmotic flow in many types of membrane cannot be obtained, while in other types the effects may be of small importance (Brydges, 1966; Barry and Hope, 1969a, b).

Similar considerations hold for practical separation processes. Sourirajan (1970) has discussed the influence of diffusion films in reverse osmosis, while the recent book by Hwang and Kammermayer (1975) has devoted considerable space to consideration of diffusion and convection processes

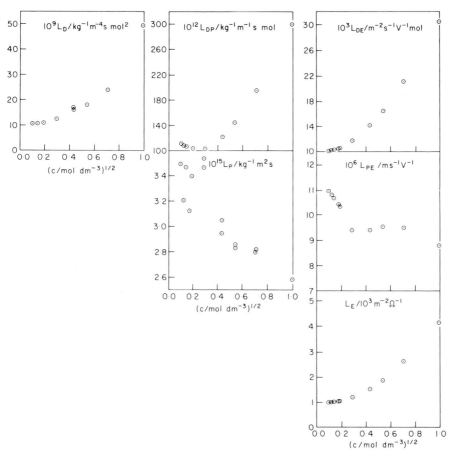

FIG. 3. Phenomenological coefficients for a phenolsulphonate ion exchange membrane as functions of concentration of aqueous NaBr solutions. (After Foley *et al.*, 1974.) The correct functional forms for the coefficients are unknown; they are plotted as functions of $c^{\frac{1}{2}}$ for convenience only.

at membrane surfaces, and their effects on practical separation processes.

Meares (1976) has recently edited a book on membrane separation processes that deals with a number of special topics, all of which involve transport phenomena.

References

Agar, J. N. (1963). *In*: "Advances in Electrochemistry and Electrochemical Engineering" (P. Delahay, ed.), p. 31. Interscience, New York and London.

Barry, P. H. and Hope, A. B. (1969a). *Biophys. J.* **9**, 700–728.

Barry, P. H. and Hope, A. B. (1969b). *Biophys. J.* **9**, 729–757.

Bikerman, J. J. (1933). *Z. physikal. Chem.* A **163**, 378–394.

Bresler, E. and Wendt, R. (1969a). *J. Phys. Chem.* **73**, 264–266.

Bresler, E. and Wendt, R. (1969b). *Science* **163**, 944–945.

Brydges, T. G. (1966). "The Measurement of Transport Numbers of Ions and Water in Ion-selective Membranes". Ph.D. Thesis, University of Western Ontario, London, Canada.

Caplan, S. R. and Mickulecky, D. C. (1966). *In:* "Ion Exchange (J. A. Marinsky, ed.). Vol. 1, Chap. 1. Marcel Dekker, Inc., New York.

Churaev, N. V. and Deryagin, B. V. (1966). *Dokl. Akad. Nauk SSSR,* **169**, 396–399; *Dokl. Phys. Chem.* **169**, 471–474.

Cohen, E. G. D. (1969). *In:* "Transport Phenomena in Fluids" (H. J. M. Hanley, ed.), Chap. 6. Marcel Dekker, New York and London.

de Groot, S. R. (1952). "Thermodynamics of Irreversible Processes". North Holland Publishing Co., Amsterdam.

de Groot, S. R. and Mazur, P. (1962). "Non-equilibrium Thermodynamics". North Holland Publishing Co., Amsterdam.

Dresner, L. (1963). *J. Phys. Chem.* **67**, 1635–1641.

Dresner, L. and Kraus, K. A. (1963). *J. Phys. Chem.* **67**, 990–996.

Flory, P. J. (1953). "Principles of Polymer Chemistry". pp. 269–282. Cornell University Press, Ithaca.

Foley, T., Klinowski, J. and Meares, P. (1974). *Proc. Roy. Soc. London* A **336**, 327–354.

Gardner, C. R. and Paterson, R. (1972). *J. Chem. Soc. Faraday* I, **68**, 2030–2040.

Glansdorff, P. and Prigogine, I. (1971). "Thermodynamic Theory of Structure, Stability and Fluctuations". Wiley-Interscience, London, New York, Sydney and Toronto.

Goldring, L. S. (1966). *In:* "Ion Exchange" (J. A. Marinsky, ed.), Vol. 1, Chap. 6. Marcel Dekker, Inc., New York.

Guggenheim, E. A. (1959). "Thermodynamics". 4th ed. Sect. 5.27. North Holland Publishing Co., Amsterdam.

Haase, R. (1963). "Thermodynamik der Irreversiblen Prozesse". Dr. Dietrich Steinkopff Verlag, Darmstadt.

Haase, R. (1969). "Thermodynamics of Irreversible Processes". Addison-Wesley, Reading, Mass.

Hanley, H. J. M., editor (1969). "Transport Phenomena in Fluids". Marcel Dekker, New York and London.

Heckmann, K., Lindemann, B. and Schnakenberg, J. (1972). *Biophys. J.* **12**, 683–702.

Helfferich, F. (1962). "Ion Exchange". Chap. 6. McGraw-Hill Book Co., New York.

Hermans, J. J. (1953). *In:* "Flow Properties of Disperse Systems" (J. J. Hermans, ed.). North Holland Publishing Co., Amsterdam.

Hertz, G. (1923). *Z. Physik* **19**, 35–42.

Hill, T. L. and Kedem, O. (1966). *J. Theor. Biol.* **10**, 399–441.

Hills, G. J., Jacobs, P. W. M. and Lakshminarayanaiah (1957). *Nature, Lond.* **179**, 96–97.

Hills, G. J., Jacobs, P. W. M. and Lakshminarayanaiah (1961). Proc. Roy. Soc. Lond. A **262**, 246–256.

Hwang, S.-T. and Kammermayer, K. (1975). "Membranes in Separations". Vol. VII (Weissberger, A., ed.) "Techniques of Chemistry" series. John Wiley and Sons, New York, London, Sydney and Toronto.

Katchalsky, A. and Curran, P. F. (1965). "Non-equilibrium Thermodynamics in Biophysics". Harvard University Press, Cambridge, Mass.

Kedem, O. (1972). In: "Reverse Osmosis Membrane Research" (H. K. Lonsdale and H. E. Podall, ed.). Plenum Press, New York and London.

Kedem, O. (1973). J. Phys. Chem. **77**, 2711.

Kesting, K. E. (1971). "Synthetic Polymer Membranes". McGraw-Hill Book Co., New York.

Klemm, A. (1953). Z. Naturforsch. **8a**, 397–400.

Kobatake, Y. (1958). J. Chem. Phys. **28**, 146–153.

Kobatake, Y. and Fujita, H. (1964a). J. Chem. Phys. **40**, 2212–2218.

Kobatake, Y. and Fujita, H. (1964b). J. Chem. Phys. **40**, 2219–2222.

Kobatake, Y. and Fujita, H. (1964c). Kolloid-Z. **196**, 58–64.

Lakshminarayanaiah, N. (1969). "Transport Phenomena in Membranes". Academic Press, New York and London.

Lakshminarayanaiah, N. (1972). In: "Electrochemistry". Vol. 2, Chap. 5, Specialist Periodical Reports. The Chemical Society, London.

Lakshminarayanaiah, N. (1974). In: "Electrochemistry". Vol. 4, Chap. 6, Specialist Periodical Reports. The Chemical Society, London.

Lamm, O. (1957). J. Phys. Chem. **61**, 948.

Lamm, O. (1947). J. Phys. Chem. **51**, 1063.

Landsberg, P. T. (1961). "Thermodynamics with Quantum Statistical Illustrations". Wiley-Interscience, New York, London, Sydney and Toronto.

Landsberg, P. T. (1972). Nature, Lond. **238**, 229–231.

Lorimer, J. W. (1976). In: "Charged Gels and Membranes II" (E. Sélégny, ed.). pp. 45–62. D. Reidel Pub. Co., Dordrecht, Holland.

Lorimer, J. W. (1978). J. Chem. Soc. Faraday Trans. II, 75–83.

Lorimer, J. W., Boterenbrood, E. and Hermans, J. J. (1956). Faraday Soc. Disc. no. 21, 141–149.

Manegold, E. and Solf, K. (1932). Kolloid-Z. **59**, 179–195.

Marinsky, J. A. (1966). In: "Ion Exchange" (J. A. Marinsky, ed.). Vol. 1, Chap. 9. Marcel Dekker, Inc., New York.

Meares, P., ed. (1976). "Membrane Separation Processes". Elsevier Pub. Co., Amsterdam, Oxford, New York.

Meares, P., Dawson, D. G., Sutton, A. H. and Thain, J. F. (1967). Ber. Bunsenges. phys. Chem. **71**, 765–775.

Meyerhoff, G. (1960). In: "Landolt-Börnstein Zahlenwerte und Funktionen". 6th ed., II. Band. 2. Teil. Springer-Verlag, Berlin-Göttingen-Heidelberg.

Michaeli, I. and Kedem, O. (1961). Trans. Faraday Soc. **57**, 1185–1190.

Mickulecky, D. C. and Caplan, S. R. (1966). J. Phys. Chem. **70**, 3049–3056.

Miller, D. G. (1960). Chem. Rev. **60**, 15–37.

Miller, D. G. (1969). In: "Transport Phenomena in Fluids" (H. J. M. Hanley, ed.), p. 421. Marcel Dekker, New York and London.

Onsager, L. (1931a). Phys. Rev. **37**, 405–426.

Onsager, L. (1931b). Phys. Rev. **38**, 2265–2279.

Pikal, M. J. (1971). J. Phys. Chem. **75**, 3124.

Rastogi, R. P., Singh, K. and Srivastava, M. L. (1969). *J. Phys. Chem.* **73**, 46–51. See also earlier papers, quoted therein.

Reichenberg, D. (1966). *In:* "Ion Exchange" (J. A. Marinsky, ed.), Vol. 1, Chap. 7. Marcel Dekker, Inc., New York.

Rice, S. A. and Nagasawa, M. (1961). "Polyelectrolyte Solutions". Academic Press, London and New York.

Robinson, R. A. and Stokes, R. H. (1959). "Electrolyte Solutions". Butterworths Scientific Publications, London, pp. 205–208.

Scattergood, E. M. and Lightfoot, E. N. (1968). *Trans. Faraday Soc.* **64**, 1135–1146.

Sourirajan, S. (1970). "Reverse Osmosis". Logos Press, London.

Spiegler, K. S. (1958). *Trans. Faraday Soc.* **54**, 1408–1428.

Spiegler, K. S. and Kedem, O. (1966). *Desalination* **1**, 311–326.

Staverman, A. J. (1952). *Trans. Faraday Soc.*, **48**, 176–185.

Staverman, A. J. (1972). *J. Electroanal. Chem.* **37**, 233–248.

Suwandi, M. S. and Stern, S. A. (1973). *J. Polymer Sci. Polymer Phys.* **11**, 663–681.

Truesdell, C. (1969). "Rational Thermodynamics". McGraw-Hill Book Co., New York, Lecture 7.

Tyrrell, H. J. V. (1961). "Diffusion and Heat Flow in Liquids". Butterworths, London.

Wagner, R. H. and Moore, L. D., Jr. (1959). *In:* "Physical Methods of Organic Chemistry" (A. Weissberger, ed.), Vol. 1, part 1. Interscience Publishing, Inc., New York.

Weinstein, J. N. and Caplan, S. R. (1973). *J. Phys. Chem.* **77**, 2710–2711.

Weiss, G. H. (1969). *In:* "Transport Phenomena in Fluids" (H. J. M. Hanley, ed.) Chap. 5. Marcel Dekker, New York and London.

Zündel, G. (1969). "Hydration and Intermolecular Interaction". Academic Press, New York.

4. Thermodynamics of Elastic Deformation Processes

L. R. G. TRELOAR

*University of Manchester Institute of Science and Technology,
Manchester, England*

I. Historical Introduction

Historically, interest in the thermodynamics of elastic deformation processes in solids dates from the early discovery by Gough (1805), later confirmed by Joule (1859), of the peculiar and intellectually challenging thermal or thermo-elastic properties of natural rubber. These "Gough–Joule" effects are two-fold, i.e.,

(i) stretched rubber, maintained under a constant tensile force, contracts (reversibly) on heating;
(ii) the extension of rubber is accompanied by a (reversible) evolution of heat.

This early interest in the thermodynamics of rubber elasticity has continued to expand with every advance in knowledge of the basic molecular constitution and detailed structure of rubber, and with the concomitant development of specific molecular theories of its mechanical and thermo-elastic properties. While considerable successes were achieved at a quite early stage of this development, notably in connection with the kinetic-statistical theory of Meyer and others, the conclusions arrived at were by no means as simple and clear-cut as might have been expected, and it has taken many years to elucidate some of the more elusive aspects of the problems involved. Even today, though the broad picture appears to be clear, there remain certain areas where a quantitative explanation of experimental observations is still lacking.

One particular difficulty which impeded progress in the early days was the confusion between the effects of crystallization, which occurs spontaneously when natural rubber is highly stretched, and the fundamental process of deformation. Though essentially a secondary process—a consequence rather than a cause of the elastic deformation—such induced crystallization produces thermal and thermodynamic effects sufficiently large to obscure the effects of the deformation process itself. At a time (1920–30) when the polymeric nature of rubber was still under discussion, and the nature of the crystallization process obscure, this basic confusion between essentially unrelated phenomena was peculiarly difficult to unravel.

The degree of success eventually achieved in the application of thermodynamics to the problem of rubber elasticity has naturally encouraged a comparable attempt to elucidate the elastic deformation processes in other polymeric materials. However, owing to the much more complex molecular structure and detailed morphology of these materials, both their elastic and their thermodynamic properties are generally less well-defined and less easily related to specific molecular mechanisms than is the case with rubber. While it is not proposed in this chapter to deal specifically with the thermodynamics of elastic deformation processes in other types of polymers, it is hoped that a discussion of the progress achieved—and the difficulties encountered—in the study of rubber elasticity may be of value in indicating the general nature of the problems involved in any such studies.

We begin with a brief account of the elementary statistical theory of rubber elasticity. We then proceed to the consideration of the relevant general thermodynamic relations and their application to the analysis of thermo-elastic properties, together with the interpretation of the results so obtained in terms of the molecular network structure. This is followed by a discussion of the more accurate network theory introduced by Flory, which takes full account of the changes in volume which accompany the application of a stress, and thus enables the contributions to the internal energy of the system arising from both *inter-* and *intra*-molecular interactions to be separately estimated.

II. Statistical Theory of Rubber Elasticity

A. CHAIN STATISTICS

The basic concepts of the statistical theory were first expounded by Meyer *et al.* (1932), and later elaborated by Guth and Mark (1934) and Kuhn (1936). The molecule is assumed to be capable of random fluctuations of form brought about by free rotation about bonds in the chain backbone structure. All such conformations are assumed to have the same internal energy. The probability of any given distance r between the ends of such a chain is defined by the *Gaussian* probability function

$$P(r) \, dr = (4b^3/\pi^{\frac{1}{2}})r^2 \, e^{-b^2r^2} \, dr \qquad (1)$$

in which the parameter b is determined by the number of bonds or rotatable "links" in the chain and the geometry of the chain structure. The *mean-square* end-to-end distance, from (1), is given by

$$\overline{r^2} = 3/2b^2 \tag{2}$$

For an idealized randomly-jointed chain of n links, in which there is no correlation between successive link orientations, we have

$$b^2 = 3/2nl^2; \qquad \overline{r^2} = nl^2 \tag{3}$$

where l is the length of the link.

The extension of a chain of this kind involves a reduction of probability, without change in internal energy, the latter being wholly kinetic, as in a gas. In thermodynamic terms, this implies a reduction in *entropy* on extension. For this reason rubber-like elasticity is sometimes referred to as *entropy* elasticity, in contrast to the elasticity of an ordinary solid, which is governed by intermolecular force fields, and associated with changes of potential energy.

Meyer at once saw the significance of the kinetic theory in providing a perfectly general and natural explanation not only of the high elastic extensibility of rubber, but also of its remarkable thermo-elastic properties. He showed that the two Gough–Joule effects referred to above were a direct consequence of the fundamental postulates of the theory, as will be shown in detail later. According to the theory, the tension in a rubber sample held at a constant stretched length should be proportional to the absolute temperature. The experiments of Meyer and Ferri (1935) showed this conclusion to be approximately borne out in the case of high extensions (Fig. 1).

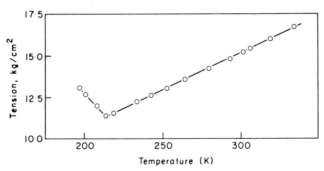

FIG. 1.   Force at constant length as function of temperature. Vulcanized rubber, 350% extension. (Meyer and Ferri, 1935.)

B. NETWORK THEORY

A vulcanized rubber is conceived of as an assembly of long-chain molecules connected together by a relatively small number of cross-linkages so as to form a three-dimensional network. The conformations of the network elements between points of cross-linkage are envisaged as being completely unhampered, and definable in terms of the Gaussian statistics. The problem is to determine the entropy, and hence the free energy or work of deformation, as a function of the state of strain.

A number of models, differing in detail, have been employed for this calculation; these all lead to substantially identical results. For the case of a simple extension, defined by an extension *ratio* λ, the tensile force f, per unit *unstrained* cross-sectional area, is given by

$$f = NkT(\lambda - 1/\lambda^2) \tag{4}$$

in which N is the number of "chains" (i.e., segments of molecules between successive points of cross-linkage) per unit volume and k is Boltzmann's constant. This result may be expressed alternatively in terms of the mean molecular weight M_c of the chains, i.e.,

$$f = \frac{\rho RT}{M_c}(\lambda - 1/\lambda^2) \tag{5}$$

where ρ is the density of the rubber and R the gas constant per mole.

The theoretical models on which Eqns. (4) and (5) are based all make use of the following assumptions:

(i) that the deformation takes place without change of volume, and
(ii) that the mean-square length of the chains in the undeformed state of the network is the same as for a corresponding set of free chains, as given by Eqn. (2).

Assumption (i) is well established experimentally as a fair approximation; assumption (ii), on the other hand, has no logical justification, and is only introduced for want of any more realistic assumption.

It is to be noted that Eqns. (4) or (5) contain essentially only a single elastic constant (NkT or $\rho RT/M_c$), which is related to the number of chains per unit volume, this in turn being a function of the degree of cross-linking. This constant, which may be shown to be equivalent to the shear modulus, is of course proportional to the absolute temperature, in accordance with the basic postulates of the kinetic theory.

III. Elementary Thermodynamic Analysis

We turn now to the general thermodynamic analysis of experimental thermo-elastic data (i.e., stress–temperature relations). The aim of this analysis is to derive the separate contributions of internal energy and entropy to the total stress. The relevant relations are derived on the basis of the first and second laws of thermodynamics. From the first law the change of internal energy dU in any process is given by

$$dU = dQ + dW \tag{6}$$

where dQ is the heat supplied to the system and dW is the work done on it by the external forces. For a *reversible* process the quantity dQ determines the change of entropy dS, in accordance with the second law, i.e.,

$$dQ = T\,dS \tag{7}$$

From (6) and (7) we have therefore, for a reversible process,

$$dU = T\,dS + dW \tag{8}$$

Furthermore, if the process is *isothermal*, the work dW is equal to the change in Helmholtz free energy dA, hence, from (8)

$$dA = dW = dU - T\,dS \tag{9}$$

Consider a specimen of length l and volume V acted on by a tensile force f under constant (e.g. atmospheric) pressure conditions. For a change of length dl the work done by the external forces is then

$$dW = f\,dl - p\,dV \tag{10}$$

where p is the pressure and dV the accompanying change in volume. The term $-p\,dV$ is normally very small compared with $f\,dl$ and may be disregarded in an elementary treatment; Eqn. (10) then becomes

$$dW = f\,dl \tag{10a}$$

Combining this with Eqn. (9) we obtain the expression

$$f = \left(\frac{\partial U}{\partial l}\right)_T - T\left(\frac{\partial S}{\partial l}\right)_T \tag{11}$$

in which the terms $(\partial U/\partial l)_T$ and $-T(\partial S/\partial l)_T$ represent the respective contributions of internal energy and entropy to the total force.

It may now be shown that the entropy term is related to the temperature coefficient of the force at constant length. From the definition of Helmholtz free energy (A), i.e.,

$$A = U - TS \tag{12}$$

we have, for *any* change (i.e. not necessarily isothermal)

$$dA = dU - T\,dS - S\,dT \tag{13}$$

whilst from (9) and (10a)

$$dU = f\,dl + T\,dS \tag{14}$$

Insertion in Eqn. (13) gives

$$dA = f\, dl - S\, dT \tag{15}$$

whence

$$\left(\frac{\partial A}{\partial l}\right)_T = f; \qquad \left(\frac{\partial A}{\partial T}\right)_l = -S \tag{16}$$

Since

$$\frac{\partial}{\partial T}\left(\frac{\partial A}{\partial l}\right)_T = \frac{\partial}{\partial l}\left(\frac{\partial A}{\partial T}\right)_l$$

it follows that

$$\left(\frac{\partial S}{\partial l}\right)_T = -\left(\frac{\partial f}{\partial T}\right)_l \tag{17}$$

In accordance with Eqn. (11) the corresponding internal energy contribution to the force is then given by

$$\left(\frac{\partial U}{\partial l}\right)_T = f - T\left(\frac{\partial f}{\partial T}\right)_l \tag{18}$$

Equations (17) and (18) have been extensively used for the thermodynamic analysis of experimental data on the temperature coefficient of the force at constant length.

A. HEAT OF EXTENSION

An alternative method of deriving the entropy contribution to the force is by the direct calorimetric determination of the heat evolved $(-dQ)$ in an isothermal extension. For a reversible process (which is assumed throughout) this is given directly by Eqn. (7). To obtain the corresponding internal energy term it is then necessary to measure the work of deformation and apply Eqn. (6). Because of the experimental difficulties this method has not attracted as much attention as the stress–temperature analysis, but it can be valuable in providing an independent check on the latter more elaborate treatment.

The value of the heat of extension may also be deduced from a measurement of the change of temperature in an adiabatic extension. If c_l is the heat capacity at constant length, the increase in temperature per unit increase in length under adiabatic (constant entropy) conditions is given by

$$\left(\frac{\partial T}{\partial l}\right)_S = -\frac{1}{c_l}\left(\frac{\partial Q}{\partial l}\right)_T \tag{19}$$

where $-(\partial Q/\partial l)_T$ is the isothermal heat of extension. The temperature rise ΔT in an extension from an initial length l_0 to a final length l is obtained by integration, i.e.

$$\Delta T = -\frac{T}{c_l}\int_{l_0}^{l}\left(\frac{\partial S}{\partial l}\right)_T dl \qquad (20)$$

Figure 2 shows the original measurements of Joule on the rise of temperature in an adiabatic extension. For comparison some more recent data by James and Guth (1943) are also included. The interpretation of these results is referred to in the following Section.

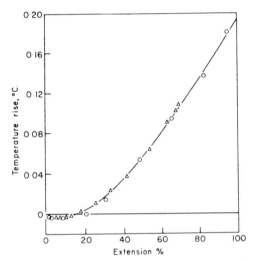

FIG. 2. Temperature change in adiabatic extension of rubber. ◯—Joule (1859). △—James and Guth (1943).

IV. Early Experiments on Rubber

A. THERMO-ELASTIC INVERSION

Reference has already been made to the early work of Meyer and Ferri (1935) on vulcanized rubber, which showed the force at constant length to be approximately proportional to the absolute temperature, thus confirming the basic postulate of the kinetic theory that the deformation process is associated essentially with a reduction of entropy. This result, however, was obtained only at high extensions; for relatively low extensions (e.g. <50% extension) the slope of the force–temperature line was lower than was to be expected theoretically, while for the lowest extensions (<10%) it was actually *negative*. This effect, known as the *thermo-elastic*

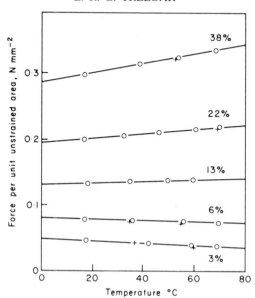

FIG. 3. Temperature dependence of force at constant length for various extensions
(Anthony *et al.*, 1942).

inversion, is illustrated in Fig. 3, taken from the later work of Anthony *et al.* (1942).

The explanation of this effect is very simple, and was correctly given both by Meyer and Ferri and by Anthony *et al.* It arises from the ordinary volume expansivity of the unstrained rubber. This implies that on raising the temperature the unstrained length of the sample increases, causing a reduction in *strain*, even though the length is held constant. The inversion point represents the value of the strain at which this effect is just sufficient to nullify the normal increase of tension with temperature (at a given value of the strain) to be expected theoretically. Meyer and Ferri also noted that if the experiment is carried out in uniaxial compression instead of extension, the increase in the unstrained length due to thermal expansivity *increases* the effective compressive strain (at constant length) and hence tends to increase the slope of the force–temperature relation. Under these conditions, therefore, the thermo-elastic inversion effect is absent.

The above complications associated with the volume expansivity of the unstrained material are reflected in the corresponding thermodynamic data. A typical example, taken from the work of Anthony *et al.* (1942), is reproduced in Fig. 4. From this it is seen that for extensions exceeding about 100% the predominant contribution to the stress arises from the entropy term $-T(\partial S/\partial l)_T$. With decreasing extension, however, the internal energy term $(\partial U/\partial l)_T$ becomes increasingly significant and ultimately

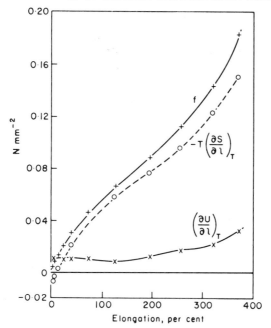

FIG. 4. Changes in internal energy (U) and entropy (S) accompanying the extension of rubber (Anthony *et al.*, 1942).

provides the major contribution. At extensions less than about 10% the entropy term, being *positive*, actually *detracts* from the stress.

Comparison with Fig. 2 reveals a direct parallel between the results of the stress–temperature analysis and the heat of extension. For small extensions there is a slight reduction of temperature (corresponding to an increase of entropy) on adiabatic extension; it is only later that this gives way to an increase of temperature. Since the temperature change involves an integration (Eqn. 20), the change in sign of the force–temperature relation $(\partial f/\partial T)_l$ corresponds not to the change in sign of the ΔT-curve, but to the change in sign of the *slope* of this curve; the thermo-elastic inversion point therefore corresponds to the minimum in the ΔT-curve, which in the case shown occurred at about 8% extension.

B. EXPERIMENTS AT CONSTANT STRAIN

From the above considerations it is plausible to assume that a more appropriate basis for the measurement of the stress–temperature coefficient would be constant *strain* rather than constant extended length. With this in mind Gee (1946) measured the relation between the force and the extension ratio λ (ratio of stretched to unstretched length) at different temperatures. Great care was taken to obtain equilibrium values of the force, which is an essential pre-condition for the application of the thermodynamic analysis, and to this end he allowed the strained sample to absorb petroleum vapour so as to break down temporary intermolecular cohesions, the vapour being subsequently dried off *in vacuo* before taking the measurement. His results (Fig. 5) indicated that under these conditions

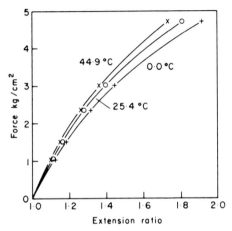

FIG. 5.   Equilibrium force-extension relations at three temperatures (Gee, 1946).

the force at a given *extension ratio* was proportional to the absolute temperature, to within the accuracy of the experiments. This experimental result is expressed by the relation

$$f - T(\partial f/\partial T)_{p,\lambda} = 0 \tag{21}$$

in which the subscripts represent constant pressure and constant strain.

A similar conclusion had been arrived at also by Anthony *et al.* (1942) by adjustment of their stress–temperature data to allow for the temperature dependence of the unstrained length of the specimen.

C. Effects of Crystallization

The experiments referred to above have been concerned with extensions which are not large enough for strain-induced crystallization to make its appearance. When this occurs, i.e. at extensions in the region of 200 to 300%, it leads to a large *reduction* in internal energy whose magnitude increases progressively up to the breaking point. This is brought out by the experiments of Wood and Roth (1944), the results of which are reproduced in Fig. 6. In a comparable study of a butadiene-styrene rubber, which does not crystallize on stretching, these effects were not present (Roth and Wood, 1944).

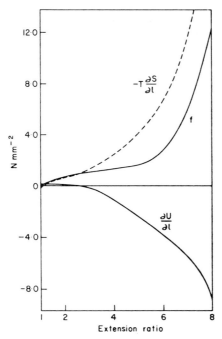

FIG. 6.   Changes in internal energy (U) and entropy (S) from data of Wood and Roth (1944).

Despite the profound influence of crystallization on the internal energy and entropy changes during extension, the available evidence strongly suggests that the basic mechanical properties of a rubber are independent of the incidence of crystallization (Treloar, 1973).

V. More Accurate Thermodynamic Analysis

For a proper understanding of the internal energy and entropy changes associated with the deformation process, and in particular of the effect of the expansivity of the unstrained material, it is necessary to distinguish between the stress–temperature coefficients at constant pressure (i.e. under normal atmospheric conditions) and at constant volume (i.e. carried out in the presence of an adjustable external pressure). Strictly speaking, the analysis given in Section 3, which ignores the term $-p\,dV$ in Eqn. (10), is valid only under conditions of constancy of volume; to be precise Eqns.

(17) and (18) should therefore be written

$$\left(\frac{\partial S}{\partial l}\right)_{V,T} = -\left(\frac{\partial f}{\partial T}\right)_{V,l} \tag{22}$$

$$\left(\frac{\partial U}{\partial l}\right)_{V,T} = f - T\left(\frac{\partial f}{\partial T}\right)_{V,l} \tag{23}$$

For constant *pressure* conditions the Helmholtz free energy A is replaced by the Gibbs free energy G, defined by

$$G = U - TS + pV \tag{24}$$

The comparable thermo-elastic relations for constant pressure conditiohs are

$$\left(\frac{\partial S}{\partial l}\right)_{p,T} = -\left(\frac{\partial f}{\partial T}\right)_{p,l} \tag{25}$$

$$\left(\frac{\partial H}{\partial l}\right)_{p,T} = f - T\left(\frac{\partial f}{\partial T}\right)_{p,l} \tag{26}$$

in which H is the heat content or enthalpy $U + pV$. It may easily be shown (Flory, 1953, p. 441) that in the present context $(\partial H/\partial l)_{p,T}$ differs insignificantly from $(\partial U/\partial l)_{p,T}$; Eqns. (25) and (26) therefore give effectively the entropy and internal energy changes accompanying a deformation at constant pressure.

Until the advent of later experimental techniques enabling the measurement of stress–temperature coefficients under constant volume conditions to be carried out, the only way of arriving at the internal energy and entropy changes at constant volume was by the application of appropriate general thermodynamic transformations to the existing constant-pressure data. Although the direct measurement of stress–temperature coefficients at constant volume was eventually successfully carried out (Allen *et al.*, 1963), it remains true that most of the experimental data relate to constant pressure conditions. A great deal of attention has therefore been given to the derivation of suitable thermodynamic relations for the conversion of constant pressure into the equivalent constant volume data, which are capable of a more direct physical interpretation.

The relevant theory has been worked out, in substantially identical terms, by Elliott and Lippmann (1945) and by Gee (1946). Gee derived the difference between $(\partial H/\partial l)_{p,T}$ and $(\partial U/\partial l)_{V,T}$ in the form

$$\left(\frac{\partial H}{\partial l}\right)_{p,T} - \left(\frac{\partial U}{\partial l}\right)_{V,T} = T\left(\frac{\partial f}{\partial l}\right)_{p,T}\left(\frac{\partial l}{\partial V}\right)_{f,T}\frac{(\partial V/\partial p)_{T,f}}{(\partial V/\partial p)_{T,l}}\left(\frac{\partial V}{\partial T}\right)_{p,l} \tag{27}$$

In order to reduce this equation to a useful form it is necessary to introduce a number of approximations. The approximations introduced by Gee were the following:

(i) the material was assumed to be isotropically compressible in the strained, as in the unstrained state, so that

$$(\partial l/\partial V)_{T,f} \approx l/3V \tag{28}$$

(ii) the isothermal compressibilities at constant force and at constant length were assumed to be equal, i.e.,

$$(\partial V/\partial p)_{T,f} \approx (\partial V/\partial p)_{T,l} \tag{29}$$

(iii) the thermal expansivity in the strained state was assumed to be the same as that of the unstrained material, i.e.,

$$\frac{1}{V}\left(\frac{\partial V}{\partial T}\right)_{p,l} \approx \beta \tag{30}$$

where β is the normal expansion coefficient.

Of these three assumptions, it is only the first which raises serious difficulties. As Gee fully realized, this assumption is only strictly true in the limit of zero strain, and is likely to become increasingly inaccurate as the strain is increased, i.e. as the material develops anisotropic mechanical properties.

On the basis of the above assumptions Eqn. (27) can be reduced to the form

$$\left(\frac{\partial H}{\partial l}\right)_{p,T} - \left(\frac{\partial U}{\partial l}\right)_{V,T} \approx \frac{1}{3}\beta l T\left(\frac{\partial f}{\partial l}\right)_{p,T} \tag{31}$$

where $(\partial f/\partial l)_{p,T}$ is the slope of the force-extension curve at the strain considered. This equation therefore provides a basis for the calculation of $(\partial U/\partial l)_{V,T}$ from properties measured under constant pressure conditions.

Of more immediate interest is the relation between $(\partial U/\partial l)_{V,T}$ and $(\partial f/\partial T)_{p,\lambda}$, the temperature coefficient of the force at constant strain. For this the following approximation was derived.

$$\left(\frac{\partial U}{\partial l}\right)_{V,T} \approx f - T\left(\frac{\partial f}{\partial T}\right)_{p,\lambda} \tag{32}$$

Comparison with Eqn. (23) shows that an experiment at constant pressure and constant extension ratio is effectively equivalent to an experiment at constant volume and constant length. Insertion of the experimental result represented by Eqn. (21) led Gee to the important conclusion that

$$(\partial U/\partial l)_{V,T} \approx 0 \tag{33}$$

i.e. that the internal energy change at constant volume is effectively zero.

Introduction of the result represented by Eqn. (33) into Eqn. (31) gives the internal energy contribution to the force under constant pressure conditions in the form

$$\left(\frac{\partial U}{\partial l}\right)_{p,T} \approx \left(\frac{\partial H}{\partial l}\right)_{p,T} \approx \frac{1}{3}\beta lT\left(\frac{\partial f}{\partial l}\right)_{p,T} \tag{34}$$

which enables it to be directly evaluated from the experimental force-extension curve. Values so obtained, as calculated by Gee, are shown in Fig. 7, together with the corresponding entropy contribution, obtained by subtraction.

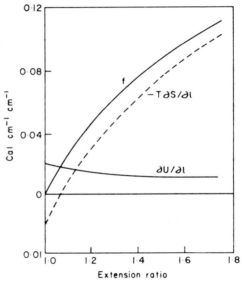

FIG. 7.   Internal energy and entropy contributions to the force, as evaluated by Gee (1946) on the basis of Eqn. (34).

The physical interpretation of these results is very simple. Since in a constant volume deformation the internal energy change is zero, it follows that the observed positive internal energy changes in a deformation at constant pressure must be due solely to the associated changes of volume. This is readily understood in terms of the statistical theory of rubber elasticity, in which two entirely separate kinds of forces are envisaged. These are (1) the network forces responsible for the rubber-like deformation, associated with the configurational entropy of the constituent chains, and (2) the intermolecular forces, not taken into account in the elementary form of the theory, which determine the *volume* of the system. These intermolecular forces are of the same kind as those in a liquid (van der Waals' forces) and are associated with changes of potential energy. The application of a tensile stress to a rubber is accompanied by a small increase of volume, which involves the performance of work against the intermolecular forces, leading to an increase of internal energy.

Before proceeding further it is necessary to add that this simple picture requires some modification in the light of the later experimental work of Allen *et al.* (1963) referred to above, which has revealed a small internal energy contribution to the stress even at constant volume, whose interpretation is discussed in Section VI. It remains true, however, that the differences between the changes of internal energy under constant pressure and constant volume conditions respectively are associated primarily with the changes of volume which are present in the former case, and to this extent Gee's conclusions remain valid.

VI. Changes of Volume on Deformation

If the changes in internal energy which accompany a deformation at constant pressure are attributed solely to the accompanying changes of volume, it should be possible to estimate the amount of the change of volume at a particular extension from the observed value of the internal energy change. The basic thermodynamic relationship between these two quantities (Gee, 1946) is

$$\left(\frac{\partial U}{\partial V}\right)_T = \frac{\beta T}{K} \tag{35}$$

where K is the volume compressibility $V^{-1}(\partial V/\partial p)_T$. Applying this formula to the present problem, Gee obtained via Eqn. (34) the increase of volume ΔV on extension in the form

$$\Delta V = \frac{K}{3} \int_{l_0}^{l} l\left(\frac{\partial f}{\partial l}\right) dl \tag{36}$$

Though the increase of volume is very small, i.e. of the order 10^{-4}, the experiments of Gee *et al.* (1950) on the direct measurement of volume changes on extension, yielded results in moderately close agreement with this equation (Fig. 8).

Gee went further and showed that even in the region where the changes in internal energy were predominantly due to crystallization, as in Roth and Wood's experiments (see Fig. 6), the same relationship applied. In this case the internal energy change is negative, as is also the associated change in volume. The reason for this is that the crystalline state involves a closer molecular packing than the amorphous state, i.e. a reduction of volume.

The normal *increase* of volume under tensile stress has its origin in the hydrostatic component of the applied stress, which is numerically equal to one third of the tensile stress. On the classical theory of elasticity the

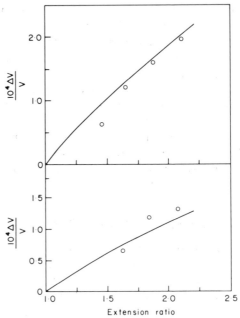

FIG. 8. Changes of volume on extension for two differently vulcanized rubbers. O—
Measured—Calculated from Eqn. (36) (Gee *et al.*, 1950).

change of volume due to a tensile force f acting on an area A is given by

$$\Delta V/V = \tfrac{1}{3}Kf/A$$

or

$$\Delta V = \tfrac{1}{3}Kfl \tag{37}$$

since $l = V/A$. It can be seen that this is the limiting form of Eqn. (36) as $l \to l_0$. It follows from this that the volume changes in rubber are no different in principle from the volume changes produced in *any* solid body on the application of a tensile stress. Rubber is unusual not in its response to the hydrostatic component of the stress (i.e. in its compressibility) but in its response to the şhear stress components (i.e. in its low value of shear modulus).

It has already been indicated that the conclusion represented by Eqn. (33) has been shown by later work to be inadequate. This automatically necessitates some modification of the above conclusions concerning the related question of the volume changes on extension. This matter is discussed more fully in Section X.

VII. Thermodynamic Measurements at Constant Volume

Reference has already been made to the direct measurements of the internal energy changes under constant volume conditions carried out by Allen *et al.* (1963). These measurements, which have led to a reappraisal of the conclusions arrived at in the foregoing discussion, will now be considered. Their full physical significance can only be understood in terms of a more precise analysis of the theoretical molecular network model which has been worked out primarily by Flory and his associates, and is discussed in the succeeding Section.

The dynamometer employed for the measurement of the stress–temperature coefficient is shown diagrammatically in Fig. 9. The rubber cylinder B was bonded to stainless steel end-pieces C attached at the top to a stiff spring E and at the bottom to a connecting rod A, the length of which determined the extension applied to the sample, the whole assembly being mounted on an Invar frame D. The length of the metal end-pieces was chosen so as to eliminate the residual expansivity of the frame, enabling the length of the sample to be held constant to within 0·01%. The stress was determined from the deflection of the spring, as measured by the sensitive transducer F. The dynamometer was contained in a steel pressure vessel

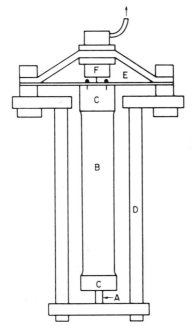

FIG. 9. Diagram of apparatus used in stress–temperature measurements. See text for explanation of symbols (Allen *et al.*, 1963).

containing mercury, to which pressures up to 150 atmospheres could be applied. (For insulation purposes the upper part containing the transducer was immersed in transformer oil floated on the mercury.)

The value of the pressure required to maintain constancy of volume while the temperature was varied was obtained from a subsidiary measurement of the "thermal pressure coefficient" of the rubber at constant volume and constant length, $(\partial p/\partial T)_{V,l}$, for various values of the extension. These measurements were made in a dilatometer. In addition the volume expansivity at constant length $V^{-1}(\partial V/\partial T)_{p,l}$, was determined. Neither of these coefficients showed any significant dependence on the state of strain.

From the measured stress–temperature coefficient at constant volume, $(\partial f/\partial T)_{V,l}$, the internal energy and entropy contributions to the force f were obtained directly by the application of Eqns. (22) and (23). The results are represented in Fig. 10 in terms of the ratio f_e/f, where f_e represents the internal energy contribution $(\partial U/\partial l)_{V,l}$. Despite the relatively large experimental scatter, these experiments established the existence of a genuine internal energy component amounting to about 20% of the total force.

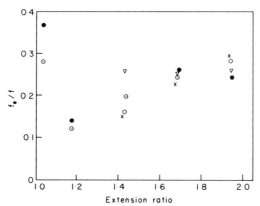

FIG. 10.   Relative internal energy contribution to the stress at constant volume (Allen *et al.*, 1963).

These direct measurements of $(\partial U/dl)_{V,T}$ were checked by an indirect derivation based on the thermodynamic relation

$$\left(\frac{\partial f}{\partial T}\right)_{V,l} = \left(\frac{\partial f}{\partial T}\right)_{p,l} + \left(\frac{\partial f}{\partial p}\right)_{T,l}\left(\frac{\partial p}{\partial T}\right)_{V,l} \tag{38}$$

From separate measurements of each of the coefficients on the right-hand side of this equation an independent value of $(\partial f/\partial T)_{V,l}$, and hence of $(\partial U/\partial l)_{V,T}$ was obtained. The comparison is shown in Table I.

TABLE I

Direct and indirect determination of $(\partial U/\partial l)_{V,T}$ at $30°C$
(Allen et al., 1963). Units, kg/cm^{-2}

Extension ratio	f kg cm^{-2}	$(\partial U/\partial l)_{V,T}$ Calc. from (38)	$(\partial U/\partial l)_{V,T}$ Directly measured
$1 \cdot 69_0$	$3 \cdot 56_4$	$0 \cdot 86_5$	$0 \cdot 91_0$
$1 \cdot 43_7$	$2 \cdot 65_2$	$0 \cdot 52_8$	$0 \cdot 52_3$
$1 \cdot 18_4$	$1 \cdot 42_3$	$0 \cdot 16_0$	$0 \cdot 20$
$1 \cdot 03_4$	$0 \cdot 31_8$	$0 \cdot 09_0$	$0 \cdot 11_8$

It should be emphasized that the general thermodynamic relation (38) is *exact*; values of $(\partial U/\partial l)_{V,T}$ calculated from it are therefore thermodynamically equivalent to values derived by the direct measurement of $(\partial f/\partial T)_{V,l}$. The close agreement between the last two columns in the above table thus provides a convincing check on the accuracy of the experimental data.

A. ORIGIN OF INTERNAL ENERGY CHANGES AT CONSTANT VOLUME

It is not possible from general thermodynamic considerations alone to deduce the mechanism responsible for the existence of a significant internal energy component of the stress at constant volume; any such deduction can only be made on the basis of a molecular or structural model. The necessary amendment of the elementary theory of the molecular network which has made this possible is considered in the following Section. In anticipation of the conclusions to be arrived at it may however be noted here that the modified theory attributes the observed changes in internal energy under constant volume conditions to energetic interactions *within* the single long-chain molecule, the presence of which is ignored in the simple statistical theory, according to which all conformations of the chain are assumed to have the same internal energy. The effects arising from these *intra*-molecular interactions are regarded as separate from, and additional to the *inter*-molecular interactions associated with changes of volume of the system.

VIII. Refinement of the Network Theory

The physical basis of the elementary network theory, leading to the force-extension relations (4) or (5), has been briefly discussed in Section II. As was seen in Section VI, the assumption of constancy of volume used in the derivation of these relations is physically unrealistic, and fails to take into account the observed volume changes on deformation and the closely related changes in internal energy. It has also been noted that the assumption that the mean-square chain vector length in the unstrained state of the network is the same as for a corresponding set of free chains is arbitrary and without any quantitative physical basis.

In the more refined treatment of the Gaussian network, due largely to Flory and his associates (Flory *et al.*, 1960; Flory, 1961) these unjustified assumptions are avoided. For this purpose the force-extension relation is written in the form

$$\frac{f}{A_i} = \frac{\nu k T}{V} \frac{\overline{r_i^2}}{\overline{r_0^2}}\left(\alpha - \frac{1}{\alpha^2}\right) \tag{39}$$

in which ν is the number of chains in the volume V, measured in the *strained* state, and f is the applied force. The symbol α represents the extension ratio referred, not to the actual unstrained length, but to the length l_i in the undistorted state at the volume V, while A_i is the corresponding undistorted cross-sectional area (at the volume V). The mean-square length of the chains in the undistorted network (at volume V) is denoted by $\overline{r_i^2}$, and that of the corresponding set of free chains by $\overline{r_0^2}$. Taking for convenience an unstrained specimen in the form of a cube, we have then

$$l_i = V^{\frac{1}{3}}, \qquad A_i = V^{\frac{2}{3}}, \qquad \alpha = l/l_i = l/V^{\frac{1}{3}} \tag{40}$$

where l is the final strained length. Introducing these relations Eqn. (39) may be transformed to

$$f = \nu k T \frac{\overline{r_i^2}}{\overline{r_0^2}}\left(\frac{l}{V^{\frac{2}{3}}} - \frac{V^{\frac{1}{3}}}{l^2}\right) \tag{41}$$

Differentiation of Eqn. (41) enables the temperature coefficient of the force to be obtained under either constant volume or constant pressure conditions. For the constant volume condition we obtain directly

$$\left(\frac{\partial f}{\partial T}\right)_{V,l} = \frac{f}{T}\left[1 - T\frac{d \ln \overline{r_0^2}}{dT}\right] \tag{42}$$

To obtain the corresponding relation for constant pressure (i.e., allowing the volume to change), we note that $\overline{r_i^2}$ is proportional to l_i^2 or $V^{\frac{2}{3}}$ while $\overline{r_0^2}$ is independent of V. Hence

$$\left(\frac{\partial f}{\partial T}\right)_{p,l} = \frac{f}{T}\left[1 - T\frac{d \ln \overline{r_0^2}}{dT} - \frac{\beta T}{\alpha^3 - 1}\right] \tag{43}$$

The difference between these two temperature coefficients is therefore given by

$$\left(\frac{\partial f}{\partial T}\right)_{V,l} - \left(\frac{df}{\partial T}\right)_{p,l} = \frac{f\beta}{\alpha^3 - 1} \tag{44}$$

From these relations the relative internal energy contribution to the force at constant volume (f_e/f) is found to be

$$\frac{f_e}{f} = 1 - \frac{T}{f}\left(\frac{\partial f}{\partial T}\right)_{V,l} = T\frac{d \ln \overline{r_0^2}}{dT} \qquad (45)$$

This very simple result enables the temperature dependence of the mean-square vector length of the free chains to be directly calculated from experimental data on the relative internal energy contribution to the force at constant volume. This in turn may be obtained either from direct measurements at constant volume, as in the experiments of Allen *et al.* (1963), or indirectly from constant pressure measurements through the application of Eqn. (44).

This modified network theory also makes possible a more accurate evaluation of the temperature coefficient of the force at constant pressure and constant *extension ratio.* The result is expressed by the equation

$$\left(\frac{\partial f}{\partial T}\right)_{p,\alpha} = \left(\frac{\partial f}{\partial T}\right)_{V,l} + \frac{f\beta}{3} \qquad (46)$$

Since $(\partial f/\partial T)_{p,\alpha}$ differs negligibly from $(\partial f/\partial T)_{p,\lambda}$, Gee's approximation (Eqn. 32), obtained on the basis of general thermodynamics, is equivalent to the relation

$$\left(\frac{\partial f}{\partial T}\right)_{p,\alpha} = \left(\frac{\partial f}{\partial T}\right)_{V,l} \qquad (47)$$

The additional term in (46) represents the error involved in this approximation. As would be expected, this error decreases and ultimately vanishes as the strain (and hence f) decrease to zero.

A. Experimental Assessment

A careful assessment of the above relations between the force-temperature coefficients at constant volume and at constant pressure has been carried out by Allen *et al.* (1971). As in the earlier work of Allen *et al.* (1963) the internal energy contribution to the force at constant volume was determined in two ways: (a) by direct measurement of $(\partial f/\partial T)_{V,l}$, (b) by calculation from measurements of $(\partial f/\partial T)_{p,l}$, $(\partial f/\partial p)_{T,l}$ and $(\partial p/\partial T)_{V,l}$, using the *exact* general thermodynamic relation (38). These values, which involve no assumptions, were compared with values of the same quantity calculated from measurements of $(\partial f/\partial T)_{p,l}$, using the formula (44) derived from the network model. Methods (a) and (b) were in close agreement with each other, and yielded the mean value $f_e/f = 0.123 \pm 0.022$. The third method, however, gave the higher value 0.18 ± 0.03. This difference, though slight in relation to the experimental error, may possibly indicate an inadequacy in the theory.

B. Possible Strain Dependence of f_e/f

The accuracy of estimates of f_e/f from Eqn. (44) tends to diminish with decreasing strain, due to the experimental difficulty of determining the unstrained length of the specimen. Any error in the strain is magnified by the presence of the quantity $\alpha^3 - 1$ in the denominator of the right-hand-side of this equation, the result becoming indeterminate as $f \rightarrow 0$. For this reason the apparent strain dependence of f_e/f in the region $\lambda = 1\cdot0$ to $\lambda = 1\cdot5$ reported by some authors (cf. Shen *et al.*, 1967) cannot be regarded as significant. Various procedures (which cannot be discussed here) have been proposed for avoiding this difficulty. The most careful work on this subject to date is probably that of Wolf and Allen (1975) which covered both extension and compression and showed no dependence on strain over the range $\lambda = 0\cdot9$ to $1\cdot7$ (Fig. 11); this yielded the result $f_e/f = 0\cdot18 \pm 0\cdot02$.

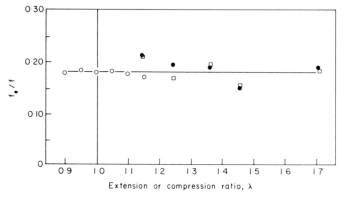

FIG. 11.   Relative internal energy contribution to the stress (f_e/f) in extension and uniaxial compression. (Wolf and Allen, 1975.)

IX. Experiments in Torsion

It was suggested long ago by Meyer and van der Wyk (1946) that the difficulties arising from the volume changes associated with a tensile stress could be avoided by working with simple shear. This was partially confirmed by their own experiments, though the accuracy achieved was rather low.

A more attractive experiment, which is basically similar to simple shear, involves the measurement of the couple in a twisted cylinder. This system has been analysed by the writer (Treloar, 1969) using a modified form of Flory's network theory. The basic equation for the couple (M) is

$$M = \frac{\pi}{2} \frac{\nu k T}{V_u} \frac{\overline{r_i^2}}{\overline{r_0^2}} \psi a_0^4 \tag{48}$$

in which V_u is the unstrained volume, a_0 the unstrained radius and ψ the torsion, defined by

$$\psi = \phi / l \tag{49}$$

where ϕ is the angular twist in the length l. By analogy with simple extension we may write, for the case of torsion at constant volume,

$$M = \left(\frac{\partial U}{\partial \phi}\right)_{V,l,T} - T\left(\frac{\partial S}{\partial \phi}\right)_{V,l,T} \tag{50}$$

where

$$\left(\frac{\partial S}{\partial \phi}\right)_{V,l,T} = -\left(\frac{\partial M}{\partial T}\right)_{V,l,\phi} \tag{51}$$

By entirely similar arguments it is found that

$$\left(\frac{\partial M}{\partial T}\right)_{V,l,\phi} = \frac{M}{T}\left[1 - T\frac{d \ln \overline{r_0^2}}{dT}\right] \tag{52}$$

giving

$$\frac{M_e}{M} = T\frac{d \ln \overline{r_0^2}}{dT} \tag{53}$$

where M_e is the relative internal energy contribution to the total couple, at constant volume. This equation is identical in form to Eqn. (45) for the case of simple extension.

The difference between the temperature coefficients of the couple at constant volume and at constant pressure is found to be

$$\left(\frac{\partial M}{\partial T}\right)_{p,l,\phi} - \left(\frac{\partial M}{\partial T}\right)_{V,l,\phi} = M\beta \tag{54}$$

Unlike the case of simple extension (Eqn. 44) this difference does not contain the strain. From the experimental standpoint this result is of considerable significance, for it means that the derivation of the equivalent temperature coefficient at constant volume from constant pressure data is not susceptible to errors arising from any inaccuracy in the measurement of the strain, as it is in the case of simple extension.

At first sight it might appear (on the basis of the classical theory of elasticity) that a torsional strain should involve no change of volume. In large-deformation theory, however [as has been pointed out by Flory *et al.* (1960)], this conclusion is invalid. A detailed study of this point (Treloar, 1969) shows that the volume changes in torsion are nevertheless of a lower order of magnitude than in the case of a tensile strain, being proportional to the *square* rather than to the first power of the strain. However, though the volume change indeed vanishes in the limit of zero torsion, its relative contribution to the internal energy remains finite, as is implied by Eqn. (54).

These advantages of torsion for thermo-elastic measurements were confirmed by the work of Boyce and Treloar (1970), illustrated in Fig. 12. The thermo-elastic inversion phenomenon is absent, and the correction to the constant-pressure data represented by Eqn. (54) is relatively slight. The numerical value of M_e/M, namely $0 \cdot 126 \pm 0 \cdot 016$, is in close agreement with that obtained from the constant-volume measurements of f_e/f by Allen *et al.* (1971), namely $0 \cdot 123 \pm 0 \cdot 022$.

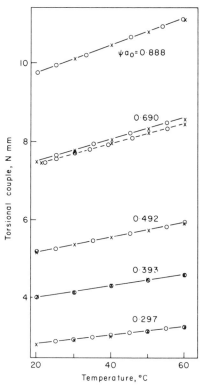

FIG. 12.   Stress–temperature relations for rubber in torsion. × —Temperature rising; ○—temperature falling. – – – Repeated after completion of set, showing degree of reproducibility. Dimensionless parameter ψa_0 represents torsional strain. (Boyce and Treloar, 1970.)

A. DIRECT CALORIMETRIC DETERMINATION OF f_e/f

The direct determination of the internal energy change on the basis of Eqn. (1) requires the calorimetric measurement of the heat of deformation ΔQ and the mechanical work ΔW. Experiments of this kind have been carried out by Allen et al. (1975), on natural rubber, using both tensile and torsional deformations. Since the experiments were carried out under constant pressure conditions, a conversion to equivalent constant volume data, as in the case of the stress–temperature coefficients, was required. The torsional data yielded the result $M_e/M = 0.20_2 \pm 0.01_1$, while the tensile data gave the value $f_e/f = 0.19 \pm 0.02$. These values, though self-consistent, are slightly higher than the values of f_e/f determined from the constant volume stress–temperature data of Allen et al. (1971), referred to above, namely 0.123 ± 0.022. It was thought that this discrepancy could be due to the deviations in the force-extension relation, and in the associated changes of volume (see below), from the forms predicted by the statistical theory. Re-calculation of the data on the basis of two alternative modifications to the theory of a semi-empirical nature worked out by Price and Allen (1973) yielded the revised values 0.13 and 0.14, respectively. In any case, these differences are not of critical importance; of greater significance is the independent confirmation of the substantial correctness of the thermo-elastic measurements.

X. Modified Treatment of Volume Changes

Reference was made in Section 5 to the inadequacy of the assumption of the isotropic compressibility of the rubber in the strained state, represented by Eqn. (28), as used in the earlier treatment of Gee. In the more precise analysis of Flory the anisotropy in the strained state is expressed in terms of the "dilation coefficient" (η), for which the following expression is derived for a rubber in simple extension (Flory, 1961)

$$\eta = \frac{d \ln V}{d \ln l} = \frac{K_l f l}{V(\alpha^3 - 1)} \tag{55}$$

in which K_l is the volume compressibility at constant length (which is substantially equivalent to K). Calculation of the total volume change ΔV involves integration of η with respect to length; the result so obtained is

$$\frac{\Delta V}{V} = K_l \frac{\nu k T}{V} \frac{\overline{r_i^2}}{r_0^2} \left(1 - \frac{1}{\alpha}\right) \tag{56}$$

The unknown factor $\overline{r_i^2}/\overline{r_0^2}$ may be eliminated by means of Eqn. (39), in which $A_i = \alpha V/l_i$, to give the more convenient form for practical application (Christensen and Hoeve, 1970)

$$\frac{\Delta V}{V} = \frac{K_i f l}{V(1 + \alpha + \alpha^2)} \tag{57}$$

In Section VI it was indicated that the observed changes of volume on extension were in approximate agreement with the original theory of Gee. The accuracy of these experiments, however, was not high. More recent work, capable of providing an assessment of the conclusions of the modified network theory, has been carried out by Allen *et al.* (1971), who obtained measured values of the dilation coefficient exceeding those predicted by Eqn. (55) by as much as 100% in some cases. Similarly Christensen and Hoeve (1970), using the integrated expression (57), obtained values of $\Delta V/V$ considerably in excess of the theoretical predictions (Fig. 13). These results were confirmed in the later work of Price and Allen (1973) already referred to, in which it was suggested that the discrepancy could be connected with the deviations of the experimental force-extension relation from the theoretical form.

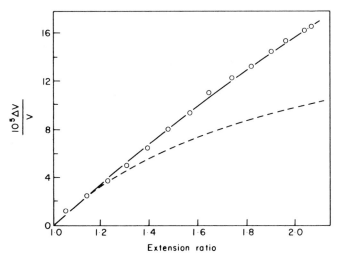

FIG. 13. Volume changes on extension of rubber. ○—Observed; – – – calculated from Eqn. (57) (Christensen and Hoeve, 1970).

XI. Temperature Dependence of Statistical Length of Chain

Data on the relative internal energy contribution to the stress and the closely related temperature dependence of $\overline{r_0^2}$ (Eqn. 45) have been obtained for a number of polymers in the rubber-like state; a selection of these data is given in Table II.

TABLE II

Values of f_e/f and temperature coefficient of chain dimensions, $d \ln \overline{r_0^2}/dT$

Polymer	f_e/f	$\dfrac{10^3 d \ln \overline{r_0^2}/dT}{K^{-1}}$	Reference
Nat. rubber	0·12	0·38	Allen *et al.*, 1971
Poly(*trans*-isoprene)	0·17	0·53	Barrie and Standen, 1967
Butyl rubber	−0·08	−0.26	Allen *et al.*, 1968
Silicone rubber	0·25	0·82	Allen *et al.*, 1969
Poly(*cis*-butadiene)	0·10	0·31	Shen *et al.*, 1971
	0·12₅	0·41	Price *et al.*, 1975
Polyethylene (above crystal melting point)	−0·42	−0·97	Ciferri *et al.*, 1961

A more comprehensive list of values has been collected by Mark (1973). A positive value of $d \ln \overline{r_0^2}/dT$ implies an increase in the statistical length of the chain with rising temperature, and *vice versa*. No simple interpretation of these temperature coefficients can be given; they are a function not only of the energy differences between different rotational positions of successive bonds, but also of more remote interactions. (For a full discussion of this subject the reader is referred to Flory, 1969.) It is seen, however, that in the typical rubbers the temperature coefficient of chain dimensions is rather small, and usually, though not necessarily, positive. To illustrate the significance of these figures, it may be noted that the value of $d \ln \overline{r_0^2}/dT$ for natural rubber ($0·38 \times 10^{-3}$) corresponds to an increase of $\overline{r_0^2}$ of 3·8% for a rise of temperature of 100°C; this is equivalent to an increase in the effective length (root-mean-square length) of only 1·9%. In polythene, however, the temperature coefficient is negative, and noticeably larger, indicating a rather strong preference, on energetic grounds, for a more extended conformation of the chain. This, of course, is in harmony with the outstanding propensity of this polymer to adopt the crystalline state, with the individual chains in the *trans* configuration in the crystal lattice.

XII. Summary and Conclusions

A. LIMITATIONS OF NETWORK MODEL

The foregoing presentation shows the elucidation of the thermo-elastic data for rubber to involve considerations going far beyond the original naive concepts in terms of which the problem was originally envisaged. Three principal stages in the development of the subject can be distinguished. The first was the demonstration that the *primary* consequence of the original statistical theory, namely that the deformation process was essentially related to a reduction of entropy, was satisfied in principle, though complications could arise due to the thermal expansivity of the unstrained rubber. The second stage was the formalization of the general thermodynamic relations between internal energy changes at constant volume and at constant pressure, leading to the tentative conclusion that the observed internal energy changes in an extension under constant pressure conditions could be attributed solely to the accompanying changes in volume. This conclusion held the field for some 14 years, but was eventually superseded by the third stage, heralded by the modified network theory of Flory and the experiments of Allen *et al.* demonstrating the presence of a significant energetic contribution to the stress, even when the volume was held constant.

The main *general* conclusion to be drawn from these historical developments, is that unambiguous conclusions regarding the mechanism of the deformation process cannot be arrived at on the basis of purely thermodynamic evidence; it is necessary in addition to have recourse to evidence of a physical or molecular character to reinforce any proposed interpretation. A consequence of this is that the criterion for "rubberlike" elasticity cannot be expressed simply in terms of a negative entropy of deformation; if this were so, we should be forced to conclude that in the case of rubber the mechanism of deformation changes at the inversion point, and that rubber itself shows "rubberlike" elasticity only for extensions exceeding about 10%! At low values of the extension the effects of the essential process of deformation are obscured by changes of volume which are subsidiary and irrelevant to this process. Again, at sufficiently high extensions the thermodynamic effects are dominated by the process of crystallization, which is entirely secondary in character.

Compared with many other polymers, a vulcanized rubber (in the amorphous state) has the advantage of a comparatively simple molecular structure which can be satisfactorily represented by a simple model which is capable of quantitative treatment by the methods of statistical thermodynamics. This model accounts not only for the mechanical and thermoelastic behaviour, but also, with minor modifications, for many other physical properties, such as for example the swelling and solution properties, and the photo-elastic behaviour. Even so, the model is by no means perfect, and substantial discrepancies from the theory which have not yet received a convincing explanation are found in the form of the stress-strain relations for various types of strain. Possibly connected with this, as has been indicated, are the anomalies in the observed changes of volume under stress. While these inconsistencies remain, a certain degree of reservation regarding the quantitative accuracy of the thermodynamic conclusions derived on the basis of the model must remain.

B. THERMODYNAMICS OF FIBRES

On the purely experimental side, the thermodynamic study of fibres encounters difficulties which are either non-existent or much less serious in the case of rubbers. First, fibres in general are much less perfectly elastic than rubbers; they exhibit the phenomena of stress–relaxation and hysteresis to a much greater degree. In these circumstances the condition that the stress shall be a unique function of the specified variables (length, volume, temperature, etc.) becomes more difficult to satisfy. Secondly, the great majority of fibres are hygroscopic systems, so that the further variables, water content and relative humidity, have to be taken into account. In the pioneer studies of Woods (1946) the fibres were immersed in water. In general, the water content will be a function of the temperature; it may also be a function of the stress (Treloar, 1953).

To these experimental difficulties must be added the more fundamental theoretical difficulty of providing a sufficiently realistic theoretical model of the structure of any particular fibre. As noted above, the derivation of internal energy and entropy changes from experimental data represents only the first stage in the problem; the interpretation of such experimental data in terms of specific mechanisms of deformation, which is the normal aim of investigations of this kind, can only usefully be pursued with the aid of a physical model. Progress towards this goal must inevitably be slow, and must take into account all the available evidence on the molecular structure and physical properties of the material.

References

Allen, G., Bianchi, U. and Price, C. (1963). *Trans. Faraday Soc.* **59**, 2493–502.

Allen, G., Kirkham, M. C., Padget, J. C. and Price, C. (1968). "Polymer Systems" (R. E. Wetton and R. W. Whorlow, eds.), p. 51–8. Macmillan, London.

Allen, G., Kirkham, M. J., Padget, J. and Price, C. (1971). *Trans. Faraday Soc.* **67**, 1278–92.

Allen, G., Price, C. and Yoshimura, N. (1975). *J. Chem. Soc. Faraday Trans. I*, **71**, 548–57.

Anthony, R. L., Caston, R. H. and Guth, E. (1942). *J. Phys. Chem.* **46**, 826–40.

Barrie, J. A. and Standen, J. (1967). *Polymer* **8**, 97.

Boyce, P. H. and Treloar, L. R. G. (1970). *Polymer* **11**, 21–30.

Christensen, R. G. and Hoeve, C. A. J. (1970). *J. Polymer Sci. A-1* **8**, 1503–12.

Ciferri, A., Hoeve, C. A. J. and Flory, P. J. (1961). *J. Amer. Chem. Soc.* **83**, 1015–22.

Elliott, D. R. and Lippmann, S. A. (1945). *J. Appl. Phys.* **16**, 50–4.

Flory, P. J. (1953). "Principles of Polymer Chemistry", Cornell, New York.

Flory, P. J. (1961). *Trans. Faraday Soc.* **57**, 829–38.

Flory, P. J. (1969). *Statistical Mechanics of Chain Molecules* (New York: Interscience).

132   L. R. G. TRELOAR

Flory, P. J., Ciferri, A. and Hoeve, C. A. J. (1960). *J. Polymer Sci.* **45**, 235–6.
Gee, G. (1946). *Trans. Faraday Soc.* **42**, 585–98.
Gee, G., Stern, J. and Treloar, L. R. G. (1950). *Trans. Faraday Soc.* **46**, 1101–6.
Gough, J. (1805). *Mem. Lit. Phil. Soc. Manchester* **1**, 288.
Guth, E. and Mark, H. (1934). *Mh. Chem.* **65**, 93–121.
James, H. M. and Guth, E. (1943). *J. Chem. Phys.* **11**, 455–81.
Joule, J. P. (1859). *Phil. Trans. Roy. Soc.* **149**, 91–131.
Kuhn, W. (1936). *Kolloidzschr.* **76**, 258–71.
Mark, J. E. (1973). *Rubber Chem. Technol.* **46**, 593–618.
Meyer, K. H. and Ferri, C. (1935). *Helv. Chim. Acta* **18**, 570–89.
Meyer, K. H. and Van der Wyk (1946). *Helv. Chim. Acta* **29**, 1842–53.
Meyer, K. H., von Susich, G. and Valko, E. (1932). *Kolloidzschr.* **59**, 208–16.
Price, C. and Allen, G. (1973). *Polymer* **14**, 576–8.
Price, C., Padget, J. C., Kirkham, M. C. and Allen, G. (1969). *Polymer,* **10**, 573–8.
Price, C., Allen, G. and Yoshimura, N. (1975). *Polymer* **16**, 261–4.
Roth, F. L. and Wood, L. A. (1944). *J. Appl. Phys.* **15**, 749–57.
Shen, M. C. (1969). *Macromolecules* **2**, 358–64.
Shen, M. C., McQuarrie, D. A. and Jackson, J. L. (1967). *J. Appl. Phys.* **38**, 791–8.
Shen, M. C., Chen, T. Y., Cirtin, E. H. and Gebhard, H. M. (1971). "Polymer Networks: Structure and Mechanical Properties" (A. J. Chomff and S. Newman, eds.), pp. 47–56. Plenum Press, New York.
Treloar, L. R. G. (1953). *Trans. Faraday Soc.* **49**, 816–23.
Treloar, L. R. G. (1969). *Polymer* **10**, 291–305.
Treloar, L. R. G. (1973), *Rep. Prog. Phys.* **36**, 755–826.
Wolf, F. P. and Allen, G. (1975). *Polymer,* **16**, 209–17.
Wood, L. A. and Roth, F. L. (1944). *J. Appl. Phys.* **15**, 781–9.
Woods, H. J. (1946). *J. Coll. Sci.* **1**, 407–19.

5. D.C. Conduction in Keratin and Cellulose

J. E. ALGIE

C.S.I.R.O., Division of Food Research
North Ryde, N.S.W., Australia

I. Introduction

From the beginning of the study of electricity the conduction properties of the naturally occurring textile fibres have been of interest. Experiments on the electricity in charged rain clouds used silken or hempen conductors and the very great increase in their conductivity when wet produced some drastic consequences. Recent interest has also been on the effect of absorbed water on the conductivity, firstly in providing a non-destructive method of measuring the water content through measurements of the conductance and secondly, in the study of the reduction of the effects of static electrification of fibres and yarns during processing. A further interest has been the relationship between conduction and molecular structure, although progress in this field has not been great.

This chapter is concerned with an examination of the theories of conduction which have been used to explain conduction in fibres, the experimental methods and a review of recent data and theories of conduction in keratin and cellulose. SI units have been used excepting where noted and the activation energies have been quoted in kilocalories/gram mole which is related to the electron volt and the kilojoule/kilogram mole as follows:

$$1 \text{ kcal/gmole} = 4 \cdot 2 \times 10^3 \text{ kjoules/kgmole}$$

$$1 \text{ electron volt} = 23 \cdot 069 \text{ kcal/gmole}.$$

II. Theories of Conduction

The electrical conductivity of polymers has been examined in the light of theories developed for conduction in ionic crystals or electronic semiconductors.

The equation for conductivity is $\sigma = \text{Nev ohm}^{-1} \text{m}^{-1}$ where N is the density of conducting units (electron, positive hole or ion) in number per

m^3, e = electronic charge in coulomb, v = mobility in m/sec per volt/m. This equation can be generalized to allow for conduction by several species of conducting units through a summation of terms

$$\sigma = \sum_{i=1}^{n} N_i e z_i v_i$$

where z_i is the valency.

Since the conductivity is dependent upon the product of the density of conducting units N and their mobility v it is necessary to determine either N or v by other methods. The mobility is usually found from measurements of the Hall effect (Shockley, 1950a) in semiconductors, although this method is difficult to apply to polymers because of their low conductivity. Ionic mobility has been measured by O'Sullivan (1947b) in cellulose containing absorbed water, by detection of the presence of ionic species using various indicators. This method is due to Sir Oliver Lodge (1886). Recently, electron and hole mobilities have been determined by injecting electrons into polymers with a pulsed electron beam (Hayashi et al., 1973).

Both the mobility and the density of conducting units are thermally activated quantities so that the activation energy for conduction will be the sum of two terms,

$$\sigma = \sigma_0 \exp\left[-(E_1 + E_2)/kT\right]$$

In addition it is found experimentally that with both electronic semiconductors and ionic semiconductors the equations relating conductivity to temperature are similar in form, i.e.

$$\sigma = A \exp\left(-E_3/kT\right) + B \exp\left(-E_4/kT\right)$$

It is thus not possible to decide upon the type of conductivity which the polymer exhibits, i.e. ionic or electronic, from conductivity measurements. It is necessary to determine whether or not mass transfer occurs in accordance with Faraday's law; if it does, then the sample is exhibiting ionic conductivity. Both electronic and ionic conductivity may occur together in the one sample.

A. IONIC CONDUCTIVITY

The following treatment is based on that of Mott and Gurney (1940). Ionic conductivity in ionic crystals occurs by virtue of defects in the crystal lattice. These are of two types, the "Frenkel defect" and the "Schottky defect". The Frenkel defect, illustrated in Fig. 1, consists of a vacancy and an interstitial atom. It can be shown by a thermodynamic argument that the number of Frenkel defects (n) is given by

$$n = (NN')^{\frac{1}{2}} \exp\left(-\tfrac{1}{2}W/kT\right)$$

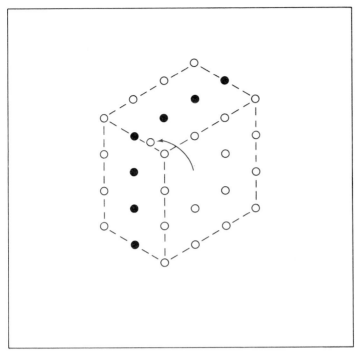

FIG. 1.   Section through an ionic crystal showing a Frenkel defect.

where N = total number of atoms, N' = total number of possible interstitial positions, W = increase in internal energy due to placing one atom in an interstitial position.

The factor $\frac{1}{2}$ in the exponential is due to the fact that two types of defect are created simultaneously, i.e. a vacancy and an interstitial atom. A Schottky defect is illustrated in Fig. 2 where a vacancy is produced by the removal of an atom from the crystal lattice to the surface. The number of Schottky defects is given by

$$n/(N - n) = \exp(-W_s/kT)$$

where W_s = the energy of formation of a Schottky defect. Both defects will occur together although one type will be predominant.

The mobility may be calculated by considering the effect on diffusion of an electric field. Figure 3 shows that the field increases the height of the potential barrier in the direction opposite to the field F by an amount $\frac{1}{2}eFa$ (where a is the width of the barrier) and decreases the potential barrier U by the same amount in the direction of the field. The probability that an ion which oscillates with a frequency ν will cross the barrier per unit time in the

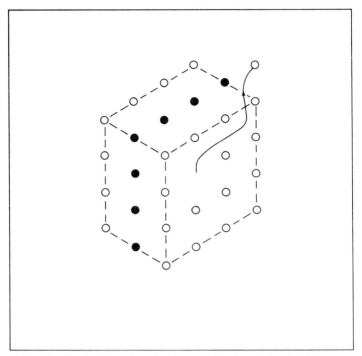

FIG. 2.   Section through an ionic crystal showing a Schottky defect.

direction of the field is

$$\nu \exp\left[-(U - \tfrac{1}{2}eFa)/kT\right]$$

and in the opposite direction is

$$\nu \exp\left[-(U + \tfrac{1}{2}eFa)/kT\right]$$

so that the mean velocity of drift is

$$u = \nu a\, 2 \sinh\left(\tfrac{1}{2}eFa/kT\right) \exp\left(-U/kT\right)$$

and when $eFa \ll kT$, which will be the case for the fields used in practice, the mobility

$$v = [e\nu a^2\, C/kT][\exp\left(-U_0/kT\right)]$$

where $\exp\left(-U/kT\right) = C \exp\left(-U_0/kT\right)$ with U_0 being the height of the potential barrier when $T = 0°K$; C is a constant. The conductivity is then

$$\sigma = (KNe^2\nu a^2/kT)(\exp\left[(-\tfrac{1}{2}W_0 - U_0)/kT\right])$$

where K is a coefficient which may be estimated. The above equation is for a single type of defect and if the conducting unit is an interstitial ion N must be replaced by $\sqrt{NN'}$.

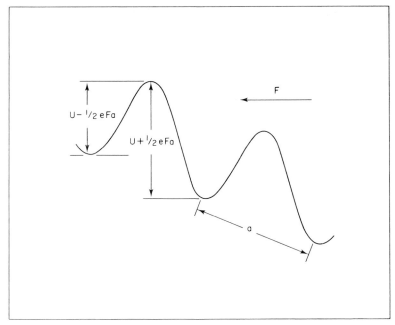

FIG. 3.   Diagram showing the effect of an electric field F in lowering the potential barrier U in the forward direction (a is the inter atomic distance).

It is found experimentally that the above equation is applicable at high temperatures but at low temperatures an impurity conductivity with a low activation energy often becomes the dominant process.

B. Electronic Conduction

From quantum mechanics has come the theory that electrons in crystals can have only certain permitted energy levels which are divided into bands separated by bands of forbidden energy. In Fig. 4 is shown the energy band diagrams for metals, insulators, intrinsic and impurity semiconductors. According to the theory of Wilson (1936), in metals there are many electrons which are free to move and which form an electron gas. The motion of these electrons is impeded by the thermal vibrations of the crystal lattice so that the conductivity of a metal decreases as the temperature rises. In contrast the conductivity of intrinsic semiconductors increases with temperature rise. At absolute zero the allowed energy bands are either filled or empty so that no conduction can occur. When the temperature is raised some electrons receive enough thermal energy to jump over the forbidden energy gap into the conduction band where they provide the conductivity.

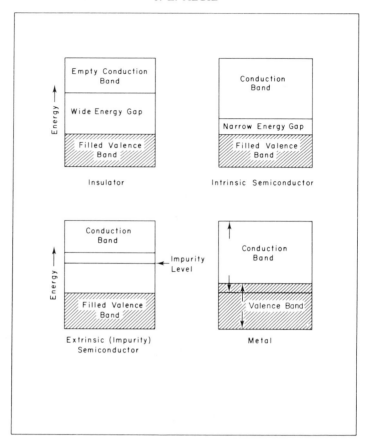

FIG. 4.   Energy band diagrams for electrons in insulators, metals and extrinsic and intrinsic
semiconductors.

Impurity semiconductors or "extrinsic" semiconductors depend upon the existence of impurity energy levels which are not far below the conduction band so that electrons can be readily raised by heat into the conduction band. The impurity centres are either due to atoms of a foreign species or to a stoichiometric excess of one constituent in an ionic crystal.

When an electron in an intrinsic semiconductor is raised into the conduction band from the valence band it leaves behind a positive "hole" which behaves in a similar fashion to an electron but with a positive charge.

Conduction will then be due to the motion of both electrons and positive holes.

For impurity semiconductors with the number of impurity centres N equal to the number of electrons in a volume V it can be shown by a thermodynamic argument (Mott and Gurney, 1940) that the number of conduction electrons is

$$n/V = (2)(N^{\frac{1}{2}}/V^{\frac{1}{2}})(2\pi mkT)^{\frac{3}{4}}/h^2[\exp(-\tfrac{1}{2}E/kT)]$$

provided that $n \ll N$. The (2) only occurs if the impurity levels contain paired electrons. In the above equation: m = mass of the electron, k = Boltzmann's constant, h = Planck's constant, E = energy required to raise an electron into the conduction band.

Since the mobility of the electrons does not vary rapidly with temperature, for $T \ll \tfrac{1}{2}E/k$ the conductivity will vary as the number of conduction electrons, i.e.

$$\sigma = \sigma_0 = \exp(-\tfrac{1}{2}E/kT)$$

This equation is for intrinsic conduction but when an intrinsic semiconductor contains impurity centres, which will usually be the case, the conductivity will have two terms,

$$\sigma = A_1 \exp(-E_1/kT) + A_2 \exp(-E_2/kT)$$

One term has a large coefficient and a large activation energy and the other term has a small coefficient and a small activation energy. This is exactly the same situation as was found in the case of ionic conductivity and a plot of log σ against $1/T$ will consist of two straight lines. The line of greatest slope, corresponding to the larger activation energy, will appear at high temperatures.

1. Narrow Conduction Band

In those cases in which the conduction band is narrow due to a small overlap of the electronic orbitals of adjacent atoms the effective mass of the electron in the conduction band will be large relative to that of a free electron (Shockley, 1950b),

$$\partial \epsilon = h^2/8a^2 m^*$$

where $\partial \epsilon$ = band width, h = Planck's constant, a = lattice constant, m^* = effective electron mass.

2. Tunnelling

Quantum theory states that when a charge carrier strikes a potential barrier there is a finite probability that it can pass through. For a rectangular barrier Kauzmann (1957) finds that the probability

$$T \simeq (16\,E(U-E)/U^2)\exp[-2(2m(U-E)a^2)^{\frac{1}{2}}/\hbar]$$

provided that

$$2m(U-E)a^2/\hbar^2 \gg 1$$

where T = probability of tunnelling, E = energy of particle, U = height of the barrier, a = width of the barrier, m = mass of the electron, $\hbar = (h/2\pi)$ where h is Planck's constant.

The probability of tunnelling is thus exponentially dependent upon the barrier width a and upon the square root of the energy difference between the height of the barrier and the energy of the electron. Tunnelling will proceed with equal probability in both directions. The effect of an applied field is effectively to raise the barrier height in the backwards direction and lower it in the forward direction by an amount $\frac{1}{2}eFa$ where e is the electronic charge and F is the field strength. Eley and Willis (1961) have calculated that the net tunnelling probability in the forward direction for a rectangular barrier is

$$T_f - T_b = \sinh\left[[(2m)^{\frac{1}{2}}/(U - E)](eFa/2)(a/\hbar)\right]$$

$$\times [(32E(U - E)/U^2)\exp[-2(2m(U - E)a^2)^{\frac{1}{2}}/\hbar]$$

where T_f = probability of tunnelling in the forward direction, T_b = probability of tunnelling in the backward direction. The effect of the field F in producing the net tunnelling in the forward direction is dependent upon the size of eFa in comparison with $(U - E)$. With the fields used in conduction experiments being only up to 10^7 volt/m, $eFa \simeq 3 \times 10^{-3}$ electron volts whilst $(U - E)$ may be 5 electron volts, the effect is consequently small.

III. Experimental Methods

Because of the high resistivity of textile fibres, particularly when dry, the effect of the insulation resistance of electrode supports is important. Also since the currents to be measured are small, careful shielding is essential to prevent interference from electric and magnetic fields. In the circuit shown in Fig. 5 the high value resistors are available up to 10^{13} ohm and the electrometer can have an input resistance of about 10^{16} ohm with a full scale sensitivity of 10^{-2} volt. This means that a sample resistance of 10^{17} ohm may be measured. Since wool has a resistivity of about 10^{16} ohm metre when dry at 20°C, a 1 cm length of fibre with a diameter of 3×10^{-3} cm will have a resistance of about 10^{23} ohm, so that in order to make a measurement on the fibres it would be necessary to place 10^6 fibres in parallel. It is more convenient to use a solid disc of another form of keratin, cut from a rhinoceros horn. Such a disc $2\cdot54 \times 10^{-2}$ cm thick and $2\cdot54$ cm in diameter has a resistance of about 5×10^{15} ohm which can be readily measured. It is advisable that a guard electrode (Fig. 6) be used to prevent interference from surface currents.

For conditions of high water content when it is necessary to make measurements quickly before products of electrolysis have time to build

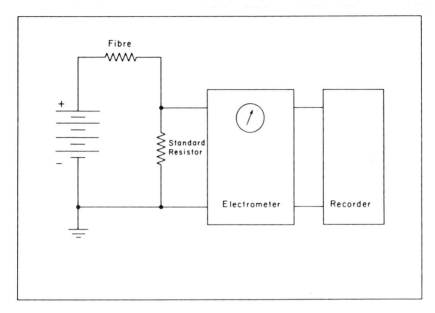

FIG. 5. Electrometer circuit for measuring the resistance of fibres.

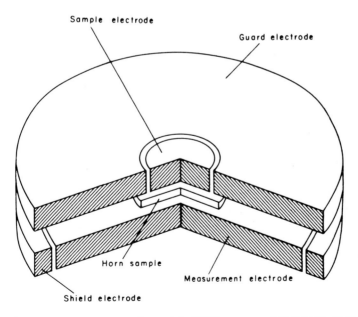

FIG. 6. Guarded electrode system. From Algie (1973), courtesy of *Kolloid Z.u.Z. Polymere.*

up, the fibre resistance may be measured with a Wheatstone bridge or a voltmeter/milliampere meter combination.

In order to avoid electrode effects Algie and Gamble (1973) have calculated the conductance from the dielectric loss at low frequencies (down to 0·1 Hz). The dielectric loss due to the d.c. conductance is

$$\epsilon'' = G_{d.c.}/2\pi f C_0$$

where $G_{d.c.}$ = d.c. conductance, f = frequency, C_0 = capacity of sample capacitor containing air.

This component of the dielectric loss produces a line with a slope -1 on a plot of log ϵ'' versus log f. It may be necessary to deduce the values of ϵ'', which are due to other loss causing mechanisms, from values of log $(\epsilon' - \epsilon_\infty)$ versus log f since the total loss $\epsilon'' = \epsilon''_{a.c.} + \epsilon''_{d.c.}$. The method outlined above was used with a clump of loose wool as a dielectric and hence gave values of the conductance of the fibre clump rather than that for single fibres. Values for the conductance of the fibre clump were obtained for a range of water contents and temperatures and these values were adjusted to keratin conductivities by comparison with literature values for single fibres at several water contents. Activation energies of the d.c. conductance can be obtained without this procedure but absolute values of the conductivity can only be obtained by comparison with data on solid keratin either as single fibres or as pieces of horn.

A. Time Effects

Textile fibres and polymers in general are characterized dielectrically by polarization processes with long relaxation times. As a consequence, the current which flows when a voltage is applied to a capacitor containing the polymer as its dielectric only slowly approaches the steady state value due to the d.c. conductance of the sample. This is shown schematically in Fig. 7. So long as the superposition principle holds for the dielectric, the charging current $i_{(t)}$ is related to the dielectric loss factor by the Fourier integral

$$\epsilon''_{(\omega)} = G_{d.c.}/\omega C_0 + 1/C_0 V \int_0^\infty i_{(t)} \sin \omega t . \, dt$$

where V is the applied voltage.

Hamon (1952) has shown that, provided that the transient current is $i_{(t)} = C t^{-n}$ where C is a constant and n is a constant such that $0·3 < n < 1·2$, then

$$\epsilon''_{(f)} = (i_t/2\pi f C_0 V) + (G_{d.c.}/2\pi f C_0)$$

where $t = 0·1/f$ sec and f = frequency in Hz. (i_t is in ampere, V in volt, C_0 in farad.)

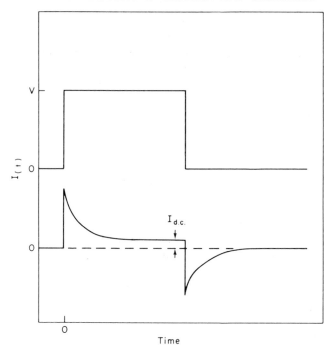

FIG. 7. Charge and discharge current as a function of time after the application of the voltage V.

If the dielectric loss factor ϵ'' due to the polarization processes is known at very low frequencies ($f < 10^{-2}$ Hz) then $i_{(t)}$ may be calculated from ϵ'' and used to determine what portion of the current at a particular time after the application of the voltage is due to the polarization processes. The calculated current must of course obey the above mentioned conditions for the Hamon approximation to be applicable.

For dry keratin at 20°C such very low frequency ϵ'' data is not available except by extrapolation from data obtained at higher temperatures. Using such extrapolated data for dry keratin at 20°C, 75% of the current flowing 20 h after the application of the voltage is due to polarization effects. For dry cellulose at 20°C (Occhini and Lawson, 1972), 3 weeks were necessary in order to obtain the true d.c. conductance.

IV. Experimental Results

A. Effect of Water Content

When dry, both cellulose and keratin are good insulators with a resistivity at room temperature of about 10^{16} ohm m. A single wool fibre 1 cm long and 30 μm in diameter has a resistance at 20°C of about 10^{23} ohm. Because of this high resistance it is difficult to measure the resistance even of bundles of fibres in the dry state. Modern methods of resistance measurement in which the resistance of the fibres is compared with standard resistances of up to 10^{13} ohm, are limited to the measurement of a fibre resistance of about 10^{16} ohm. Consequently the early workers did not measure the resistivity of wool fibres below water contents of about 12%. A factor to be considered is the long time required for the decay of the polarization current. When keratin is dry the most probable relaxation time of the relaxation process which has been ascribed to interfacial polarization (Algie, 1973) is $\tau \simeq 2 \times 10^8$ sec at 20°C so that if 10 times τ is taken as a guide for the time (after the application of the voltage to the specimen) at which the current should be measured then that time at 20°C is about 63 years. It is obvious from these considerations that it is impossible to measure the true d.c. conductivity of dry keratin. Algie (1973) used a time of 20 h after application of the voltage for the measurement of the d.c. conductance of a thin disc of rhinoceros horn and, as stated in the previous section, 75% of the current is due to the polarization. For the above reasons Hearle's (1953a) values for the conductivity at low water contents (Fig. 8) are suspect.

For keratin, at higher water contents, there is a spread of the values obtained by various workers (Marsh and Earp, 1933; King and Medley, 1949b; Hersh and Montgomery, 1952; Hearle, 1953a; Murphy, 1960a, b; Algie and Gamble, 1973) although the slope in Fig. 8 of the log conductivity versus water content is approximately the same for all the data. As outlined in the experimental methods section, Algie and Gamble's (1973) data has been determined from the component of the dielectric loss factor ϵ'' which is due to the d.c. conductance $\epsilon'' = G_{d.c.}/\omega C_0$ and should be free from electrode effects and polarization effects.

In the case of cellulose (Fig. 9), (Hersh and Montgomery, 1952; Hearle, 1953a; Murphy, 1960b), most of the data was obtained with cotton although the value for dry cellulose was obtained by Murphy (1960a) from experiments with paper. Hearle's data show a higher conductivity, probably because the results are for cotton as received and hence not freed from conducting salts. He showed that purification by washing in distilled water decreased the conductivity by about one decade. Hersh and Montgomery's (1952) data shows a higher conductivity than that of Murphy's (1960b) primarily due to the higher temperature of 30°C in comparison with 25°C. The agreement between the three sets of data is reasonable.

The most remarkable fact displayed by the data in Figs 8 and 9 is the very great effect which absorbed water has on the d.c. conductivity. For keratin the conductivity increases with water content by a factor of 10^9 from 0% to 20% water content; whilst for cellulose, the increase in conductivity is by a factor of 10^{14} from 0% to 20% water content.

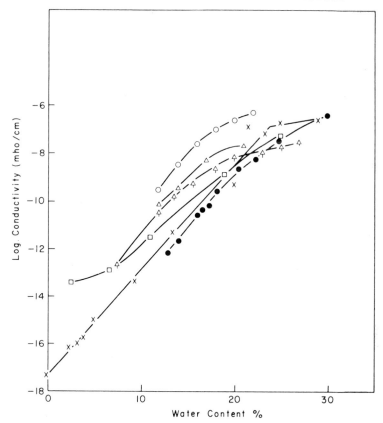

FIG. 8. Log conductivity as a function of water content. × Algie and Gamble (1973) wool 20°C; ○ Marsh and Earp (1933) wool 25°C; △ Hersh and Montgomery (1952) wool 30°C; ⇑ King and Medley (1949b) horn 25°C; ● Murphy (1960) wool 25°C; □ Hearle (1953) wool 20°C.

The explanation of these phenomena has intrigued the theoreticians for many years and will be discussed later.

B. EFFECT OF TEMPERATURE

The conductivity of both cellulose and keratin is thermally activated, i.e. it increases with temperature and may be represented by an Arrhenius equation

$$\sigma = \sigma_0 \exp(-E/RT)$$

where E is the activation enthalpy. Figure 10 shows a plot of log R_s versus $10^3/T$, where R_s is the mass specific resistivity quoted by Hearle (1953b); apart from the values at high temperatures (80°C) which may be affected by a small loss of water from the sample, the data is a reasonable fit to a

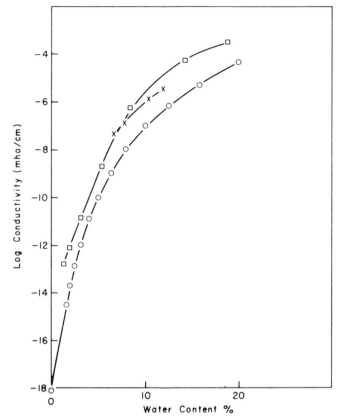

FIG. 9. Log conductivity for cotton as a function of water content. □ Hearle (1953) 20°C;
× Hersh and Montgomery (1952) 30°C; ○ Murphy (1960) 25°C.

series of straight lines. The activation enthalpies calculated from the slopes
of these lines is given in Table I whilst data for keratin is given in Table II.

The activation enthalpies for keratin determined by various workers are
in reasonable agreement at high water contents except for the data of
Baxter (1943) which is suspect at all water contents. The data of King and
Medley (1949b) appears to give values which are high at low water
contents in comparison with those of Algie and Gamble (1973) which were
obtained over a much larger temperature range and furthermore are values
obtained after the polarization current had virtually disappeared, i.e. after
20 h at high temperatures.

The conductivity of dry keratin (Algie, 1973) increases rapidly above
a temperature of about 150°C with an activation enthalpy of about
40 kcal/gmole, and this seems to be associated with a transition as indicated

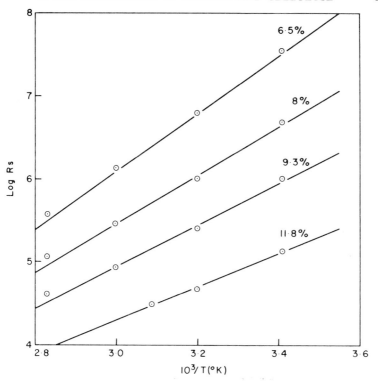

Fig. 10. Log Mass Specific Resistance R_s as a function of $10^3/T$ (T in °K) for cotton at various water contents. From data of Hearle (1953).

TABLE I

Activation enthalpy of conduction in cotton

Water content %	Activation enthalpy kcal (g mol)$^{-1}$
0	24.6[a]
6·5	16.1
8·0	13.4
9·3	11·6
11·8	9·2

[a] Ochinni and Lawson (1972).

TABLE II

Activation enthalpy of conduction in keratin

Water content %	Activation enthalpy kcal (g mole)$^{-1}$			
	King and Medley (1949b)	Hearle (1953b)	Baxter (1943)	Algie and Gamble (1973) Algie (1973)
0			25·3	24
3			32·0	
3·6	31			
7·2	29·2			
9·0			30·6	
9·5				26
10·8	26·5			
12·9		24·6		
13·5				24
14·5	22·8	22·2		
16·0			30·6	
18·1	20·1	18·1		
19·6		17·7		
21·6	16·4			15
23·3				15
25·2	13·7			
28·5			31·0	

by the corresponding sudden increase in the activation enthalpy of the relaxation process ascribed to the motion of main chains.

Dry cellulose has been studied extensively by Murphy (1960a) who found an increase in the activation enthalpy of conduction above a temperature of about 60°C. He derived an equation for the conductivity,

$$\sigma = 4\cdot5 \times 10^2 \exp\left(-30\cdot7/RT\right) + 3\cdot55 \times 10^{-10} \exp\left(-10\cdot6/RT\right)(\text{ohm cm})^{-1}$$

and treated the conductivity as being similar to that of an ionic crystal.

Occhini and Lawson (1972) claim that the apparent decrease in the activation enthalpy which Murphy (1960a) found at 60°C is due to the fact that the currents determined by Murphy were mainly displacement currents due to polarization processes. Occhini and Lawson found no change of slope down to 20°C and determined the activation enthalpy to be about 25 kcal/gmole.

C. TIME AND VOLTAGE EFFECTS

When a d.c. voltage is applied to a sample of cellulose or keratin the current is initially high then falls away with time. For samples which are dry or at low relative humidities (<30%) the major part of this fall of current is due to a decrease with time of the polarizing current as the effective capacitance of the sample plus electrodes becomes charged. The theory of this process has been outlined in the experimental section. For keratin the most probable relaxation times (at various water contents and temperatures) due to the slowest relaxation process, the interfacial polarization, are known (Algie, 1973).

Also, the magnitudes of the relaxation processes are known from data at high temperatures. Using these extrapolated values for the magnitudes of the dielectric loss factors ϵ''_m of the two slowest relaxation processes Ω and α', their extrapolated relaxation times and their Cole–Cole parameters β (see Chapter 18), it is possible to calculate the polarization current which they would produce. Figure 11 shows actual values of the discharge current for dry keratin at 22°C. During the charging process the current at 20 h was $1\cdot3 \times 10^{-14}$ amp; however the calculated polarization current from the α' process is $0\cdot8 \times 10^{-14}$ amp and from the interfacial polarization Ω, assuming $\epsilon''_m = 100$, Cole–Cole parameter $\beta = 0\cdot73$ and $f_m = 10^{-10}$ Hz, the polarization current is $0\cdot2 \times 10^{-14}$ amp. This gives a total current at 20 h of $1\cdot0 \times 10^{-14}$ amp due to polarization. The conductivity calculated by Algie (1973) on the basis of the 20 h value, i.e. $1\cdot3 \times 10^{-14}$ amp is therefore too high by a factor of approximately four. It is worth noting two facts, (i) the current due to the slowest relaxation process provided only a small correction, so that although theoretically one should wait about 63 years, in

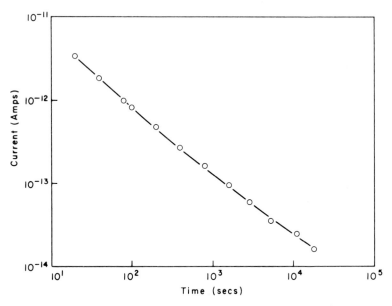

FIG. 11.   Discharge current versus time for dry rhinoceros horn at 20°C on log scales.

practice the effect may be small; (ii) conductivities calculated on the basis of short times as recommended by some workers could give highly erroneous values, e.g. at 20 sec the calculated conductivity would be high by a factor of 300 whilst at 2 min it would be high by a factor of 50.

At 20°C the relaxation time τ of the α' process decreases from $1 \cdot 6 \times 10^9$ sec for a dry sample to $3 \cdot 2 \times 10^{-3}$ sec for a sample containing $29 \cdot 1\%$ H_2O. Thus it is necessary to wait a long time for a true equilibrium to be achieved for dry specimens but only milliseconds for samples nearly saturated with water. Unfortunately there are other effects occurring at high relative humidities due to the similarity between a water-containing fibre and an electrolytic cell.

This has been studied extensively in cotton by Murphy (1929), Hearle (1953c) and Cusick and Hearle (1955), and in cellophane by O'Sullivan (1947a).

Murphy found that for cotton at $88 \cdot 4\%$ R.H., 25°C, the contact resistance with brass and platinum electrodes was negligible. Furthermore the increase of resistance with time was greatest near the anode section for brass electrodes but greatest in the centre section of the thread for platinum electrodes. Murphy attributed this to the effect of different electrode reactions for brass and platinum electrodes so that the resulting products of electrolysis would have different effects on the conductivity.

Hearle (1953c) also studied the relationship between conductivity and electrode materials. Figure 12 shows the increase of resistance with time for cotton at $10 \cdot 3\%$ water content at 20°C and for electrodes of copper, zinc, tin, aluminium and platinum. Whilst tin, aluminium and platinum produce only a small increase of resistance with time, the effect of copper and zinc electrodes is such that the resistance increased by about 50% in 2 minutes in comparison with the value at 2 seconds. Even with platinum electrodes the time effect can be large. A 2 cm length of cotton at $88 \cdot 4\%$ R.H., 25°C, with 370 volts applied, increased its resistance from 272×10^6 ohm to 7500×10^6 ohm over a period of 59 h (Murphy, 1929). It would be expected that if the time effects were due to electrolytic processes then they would be larger if the current through the fibres was increased and this in fact occurs. The current can be increased at a fixed voltage by increasing the water content of the sample and the time effect increases as shown in Fig. 13. At a fixed water content the current can be increased by increasing the voltage and the time effect, although not proportional to the voltage, increases as shown in Fig. 14. The addition of salts such as KCl also increases the time effect although it would be expected that only part of the increase is due to the increased current.

It is clear that there are two different processes causing the time effects, one of which, namely that due to dielectric polarization, requires that the true conductivity be determined after a time large in comparison with the

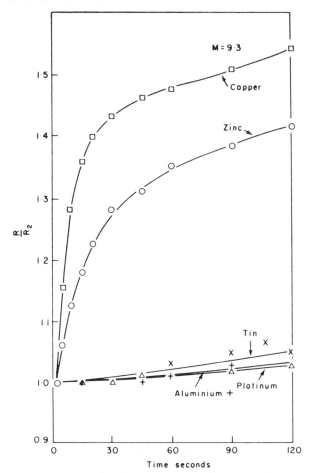

FIG. 12.   Resistance of cotton divided by the resistance at 2 sec as a function of time for various electrode materials, 20°C (moisture content 9·3%). From Hearle (1953c) courtesy of *Textile Research Journal.*

most probable relaxation time of the slowest dielectric relaxation process, whilst the other, due to electrolytic processes, requires that the true conductivity be determined at short times before the products of electrolysis disturb the conductivity.

These conflicting requirements imply that at some water contents it will be impossible to determine the true conductivity by the usual d.c. methods and an a.c. method as used by Algie and Gamble (1973) is more likely to give reliable results.

The mobility of various ions in cellophane sheets containing salts was measured by O'Sullivan (1947c) using the method of Lodge (1886).

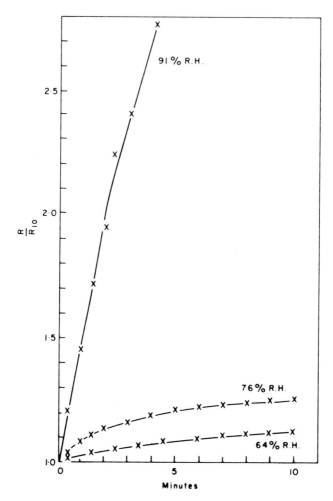

FIG. 13. Resistance of cotton divided by the resistance at 10 sec after application of 1 k volt as a function of time at various relative humidities, 30°C. (Tin electrodes, specimen length 0·8 cm.) From Cusick and Hearle (1955) courtesy of *Textile Research Journal.*

Hydrogen and hydroxyl ions increased in mobility by a factor of 10^3 from a water content of 11% to one of 40%. This corresponds to a large percentage of the increase in conductivity over that range of water content, namely 3×10^4. The mobility of hydrogen ions at 40% water content, corresponding to 92% R.H., 20°C, was $8·0 \times 10^{-9}$ m^2 volt^{-1} sec^{-1} which is only about 2% of that in bulk aqueous solution, namely $3·3 \times 10^{-7}$ m^2 volt^{-1} sec^{-1}.

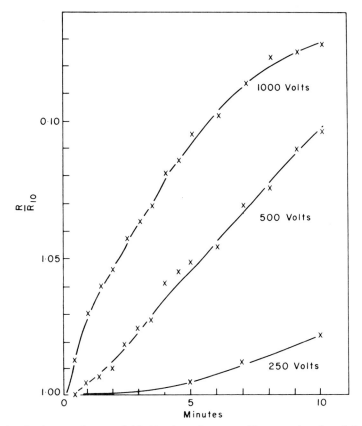

FIG. 14. Resistance of cotton divided by the resistance at 10 sec as a function of time after the application of various voltages to a specimen 0·8 cm long. 64% R.H. 30°C. From Cusick and Hearle (1955) courtesy of *Textile Research Journal*.

D. EFFECT OF TENSION OR STRAIN

Hearle (1953c) found that the effects of tension on the d.c. resistance of cotton in the laboratory atmosphere were very small, i.e. barely detectable. For cotton yarns there was a tendency for the resistance to increase with tension whilst for viscose rayon filament there was a tendency for the resistance to decrease.

Algie (1959, 1964b) has examined the effects of strain on horsehair at 90% R.H., 20°C, and was able to show that the change of resistance with time after the strain was applied, was due in part to an increase in resistance proportional to the strain for strains greater than 1%, and a decrease in resistance due to the entry of extra water vapour into the fibre. When water was prevented from entering the fibre the increase of resistance with strain was what would be expected if some water left a conducting network of water molecules in the amorphous sections of the fibre and became absorbed on new sites for water absorption, made available by the opening out of the alpha helical crystalline portion of the fibre, due to the strain.

E. CONDUCTANCE EFFECTS PRODUCED BY SUDDEN INCREASES IN THE RELATIVE HUMIDITY

Two series of experiments have been described by Algie (1964a) and Algie and Watt (1965) on this subject. Although the experiments were undertaken to obtain information as to the nature of the absorption process, they did provide data of value to an understanding of d.c. conduction.

It was found that following an increase in the relative humidity from 0% to values in the range of 18% to 80% at 55°C the conductance initially increased then fell to a lower final value, i.e. there was an overshoot of conductance. The results for an $0 \rightarrow 97\%$ R.H. step however were different; instead of a decrease to a final value there was a slight increase in the conductance even though the actual water content was falling. Results obtained at 35°C were similar. Dielectric constant (ϵ') values at 1592 Hz were also measured during step changes of the relative humidity (Algie, 1964a). Very high values of ϵ' were recorded following an R.H. change from 0 to 5·8%, after 1 h the ϵ' was about 9 and almost equal to the value obtained for a $0 \rightarrow 50\%$ R.H. step. By this time the water content should have reached an equilibrium condition for both the $0 \rightarrow 5·8\%$ and the $0 \rightarrow 50\%$ step. However, the conductance for the $0 \rightarrow 5·8\%$ R.H. step was more than 10^3 less than that for the $0 \rightarrow 50\%$ R.H. step indicating that theories in which the dielectric constant of the keratin is supposed to influence the conductance through its effect on ionic dissociation are not acceptable.

V. Theories of Conduction

The early theories of conduction have been reviewed by Hearle (1952). These were concerned mainly with the concept of ionic migration through capillary channels which were thought to exist inside fibres. With the advent of synthetic fibres which have similar conduction properties to the natural fibres but which clearly do not contain capillaries, such theories lost favour. In an attempt to explain the exponential dependence of the conductance on the water content, various theories were proposed linking the increase in dielectric constant with water content and the concurrent increase in the conductance. The first of these is due to King and Medley (1949b). They initially showed that the conduction process was ionic in keratin for water contents greater than 15% by electrolysis experiments (King and Medley, 1949a) and then developed a theory based on Bjerrum's (1926) relation between the dielectric constant and the dissociation constant. For solutions of electrolytes in solvents of low dielectric constant Fowler and Guggenheim (1952) derived the following relationship between the fraction of associated ion pairs α and the dielectric constant ϵ'.

$$\alpha = N(\pm)/V \int_a^q (4\pi r^2)[\exp(z^2 e^2/\epsilon' kTr)]\, dr$$

where $\alpha \ll 1$, $N\pm$ = the total number of ions of each type in a volume V, e = the electronic charge, z = ionic valency, k = Boltzmann's constant, T = absolute temperature, a = closest distance of ionic approach, q = closest distance of approach for free ions, ϵ' = dielectric constant.

Even when α is not very much less than 1 King and Medley (1949b) state that

$$N[(1/\alpha)^2/\alpha] = V \cdot \left[\int_a^q \exp[z^2 e^2/\epsilon' kTr]4\pi r^2 \cdot dr\right]^{-1}$$

They assumed that the conductivity was proportional to the amount of dissociation $(1 - \alpha)$ and calculated the value of the integral for a range of values of ϵ' corresponding to various water contents. The values of ϵ' were for a frequency of 10^4 Hz. Calculated values of $\log_{10}(1 - \alpha)$ against water content had a smaller slope than the log conductivity relation. They attributed this to the use of dielectric constant values which did not rise quickly enough with increasing water content. Values of ϵ' obtained at lower frequencies would possibly provide a better fit; however they claimed that the use of ϵ' obtained at "relatively high frequencies" would tend to compensate for the saturation effect in strong ionic fields which would tend to lower the value of the effective dielectric constant. King and Medley concluded that ionic dissociation controlled by the dielectric constant provides an adequate description of the conduction process.

Hearle (1953d, 1957) has developed an elementary theory of the effect of the dielectric constant on the conductivity. He applied the Law of Mass Action to the equilibrium between ions and ion pairs

$$A^+B^- \overset{U}{\underset{(1-\gamma)}{\rightleftharpoons}} \underset{\gamma}{A^+} + \underset{\gamma}{B^-}$$

where γ is the degree of dissociation, U = energy of dissociation and obtained

$$\gamma^2/(1 - \gamma) = K \exp(-U/kT)$$

K is almost constant for a given material and ion content and is apporoximately equal to

$$\frac{\text{weight of material}}{\text{weight of ions}}$$

He assumed that the conducting ions were from an impurity such as KCl with a concentration of about 1%. Assuming that the ions are surrounded by a medium of dielectric constant ϵ'

$$U = U_0/\epsilon'$$

where U_0 = energy required to separate the ions in a vacuum. He further assumed that $\gamma \ll 1$ which is equivalent to assuming that ϵ' is not greater than 20. Then

$$\gamma^2/(1 - \gamma) \simeq \gamma^2$$

and it follows that

$$\gamma = (K)^{\frac{1}{2}} \exp{(-U_0/2\epsilon'kT)}$$

The resistance

$$R \propto K^{-\frac{1}{2}} \exp{(U_0/2\epsilon'kT)}$$

Taking logarithms,

$$\log R = \Psi/\epsilon' + \chi$$

where

$$\psi = U_0 \log e/2kT$$

$$\chi = \text{a constant}$$

The above equation was applied to cotton and keratin and values of ψ and χ could be obtained which fitted Hearle's experimental results.

A similar equation can be deduced for the conductivity:

$$\log \sigma = \log{[(A'/v')(2\mu v)]} - U_0 \log e/2kT\epsilon'$$

where $A'/v' = \dfrac{\text{total volume}}{\text{volume occupied by ions}}$, μ = mobility (m^2 . volt^{-1} . sec^{-1}), v = number of molecules per m^3.

Applying this equation to Algie's results on keratin in Fig. 8 for ϵ' obtained by King at 10^6 Hz gives

$$\log_{10} \sigma = -88/\epsilon' + 2 \cdot 5$$

when $A'/v' = 100$, i.e. for an electrolyte content of 1% the mobility, $\mu = 0 \cdot 44 \times 10^{-4} \text{ m}^2 \text{ sec}^{-1} \text{ volt}^{-1}$.

It should be noted that the ionic mobility in this simple theory is considered to be constant for the range of water contents over which $\log \sigma \propto 1/\epsilon'$. The value 0.44×10^{-4} m^2 volt^{-1} sec^{-1} for the ionic mobility in keratin may be compared with the values for the ionic mobility in salt impregnated cellulose sheet found by O'Sullivan (1947c). For the potassium ion at 10% water content $\mu = 1.7 \times 10^{-12}$ m^2 volt^{-1} sec^{-1}. For the hydrogen ion at 11% water content $\mu = 7.2 \times 10^{-8}$ m^2 volt^{-1} sec^{-1}. For the hydrogen ion at 40% water content $\mu = 7.8 \times 10^{-9}$ m^2 volt^{-1} sec^{-1}. The discrepancy between these figures and the 0.44×10^{-4} m^2 volt^{-1} sec^{-1} determined from Hearle's theory is so great that the theory is suspect on that account. A much greater difficulty of all the theories based on the dielectric constant is that it is not clear what dielectric constant should be used, i.e. at what frequency the dielectric constant should be determined. The author is of the opinion that since the dissociation equilibrium involves the breaking and recombining of ionic bonds, a diffusion type of process will be involved, and the ionic mobilities are so small that a dielectric constant determined at low frequencies, e.g. $f < 10^2$ Hz, would be appropriate.

However, the theories will only fit the experimental data if a high frequency (10^6 Hz) ϵ' is used. Various reasons have been proposed to explain this. King and Medley (1949b), whilst recognizing that "dielectric constant values measured at a lower frequency (zero frequency preferably)" may give a better fit to their data, later claim that hydration effects would tend to lower the internal dielectric constant so that the lower values at 10^4 Hz would be more suitable. Taylor (1961) stated that the high frequency dielectric constant is "probably the correct one to use, corresponding to the rapid motion of free ions". Even if the high frequency dielectric constant is the correct one to use, the theory does not appear to apply to keratin whilst it is absorbing water. Algie (1964a; Algie and Watt, 1965) has shown that following a step in the relative humidity from $0 \rightarrow 5\%$ R.H. there is a transient increase in the dielectric constant (measured at 1590 Hz) which does not produce a comparable increase in the conductivity. One hour after the step change the dielectric constant is equal to that produced by 50% R.H., i.e. $\epsilon' = 9.5$, although the conductance was at least 10^4 less than that for a wool fibre at 50% R.H. This failure of a correspondence between ϵ' and the conductance suggests that some other parameter which is affected by water content may be responsible for the change in conductance with water content. This matter will be considered later but first let us examine other theories of conductance which are related to the dielectric constant. Taylor (1961) has extended King and Medley's ideas and applied his theory to dry cytochrome which he believes shows ionic conductivity.

Haemoglobin has been studied by Rosenberg (1962b) and he found that the activation energy depended upon the water content whilst the preexponential factor remained constant. This led him to develop a theory in which the activation energy is considered to be a function of the dielectric constant.

This theory is inapplicable to keratin because in keratin between 0% and 15% water content the activation energy remains practically constant (Table II) whilst the conductivity increases by 10^7.

Murphy (1960a) had a different approach and applied the theory of ionic conduction in ionic salt crystals to conduction in dry cellulose. His experimental results could be fitted to an equation

$$\sigma = 4 \cdot 5 \times 10^2 \exp\left[(-30 \cdot 7 \times 10^2)/RT\right] + 3 \cdot 55$$
$$\times 10^{-10} \exp\left[(-10 \cdot 6 \times 10^3)/RT\right]$$

where σ is the conductivity in $\text{ohm}^{-1}\,\text{cm}^{-1}$. This equation has the same form as that for an ionic crystal as outlined in the Theory section. However, it has been shown by Occhini and Lawson (1972) that the change in slope obtained by Murphy in the conductivity versus $1/T$ plot was probably due to not allowing sufficient time for polarization currents to cease. The fact that Murphy's equation is in error does not indicate that the conductivity may not be similar to that in ionic crystals, it merely throws doubt upon the detailed calculations based upon the theory.

For cellulose and proteins containing water, Murphy (1960b) has developed a theory of conduction based upon the notion that conduction takes place along chains of water molecules. These water molecules are being continuously absorbed and desorbed on n sites which are distributed between ion generating sites which themselves are uniformly distributed along a cellobiose chain (for example). The fraction of the time that a particular site is occupied is assumed to be proportional to the water content expressed as a proportion of the saturation water content, i.e. α/α_0 where α = water content, α_0 = saturation water content.

The probability that the n sites are occupied simultaneously is proportional to $[\alpha/\alpha_0]^n$. He then assumes that an elementary contribution to the conductivity occurs when the chain of water molecules is complete, i.e. the n sites are occupied simultaneously. The conductivity is then given by $\sigma = \sigma_s(\alpha/\alpha_0)^n$ where σ_s = the conductivity at saturation.

For this equation to be satisfied by the experimental results there should be a linear relation between $\log \sigma$ and $\log(\alpha/\alpha_0)$ or $\log$ (water content) with a slope of n. For keratin with water contents from 0 to 12%, Fig. 15 shows that the relationship is non-linear so that the theory is not supported by the experimental results. In addition there is the implicit assumption in the theory that only one water molecule may occupy a site thus giving a continuous chain of water molecules only at saturation, i.e. 100% R.H. From specific heat measurements (Haly and Snaith, 1969) it is known that the water absorbed by keratin can be frozen for water contents greater

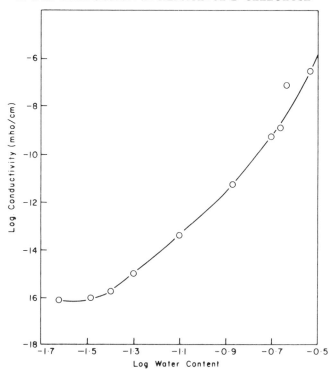

FIG. 15. Log conductivity of wool as a function of log water content at 20°C. From data of Algie and Gamble (1973).

than 22%. This implies that clusters of water molecules are formed for water contents greater than 22% so that the sites are all occupied for a water content of about 20%. It follows that the linear relationship between log σ and log (water content) should not exist for water contents greater than 22% and in fact σ should reach a limiting value. Murphy's (1960b) data does not support this, as he shows a linear relationship from 12% to 30% water content for wool.

In the case of cellulose, log σ is linear with log (water content) from 1·6% to 20% water content at a temperature of 25°C with a slope of about 9. However, Hearle (1953b) found that the activation energy of conduction decreased with increase of water content as shown in Table I. The effect of this is that at a temperature of 100°C the calculated slope n has dropped to 5·7. Murphy (1960b) states that n is the number of sites for water absorption between ion generating sites but it is difficult to see how the number of water absorption sites could change with temperature especially as they are supposedly related to the secondary crystalline structure. It follows that some or all of the initial premises involved in this theory must be inapplicable to keratin and cellulose.

Proteins other than keratin have been extensively studied by Eley and his co-workers (Eley *et al.*, 1953; Cardew and Eley, 1959; Eley and Spivey, 1960b); because of the fact that the activation energies were similar for a range of proteins, Eley and Spivey (1960b) concluded that the conduction was electronic and intrinsic, i.e. not due to impurities. The early ideas, on electronic conduction in proteins, of Szent–Györgyi (1941) and Evans and Gergly (1949) were used by Eley to support his conclusions. In particular Evans and Gergly calculated that the energy gap between the highest filled band and the lowest unfilled band for the hydrogen bonded system $(N-H \ldots O=C)_n$ was 4·23 ev (electron volts) for pyrimidal N, 3·05 ev for trigonal N and 4·76 ev for NH isoelectronic with O. Since Eley and Spivey (1960b) found that the activation energy (ΔE) in the equation

$$O\sigma = \sigma_0 \exp(-\Delta E/2kT)$$

was 2·97 ev for globin, 2·75 ev for haemoglobin, 2·92 ev for glycine and 3·12 ev for polyglycine, they suggested that the conduction was due to excited π electrons in the $N-H \ldots O=C$ system. The above values of ΔE are close to that calculated by Evans and Gergly for trigonal nitrogen.

Suard-Sender (1965) has calculated molecular orbitals, for the poly-peptide system $(N-H-C-O)_n$ bound by hydrogen bonds, using a self consistent field (SCF) method. She found that for polypeptides the n-π^* transition had an energy gap of 5 ev for the triplet and 5·5 ev for the singlet state. π-π^* was higher at 6 ev for the triplet and 7·5 ev for the singlet state. Accordingly she claimed that, contrary to Eley's proposals, the conduction is not intrinsic semi-conduction but possibly due to impurities. Further-more it was not due to the π electrons and the conduction cannot be explained by the presence of H bonding, in fact from the point of view of electronic conduction polypeptides are insulators.

Another difficulty facing the electronic theory is the large *calculated* values for the mobility. Cardew and Eley (1959) quoted values of $7 \, m^2 \, volt^{-1} \, sec^{-1}$ at 400°K for globin and $2\cdot3 \, m^2 \, volt^{-1} \, sec^{-1}$ for haemo-globin assuming an effective electron mass equal to that of a free electron. To account for these high mobilities an electron tunnelling mechanism was invoked (Eley *et al.*, 1953; Eley and Spivey, 1960b). However this gave a value for the pre-exponential factor (which includes the mobility) of 69·7, which is 10^3–10^4 times smaller than that calculated from the experimental values (Eley and Spivey, 1960b). In this calculation the effective electron mass was taken to be 40 times the free electron mass since the energy band was assumed to be only 0·1 ev wide. To explain the discrepancy in the pre-exponential factors, Eley stated that the calculated value could be increased if the activated electron, after passing through a barrier, could travel through a large number of hydrogen bridges before striking another barrier. This seems unlikely if Suard-Sender's (1965) comments on H-bonding chains are correct. Furthermore Eley and Willis (1962) point out that the above mechanism would lead to a lowered mobility due to a decreased frequency of striking the barrier.

Electron tunnelling as a basis for semiconduction in proteins has been examined theoretically by Flax and Flood (1971). They employed a potential barrier which was rectangular with a rounded top and concluded that the resulting conduction is several orders of magnitude less than that observed experimentally. They also criticized the treatment of electron tunnelling by Kemeny and Rosenberg (1970), claiming that they failed to take into account tunnelling in the backwards direction. Backward tunnelling severely reduces the net conduction, as the presence of an electric field (up to 10^7 volts m^{-1}) only slightly distorts the potential barrier in favour of the forward conduction. The probability of tunnelling in both directions will be almost the same so that the net current will be small.

Throughout the discussion to this point, theories involving either electronic conduction or ionic conduction have been examined without consideration of the evidence for and against these different forms of conduction. We will now consider this point.

As outlined in the Theory section, the form of the current as a function of temperature is similar in both ionic and electronic conduction, the only difference is that with ionic conduction alone mass transfer occurs according to Faraday's law. King and Medley (1949a) conducted electrolysis experiments on keratin with water contents greater than 15%. Sufficient hydrogen was evolved to account for 95% of the electricity which flowed in the case of keratin with 15% water content, however simple electrolytic decomposition of the absorbed water did not occur as only a small quantity of oxygen was produced and some other gas, thought to be CO_2, was also evolved.

Conduction in keratin for water contents greater than 15% is thus ionic.

Cellulose containing water has been studied by O'Sullivan (1947b); he measured the mobilities of various ions including H^+ and OH^- in cellulose with absorbed salts. The mobilities decreased markedly with decreasing water content and the decrease was almost sufficient to account for the decrease in conduction. Electrolysis experiments were carried out on dry cellulose by Murphy (1963). In order to obtain sufficient conduction the temperature employed was high, being 130° or 160°C; however the gases evolved by electrolysis were clearly differentiated from those produced by thermal decomposition.

The gases evolved were H_2, CO and CO_2 in amounts predicted by Faraday's laws except for the high field strengths where an excess of gas was produced. Murphy concluded that the conduction process was ionic and mainly protonic. Since the activation energy of conduction is constant from 120°C down to 20°C it is likely that the same mechanism is involved at low temperatures as at high temperatures so that ionic conduction is probably the mechanism for conduction in dry cellulose even at room temperature.

It is possible to use a similar argument for conduction in dry keratin, namely that the activation energy of conduction is practically the same from 15% to 0% water content and thus the conduction mechanism is similar at 0% and 15%, implying ionic conduction in dry keratin. In the absence of direct experimental evidence this argument cannot be proved and there is one piece of evidence against it. Rosenberg (1962a) conducted an experiment on another protein, haemoglobin, which contained 7·5% H_2O. A voltage was applied to the sample which was in a sealed Teflon enclosure and the current flowing was monitored. He claimed that if the conduction was ionic, electrolysis of the absorbed water would have occurred and the current would have decreased with time. This did not happen, so that according to him the conduction could not be ionic. The one vital assumption which Rosenberg made is that if the conduction is ionic, electrolysis of the absorbed water would take place. Experimental evidence suggests that this is not so. King and Medley (1949a) found that very little O_2 was evolved, with H_2 being the main gas, but some small quantities of CO_2 were also present following electrolysis of keratin. Haemoglobin with 18% H_2O was electrolysed by Maričić et al. (1964) and they found that only H_2 was evolved with a current efficiency of from 26% to 112%. They also stated that ordinary electrolysis of the absorbed water did not occur, as Rosenberg (1962a) had hypothesized, since the relationship between 1/current and time was non-linear.

A direct electrolysis experiment on haemoglobin with about 8% H_2O should settle the question; however the resistance of the sample is very high and hence currents are low and the rate of evolution of gas would be small. Maričić et al. (1964) have performed such an experiment on haemoglobin containing 9·17% H_2O using a constant voltage of 150 V at 35°C.

They detected no gas after 7 days. Using the fact that the conductivity of haemoglobin containing 9·17% H_2O is about 10^3 times less than that for a water content of 18% (see Rosenberg and Postow, 1969) one can calculate that to obtain the same amount of gas as was obtained with a water content of 18%, the electrolysis would have to be carried out for about 10^3 days. Seven days is consequently too short a time if one is to be certain that no gas was being evolved.

Finally it is not necessarily true that the results and mechanisms of conduction for haemoglobin are the same as those of keratin since dry keratin exhibits strong time dependence of the conduction whereas haemoglobin and the other proteins studied by Eley and Spivey (1960b) and Cardew and Eley (1959) do not show a time dependence.

Eley and Thomas (1968) claim that a time dependent conductivity is evidence of ionic or protonic conduction. However, dielectric studies of dry keratin (Algie, 1973) indicate that the time dependent conduction is due in part to polarization processes associated with the relaxation of main polymer chains and in part to interfacial polarization which only implies inhomogeneity of the dielectric constant and conductivity, and does not specify a conduction mechanism.

The conclusion from these arguments is that although keratin is an ionic conductor for water contents greater than 15% the conduction process for lower water contents is not yet determined.

In regard to the ionic conductivity of keratin at high water contents, Algie *et al.* (1960) have claimed that the conduction is protonic following the earlier proposals of Baker and Yager (1942) for protonic conduction in the polyamides. Eley and Spivey (1960a) have presented a similar argument for protonic conduction in hydrated haemoglobin.

The theories of electronic conduction which have been discussed above have been in most cases developed for regular crystalline materials and it is doubtful whether they are applicable to polymeric materials. In both keratin and cellulose the absorbed water is largely confined to the non-crystalline regions so that the conductivity, which is so influenced by water, must occur in these non-crystalline or amorphous regions. It seems possible that the special attributes of amorphous polymeric materials should influence the conductivity. Fuoss (1939) found that there was a relationship between the frequency of maximum dielectric loss f_m in plasticized polyvinyl chloride and both the Young's modulus and the d.c. conductance. In particular for a frequency of 60 Hz the d.c. conductance was the same for different compositions when the temperature was that for the peak dielectric loss. The contents of diphenyl plasticizer ranged from 1% to 20% and the temperatures at which the frequency of maximum loss was 60 Hz ranged from 91°C to 40°C. It is now well established (McCall, 1969) that for polymers the electrical relaxation processes which give rise to the dielectric loss are a reflection of mechanical relaxation processes due to the microbrownian motion of polymer chains or parts of chains. It follows that Fuoss's data implies that the dominant factor in producing the d.c. conductivity is the rate at which the polymer chains can relax. Hence the increase in the conductivity at a fixed temperature with an increase in plasticizer content is due to the increased rate at which the polymer chains can perform their microbrownian motions. Fuoss (1941) suggested that the conduction was ionic and that the enhanced chain motion due to the plasticizer facilitated the diffusion of ions. King and Medley (1947) recognized that water and formic acid acted as plasticizers in keratin, with formic acid being most effective on a molar basis. They later showed (1949b) that the conductivity of keratin was increased more by formic acid than by water and they consequently felt that Fuoss's concept was correct and applicable to keratin.

Various authors (Eley and Spivey, 1960a; Murphy, 1963) have since referred to the possibility that conduction may be assisted by the motion of polymer chains, without making this the dominant factor. It is the author's view (Algie, unpublished observations) that the primary process is the plasticization of keratin by water or formic acid; this presumably occurs by a weakening and/or breaking of inter-chain hydrogen bonding. The increase in the mechanical relaxation rate due to an increase in the micro-brownian motion of main chains in the amorphous regions then follows. Since the keratin is polar the motion of the main chains exhibits dielectric relaxation effects and there is an increase in the frequency f_m of maximum dielectric loss and an associated (Algie and Gamble, 1973) increase in the dielectric constant, as these are different manifestations of the same molecular process. The mobility of charge carriers either ionic or electronic is considered to be proportional to the rate at which main chains in the amorphous regions assume configurations which permit the transfer of a charge carrier from one chain to another. This rate will be proportional to the rate of main chain motion in the amorphous regions.

The experimental results in Fig. 8 show that up to about 20% water content log σ is proportional to the water content and also to log f_m, for the mechanical relaxation of the main chains in the amorphous regions is approximately linear with water content (Fig. 16). Activation energies of the main chain motion are about 24 kcal (g mole)$^{-1}$ from dielectric data in the dry state (Algie, 1973) and also about 24 kcal (g mole)$^{-1}$ mechanical relaxation data in the saturated state, i.e. 33% H_2O at 20°C (Feughelman, 1958).

The conductivity has a similar activation energy of about 24 kcal(g mole)$^{-1}$ (see Table II) for water contents up to about 20%. However for higher water contents the activation energy of conduction begins to fall until at 30% H_2O it is about 15 kcal(g mole)$^{-1}$. Also the increase in conductance with water content is very much less for water contents greater than 20%. These two facts suggest that a different mechanism of conduction is involved at the higher water contents probably due to the fact that there is sufficient water available to allow the formation of a water network (Algie, 1964b). More than 23% H_2O has to be absorbed by keratin before the water can be frozen (Haly and Snaith, 1969) which implied that a water network is beginning to be formed. To explain the fact that the conductivity no longer increases at the same rate as the mechanical relaxation at high water contents (greater than 20%) it is postulated that chain to chain contacts can no longer be made to the same extent due to the swelling and the interpositioning of a layer of water molecules between chains. Conduction in the high water content region is

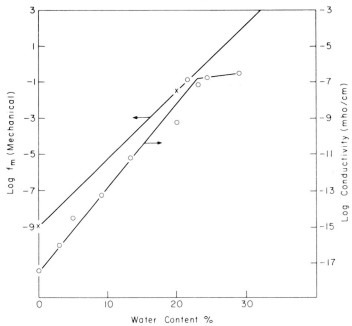

FIG. 16. Log f_m (mechanical) ($\times$) and log conductivity ($\bigcirc$) of wool as a function of water content, 20°C. Data from Algie and Gamble (1973) and Algie (unpublished observations).

assumed to be similar to that proposed for ice as suggested by Riehl (1957) for hydrated gelatine.

A direct relationship between the rate of chain motion, as determined dielectrically, and conduction is shown in Fig. 17 for a specimen of dry keratin which was in a state of internal stress. The conductivity and the relaxation rate f_m were both high and both dropped by the same extent when the specimen had been annealed by heat treatment.

The theory also provides an explanation for the overshoot in conductance which follows a rapid increase in the relative humidity (R.H.) surrounding a wool fibre (Algie, 1964a; Algie and Watt, 1965). This overshoot is probably a reflection of the overshoot in the torsional modulus at frequencies of $1 \rightarrow 100$ Hz found by Nordon (1962) after a similar rapid increase in R.H.

It is implicit in the theory that the dissociation into ions is saturated, i.e. that the internal dielectric constant is high enough for this to occur. If it is accepted that the frequency at which the dielectric constant governing dissociation should be measured is related to the mobility of the ions, then, since the dissociation and recombination of ions is controlled by a diffusion process, the frequency will be low (less than 100 Hz) and will become lower as the ionic mobility decreases. It has already been mentioned that in

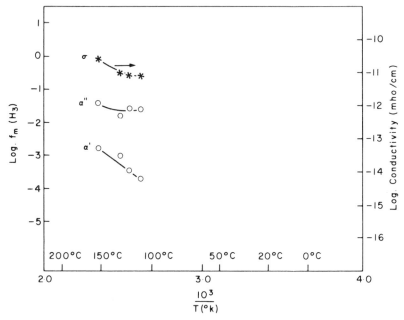

FIG. 17. Log conductivity and log f_m for various relaxation processes in dry unannealed rhinoceros horn as a function of $10^3/T$ (T in °K). Data from Algie (unpublished observations).

cellulose the ionic mobility (O'Sullivan, 1947b) decreased with decreasing water content almost as fast as the conductivity, which implies that in the dry state the ionic mobility will be very small (less than 10^{-14} m^2 volt^{-1} sec^{-1}) and hence the appropriate frequency for the dielectric constant would be correspondingly low. A dielectric constant of 50 or more, changing little with water content (due to the changing measurement frequency), could then be the appropriate value.

In the absence of definitive measurements of the mobility of charge carriers in keratin the above theory fits the available experimental data and is open to test.

References

Algie, J. E. (1959). *Text Res. J.* **29**, 1–6.
Algie, J. E. (1964a). *Text Res. J.* **34**, 477–486.
Algie, J. E. (1964b). *Text. Res. J.* **34**, 273–274.
Algie, J. E. (1973). *Kolloid Z.u.Z. Polymere* **251**, 305–309.
Algie, J. E. and Gamble, R. A. (1973). *Kolloid Z.u.Z. Polymere* **251**, 554–562.
Algie, J. E. and Watt, I. C. (1965). *Text Res. J.* **35**, 922–929.
Algie, J. E., Downes, J. G. and Mackay, B. H. (1960). *Text. Res. J.* **30**, 432–434.
Baker, W. O. and Yager, W. A. (1942). *J. Am. Chem. Soc.* **64**, 2171–2177.

Baxter, S. (1943). *Trans. Faraday Soc.* **39**, 207–214.

Bjerrum, N. (1926). *Math.-fys, Meddr.* **7** (9), 2–48.

Cardew, D. H. and Eley, D. D. (1959). *Discuss. Faraday Soc.* **27**, 115–128.

Cusick, G. E. and Hearle, J. W. S. (1955). *Text. Res. J.* **25**, 563–566.

Eley, D. D. and Spivey, D. I. (1960a). *Nature, Lond.* **188**, 725.

Eley, D. D. and Spivey, D. I. (1960b). *Trans. Faraday Soc.* **56**, 1432–1442.

Eley, D. D. and Thomas, P. W. (1968). *Trans. Faraday Soc.* **64**, 2459–2462.

Eley, D. D. and Willis, M. R. (1962). *In:* "Symposium on Electrical Conductivity in Organic Solids" (H. Kallmann and M. Silver eds.), pp. 257–275, Interscience, New York and London.

Eley, D. D., Parfitt, G. D., Perry, M. J. and Taysum, D. H. (1953). *Trans. Faraday Soc.* **49**, 79–86.

Evans, M. G. and Gergly, J. (1949). *Biochim. biophys. Acta* **3**, 188–197.

Feughelman, M. (1958). *J. Text. Inst.* **49**, T361–T378.

Flax, L. and Flood, D. (1971). N.A.S.A. Tech. Note NASA TN D-6559.

Fowler, R. and Guggenheim, E. A. (1952). "Statistical Thermodynamics", Cambridge University Press, London.

Fuoss, R. M. (1939). *J. Am. Chem. Soc.* **61**, 2334–2340.

Haly, A. R. and Snaith, J. W. (1969). *Biopolymers* **7**, 459–474.

Hamon, B. V. (1952). *Proc. Inst. elect. Engrs.* IV, **99**, 151–155.

Hayashi, K., Yoshino, K. and Inuishi, Y. (1973). *Japan J. appl. Phys.* **12**, 754–755.

Hearle, J. W. S. (1952). *J. Text. Inst.* **43**, P194–P223.

Hearle, J. W. S. (1953a). *J. Text. Inst.* **44**, T117–T143.

Hearle, J. W. S. (1953b). *J. Text. Inst.* **44**, T144–T154.

Hearle, J. W. S. (1953c). *J. Text. Inst.* **44**, T155–T176.

Hearle, J. W. S. (1953d). *J. Text. Inst.* **44**, T177–T198.

Hearle, J. W. S. (1957). *J. Text. Inst.* **48**, P40–P51.

Hersh, S. P. and Montgomery, D. J. (1952). *Text. Res. J.* **22**, 805–818.

Kauzmann, W. (1957). "Quantum Chemistry", p. 197. Academic Press, Inc., New York and London.

Kemeny, G. and Rosenberg, B. (1970). *J. chem. Phys.* **52**, 4151–4153.

King, G. and Medley, J. A. (1947). *Nature, Lond.* **160**, 438.

King, G. and Medley, J. A. (1949a). *J. Colloid Sci.* **4**, 1–8.

King, G. and Medley, J. A. (1949b). *J. Colloid Sci.* **4**, 9–18.

Lodge, O. (1886). Rep. British Assoc. No. 56. 389.

McCall, D. W. (1969). *In:* "Molecular Dynamics and Structure of Solids" (Carter, R. S. and Rush, J. J., eds.), pp. 475–537. Special Publs. U.S. Nat. Bur. Stand No. 301.

Marsh, M. C. and Earp, K. (1933). *Trans. Faraday Soc.* **29**, 173–193.

Maričić, S., Pifat, G. and Pravdić, V. (1964). *Biochim. biophys. Acta* **79**, 293–300.

Mott, N. F. and Gurney, R. W. (1940). "Electronic Processes in Ionic Crystals", Clarendon Press, Oxford.

Murphy, E. J. (1929). *J. phys. Chem. Ithaca* **33**, 509–532.

Murphy, E. J. (1960a). *J. Phys. Chem. Solids* **15**, 66–71.

Murphy, E. J. (1960b). *J. Phys. Chem. Solids* **16**, 115–122.

Murphy, E. J. (1963). *Can. J. Phys.* **41**, 1022–1035.

Nordon, P. (1962). *Text. Res. J.* **32**, 560–571.

Occhini, E. and Lawson, W. G. (1972). *Rep. Conf. elect. Insul.* **1971**, 212–218.

O'Sullivan, J. B. (1947a). *J. Text. Inst.* **38**, T271–T290.

O'Sullivan, J. B. (1947b). *J. Text. Inst.* **38**, T291–T297.

O'Sullivan, J. B. (1947c). *J. Text. Inst.* **38**, T298–T306.

Riehl, N. (1957). *Kolloid Z.* **151**, 66–72.

Rosenberg, B. (1962a). *Nature, Lond.* **193**, 364–365.

Rosenberg, B. (1962b). *J. chem. Phys.* **36**, 816–823.

Rosenberg, B. and Postow, E. (1969). *Ann. N. Y. Acad. Sci.* **158**, 161–190.

Shockley, W. (1950a). "Electrons and Holes in Semiconductors", p. 215, D. Van Nostrand Co., New York.

Shockley, W. (1950b). "Electrons and Holes in Semiconductors", p. 398, D. Van Nostrand Co., New York.

Suard-Sender, M. (1965). *J. Chim. phys.* **62**, 79–98.

Szent-Györgyi, A. (1941). *Nature, Lond.* **148**, 157–159.

Taylor, C. P. S. (1961). *Nature, Lond.* **189**, 388–389.

Wilson, A. H. (1936). "The Theory of Metals", Cambridge University Press, London.

6. Dielectric Properties of Keratin and Cellulose

J. E. ALGIE

C.S.I.R.O., Division of Food Research
North Ryde, N.S.W., Australia

I. Introduction

This chapter has been restricted to the dielectric properties of keratin and cellulose since the synthetic textile polymers have been dealt with extensively elsewhere. The impetus for research on the dielectric properties has come partly from practical and partly from theoretical considerations. Moisture meters based on the relationships between the dielectric constant or loss and the water content of fibres have been produced and their accuracy and the factors affecting that accuracy have had to be examined.

The use of high frequency electric fields to heat bales of wool and to cure the polymeric glues used in the manufacture of plywood had led to an examination of the dielectric properties of wool and wood at high frequencies as functions of water content and temperature.

Microwave frequencies have been employed by those interested in the effects of water as the frequency of maximum dissipation of energy for free water occurs in this frequency band. The effect which the long chain polymeric nature of the textile fibres molecular structure has upon the dielectric properties have also been studied and in conjunction with mechanical experiments have led to a better understanding of the relationships between structure and those physical properties which are frequency dependent. Relationships between d.c. conductivity, mechanical and dielectric properties have been postulated and a better understanding of these phenomena is emerging. Some problems remain to be solved and further study of the electrical properties of fibres should prove fruitful.

II. Theory of the Dielectric Constant

Consider a capacitor consisting of two parallel metal plates separated by an insulator, i.e. a dielectric material. If a d.c. voltage is applied to the plates, charges within the dielectric move a small amount and the dielectric becomes polarized. The rate of movement of the charge may be fast when it is due to the displacement in individual atoms of electrons with respect to the nucleus (electronic polarization) or due to the movement of atoms with respect to each other (atomic polarization). Slower rates of movement are associated with the accumulation of charge at interfaces between portions of the dielectric with different conductivities and dielectric constants (interfacial polarization) and a slow rate is also due to the orientation of molecular groups which have permanent charges and thus constitute electrical dipoles (dipolar orientational polarization). If the d.c. voltage is replaced by an alternating voltage of frequency f the fast polarization processes (atomic polarization and electronic polarization) produce effects at very high frequencies (infrared and ultraviolet light frequencies). Because the orientational and interfacial polarization processes each occur at a less rapid rate, at very high frequencies they contribute little to the polarization and hence to the related dielectric constant; at low frequencies however the moving charges can follow the electric field and contribute to the low frequency dielectric constant. Figure 1 shows how the frequency

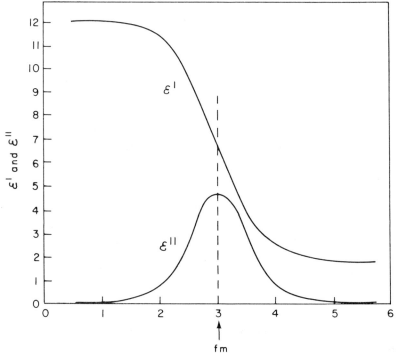

FIG. 1.   Debye curves for ϵ' and ϵ'' as a function of log frequency for $\epsilon_\infty = 2$, $\epsilon_s = 12$ and $\tau_0 = 1\cdot59 \times 10^{-4}$ sec.

dependent relative dielectric constant ϵ' varies with log frequency for an idealized orientational polarization.

Since the polarization processes dissipate energy, the dielectric constant is a complex quantity $\epsilon_0\epsilon^*$ and $\epsilon_0\epsilon^* = \epsilon_0(\epsilon' - j\epsilon'')$ where ϵ_0 is the dielectric constant or permittivity of space, ϵ^* is the complex relative dielectric constant, ϵ' is the relative dielectric constant and ϵ'' is the relative dielectric loss. For a capacitor containing a plane parallel sided slab of dielectric the capacitance C is $\epsilon'C_0$ where C_0 is the capacitance with a vacuum between the plates and the conductance is $G_{a.c.} = \epsilon''(2\pi f C_0)$. If the dielectric has d.c. conductance then $G_{dc} + G_{ac} = \epsilon''(2\pi f C_0)$ and f is the frequency of the applied a.c. voltage. Orientational polarization (Debye, 1912) may occur at frequencies up to about 10^{11} Hz whilst interfacial polarization occurs at frequencies up to about 10^6 Hz. In the remainder of this discussion the electronic and atomic polarization processes will not be considered further, excepting that their contribution to ϵ' will be denoted ϵ_∞.

A. PHENOMENOLOGICAL THEORY

Using Hopkinson's (1877) principle of linear superposition Von Schweidler (1907) showed that for a voltage independent dielectric the behaviour of $\epsilon'_{(\omega)}$ and $\epsilon''_{(\omega)}$ as functions of the angular frequency $\omega(= 2\pi f)$ can be determined from the charging current $i_{(t)}$ which flows when a step voltage V is applied to a capacitor containing that dielectric. The discharge current may be used also, provided that the capacitor was previously charged for a time sufficiently long for the current to have reached the steady value due to the direct current conductance $G_{d.c.}$ (see Fig. 2). The relaxation function $\phi(t)$ is defined as the current $i_{(t)}$ per unit voltage, i.e.

$$\phi_{(t)} = i_{(t)}/V$$

then

$$\epsilon'_{(\omega)} - \epsilon_\infty = C_0^{-1} \int_0^\infty \phi_{(t)} \cos \omega t \, dt \qquad (1)$$

$$\epsilon''_{(\omega)} = G_{dc}/\omega C_0 + C_0^{-1} \int_0^\infty \phi_{(t)} \sin \omega t \, dt \qquad (2)$$

where the voltage V is applied at time $t = 0$. Note $i_{(t)}$ does not include the d.c. current.

If the charging current $i_{(t)}$ is an exponential function and

$$\phi_{(t)} = [(\epsilon_s - \epsilon_\infty)/\tau_0] \exp(-t/\tau_0) \qquad (3)$$

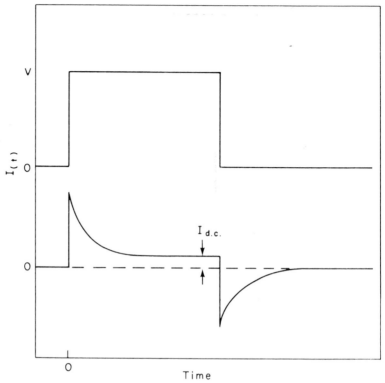

FIG. 2. Charge and discharge current as a function of time after the application of the voltage V.

then substituting (3) into (1) and (2) and integrating gives

$$\epsilon'_{(\omega)} - \epsilon_\infty = (\epsilon_s - \epsilon_\infty)/(1 + \omega^2 \tau_0^2) \qquad (4)$$

$$\epsilon''_{(\omega)} = (\epsilon_s - \epsilon_\infty)\omega\tau_0/(1 + \omega^2 \tau_0^2) \qquad (5)$$

where ϵ_s is the dielectric constant at zero frequency.

Equations (4) and (5) are usually referred to as the Debye equations. They represent the response of a dielectric with a polarization process, represented by a single relaxation time τ_0, to a sinusoidal voltage of angular frequency (ω). Figure 1 shows the Debye curves for $\epsilon'_{(\omega)}$ and $\epsilon''_{(\omega)}$ as a function of frequency for $\epsilon_\infty = 2$, $\epsilon_s = 12$ and $\tau_0 = 1\cdot59 \times 10^{-4}$ sec. The maximum dielectric loss ϵ''_m occurs at a frequency ω_m where $\omega_m = \tau_0^{-1} = 2\pi f_m$. In solids and in particular in polymers the orientational polarization processes cannot be represented by a single relaxation time and various empirical distributions of the relaxation time have been devised in an attempt to describe the form of $\epsilon'_{(\omega)}$ and $\epsilon''_{(\omega)}$ curves. Wagner (1913) and

Yager (1936) used a Gaussian normal distribution function

$$f_{(s)} = b\pi^{-\frac{1}{2}} \exp(-b^2 s^2) \tag{6}$$

where

$$s = \ln(\tau/\tau_0)$$

and

$$\int_{-\infty}^{\infty} f(s)\, ds = 1$$

b is the parameter which specifies the width of the distribution. Fuoss and Kirkwood (1941) found that the dielectric loss factor of some polymeric substances could be represented by

$$\epsilon''_{(\omega)} = (\epsilon_s - \epsilon_\infty)(\omega\tau)^\alpha/(1 + (\omega\tau)^{2\alpha})$$

This is associated with a distribution function

$$f_{(s)} = \alpha\pi^{-1}((\cos\alpha\pi/2)(\cosh\alpha s))/(\cos^2\alpha\pi/2 + \sinh^2\alpha s) \tag{7}$$

Cole and Cole (1941) postulated that their data could be fitted to an equation

$$\epsilon^*_{(\omega)} = \epsilon'_{(\omega)} - j\epsilon''_{(\omega)} = \epsilon_\infty + (\epsilon_s - \epsilon_\infty)/(1 + (j\omega\tau)^n)$$

from which, separating the real and imaginary parts

$$\epsilon'_{(\omega)} = \epsilon_\infty + (\epsilon_s - \epsilon_\infty)[1 + (\omega\tau_0)^n \cos n\pi/2]/[1 + 2(\omega\tau_0)^n(\cos n\pi/2)$$
$$+ (\omega\tau_0)^{2n}] \tag{8}$$

$$\epsilon''_{(\omega)} = (\epsilon_s - \epsilon_\infty)[(\omega\tau_0)^n \sin n\pi/2]/[1 + 2(\omega\tau_0)^n \cos(n\pi/2) + (\omega\tau_0)^{2n}] \tag{9}$$

These equations are associated with a distribution function

$$f_{(s)} = (2\pi)^{-1} \sin n\pi/(\cosh ns + \cos n\pi) \tag{10}$$

where n is the parameter describing the breadth of the distribution. In Fig. 3 the Cole and Cole expressions for $\epsilon'_{(\omega)}$ and $\epsilon''_{(\omega)}$ have been plotted for $n = 0.5$; the curve for $\epsilon''_{(\omega)}$ is much broader than the Debye curve (Fig. 1). From the Cole and Cole (1941) expression (9), the maximum dielectric loss factor is

$$\epsilon''_m = ((\epsilon_s - \epsilon_\infty)/2)(\tan(n\pi/4)) \tag{12}$$

when

$$\omega = \tau_0^{-1}$$

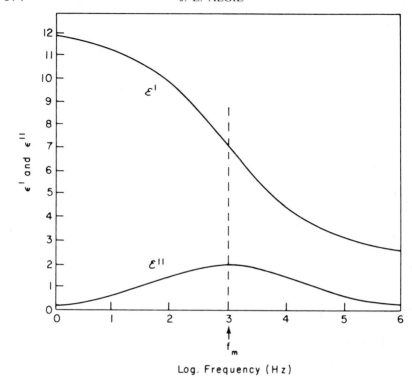

Log. Frequency (Hz)

FIG. 3.   Cole and Cole curves for ϵ' and ϵ'' as a function of log frequency for the distribution
parameter $n = 0 \cdot 5$, $\epsilon_\infty = 2$, $\epsilon_s = 12$ and $\tau_0 = 1 \cdot 59 \times 10^{-4}$ sec.

It is clear from Eqns. (8), (9) and (12) that if $\epsilon''_{(\omega)}$ is known and it fits the
Cole and Cole expression (9) then n, τ_0 and $(\epsilon_s - \epsilon_\infty)$ are known and hence
using (8), $\epsilon'_{(\omega)}$ can be calculated. $\epsilon''_{(\omega)}$ and $\epsilon'_{(\omega)}$ are thus not independent
functions; this can be seen also from the fact that both $\epsilon'_{(\omega)}$ and $\epsilon''_{(\omega)}$ can be
calculated from the decay function $\phi_{(t)}$ using Eqns. (1) and (2). Also direct
relationships between $\epsilon'_{(\omega)}$ and $\epsilon''_{(\omega)}$ have been derived by Gross (1941) and
they are similar to those derived for optical frequencies by Kramers (1927)
and Kronig (1926).

B. INTERFACIAL POLARIZATION

Many polymeric dielectrics contain inhomogeneities such as crystalline
regions surrounded by amorphous regions and the differences in the
electrical properties of these regions is accentuated by the presence of
plasticizers such as water which may be preferentially absorbed in the
amorphous regions. Maxwell (1892) showed that an inhomogeneous
dielectric consisting of two parallel sided slabs of materials with different

ratios of the relative dielectric constant to the d.c. conductivity ϵ'/σ, had a complex relative dielectric constant $\epsilon^*_{(\omega)}$ which was described by equations similar to the Debye equations. This treatment was extended by Wagner (1914) to a system consisting of an insulator containing a small quantity (up to 10%) of semi-conducting spheres, again a Debye type function was obtained

$$\epsilon' = \epsilon_\infty(1 + (K/(1 + \omega^2\tau_0^2)))$$ (13)

where

$$\epsilon_\infty = \epsilon'_1(1 + (3q(\epsilon'_2 - \epsilon'_1)/(2\epsilon'_1 + \epsilon'_2)))$$

$$K = 9q\epsilon'_1/(2\epsilon'_1 + \epsilon'_2)$$

$$\tau_0 = (2\epsilon'_1 + \epsilon'_2)/4\pi\sigma_2$$ (14)

ϵ'_1 = the relative dielectric constant of the insulating phase, ϵ'_2 = the relative dielectric constant of the material of the spheres, σ_2 = the conductivity (in e.s.u.) of the material of the spheres, q = the fraction of the conducting material by volume.

Sillars (1937) has derived similar functions for the dielectric constant when the conducting material is in the form of ellipsoids rather than being restricted to spheres. The effect of long thin needle shaped inclusions of conducting material rather than spheres was to increase the dielectric loss and to reduce the frequency at which the maximum loss occurred.

Artificially constructed inhomogeneous dielectrics have been tested by Sillars (1937) and Hamon (1953), and the results showed reasonable agreement with the theory.

The assumptions made in the derivation of Eqns. (13) and (14) are that the insulating material is a low loss dielectric, which implies that it is a nonpolar material so that ϵ'_1 is constant over the frequency range of the interfacial polarization, also that ϵ'_2 is constant over that frequency range, and furthermore that the conducting inclusion has the properties of that material in bulk. For polar polymers containing water none of these assumptions are true, so that it is difficult to apply Eqns. (13) and (14) to keratin and cellulose. It should be noted that it is not possible to determine from the shape of the $\epsilon'_{(\omega)}$ and $\epsilon''_{(\omega)}$ against log ω curves whether the polarization is due to inhomogeneities or to dipolar orientation. The shape of $\epsilon'_{(\omega)}$ and $\epsilon''_{(\omega)}$ curves in the case of interfacial polarization can be modified greatly if there is a distribution of conductivities (σ_2) (Wagner, 1914) or a distribution of shapes of included particles (Sillars, 1937).

C. MOLECULAR THEORY

The phenomenological theory has been concerned with the description of dielectric data in terms of observed quantities, e.g. the relaxation time τ_0 is the macroscopic relaxation time and its relationship to molecular relaxation times has not been investigated. Consequently the phenomenological theory can tell us nothing about molecular processes. Debye (1929) proposed a molecular theory for the polarization of polar liquids in alternating electric fields. He assumed that some of the polarization was due to the orientation of permanent molecular dipoles by the field. This involved small rotations of the dipoles which were being subjected to Brownian motion. Onsager (1936) corrected Debye's theory by introducing the reaction field due to the polarizing influence of the molecule on the surrounding medium due to its own presence. Hill (1961) extended Onsager's theory to alternating fields and found that $\tau_0 \simeq \tau_m$ where τ_m is the molecular relaxation time.

A similar result was obtained by Fatuzzo and Mason (1967) in that although τ_0 was found to be distributed for a single molecular relaxation time τ_m, in most cases it would be difficult to observe that distribution experimentally. The macroscopic relaxation time τ_0 was found to be

$$\tau_0 \simeq 2\tau_m \epsilon_s / (\epsilon_s + \epsilon_\infty)$$

somewhere between τ_m and $2\tau_m$.

Kauzmann (1942) considered that a more realistic approach was to treat the reorientation of the dipoles as being due to jumps over a potential barrier from one equilibrium position to another. He applied Eyring's Reaction Rate Theory (Glasstone et al, 1941) to the dielectric relaxation and, assuming that all possible directions of reorientation are equally possible in the absence of a field and that the jumps involve only small changes in direction, he obtained the equations originally derived by Debye (1929). Because of the necessary assumptions these two molecular theories are only applicable to unassociated liquids. For polymers the situation is much more complicated as the long molecular chains may be arranged in a regular crystalline fashion or in a random amorphous manner. In addition the main chains may have side chains attached and these can be of different types, capable of motions of their own and in some cases bonding through salt linkages to the side chains of adjacent main chains. Even covalent bonding inter or intra chain may occur, such as the bonding due to cystine in keratin.

With such a situation it is easy to see why there is no satisfactory molecular theory of the dielectric properties of polymers. In the special case of simple amorphous polymers some progress has been made and the theories have been reviewed by McCrum et al. (1967). One important aspect of the theories based upon the motions of chain segments such as those of Kirkwood and Fuoss (1941) and of Yamafuji and Ishida (1962) is that they predict a distribution of relaxation times, thus accounting for the broadening of experimental curves of ϵ'' versus log f in comparison with the single relaxation time Debye curve.

III. Dielectric Properties of Polymers

Much of our knowledge of the dielectric relaxation processes occurring in polymers has come from studies on homologous series of polymers and intercomparisons of data obtained in dielectric, mechanical and nuclear magnetic resonance experiments. McCall (1969) has reviewed this data and has described the following types of molecular motion and their associated dielectric relaxations:

(a) primary main chain motion in crystalline regions;
(b) primary main chain motion in amorphous regions;
(c) secondary main chain motion in crystalline regions;
(d) secondary main chain motion in amorphous regions;
(e) side chain motions;
(f) impurity motions.

The above listing is largely self explanatory but there are some important features which may be noted. Both the main chain motions may be affected by plasticizers, the motion in crystalline regions being associated with a type of melting whilst the motion in the amorphous regions is connected with the glass to rubber transition. In a purely amorphous polymer, increasing temperature changes the polymer from a rigid glass to a viscous rubber whilst in a semi-crystalline polymer it is the amorphous regions which may experience such a transition.

Secondary main chain motions are thought to be due to the damped oscillation of chain segments about their equilibrium positions. The side chain motions are due to the rotation of polar side chains and have small activation energies $(2 \rightarrow 20 \text{ kcal (g mole)}^{-1})$.

Impurity motions are those of low molecular weight materials such as water.

In the literature the relaxation processes occurring in a particular polymer are often denoted α, β, γ, σ in order of the increasing frequency (f_m) at which they occur for a fixed temperature, although there has been a tendency for main chain motions to be denoted α_a or α_c for the amorphous and crystalline regions respectively; also multiple side chain relaxations have been labelled γ', γ'' and so on. Yamafuji and Ishida (1962) have called the local mode relaxation process the β relaxation.

It is clear that in many cases it will not be possible to unequivocally assign a particular relaxation to one of the above categories and this is especially so in complex systems such as keratin, which is not only semi-crystalline but appears to consist of several polymeric species.

IV. Experimental Methods

The methods used for determining the dielectric properties of materials have been reviewed by McCrum *et al.* (1967) and Hill *et al.* (1969). Since those texts contain detailed surveys, we will only consider the measurement techniques in a general way with some particular emphasis on methods especially suited to fibres and the recent developments in time–domain techniques.

At frequencies up to 10^8 Hz the sample dimensions are sufficiently small compared with the wavelength of the applied electric field that the capacitance and conductance of a parallel plate capacitor containing the sample can be considered to be lumped constants. For frequencies greater than 10^8 Hz the capacitance and conductance must be considered to be distributed in space and the methods employed for determining the complex relative dielectric constant ϵ^* depend upon the propagation of the electric field through the sample.

A. BRIDGE METHODS

The problem of determining the capacitance and conductance of a parallel plate capacitor completely filled with the sample is most conveniently solved by comparing that capacitance and conductance with standard capacitances and conductances using bridge circuits. For frequencies from 10^{-2} Hz to 10^2 Hz Scheiber (1961) and Nakajima and Saito (1958) have described bridges which are satisfactory for use with polymer samples. Considerable patience is required for balancing the bridges at 10^{-2} Hz. Transformer ratio arm bridges (Cole and Gross, 1949) are very suitable for the frequency range 10^2 Hz to 10^6 Hz as they are not affected by stray capacitances and consequently can be used with guarded sample capacitors (Fig. 4). The guard ring eliminates edge effects and thus simplifies the calculation of the relative dielectric constant. Transformer ratio arm bridges are available commercially and one such is described in the General Radio Experimenter (1962).

The relative permittivity ϵ' and the relative dielectric loss ϵ'' can be calculated from the capacitance and conductance using the following relationships

$$\epsilon' = C/C_0$$

$$\epsilon'' = G/\omega C_0$$

where G = conductance (ohm)$^{-1}$, C = capacitance (farad), C_0 = capacitance with the sample replaced by air or a vacuum.

A resonance technique may be used to 10^9 Hz by the use of a resonant re-entrant cavity (Parry, 1951).

B. TRANSMISSION METHODS

For frequencies greater than 10^8 Hz the propagation characteristics of the sample are determined by placing it in a shorted transmission line or wave

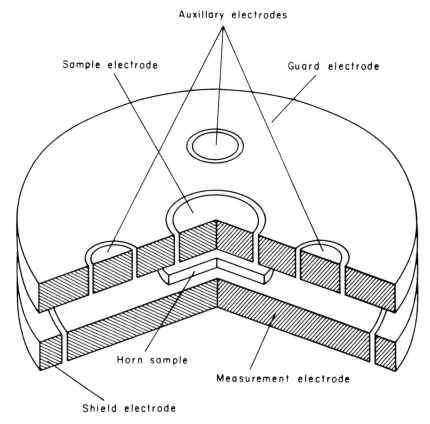

FIG. 4. Guarded capacitor for solid polymeric samples. Auxiliary electrodes are used for determining sample thickness. From Algie (1973) courtesy of *Kolloid Z.u.Z. Polymere.*

guide and measuring the magnitude and location of the standing wave due to the interference of the reflected and transmitted wave. Von Hippel (1954) has given details of the method, the equations relating the propagation factor to the standing wave and the relationship between the dielectric constant and the propagation factor.

C. MEASUREMENT TECHNIQUES FOR FIBRES

At low frequencies the difficulty with fibres is that the sample capacitor is not completely filled by the fibres, so that an air/fibre mixture is measured. When the fibres are placed parallel to the plates of the sample capacitor there is no satisfactory equation for determining the true dielectric properties of the material from which the fibres are made. Hearle (1954) has examined this problem in detail. Useful information can be obtained however (Hearle, 1954, 1956; Algie, 1964b, 1966, 1967, 1969; Algie and

Gamble, 1971) and the relaxation times of the various relaxation processes can be determined (Algie and Gamble, 1973).

To overcome the above problem Errera and Sack (1943) and De Luca *et al.* (1938) used an immersion technique whereby the sample capacitor is filled with a mixture of liquids whose dielectric constant is changed until the introduction of the fibres makes no effect on the capacitance. When this is achieved the dielectric constant of the fibres is equal to that of the liquid. The system has two major faults: (a) it is difficult to find liquids which are perfectly dry and which do not penetrate the fibres, (b) for every frequency and temperature the whole procedure must be undertaken.

Another technique is to make use of a formula derived by Mack (1955) which relates the dielectric constant of a dielectric cylinder, small in relation to the sample capacitor, to the change in capacity produced by the introduction of that cylinder between the plates. Ward (1961) and Algie (1964a) have used this method and Fig. 5 shows the capacitor assembly used by Algie (1964a).

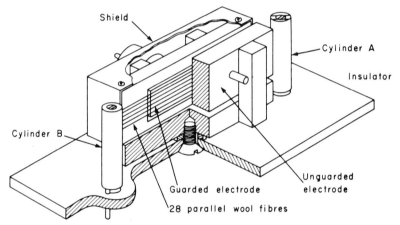

FIG. 5.   Capacitor for use with single fibres. From Algie (1964a) courtesy of *Textile Research Journal.*

The equation for the capacitance change is

$$\delta C \simeq (\epsilon' - 1)r^2/((\epsilon' + 1)2D^2)$$

where δC = capacitance change due to the insertion of the fibre, r = radius of the fibre, D = distance between capacitor plates. The above equation only applies if $r < D/10$ and the length of the fibre and hence the length of the capacitor's plates is large compared to D.

A feature of the equation is that due to the term $(\epsilon' - 1)/(\epsilon' + 1)$, δC is not very sensitive to ϵ', particularly when ϵ' becomes greater than about

10. Also $\delta C \propto r^2$ so that the radius of the fibre must be precisely known. Because of these factors the precision of the method is not high and great care is required in the design and construction of the sample capacitor.

If the fibres are arranged perpendicular to the plates of the sample capacitor, i.e. as many short lengths, the axial dielectric constant can be determined since each short length of fibre can be considered to form a capacitor in parallel with all the others and the air capacitors are also in parallel and can be easily allowed for. Balls (1946) and Ishida *et al.* (1959) have used the method on cotton fibres.

At very high frequencies $(3 \times 10^9 \rightarrow 3 \times 10^{10}\,\text{Hz})$ cavity resonators can be used (see Fig. 6). Shaw and Windle (1950) described the method for a frequency of $3 \times 10^9\,\text{Hz}$. Since a bundle of fibres had to be used, a mixture equation had to be employed to determine the dielectric constant of the material itself. However, at $26 \times 10^9\,\text{Hz}$ a single fibre when placed in an appropriate cavity resonator produces a measurable change in the resonant frequency, enabling the axial dielectric constant to be determined. From the bandwidth of the resonance curve the dielectric loss factor also can be obtained.

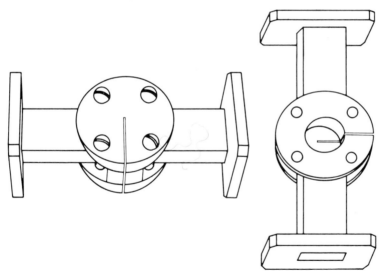

FIG. 6.   Cavity resonator for use with a single fibre at 26,000 MHz. From Windle and Shaw (1955) courtesy of *Textile Research Journal.*

Windle and Shaw (1955) have described the technique and applied it to wool fibres (Windle and Shaw, 1956). The frequency change due to the insertion of the fibre in the cavity is given by:

$$\Delta f = 1{\cdot}855\,(\epsilon' - 1)\,f_0 r^2/d$$

where Δf = frequency change, ϵ' = dielectric constant of fibre, r = fibre radius, a = cavity radius, f_0 = resonant frequency of the empty cavity.

D. MEASUREMENTS IN THE TIME DOMAIN

Up till now the measurement methods have determined the dielectric constant as a function of frequency by varying the frequency of the applied electric field, but it is also possible to measure the response of the sample to a step voltage or pulse as a function of time. The dielectric constant as a function of frequency can be determined by Fourier transforming the information obtained as a function of time or, as often stated, in the time domain.

In the Dielectric Theory section Eqns. (1) and (2) show how the relative permittivity and relative dielectric loss can be obtained from the charge or discharge current which flows after the application or removal of a step voltage. Hamon (1952) developed a simple transform which enables the relative dielectric loss $\epsilon''_{(\omega)}$ to be deduced as a function of frequency directly from the charge or discharge current $i_{(t)}$,

$$\epsilon'' = i_t/2\pi f C_0 V$$

where i_t = the current at time t and $t = 0\cdot1/f$. (In this equation i_t is in amperes and V is in volts.) It follows that the value of i at $t = 10\,\text{sec}$ provides a value of ϵ'' for $f = 10^{-2}$ Hz. The above transform is only applicable if $i_{(t)} = At^{-n}$ where A is a constant and $0\cdot3 < n < 1\cdot2$. However, Hamon (1952) showed that even if the curve of log $i_{(t)}$ against log t had a change of slope, but that both slopes had $0\cdot3 < n < 1\cdot2$, satisfactory results were obtained. In the case of a Debye curve for a single time constant the transform correctly showed the frequency location of the peak in ϵ'' but for $\omega\tau_0 < 1$ gave values of ϵ'' which were incorrect. The use of the discharge current has the big advantage that the d.c. conductance is not included and so the dielectric loss due to this term does not appear. It is necessary when measuring the discharge current that the voltage should have been applied for a time which is large with respect to the most probable relaxation time τ_0 of the low frequency peak. This and other details have been discussed by Baird (1968). The usual frequency range covered by the technique is 10^{-5} Hz to 10^{-2} Hz if $i_{(t)}$ is measured by an electrometer plus recorder. Williams (1963) has given a suitable circuit. It is possible to increase the upper frequency to about 10^2 Hz by photographing $i_{(t)}$ on a cathode ray oscillograph taking care that the time constant RC is small in relation to the minimum time interval after which $i_{(t)}$ is to be measured. Also, care should be taken to avoid spurious oscillatory currents due to the inductance and stray capacitance of the leads. Through the use of a pulse train Hyde (1970) has extended the frequency to 10^6 Hz.

The Hamon approximation has been investigated by Nakajima (1960) for the case of dielectrics which follow Cole and Cole (1941) expressions and that of Davidson and Cole (1950). He found that a good agreement was obtained to the values of ϵ'' obtained by the Hamon approximation and the Cole–Cole function for $\omega\tau_0 > 1$ but that good agreement was only obtained for $\omega\tau_0 < 1$ for a Cole–Cole distribution parameter $0 < n < 0.4$. However the frequency f_m of maximum dielectric loss was approximately correct for even $n = 1$, i.e. for a Debye curve. Williams (1962) has also examined the use of the Hamon approximation to materials obeying the Cole–Cole function and reached the same conclusions as Nakajima. Williams claimed that it is always preferable to evaluate the Fourier integral and Block *et al.* (1972) have examined the requirements for such an integration. The general computational requirements in dielectric measurements have been reviewed by Dev *et al.* (1972).

Time domain methods have been extended up to 13×10^9 Hz through the use of a very rapidly rising pulse (rise time $\simeq 35 \times 10^{-12}$ sec) and a sampling cathode ray oscilloscope which converts a high frequency signal into a low frequency one. The pulse is propagated down a coaxial line, part of which is filled with the dielectric under investigation. The reflection from or transmission through the dielectric can be analysed to provide the dielectric constant in the frequency domain. Fellner-Feldegg (1969) described a method for measuring the properties of dielectrics in the above manner. This publication and the Hewlett-Packard Application Note. 118 contain errors as pointed out by Whittingham (1970) and Van Gemert (1971). A corrected method suitable for Debye dielectrics with negligible conductivity has been presented by Fellner-Feldegg and Barnett (1970).

However it is unlikely that it would be known beforehand whether or not the dielectric was an ideal Debye dielectric and, as shown by Van Gemert and De Graan (1972), the TDR response is markedly different for dielectrics which can be represented by a distribution of relaxation times such as that due to Cole and Cole. Nicolson and Ross (1970) describe a technique suitable for thin samples in which various pulses, incident and reflected are measured and Fourier transformed into their frequency spectra. From ratios of these quantities the complex permittivity can be calculated. Suggett *et al.* (1970) also suggest the use of the Fourier transform to transform the complex reflection coefficient in the time domain into the complex reflection coefficient in the frequency domain and thence calculate the relative complex dielectric constant from

$$\epsilon^* = (1-\rho)^2/(1+\rho)^2$$

where ρ is the reflection coefficient. Various other experimental methods using (a) direct reflection, (b) the ratio of transmission coefficients using different lengths of dielectric filled line and (c) the ratio of the first-echo reflection coefficient to the transmission coefficient have been described by Loeb *et al.* (1971). They also discuss the origin of errors and ways of eliminating timing errors which can be large. Sources of error are also discussed by Fellner-Feldegg in Hewlett-Packard Application Note 153 which supersedes Note 118. Loeb *et al.* (1971) give an example of the effect of timing errors in the direct reflection method which is worth quoting; for water at 278°K and a frequency of 10×10^9 Hz the calculated phase shift is about 5° but a timing error of only 1×10^{-12} sec produces a phase shift at 10×10^9 Hz of 3·6°. Since sampling oscilloscopes have timing jitter of up to 15×10^{-12} sec some method has to be used to reduce this jitter. It appears that some sort of signal averaging equipment is essential if accurate data is to be obtained.

Loeb *et al.* (1971) claim that, using the above plus a timing error minimization procedure which they describe, a timing accuracy for the direct reflection method of $0 \cdot 1 \times 10^{-12}$ sec can be achieved. Figure 7 shows an experimental arrangement for TDR reflection measurements. A typical waveform to be observed on the sampling oscilloscope is shown in Fig. 8.

Although TDR has not been used with fibres it has been employed to determine the dielectric properties of powdered lysozyme by Harvey and Hoekstra (1972) and of maple wood by Nicholson and Ross (1970).

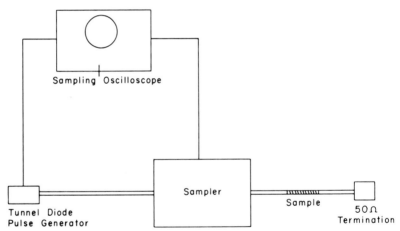

FIG. 7. An experimental arrangement for measuring the Time Domain Reflectometry reflection response from a dielectric sample.

V. Dielectric Properties of Cellulose

As is the case with most polymers, cellulose has a dielectric constant which is far from constant. It is affected by the frequency of measurement, the temperature of the sample, the presence of plasticizers of which water is

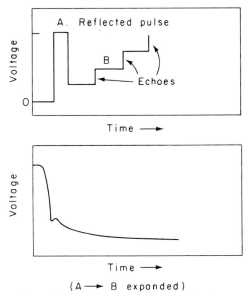

FIG. 8.   Waveform observed on the sampling oscilloscope for the arrangement in Fig. 7.

the most ubiquitous, the accessibility of water (which is related to the crystallinity), the density and the anisotropy of the macromolecular structure. In addition, since it is difficult to deduce the dielectric constant of fibrous materials from measurements of fibre/air mixtures, more tractable forms of cellulose such as wood and paper have been included in this survey.

For wood alone, Hearmon and Burcham (1954) have reviewed the literature up to 1954 and Lin (1967) up to 1967.

A. ϵ' AND ϵ'' AS A FUNCTION OF FREQUENCY

The total range of frequencies covered in dielectric experiments on cellulose is rather wide, 10^{-5} to 10^{10} Hz, although only one worker (Riaux, 1965) has examined the low frequency region 10^{-5} to 10^{-2} Hz and seven (Takeda, 1951; Trapp and Pungs, 1956; Le Petit, 1963; Tong, 1964; Chéne et al., 1965; Brady, 1968; Nanassy, 1970) have covered the higher frequencies about 10^9 or 10^{10} Hz. At low frequencies the use of the discharge current technique has not become widespread in spite of its simplicity. Consequently most workers have employed frequencies from 10^2 to 10^7 Hz. Figures 9 and 10 show results obtained by Ishida et al. (1959) in which the real relative dielectric constant and loss factor have been measured for small lengths of dry cotton sliver aligned parallel to the electric field. In the frequency and temperature range covered by these

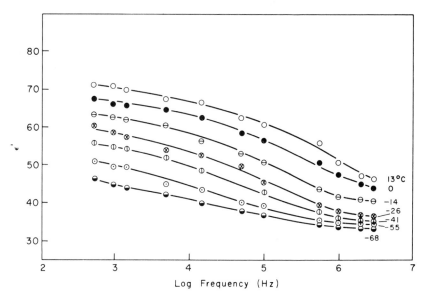

FIG. 9. ϵ' as a function of log frequency for dry cotton silver at the temperatures shown. From Ishida *et al.* (1959) courtesy of *Journal of Applied Polymer Science*.

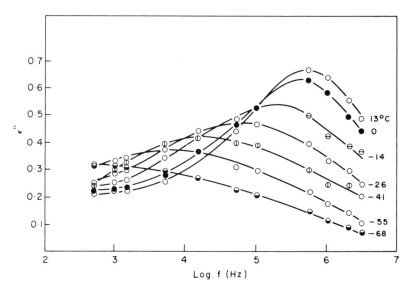

FIG. 10. ϵ'' as a function of log frequency for dry cotton silver at the temperatures shown. From Ishida *et al.* (1959) courtesy of *Journal of Applied Polymer Science*.

experiments (5×10^2 Hz to 3×10^6 Hz and -60 to 20°C) a clearly defined single dielectric loss peak can be seen. It is broader than that for a single relaxation time τ and can be described by a Cole and Cole distribution function in which the distribution parameter n increases with temperature from 0·35 to 0·55; this means that the loss peak becomes narrower with increasing temperature.

Another set of data showing the effects of frequency on ϵ' and ϵ'' of dry wood is due to Nanassy (1970) (Figs 11 and 12). Two loss peaks and the high frequency section of a third peak occurring at low frequencies can be seen. Each overlapping peak is due to a specific dielectric absorption process and this will be discussed later. The single peak occurring at about 10^6 Hz for a temperature of 20°C is evidently related to that in Fig. 10, indicating that wood and cotton are similar materials from a dielectric viewpoint.

B. ϵ' AND ϵ'' AS A FUNCTION OF TEMPERATURE

Effects of temperature have been investigated in the range -269°C to 250°C. Allan and Kuffel (1968) measured ϵ' and $\tan \delta = \epsilon''/\epsilon'$ for dry paper and paper containing 8% water content for temperatures from -269°C to 22°C and frequencies from 47 Hz to $8·95 \times 10^4$ Hz. Only one loss peak was observed and the relative ϵ' approached that for high frequencies ϵ_∞ at the lowest temperatures. Activation enthalpies and entropies were deduced and the shift of the frequency of maximum $\tan \delta$ with temperature is shown on the transition map Fig. 13.

Figures 10 and 12 show that the shift of f_m to higher frequencies with increasing temperature occurs for each of the loss peaks, indicating that the processes responsible are thermally activated. Table I lists the activation enthalpies for the three peaks as determined by various workers. The spread in values may be due to small differences between the various cellulosic materials, although Ishida et al. (1959) find the same activation enthalpies for cotton, cuprammonium rayon and viscose rayon, which suggests that some of the differences may be instrumental.

Another effect which an increase in temperature produces is an increase in the magnitude of each of the loss peaks (Fig. 12). This was also observed by Tsutsumi (1967), Trapp and Pungs (1956) and Ishida et al. (1959) (Fig. 10), for the low and high frequency peaks. The middle frequency peak only occurs in the results obtained by Nanassy (1970); following Calkins (1950) he considers that it is due to hemicellulose or lignin.

Tan δ for a range of dry cellulosic materials has been obtained by Mikhailov et al. (1969) for temperatures from -160°C to 250°C. Their results for a frequency of 10^4 Hz are shown in Fig. 14.

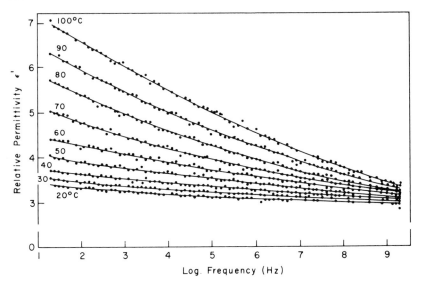

FIG. 11. ϵ' as a function of log frequency for end grain oriented wood at various temperatures. From Nanassy (1970) courtesy of *Wood Science and Technology.*

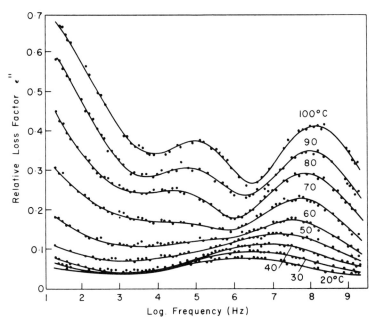

FIG. 12. ϵ'' as a function of log frequency for end grain oriented wood. From Nanassy (1970) courtesy of *Wood Science and Technology.*

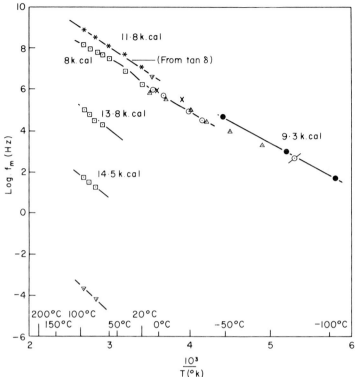

FIG. 13.   Transition map showing log frequency of maximum loss (f_m) as a function of $10^3/T$ (where T is in °K). (Δ Ishida *et al.* (1959), □ Nanassy (1970), ○ Seidman and Mason (1954), × Stoops (1934), ∇ Riaux (1965), ● Allan and Kuffel (1968), ✱ Trapp and Pungs (1956), ∅ Zelenev and Glazkov (1972).

TABLE I

Activation enthalpy of three dielectric loss peaks in dry cellulose (see Fig. 13)

Author	High Frequency	Mid Frequency	Low Frequency
Trapp and Pungs (1956)	11·8 kcal (g mole)$^{-1}$		
Nanassy (1970)	8·0	13·8	14·5
Allan and Kuffel (1968)	9·3		
Seidman and Mason (1954)	9·3		
Ishida *et al.* (1959)	4 at 20°C		
	13 at −20°C		

The peak at a temperature of −80 to −60°C is related to the high frequency peak in Fig. 12 whilst the increase in tan δ at high temperatures is related to the low frequency peak and is due to either a Maxwell–Wagner interfacial polarization or the onset of dipolar orientation due to the glass–rubber transition.

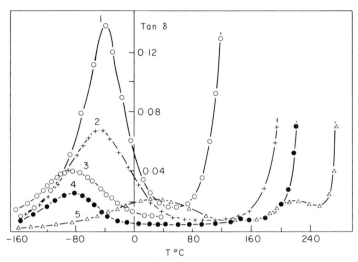

FIG. 14.   tan σ for a frequency of 10^4 Hz as a function of Temperature. 1 cyanoethyl cellulose, 2 cellulose hydrate, 3 methyl cellulose, 4 methyl cellulose (less methylation than 3), 5 cellulose triacetate. From Mikhailov *et al.* (1969) courtesy of *Vysokomol Soyed.*

C. ϵ' AND ϵ'' AS A FUNCTION OF WATER CONTENT

Trapp and Pungs (1956) have made the most extensive study of the effects of water content on the dielectric constant and loss of wood. Figure 15 shows ϵ' as a function of frequency for water contents from 0% to 100% and at a temperature of 20°C. Both the high frequency and low frequency peaks shift to higher frequencies with increasing water content and their magnitude also increases. This is similar to the effects of temperature and just as it is frequently possible to superimpose curves of ϵ' and ϵ'' for different temperatures by shifting them along the frequency axis, Tsuge and Wada (1962) have shown that a similar superposition for ϵ' and ϵ'' obtained at different water contents, may be obtained by shifting along the frequency axis together with an amplitude adjusting factor to allow for the effect of water content on the magnitude of both ϵ' and ϵ'' curves. Their results were obtained at a sample temperature of 35°C, for cellophane with water contents from 2·3 to 17% in the frequency range 10^2 Hz to $1·2 \times 10^5$ Hz. This corresponds to the range of parameters associated with the low frequency peak.

Another study, in which wood with water contents from 0 to 15% was examined in the frequency range 30 Hz to $2·5 \times 10^6$ Hz at temperatures of −70° to 40°C, is that of Tsutsumi (1967). At higher temperatures, up to 100°C, results were only obtained at low water contents. Both the low and high frequency peaks were studied and the dispersion $\epsilon'_s - \epsilon_\infty$ for the low frequency peak was 100 $\simeq$ 150 whilst that of the high frequency peak was less than 5.

D. TRANSITION MAPS

Transition maps, which are plots of the log of the frequency of maximum absorption f_m as a function of T^{-1}, where T is in degrees Kelvin, are useful for obtaining an overall view of the various dielectric absorption bands, their activation enthalpies ΔH and the effect of sorbed materials such as water.

Figure 13 is a transition map for dry cellulose. There are four absorption bands; the one at high frequencies studied by many workers, the mid frequency band which only appears in the work of Nanassy (1970), the low frequency band (Muus, 1953; Tsuge and Wada, 1962; Nanassy, 1970) and a very low frequency band found by Riaux (1965). The effects of water in shifting f_m can be seen in Figs 15 and 16.

In Fig. 13 the separation of the data into two approximately parallel lines is due to the use of the frequency of the peak of tan δ for the indicated points and the frequency of the peak of ϵ'' for the others. Theoretically the peak of tan δ should occur at a higher frequency than the peak of ϵ''. It is interesting to observe that the various forms of cellulose, wood, paper, cotton, two forms of rayon and cellophane, all have dielectric properties in regard to the relationship between f_m and T which are similar. This implies that the basic process responsible for the high frequency dielectric absorption band is the same in each case. Mikhailov et al. (1969) studied the properties of various forms of cellulose, cotton, wood, cellulose hydrate, methylated cellulose with various degrees of methylation, cellulose triacetate, cyanoethyl cellulose, xylan and triphenyl methyl cellulose ester. He used dielectric and nuclear magnetic resonance techniques and was able to deduce that (1) the high frequency relaxation processes observed dielectrically were the same as those in the NMR data, (2) by deuteration of the sample, that rotation of the whole CH_2OH group rather than just the OH group was responsible for the relaxation, (3) that the CH_2OH groups were those in the amorphous regions.

Confirmation for (3) comes from the fact that the high frequency relaxation process is plasticized by water and it is known that the water is only absorbed in the amorphous regions (Hermans, 1946).

The mid-frequency band has been ascribed to lignin or hemicellulose by Nanassy (1970). Whilst the low frequency band, which must also be due to a process in the amorphous regions for the reasons stated above, has been ascribed to (1) segmental motion of cellulose chains (Tsuge and Wada,

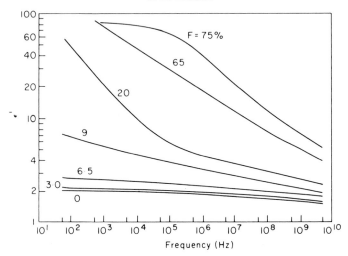

Fɪɢ. 15. ϵ' as a function of log frequency at a temperature of 20°C for wood with various water contents (F%). From Trapp and Pungs (1956) courtesy of *Holzforchung*.

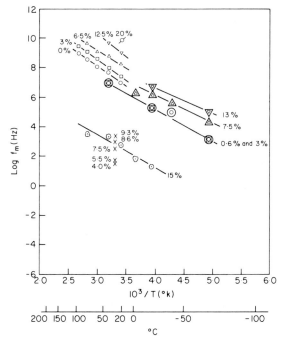

Fɪɢ. 16.   Transition map for cellulose at various water contents showing log f_m as a function of $10^3/T$ (where T is in °K). Single symbols top left Trapp and Pungs (1956), double symbols Tsutsumi (1967), × Tsuge and Wada (1967).

1962), (2) OH groups in the crystalline regions (in the dry state) (Muus, 1953), (3) Maxwell–Wagner (Wagner, 1914) interfacial polarization by Trapp and Pungs (1956) or a modification due to Daenzer (1934).

Mechanical data on cellulose (Stratton, 1973) does not indicate segmental motion in the correct temperature range for (1) to be correct, furthermore the values of the low frequency dispersion $\epsilon_s' - \epsilon_\infty = 100 \rightarrow 150$ are very large for a segmental motion. (2) is not likely to be correct because f_m increases approximately linearly with water content and Tsuge and Wada (1962) have shown that curves of ϵ'' as a function of frequency at various water contents can be transposed onto a single "master" curve which indicates that a common relaxation process is involved. It is most likely that it is an interfacial polarization process with the increase in f_m being related to the increase in d.c. conductivity with water content.

Riaux (1965) has claimed that the very low frequency absorption process is also due to a Maxwell–Wagner interfacial polarization. Since little is known about this absorption band further evidence is required before a decision can be made.

E. EFFECTS OF DENSITY, CRYSTALLINITY AND ANISOTROPY

The dielectric constant ϵ' at 2 M Hz for 30 varieties of wood has a linear relationship with the density at water contents of 0, 5, 10 and 15% according to Skaar (1948). Densities from 0·3 to 0·95 cm^{-3} were determined in the investigation. Peterson (1949, 1960) also found a proportionality between ϵ' and density for wood at a frequency of 5 M Hz. These results indicate the basic similarity between the cellulose molecules which comprise the various timbers.

Over and above the effects of density a relationship has been found between ϵ' and the accessibility or the related crystallinity of various forms of cellulose. Kane (1955) studied cotton, cellophane, wood pulp and viscose rayon in the dry state at a frequency of 10 K Hz and a temperature of 30°C. An immersion technique was employed to enable the dielectric constant of the material itself, rather than that of a cellulose/air mixture, to be determined. The relationship between ϵ' and accessibility was linear and shows that the polarization is occurring in the amorphous regions of the cellulose.

Calkins (1950) and later Verseput (1951) used high pressures on a range of cellulosic materials—cotton, wood fibres, cellophane, viscose rayon, cuprammonium rayon—in an attempt to obtain consistent values of the dielectric constant. They also found a linear relationship between ϵ' and accessibility.

ϵ' is also dependent upon the direction of the electric field in relation to the direction of the crystallites. Thus in dry cotton at 1·755 M Hz, Balls (1946) found that ϵ' along the fibre was 6 whilst in the transverse direction it was only 3. Even though the value of 3 may be in error due to the method used to deduce the real ϵ' from that of a cotton/air mixture, it is likely that there is some anisotropy. Brown, quoted by Seidman and Mason (1954), found that the transverse and axial ϵ' for cellulose crystallites was 5·27 and 7·19 at a frequency of 300 K Hz.

In the microwave region, at 9000 M Hz, Chéne *et al.* (1965) measured the axial and transverse ϵ' of paper. For dry paper there was a small anisotropy which decreased when the paper was purified by the removal of lignin. The effect of water was to increase the anisotropy which they thought was due to anisotropy in the water itself. Norimoto and Yamada (1973) have examined the relationship between the dielectric properties and the non-crystalline fraction of cellulose both theoretically and experimentally.

VI. Dielectric Properties of Keratin

A. ϵ' AND ϵ'' AS A FUNCTION OF FREQUENCY

A composite graph of ϵ' for keratin at 25°C as a function of log frequency is shown in Fig. 17. The low frequency data (500 Hz − 10^6 Hz) are taken from the work of King (1947) on cow-horn keratin whilst the data of Laird (1952) on horn keratin and Windle and Shaw (1954, 1956) on wool have been used for the high frequencies (3×10^9 Hz → $2·6 \times 10^{10}$ Hz). There is very little evidence of dispersion in dry keratin in this graph and it is not

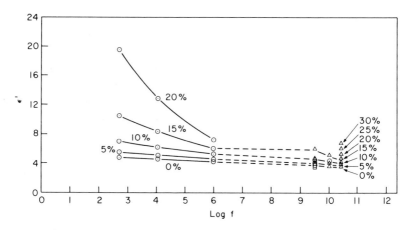

FIG. 17. ϵ' as a function of log frequency for wool and horn at 25°C and stated water contents. ⊙ King (1947) horn, △ Windle and Shaw (1954, 1956) wool, ⊡ Laird (1952) horn. Figure from Algie (1964a) courtesy of *Textile Research Journal.*

until the dielectric loss ϵ'' is examined that absorption bands can be observed. Figure 18 shows α' and β absorption bands as well as a band due to a small amount of water. (This water can only be removed by vacuum drying at a temperature of about 140°C.)

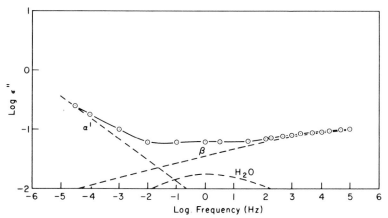

FIG. 18.   Log dielectric loss ϵ'' as a function of frequency for dry rhinoceros horn at a temperature of 20°C. Showing the analysis into component curves α', β and H_2O.

From measurements made on dry rhinoceros horn at high and low temperatures, Algie (1973) has shown that ϵ'' for dry horn keratin can be analysed into seven absorption bands, which he has labelled Ω, α', α'', β, γ', γ'' and δ in order of increasing frequency of the f_m of each band. This terminology has been discussed in the review by McCall (1969). The Ω absorption is thought to be due to a type of interfacial polarization, the α' and α'' due to the motion of main chains in amorphous regions, the β due to localized motion of main chains, γ' and γ'' due to side chain motions whilst σ may be due to the motion of end groups. It should be understood that these assignments have not been proven and arise only by analogy with dielectric and mechanical data obtained on polymers and homologous series of polymers. There are two exceptions to this general statement and they are (i) the Ω absorption has been labelled an interfacial polarization because of its large magnitude $\epsilon''_{m\Omega} \simeq 200$ and the fact that its f_m increases with temperature and water content in the same way as the d.c. conductivity; (ii) the α' absorption is thought to be due to the motion of main chains because it appears to be related to a mechanical absorption band observed by Menefee and Yee (1965) and exhibits a transition at about 150°C which is probably related to a type of glass to rubber transition.

The seven absorption bands or relaxation processes are displayed on the transition map Fig. 19. When considering ϵ' as a function of frequency it is of interest to calculate the static dielectric constant ϵ_s and the dielectric

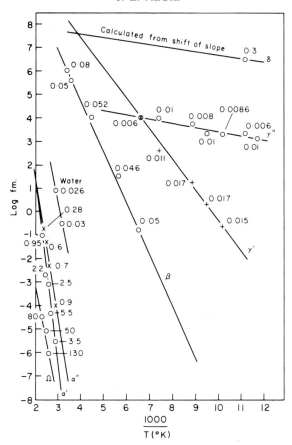

FIG. 19.   Transition map showing log f_m as a function of $10^3/T$ (T in °K). The numbers indicate the amplitude ϵ_m''. Algie (1973) courtesy of *Kolloid Z.u.Z. Polymere.*

constant at infinite frequency ϵ_∞ for a temperature of 20°C. Assuming that there are no other relaxation processes i.e. other than the seven already described, ϵ_s can be calculated by adding the contributions to ϵ' from each of the relaxation processes. This involves the extrapolation of the high temperature curves for Ω and α' to 20°C and the assumption that $\epsilon_{m\Omega}''$ and $\epsilon_{m\alpha'}''$ are the same at 20°C as at higher temperatures. Taking $\epsilon_{m\Omega}''$ as 100 and $\epsilon_{m\alpha'}''$ as 3, ϵ_s is found to be about 334 for frequencies below 10^{-14} Hz. By a similar process and taking into account the β, γ', γ'' and σ absorption bands, ϵ' at 10^{10} Hz is about 3·20 and at 10^{12} Hz is about 3·14. The calculated value of 3·20 for dry horn keratin at 10^{10} Hz, 20°C can be compared with the value of 3·4 at $2·6 \times 10^{10}$ Hz, 25°C obtained by Windle and Shaw (1956) for a single dry wool fibre.

This comparison indicates that wool and horn are dielectrically similar. Further evidence is obtainable at low frequencies in that King (1947) found $\epsilon' = 4\cdot2$ for dry horn keratin at 10^6 Hz, 25°C, in agreement with the data obtained by Errera and Sack (1943) on dry wool fibres. Algie (1964a, 1973) obtained a value of $\epsilon' = 4\cdot33$ for dry rhinoceros horn at $1\cdot59 \times 10^3$ Hz, 20°C, and $\epsilon' = 4\cdot2$ for dry wool fibres at $1\cdot59 \times 10^3$ Hz, 35°C.

B. ϵ' AND ϵ'' AS A FUNCTION OF TEMPERATURE

Algie (1973) has measured the dielectric properties of vacuum dried rhinoceros horn over the temperature range of −188°C to 200°C. ϵ' was determined in the frequency range 10^{-1} Hz to 10^5 Hz whilst ϵ'' was determined for frequencies from 10^{-6} Hz to 10^6 Hz.

So far as ϵ' is concerned the effects of temperature, as shown in Fig. 20, primarily indicate the increase of ϵ' at low frequencies due to the shift upwards in frequency of the α' peak with increasing temperature.

The effects of temperature on the dielectric loss ϵ'' are more interesting and the transition map Fig. 19 shows the way in which f_m for each of the absorption bands shifts upwards with increasing temperature. There is little effect of temperature on the magnitude ϵ''_m of the peaks or on the width of

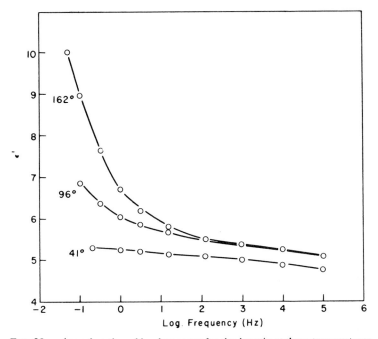

FIG. 20. ϵ' as a function of log frequency for dry keratin at three temperatures.

the peaks signified by the Cole and Cole parameter n which represents the distribution of relaxation times.

Typical results are shown in Figs 21 and 22. These figures portray the analysis of the ϵ'' versus frequency data into the component relaxation processes Ω, α', α'', β, γ', γ'' and σ. The amplitudes, Cole–Cole parameters n and activation enthalpies for each relaxation process are listed in Table II.

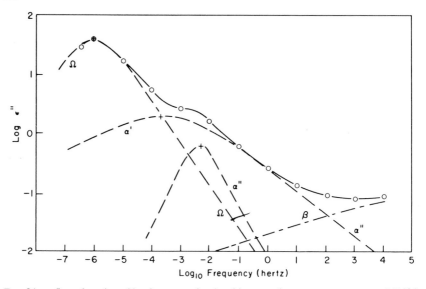

FIG. 21. ϵ'' as a function of log frequency for dry rhinoceros horn at a temperature of 132°C. Showing component curves Ω, α', β and α''. From Algie (1973) courtesy of *Kolloid Z.u.Z. Polymere*.

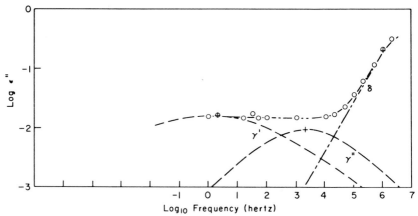

FIG. 22. ϵ'' as a function of log frequency for dry rhinoceros horn at a temperature of −168°C. From Algie (1973) courtesy of *Kolloid Z.u.Z. Polymere*.

TABLE II

Dielectric data for dry horn keratin

Relaxation process	Activation enthalpy	Cole–Cole distribution parameter n	Amplitude ϵ_m''
Ω	26 kcal (g mole)$^{-1}$	0·73	≈ 100
α'	32	0·38	≈ 3
α''	28	0·84	$\approx 0·7$
β	10	0·18	$\approx 0·05$
γ'	6	0·36	$\approx 0·015$
γ''	0·9	0·38	$\approx 0·01$
δ	0·6	1·0	$\approx 0·3$

C. EFFECT OF WATER CONTENT

Water increases the dielectric constant of wool and horn. King (1947) examined cow horn containing from 0% to 20% water and his results are drawn in Fig. 17. Also in this figure are the results of Windle and Shaw (1954, 1956) which indicate a dispersion region at high frequencies due to the presence of water in wool fibres. Hearle (1954, 1956) studied the dielectric constant and power factor of wool/air mixtures for water contents from 1·3% to 17·8% at 70°F, for frequencies from 50 Hz to 10^6 Hz. Single wool fibres were measured by Algie (1964a) at a frequency of $1·59 \times 10^3$ Hz, 35°C, and ϵ' was determined for water contents from 0% to 25%. Within experimental error, the results agreed with those obtained by King (1947) with horn keratin. Algie and Gamble (1973) have studied the effects of water and temperature on the dielectric properties of a wool/air mixture over a frequency range of 10^{-2} to 2×10^6 Hz. Figure 23 shows the increase in f_m for the Ω and α' peaks with increase in water content. The d.c. conductivity is also plotted on this figure to enable comparison with the data for the Ω absorption. There is a strong similarity between the two curves, they have the same slope and a change of slope at the same water content. Furthermore the activation enthalpies of the d.c. conductivity and the Ω absorption are similar at various water contents. It is reasonable to conclude that there is a functional relationship between the two processes. Interfacial polarization as described by Wagner (1914) provides such a relationship in that for spherical conducting particles in a non-conducting matrix the dielectric loss factor is given by

$$\epsilon'' = 9q[(\epsilon_1')^2/(2\epsilon_1' + \epsilon_2')][(f/f_m)/(1 + (f/f_m)^2)]$$

where

$$f_m = 1·8 \times 10^{12}\sigma_2/(2\epsilon_1' + \epsilon_2')$$

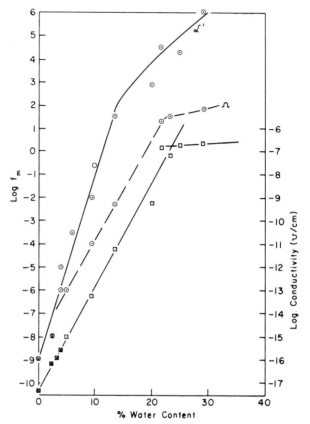

FIG. 23. Log f_m as a function of water content and d.c. conductivity as a function of water content for wool and horn ($\square$) at a temperature of 20°C. From Algie and Gamble (1973) courtesy of *Kolloid Z.u.Z. Polymere.*

q = volume fraction of conducting material, σ_2 = conductivity (ohm^{-1} cm^{-1}) of conducting material, ϵ_1' = relative dielectric constant of the matrix, ϵ_2' = relative dielectric constant of the conducting material.

Since the conductivity σ_2 has a much greater change with water content than ϵ_1' and ϵ_2' then $f_m \propto \sigma_2$. Also

$$\epsilon_m'' = \tfrac{9}{2}q(\epsilon_1')^2/(2\epsilon_1' + \epsilon_2')$$

so that $\epsilon_m'' \propto q$ provided that the term

$$(\epsilon_1')^2/(2\epsilon_1' + \epsilon_2')$$

does not vary significantly with water content.

Both these relationships $f_m \propto \sigma_2$ and $\epsilon_m'' \propto q$ are supported by the experimental data although it should be noted that σ_2 the conductivity of the conducting material is assumed to vary in the same way as the overall conductivity of the wool sample and the quantity of conducting material is assumed to vary with the water content. It follows that it has been assumed that the conducting material exists as a keratin–water complex surrounded by non-conducting regions which may be the crystalline regions.

Close agreement between the experimental data and the Maxwell-Wagner equations are only applicable to small quantities of the conducting material (less than 10%) which are sufficiently separate for no interaction to occur. It is sufficient that the data shows the general trends to be expected for an interfacial polarization.

References

Algie, J. E. (1964a). *Text. Res. J.* **34**, 477–486.

Algie, J. E. (1964b). *Text. Res. J.* **34**, 1026–1031.

Algie, J. E. (1966). *Text. Res. J.* **36**, 317–322.

Algie, J. E. (1967). *Text. Res. J.* **37**, 224–227.

Algie, J. E. (1969). *Text. Res. J.* **39**, 213–216.

Algie, J. E. (1973). *Kolloid-Z.u.Z. Polymere* **251**, 305–309.

Algie, J. E. and Gamble, R. A. (1971). *Text. Res. J.* **41**, 362–363.

Algie, J. E. and Gamble, R. A. (1973). *Kolloid-Z.u.Z. Polymere* **251**, 554–562.

Allan, R. N. and Kuffel, E. (1968). *Proc. Instn. elect. Engrs.* **115**, 432–440.

Baird, M. E. (1968). *Rev. mod. Phys.* **40**, 219–227.

Balls, W. L. (1946). *Nature, Lond.* **158**, 9–11.

Block, H., Groves, R., Lord, P. W. and Walker, S. M. (1972). *J. chem. Soc. Faraday Trans. II.* **68**, 1890–1896.

Brady, M. M. (1968). *J. microwave Power* **3**, 194–197.

Calkins, C. R. (1950). *TAPPI* **33**, 278–285.

Chène, M., Coumes, A. and Lafaye, F. (1965). *C.R. Acad. Sci. Paris* **260**, 3632–3635.

Cole, K. S. and Cole, R. H. (1941). *J. chem. Phys.* **9**, 341–351.

Cole, R. H. and Gross, P. M. (1949). *Rev. scient. Instrum.* **20**, 252–260.

Daenzer, H. (1934). *Annln. Phys. ser. 5* **20**, 463–480.

Davidson, D. W. and Cole, R. H. (1950). *J. chem. Phys.* **18**, 1417.

Debye, P. (1912). *Phys. Z.* **13**, 97–100.

Debye, P. (1929). "Polar Molecules". Dover Publications, Inc., New York.

De Luca, H. A., Campbell, W. B. and Maas, O. (1938). *Can. J. Res. Sect. B,* **16**, 273–288.

Dev, S. B., North, A. M. and Pethrick, R. A. (1972). *Adv. Mol. Relaxation Processes* **4**, 159–191.

Errera, J. and Sack, H. S. (1943). *Ind. Engng. Chem. analyt. Edn.* **35**, 712–716.

Fatuzzo, E. and Mason, P. R. (1967). *Proc. phys. Soc.* **90**, 729–740.

Fellner-Feldegg, H. (1969). *J. phys. Chem.* **73**, 616–623.

Fellner-Feldegg, H. and Barnett, E. F. (1970). *J. phys. Chem.* **74**, 1962–1965.

Fuoss, R. M. and Kirkwood, J. G. (1941). *J. Amer. chem. Soc.* **63**, 385–394.

Glasstone, S., Laidler, K. J. and Eyring, H. (1941). "The Theory of Rate Processes". McGraw-Hill, New York.

Gross, B. (1941). *Phys. Rev.* **59**, 748–750.

Hamon, B. V. (1952). *Proc. Instn. elect. Engrs.* **99**, 151–155.

Hamon, B. V. (1953). *Aust. J. Phys.* **6**, 304–315.

Harvey, S. C. and Hoekstra, P. (1972). *J. phys. Chem.* **76**, 2987–2994.

Hearle, J. W. S. (1954). *Text. Res. J.* **24**, 307–321.

Hearle, J. W. S. (1956). *Text. Res. J.* **26**, 108–111.

Hearmon, R. F. S. and Burcham, J. N. (1954). C.S.I.R.O. Div. Forest Products Special Report No. 8, 1–19.

Hermans, P. H. (1946). "Contributions to the Physics of Cellulose Fibres". Elsevier Publishing Co., Inc., Amsterdam.

Hill, N. E. (1961). *Proc. phys. Soc.* **78**, 311–312.

Hill, N. E., Vaughan, W. E., Price, A. H. and Davies, M. (1969). "Dielectric Properties and Molecular Behaviour". Van Nostrand Reinhold Company, London.

Hopkinson, J. (1877). *Phil. Trans. R. Soc.* **166**, 489–494.

Hyde, P. J. (1970). *Proc. Instn. elect. Engrs.* **117**, 1891–1901.

Ishida, Y., Yoshino, M. and Takayanagi, M. (1959). *J. appl. Polymer Sci.* **1**, 227–235.

Kane, D. E. (1955). *J. Polymer Sci.* **18**, 405–410.

Kauzmann, W. (1942). *Rev. mod. Phys.* **14**, 12–44.

King, G. (1947). *Trans. Faraday Soc.* **43**, 601–611.

Kirkwood, J. G. and Fuoss, R. M. (1941). *J. chem. Phys.* **9**, 329–351.

Kramers, H. A. (1927). Atti Congr. dei Fisici, Como, p. 545 quoted by Kronig (1926).

Kronig, R. de L. (1926). *J. opt. Soc. Am.* **12**, 547–557.

Laird, E. R. (1952). *Can. J. Phys.* **30**, 663–667.

Le Petit, J. P. (1963). *Journal de Physique* **24**, 409–410.

Lin, R. T. (1967). *Forest Prod. J.* **17**, No. 7, 54–66.

Loeb, H. W., Young, G. M., Quickenden, P. A. and Suggett, A. (1971). *Ber. Bunsen-Gesellschaft für phys. Chem.* **75**, 1155–1165.

McCall, D. W. (1969). *In* Natn. Bur. Stand. Special Pub. No. 301 "Molecular Dynamics and Structure of Solids" (R. S. Carter and J.J. Rush, eds.).

McCrum, N. G., Read, B. E. and Williams, G. (1967). "Anelastic and Dielectric Effects in Polymeric Solids". John Wiley and Sons, London.

Mack, C. (1955). *Shirley Inst. Mem.* **28**, 91–102.

Maxwell, J. C. (1892). "A Treatise on Electricity and Magnetism", pp. 452–456. Oxford University Press, London.

Menefee, E. and Yee, G. (1965). *Text. Res. J.* **35**, 801–812.

Mikhailov, G. P., Artyukhov, A. I. and Shevelev, V. A. (1969). *Vysokomolek. Soedin,* **A11**, No. 3, 553–563.

Muus, L. T. (1953). Trans. Danish Acad. Techn. Sci. ATS No. 4, pp. 1–128.

Nakajima, T. (1960). *Japan Electrotech. Lab. Bull.* **24**, No. 10, 35–40.

Nakajima, T. and Saito, S. (1958). *J. Polymer Sci.* **31**, 423–437.

Nanassy, A. J. (1970). *Wood Sci. and Technology* **4**, 104–121.

Nicolson, A. M. and Ross, G. F. (1970). I.E.E.E. Trans. Inst. and Meas. IM19, 377–382.
Norimoto, M. and Yamada, T. (1973). *Wood Res.*, *Kyoto* No. 54, 19–30.
Onsager, L. (1936). *J. Am. chem. Soc.* **58**, 1486–1493.
Parry, J. V. L. (1951). *Proc. Instn. elect. Engrs. Pt. 3*, 303–311.
Peterson, R. W. (1949). *Brit. Columbia Lumberm.* **33**, 99–100.
Peterson, R. W. (1960). Forest Products Laboratory Can. Tech. Note. 16.
Riaux, M. E. (1965). *C.R. Acad. Sc. Paris*, **261**, 2845–2847.
Scheiber, D. J. (1961). *J. Res. natn. Bur. Stand.* **65C**, 23–42.
Seidman, R. and Mason, S. G. (1954). *Can. J. Chem.* **32**, 744–762.
Shaw, T. M. and Windle, J. J. (1950). *J. appl. Phys.* **21**, 956–961.
Sillars, R. W. (1937). *J. Instn. elect. Engrs.* **80**, 378–394.
Skaar, C. (1948). New York State College of Forestry Tech. Pub. No. 69.
Stoops, W. N. (1934). *J. Am. chem. Soc.* **56**, 1480–1483.
Stratton, R. A. (1973). *J. Polymer Sci. Polymer Chem. Edn*, **11**, 535–544.
Suggett, A., Mackness, P. A., Tait, M. J., Loeb, H. W. and Young, G. M. (1970). *Nature, Lond.* **228**, 456–457.
Takeda, M. (1951). *Bull. chem. Soc. Japan* **24**, 169–173.
Tong, D. P. (1964). *Svensk Papp-Tidn.* **67**, 686–689.
Trapp, W. and Pungs, L. (1956). *Holzforschung* **10**, No. 5, 144–150. Translation available from U.S. Dept. Agriculture Forest Products Lab Report No. 634.
Tsuge, K. and Wada, Y. (1962). *J. phys. Soc. Japan* **17**, 156–164.
Tsutsumi, J. (1967). *Bull. Kyushu Univ. Forests* No. 41, 109–169.
Van Gemert, M. J. C. (1971). *J. phys. Chem.* **75**, 1323–1324.
Van Gemert, M. J. C. and De Graan, J. G. (1972). *Appl. scient. Res.* **26**, 1–17.
Verseput, H. W. (1951). *TAPPI* **34**, 572–576.
Von Hippel, A. R. (1954). "Dielectric Materials and Applications". Technology Press M.I.T. and John Wiley & Sons, New York.
Von Schweidler, E. R. (1907). *Ann. Physik Annln. Phys. ser. 4* **24**, 711–770.
Wagner, K. W. (1913). *Annln. Phys. ser. 4* **40**, 817–855.
Wagner, K. W. (1914). *Arch. Elektrotech.* **2**, 371–387.
Ward, F. S. (1961). *Brit. J. appl. Phys.* **12**, 450–455.
Williams, G. (1962). *Trans. Faraday Soc.* **58**, 1041–1044.
Williams, G. (1963). *Polymer* **4**, 27–34.
Windle, J. J. and Shaw, T. M. (1954). *J. chem. Phys.* **22**, 1752–1757.
Windle, J. J. and Shaw, T. M. (1955). *Text. Res. J.* **25**, 856–870.
Windle, J. J. and Shaw, T. M. (1956). *J. chem. Phys.* **25**, 435–439.
Whittingham, T. A. (1970). *J. chem. Phys.* **74**, 1824.
Yager, W. A. (1936). *Physics* **7**, 434–450.
Yamafuji, K. and Ishida, Y. (1962). *Kolloid Z.u.Z. Polymere* **183**, 15–37.
Zelenev, Yu. V. and Glazkov, V. I. (1972). *Vysokomolek. Soedin.* **A14**, 16–22.

7. Solid–Liquid Interactions: Inter- and Intracrystalline Reactions in Cellulose Fibers

STANLEY P. ROWLAND

*Southern Regional Research Center, Agricultural Research Service,
U.S. Department of Agriculture, New Orleans, Louisiana, U.S.A.*

Increasing sophistication of research and instrumentation has made possible studies that delve progressively deeper into the character of surfaces of solid polymers with which liquids react under a variety of circumstances. The ultimate surface is the side of a segment of a polymer chain, which eludes direct characterization. Electron microscopy brings into clear focus microstructural features having dimensions down to a few hundred angstroms and, with staining, it permits dimensional measurements to less than 100 Å with fair accuracy. The former range is about two orders of magnitude higher than the width of extended polymer chains (e.g., 3–8 Å). Based on Dreiding stereomodels, the largest dimension of single molecules of typical dyes, N-methylol reagents, and water are about 30–45, 9, and 2 Å, respectively. Thus, solid surfaces that correspond to the sizes of these molecules are substantially smaller than can be resolved clearly and visually by electron microscopy; information at these dimensions evolves from physical and chemical measurements, which will be discussed in this chapter.

Information in this chapter is organized with concern for the nature of the interaction between liquids (i.e. water, reagents, polymers, swelling agents) and surfaces of cellulose fibers. Answers are sought to questions such as (1) where are the surfaces? (2) how are they reached? (3) does the agent provide unique information on the nature of these surfaces?

Certain points are evident. In highly disordered celluloses in which hydroxyl groups are completely accessible to small molecules, surfaces contacted by the liquid or reagent must be constituted of essentially all of the hydroxyl-bearing sides of the D-glucopyranosyl units in the molecular chains. The chains must be in substantial disarray from normal linear and lateral packing order. At the other extreme, large molecules, such as those polymers commonly applied to fibers from emulsion, will find the exterior of the fiber and possibly some of the large pores as accessible solid surfaces. We are concerned with surfaces between these extremes, primarily as they relate to the cotton fiber of commerce. About 95% of the cellulose of this fiber lies in the secondary wall, made up of lamellae, macrofibrils, microfibrils, and elementary fibrils. A frequency distribution of the thickness of lamellae shows distinct peaks at intervals of about 350 Å, beginning at about 700 Å (Grant and deGruy, 1972). Macrofibrils are about 1000 Å in diameter and are composed of a great many microfibrils, which seem to be about 200 Å in width and a single elementary fibril in thickness (Peterlin and Ingram, 1970). Elementary fibrils are about 50 Å wide. It is these, alone or in a larger composite, that provide the surface for interactions with liquids.

Contact of liquids with surfaces of cellulose fibers brings forth considerations of accessibility, porosity, and interaction, the last including adsorption, substitution, alteration, and degradation. Several reviews are available (Ward and Morak, 1964; Nissan, 1967; Ranby, 1969; Segal, 1971; Lin, 1972). The solid surface at which a particular action occurs depends upon the chemical natures of the liquid and fiber and also upon physical accessibilities of the macro- and micro-structural surfaces on and within the fiber. The polarity, molecular size, and shape of the ingressing liquid have a major bearing on the interface that becomes the pertinent surface for interaction. The objective, therefore, is to consider these factors to the extent that new and useful information is available.

I. Intercrystalline Interactions

A. Accessibility

Interactions of liquids with solid surfaces of cellulose fibers are facilitated by substantial accessibilities of internal surfaces, which increase progressively from cotton to wood pulp to mercerized cotton to regenerated cellulose (Table I) (Jeffries et al., 1969). Numerous methods, some of which are summarized in Table I, have been used to estimate the percentage of disordered material in various celluloses or the extents to which surfaces, D-glucopyranosyl units, and hydroxyl groups are accessible. Child and Jones (1973) found an almost linear relationship between accessibility from prolonged hydrogen–deuterium exchanges, measured by broad-line nuclear magnetic resonance, and disorder, i.e. 100 minus percent crystallinity. Crystallinity is a consequence of ordered lateral packing of cellulose chains; high lateral order results in low accessibility. However, chemical reactions are the basis for estimating disorder for cotton ranging from as low as 8% to as high as 43%, indicating that different reagents reach and measure different internal surfaces.

TABLE I

Percentage of disordered material in various celluloses
(approximate average values from published literature[a])

Technique	Cotton	Wood pulp	Mercerized cotton	Regenerated cellulose
X-ray diffraction	27	40	49	65
Density	36	50	64	65
Deuteration	42	55	59	72
Moisture regain (sorption ratio)	42	49	62	77
Hailwood Horrobin	33	45	50	65
Non-freezing water	16	—	23	48
Acid hydrolysis	10	14	20	28
Alcoholysis	10	15	25	—
Periodate oxidation	8	8	10	20
Nitrogen dioxide oxidation	23–43	—	—	40–57
Formylation	21	31	35	63
Iodine sorption	13	27	32	52

[a] Values summarized by Jeffries *et al.* (1969).

A single method for measuring accessibility or disorder may provide more than one answer, depending upon the particular sample of cellulose and the manner in which the data are handled. The periodate oxidation method, which measures accessible vicinal hydroxyl groups O(2)H and O(3)H, provides an example. Kinetic analysis of oxidation data into rapid and slow components of reaction results in an estimate that 10·3% (Jeffries *et al.*, 1968) or 11·7% (Cousins *et al.*, 1964) is readily accessible, whereas analysis of the effect of the oxidation upon decrease in crystallinity results in estimates that 40% of the D-glucopyranosyl units are not in crystalline regions (Rowland and Cousins, 1966). Thus, with a single reagent, a small fraction of cellulose surfaces is readily accessible and a much larger fraction is non-crystalline.

Recently it was shown that accessibility based on $D_2^{18}O$ and mass spectrometry (Guthrie and Heinzelman, 1974) yields higher values than some methods based on D_2O and infrared spectroscopy (Rousselle and Nelson, 1971; Mann, 1971). The former method measures total exchangeable hydroxyl hydrogens, but the method of Dechant (1967), utilized by Rousselle and Nelson (1971), measures the more readily accessible hydroxyl hydrogens that re-exchange easily with water. Total exchangeable hydroxyl hydrogens are significantly lower for purified cotton (48·8%) than for raw fiber (52·2%), a fact attributed to the presence and high deuterium exchange of sugars, pectins, etc. in the raw fiber and to crystallization of cellulose during purification (Guthrie and Heinzelman, 1974). A high correlation exists between accessibilities measured by moisture regain and by total hydroxyl hydrogens exchangeable to $D_2^{18}O$, but this correlation does not exist for formaldehyde-treated cottons for which moisture regain shows much higher accessibilities than does deuterium exchange. Evidently moisture sorption is not specific to hydroxyl groups on cellulosic surfaces as is often assumed (Guthrie and Heinzelman, 1974).

Among conventional accessibility measurements, there are few instances that provide a reasonable basis for an assessment of the effect of molecular weight of reagent or solute over an order of magnitude range. Data listed in Table II were obtained from reactions or adsorptions on water-swollen cellulose from dilute solutions of agents. On the basis that a maximum of $9 \cdot 1$ m^2/g was measured for adsorbed dye on mercerized cotton, Johnson et al. (1969) estimated that this fixation occurred on lamellar surfaces only, and that these surfaces were limited to a total area of 17 m^2/g. The surface area estimated for elementary fibrils, proposed to be accessible to water vapor, was 700 m^2/g, and that of microfibrils, proposed to be accessible to water and other small molecules, was 260 m^2/g. The estimate was based on elementary fibrils 35 × 35 Å, microfibrils 175 × 75 Å, and lamellar thickness 1000 Å. The ratio of measured accessibilities of total $D_2^{18}O$, chemical reagent, and dye for both purified cotton and mercerized cotton is similar in magnitude to the ratio of surfaces estimated above for elementary fibrils, microfibrils, and lamellae.

TABLE II

Accessibility and molecular weight

Measurement	Purified Cotton	Mercerized Cotton
Total $D_2^{18}O$ accessibility[a]	48·8%	63·9%
Readily accessible, D_2O[b]	36·0–37·2%	48·1%
Readily accessible, N,N-diethylaziridinium chloride[c] (mol. wt. 136)	19–21%	32–33%
Readily accessible, Diphenyl Fast Red 5BL[d] (mol. wt. 676)	5·4 m^2/g (2·5%)[e]	9·9 m^2/g (4·5%)[e]

[a] Guthrie and Heinzelman (1974). [b] Rousselle and Nelson (1971). [c] Bose et al. (1971). [d] Johnson et al. (1969). [e] Percentages calculated from an estimated 220 m^2/g of surface for completely accessible cellulose chains.

B. ROLE OF WATER

Both moisture transport and physical data on never-dried fibers indicate independently of each other that the cotton fiber in this unique state has unusually high mobility (Ingram et al., 1974). Never-dried fibers contain almost 100% more water at a given level of relative humidity than dried and rewetted fibers. Extremely long fracture tips are observed to break upon deformation; this is attributed to the plasticizing effect of water between fibrillar surfaces. The unique mobility of the never-dried fiber is irretrievably lost on drying or stretching, both of which cause a substantial fraction of elementary fibrils, previously separated from one another by water, to be irreversibly hydrogen-bonded together. In spite of the high moisture content of never-dried cotton, Morosoff (1974) concluded that it was at least as crystalline, and perhaps slightly more crystalline, than rewetted cotton. High accessibility of surface of elementary fibrils of never-dried cellulose is also evident from sorption studies (Patel *et al.*, 1975) and disaggregation of chemically modified, never-dried wood pulp cellulose into long, rod-like particles with lateral dimensions of 35–45 Å (Lepoutre and Robertson, 1974). These particles were observed in transmission electron micrographs of films from colloidal dispersions obtained from graft polymerizing acrylonitrile onto this cellulose and subsequently hydrolysing the product to poly(sodium acrylate-acrylamide)-cellulose copolymer.

When never-dried cotton was dried, crystallite size and/or perfection decreased in response to stresses induced on collapse of the cylindrical fibers into a convoluted shape (Morosoff, 1974). Reorganization takes place on rewetting, leading to larger and/or more perfect crystals in the first and second rewetted states than in the never-dried state; this effect is greater for the second-dried than for the first-dried state. Increases in crystallinity for scoured and purified cottons going from the dry to the humid state were observed by Creely and Tripp (1971). Differences in crystallinity due to changes in relative humidity from 0 to 100% are about 10% as measured by the crystallinity index. It was suggested that drying decreased X-ray crystallinity because of stresses induced in the cellulose structure; conversely, moisture in the cellulose structure plasticizes and allows a relaxation to a more ordered state. Cellulose decrystallized by dry ball milling regains crystallinity on exposure to moisture or water (Caulfield and Steffes, 1969). The recovery of crystallinity in partially decrystallized cellulose is greater on exposure to atmospheres of higher moisture content. The higher moisture content presumably plasticizes the longer segments of cellulose chains, facilitating ordered alignment. Although partially decrystallized cellulose regains the cellulose I lattice even at low levels of humidity, completely decrystallized cellulose recovers the cellulose II lattice, but only upon exposure to 100% humidity. A phenomenon that appears to be closely related to those noted above is the increase in lateral order that occurs upon hydrolysis of fibrous cotton to hydrocellulose. Supplementary to the well known fact that hydrocelluloses are higher in crystallinity than fibrous cellulose, it has now been established that hydrocelluloses I and II reach a maximum in lateral order at a particular stage of hydrolysis (Rowland *et al.*, 1971b; Rowland and

Roberts, 1974). These research workers proposed that increases in lateral order occurred as molecular chains were cleaved, relieving stresses in elementary fibrils. A major effect may be from the plasticizing contribution of water during hydrolysis.

Calorimetric data from wetting of cotton are in good agreement with the view (Morrison and Dzieciuch, 1959) that the interaction of water with cellulose is a surface phenomenon (Iyer and Baddi, 1969). Weatherwax (1974), utilizing the BET equation (Brunauer et al., 1938), the "t" procedure (Hagamassy et al., 1969), and the cluster theory (Zimm and Lundberg, 1956), which constitute three conceptually different approaches for measuring surface area, demonstrated the existence of a large solid–water interfacial area of transient pores in swollen cellulosic materials. The surface area of these transient pores is consistent over most of the relative humidity range. It is interpreted as evidence of a parallel plate or concentric laminar structure of pores at the submicroscopic level.

The idea that water should be regarded as an equilibrium mixture of strongly hydrogen-bonded clusters and monomeric molecules (Frank and Wen, 1957) in the vicinity of cellulose surfaces has been considered in detail by Goring (1966) and Dobbins (1970). This concept was called upon to explain the thermal behavior of cellulose in the presence of water. Expansion of cellulose is much greater in water than in other confining liquids, such as toluene or mercury. The change with temperature in apparent specific volume of cellulose in water was interpreted by Neal and Goring (1969) in terms of accessibility of water to cellulose surfaces. Ramiah and Goring (1965) proposed that these hydroxyl-bearing surfaces act as structure breakers for water clusters, enhancing concentration of unbonded water, which has a higher coefficient of thermal expansion than clusters have. Thus an increase in thermal expansion of cellulose (and polysaccharides) occurs in the wet state. The amounts of destructured water estimated for cotton linters, cellophane, and glucose were 0·03, 0·07, and 0·25 g/g, respectively, at 20°C (Neal and Goring, 1969).

Results of application of nuclear magnetic resonance (NMR) spectroscopy to cellulose have furthered the concepts of bound water (Pittman and Tripp, 1971; Carles and Scallan, 1973). Instrumentation and approaches have been varied, and different interpretations of data concerning water in cellulose have been explored (Barker and Pittman, 1971).

A broad-line NMR spectrum of moist cellulose exhibits a narrow component, corresponding to protons in rapid motion (presumably free water), and a broad component, associated with protons of shorter spin–spin relaxation time (presumably protons of cellulose and bound water). Child (1972), using pulsed NMR, found that the molecular motion of sorbed water molecules at conventional moisture regain levels depended on the physical state of the cellulose, particularly the degree of crystallinity.

More sites of high binding energy were indicated for hydrocellulose (7·3% regain) than for cotton linters (8·9% regain), and the absence of a spin–spin transition characteristic of sorbed water at 0°C was observed only for the hydrocellulose.

C. PORES

1. Size Distributions

Pore size distributions have been calculated for cellulosic fibers from nitrogen adsorption isotherms by the method of Cranston and Inkley (1957), using the Kelvin equation (Gregg and Sing, 1967). Whether the fiber is natural, purified, mercerized, or chemically modified cotton, the pattern of distribution is, on the whole, similar to those shown in Fig. 1, and the peak in the distribution occurs for pores of 25 to 40 Å (Porter and Rollins, 1972a,b). Harris and Whitaker (1963) suggested that the Kelvin equation leads to erroneous results when applied to pores of radius below 20 Å. It is evident from curves such as those in Fig. 1 that as one proceeds from larger towards smaller pores, the $\Delta V/\Delta D$ increases at least to and possibly beyond the range of diameter at which the Kelvin equation becomes suspect.

Pore surface area for the cotton fiber, calculated by the BET method (Brunauer *et al.*, 1938) from adsorption isotherms, is around 1 m^2/g as

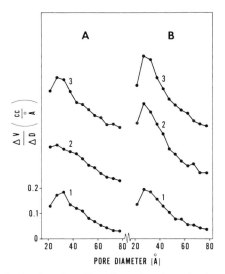

FIG. 1. Pore size distribution in cotton fibers from the adsorption branch of the isotherm: (A) dewaxed fibers; (B) dewaxed and scoured fibers; (1) Rowden variety; (2) Deltapine variety; (3) Pima variety. Ordinate values are shifted 0·2 units for successive distributions. Reproduced with the permission of John Wiley & Sons, Inc. (Porter and Rollins, 1972a).

measured with nitrogen; the values are 20·1, 7·3, 18·3, and 137 m²/g from sorption of vapors of methanol, ethanol, acetic acid, and water, respectively (Klenkova and Ivashkin, 1963), illustrating the effect of molecular size but also differing polarities and abilities of molecules to open pores by disrupting interfibrillar hydrogen bonds. Surface areas of cotton fibers that are swollen in water and solvent exchanged to nitrogen gas are about 20 m²/g (Porter and Rollins, 1972a). Collapse of some of the pores during solvent exchange has long been suspected; in the case of fibers from sprucewood, the pore volume measured by nitrogen adsorption after solvent exchange was only 20% of the volume of water removed (Stone and Scallan, 1966). This is the approximate relationship between surfaces measured with nitrogen on water-swollen, solvent-exchanged fibers (20 m²/g) and with water as the adsorbate gas (137 m²/g).

An investigation of Aggebrandt and Samuelson (1964) of penetration of water-soluble polymers into cellulose was the forerunner of more advanced methods. The solute exclusion technique of Stone and Scallan (1968) measures pore sizes in the water swollen state by means of solute molecules that serve as "feeler gauges". If all the water originally associated with a porous body is accessible to a (small) molecule, it will contribute to dilution of the solution by virtue of an increased volume available to the solute (case A, Fig. 2). As solutions of progressively larger molecules are used (cases B and C), smaller pores, and finally all of the pores, become inaccessible to the solute molecules. The water in the inaccessible pores is then unavailable for dilution or for occupancy by the solute. Differences in concentration are the basis for the solute exclusion technique (Stone and Scallen, 1967), and differences in elution volumes of solutes passing

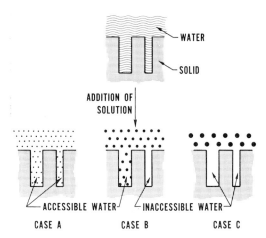

FIG. 2. Illustration of the principle of solute exclusion with molecules of three sizes and pores of two sizes. Reproduced with the permission of *Tappi* (Stone *et al.*, 1969a).

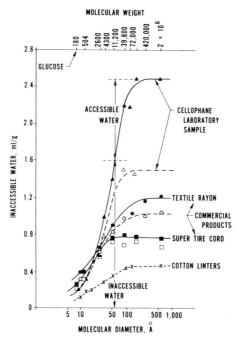

FIG. 3. The volume of water that is inaccessible to molecules of increasing size. Solid line = never dried, broken line = dried and reswollen cellulose. The curve for cotton linters (Stone *et al.*, 1969b) was added to the original figure (Stone *et al.*, 1969a). Reproduced with the permission of *Tappi*.

through a bed of porous material in a column are the basis for the gel permeation technique of Martin and Rowland (1967). Results from the solute exclusion technique are illustrated in plots of inaccessible water as a function of molecular weight or molecular diameter of the solute molecule (Fig. 3) (Stone *et al.*, 1969a). Useful information from such curves includes (a) total water within the cellulose (i.e. water inaccessible to large solutes), (b) maximum pore size (in terms of molecular weight or molecular diameter of solute), and (c) median pore size. Data of Fig. 3 show that accessible water decreases as molecular size of solute increases, and total accessible water decreases as each never-dried cellulosic substrate (solid line) is dried and reswollen (broken line). The curve for cotton linters (Stone *et al.*, 1969b) shows that this fiber has a wider distribution of pore sizes than some regenerated fibers have in spite of lower total internal volume.

On the basis of results such as those illustrated in Fig. 3 and calculations relating pore diameters to internal surface areas, Stone and Scallan (1968) proposed a structural model for the water-swollen wood cellulose fiber. This model consists of a predominance of larger pores between concentric lamellae, with decreasing distance between lamellae on progression from the external surface to the lumen surface of the fiber, and a predominance of smaller pores in lamellae between fibrils. From data on undried, bleached, pine kraft fibers, total surface area was estimated at $1030 \text{ m}^2/\text{g}$, a value about one order of magnitude higher than that estimated from water vapor sorption but generally similar to that estimated by Johnson *et al.* (1969) for the total potentially accessible surface in the cotton fiber. In the estimate of Stone and Scallan (1968), pores inaccessible to maltose accounted for about 11% of the total volume, but accounted for about 39% of the total surface area accessible to water. The significance of rapidly increasing surface area per unit of pore volume that results from decreasing pore diameter is evident in the results of Nelson and Oliver (1971). They related acetylation reactivity to (a) total pore volume of cellulose, (b) total surface area of pores larger than 20–25 Å, the lower limit of accessibility for the acetylation reagent, and (c) the volume of pores between 25 and 75 Å diameter, which provides the predominance of internal surfaces accessible to the reagent.

2. Shapes/Molecular Accommodations

Pores visible in the swollen cotton fiber by electron microscopy are perceptible as parallel plate type openings between lamellae (Rollins *et al.*, 1966). Smaller pores, representing spaces between microfibrils and between elementary fibrils, are probably characterized by more equilateral dimensions. Information concerning pore shapes might be obtained by use of molecular "feeler gauges" selected from homologous series of polymers characterized by different shapes. Unfortunately, precise information concerning shapes and volumes occupied by polymers in solution is insufficient for the purpose.

Rowland and Bertoniere (1976) examined the pore structure of cotton with two primary molecular probes, polysaccharides and polyethylene glycols (Fig. 4), each of which is free of sorption to cellulosic surfaces (Stone and Scallan, 1968; Nelson and Oliver, 1971). These solutes exhibit strong donor and acceptor hydrogen bonding (saccharides) and primarily acceptor bonding (polyethylene glycols). Figure 4 shows the different relationships between A_w, the fraction of total internal water accessible to the solute, and molecular weight of solute. Comparisons based on molecular size do not eliminate the difference. The implications of these and additional information are that unavailability of internal water to function as a solvent for saccharides is based solely on the inadequacy of small pores to accommodate the solute (i.e., saccharides show no evidence for nonsolvent or bound water in an accessible pore) and that bound water is still available as solvent water to a saccharide, since it has the same interaction with

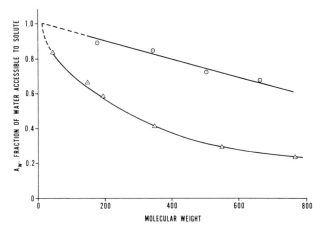

FIG. 4. The fraction of total water in pores that is accessible to solutes: ○ = mono-, di-, tri-, and tetrasaccharides; △ = mono-, tri-, and tetraethylene glycols and commercial "Carbowaxes".

water as a cellulosic surface. Since polyethylene glycols have lower interactions with water, these solutes confront an increasing shell of nonsolvent or bound water at the water-cellulose interface in pores as the molecular weight of these solutes increases, especially for the first members of the series.

D. FIBRILLAR SURFACES

Hebert *et al.* (1972) measured lateral distances between accessible surfaces of elementary fibrils of cotton and determined the average value to be 33 Å. The estimate was made by electron microscopic measurements on fiber bundles sectioned parallel to the fiber axis after reaction was carried out to introduce acrylate units into the cotton and OsO₄ was allowed to react with the acrylate double bonds. Earlier estimates of the widths of elementary fibrils ranged from 30 Å to 150 Å (Warwicker *et al.*, 1966), the more generally accepted values falling at about 30–60 Å.

Information on total accessible surfaces of elementary fibrils was obtained by Haworth *et al.* (1969) in studies of successive methylations of cotton cellulose with dimethylsulfate-dimethylsulfoxide-2N sodium hydroxide. The degree of substitution (DS) increased rapidly during the first five methylations, but quite slowly after the tenth methylation. Infrared deuteration and moisture regain accessibilities of samples during the first five treatments showed no change in the amount of hydrogen-bond order or accessible material, but significant changes occurred during the later, slower phase of reaction. The methylation was considered to proceed initially at accessible hydroxyl groups, and subsequently by slow penetration of ordered material with consequent disruption of the elementary fibrillar structure. The conversion of D-glucopyranosyl units in the cotton cellulose to substituted D-glucopyranosyl units throughout successive methylations conformed to the general pattern of increase in DS, i.e. a rapid increase in mole fraction of individual mono- and disubstituted D-glucopyranosyl units during early methylation followed by a leveling off, as illustrated in Fig. 5. The simultaneous pattern of change in mole fraction

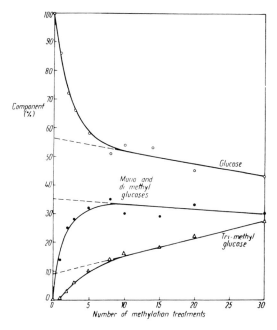

FIG. 5. Extents of mono-, di-, and trisubstitutions of D-glucopyranose units of cotton measured by gas-liquid chromatographic measurements of substituted and unsubstituted D-glucopyranoses (Haworth *et al.*, 1969). Reproduced with the permission of *Carbohydrate Research*.

of unsubstituted D-glucopyranosyl units was a rapid decrease and a leveling off. By extrapolations back to zero time, it was estimated that 56% of the D-glucopyranosyl units were beneath, or that 44% were on the outer surfaces of the elementary fibrils. Considering this evidence and the proportions and rates of formation of the various substituted D-glucopyranosyl units, Haworth *et al.* (1969) deduced that a cross section of the elementary fibril was best represented by a rectangle composed of eight molecular chains on each of two sides and ten molecular chains on each of the other two sides. The size of a cross section of this elementary fibril would be about 40×50 Å.

Evidence of aggregation of elementary fibrils was found by Haworth *et al.* (1969) when methylations were conducted in media less capable of penetrating between elementary fibrillar units. Methylation of cotton cellulose with diazomethane in dry ether was extremely slow, due to almost complete aggregation of elementary fibrils in the dry state and inability of dry ether to cause them to separate. In water-saturated ether, methylation proceeded more rapidly. Data from repeated methylations showed that the reaction was confined to 13% of the total hydroxyl groups situated on 27% of the *D*-glucopyranosyl units, compared to 26% and 44%, respectively, for repeated methylation with dimethyl sulfate, as described above. Restriction of this methylation to the surfaces of 2×2 aggregates of the model elementary fibrils described above approximates the actual results.

Until recently (Wade *et al.*, 1968), there was no experimental basis or justification for questioning the general unstated assumption that O(2)H, O(3)H, and O(6)H are equally available for reaction on accessible surfaces in the cotton fiber. In a series of studies, Rowland *et al.* (1969a, 1971a, b, 1974; Rowland and Roberts, 1974) used *N,N*-diethylaziridinium chloride (DAC) to probe reactions with various forms of cotton cellulose. They found a selective availability of hydroxyl groups at C(2), C(3), and C(6) of the *D*-glucopyranosyl units of cotton cellulose. This was measured by comparison of the distributions of 2-(diethylamino)ethyl substituents introduced into fibrous cellulose and into disordered celluloses. The term disordered cellulose refers to cellulose that is decrystallized and highly accessible; all three types of hydroxyl groups in disordered cellulose may be expected to be free of intra- and inter-molecular hydrogen bonds and to be equally available for reaction. Distribution of substituents developed under specific conditions from reactions of disordered celluloses affords a measure of relative rate constants for the three types of hydroxyl groups; therefore, a different distribution of substituents, resulting from reaction of a crystalline fibrous cellulose, provides a basis for estimating relative availabilities of hydroxyl groups in that substrate. Relative availabilities of O(2)H, O(3)H, and O(6)H were found to be $1\cdot00:0\cdot32:0\cdot77$ for fibrous cellulose (lattice I) (Rowland *et al.*, 1971b) and $1\cdot00:0\cdot84:0\cdot85$ for mercerized fibrous cellulose (lattice II) (Rowland and Roberts, 1974). It was proposed that these are the relative availabilities of the three types of hydroxyl groups on the surfaces of elementary fibrils, and that they reflect the conformation and order of molecular segments of cellulose chains on the surfaces of elementary fibrillar units.

Hydrocelluloses from native and from mercerized fibers differed from one another in perfection of lateral order and in availability of hydroxyl groups (Rowland *et al.*, 1971a, 1974). The term Exemplar Hydrocellulose (EHC) was used to designate hydrocellulose of higher perfection than those resulting from longer or shorter periods of hydrolysis. EHC-I, from

40 min hydrolysis of native cotton with 2·5 N hydrochloric acid at reflux, exhibited availabilities of O(2)H, O(3)H, and O(6)H in the ratio 1·00:0·29:0·54, reflecting a substantially higher state of order than those of the native cellulose or other hydrocelluloses-I.

Surfaces of the idealized elementary fibril that are proposed to account for the relative availabilities of hydroxyl groups of EHC-I are shown in a cross-sectional view in Fig. 6. Hydrogen bonds tie up O(6)H's on the left and upper surfaces, but O(6)H's on right and lower surfaces are free of hydrogen bonds and protrude from the surfaces. When this consideration of hydrogen bonding is carried to the D-glucopyranosyl units immediately behind those in the projection of Fig. 6, the D-glucopyranosyl units are rotated through 180°; the situation then is that O(6)H's on the left and upper surfaces protrude from the surface and O(6)H's on the right and lower surfaces are hydrogen-bonded back into the body of the elementary fibril. Hydrogen bonds that tie up O(6)H are believed to involve the bridge oxygen of an adjacent molecular chain, O(6)H . . . O(1″), as proposed by Frey-Wyssling (1955). Low availability of O(3)H's for reaction is attributed to

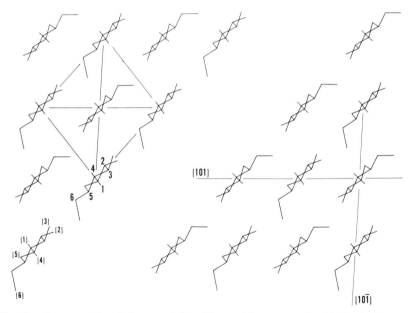

FIG. 6.   A cross section of the upper left and lower right corners of an idealized elementary fibril of native cotton constructed with O(6)H . . . O(1″) hydrogen bonds as proposed by Frey-Wyssling (1955). The unit cell is shown in the upper left portion. The direction of the molecular chains is perpendicular to the plane of the projection. The projections of D-glucopyranosyl units on the left and right sides lie in 10$\bar{1}$ planes and those on top and bottom surfaces lie in 101 planes. Numbers in open and in brackets refer to carbon and oxygen atoms, respectively (Rowland and Roberts, 1972a).

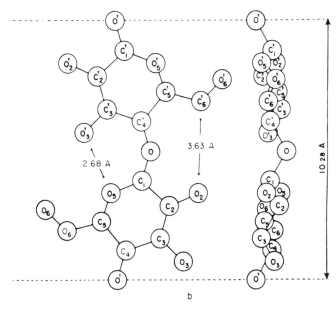

FIG. 7. The bent conformation of the cellobiose segment of the crystalline cellulose chain. Numbers identify specific carbon and oxygen atoms (Marchessault and Sarko, 1967).

hydrogen bonds, O(3)H . . . O(5'), in the Hermans conformation (Hermans, 1943) of sequential D-glucopyranosyl units in a two-fold screw axis (Fig. 7). The crux of the picture proposed above to describe perfectly ordered surfaces of the elementary fibril of native cellulose is: (a) all O(2)H's on accessible surfaces are free of hydrogen bonds and are readily available for reaction, (b) all O(3)H's are tied up in hydrogen bonds and are unavailable for reaction, and (c) half of the O(6)H's are tied up in hydrogen bonds and are unavailable for reaction. The difference between the experimentally measured relative availability of hydroxyl groups for EHC-I $(1·00:0·29:0·54)$ and the relative availability predicted from the unit cell structure of native cellulose $(1·00:0·00:0·50)$ is attributed to imperfections in accessible surfaces of EHC-I resulting from disruptions of hydrogen bonding arrangements.

Relative availabilities of O(2)H, O(3)H, and O(6)H on surfaces of EHC-II from completely mercerized cotton fibers were found experimentally to be $1·00:0·26:0·72$ (Rowland et al., 1974). These results were related to theory through the idealized structure shown in Fig. 8. The projections of D-glucopyranosyl units shown in this figure were constructed by model building and diffraction studies. On upper and lower surfaces, all O(2)H's and O(6)H's are free of hydrogen bonds, protrude from the surface, and are available for reaction. On left and right surfaces, all O(2)H's are

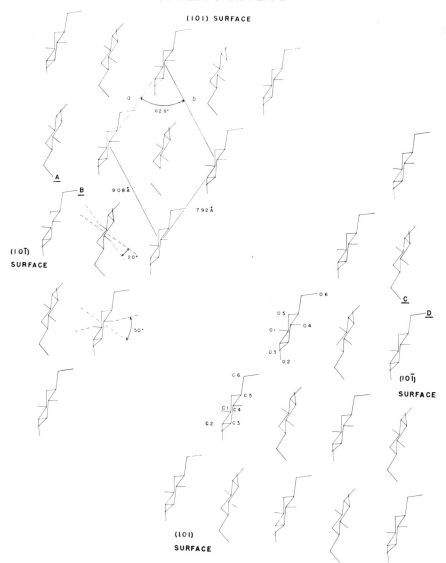

FIG. 8. Cross-sectional view of the upper left and lower right corners of an idealized elementary fibril of cellulose II. The unit cell is shown in the upper left portion. Surfaces and chain directions correspond to those in Fig. 6 (Rowland *et al.*, 1974).

available for reaction, but only one out of every two O(6)H's is free of hydrogen bonds and available for reaction. Again, all O(3)H's are expected to be tied up in hydrogen bonds to O(5′). The experimentally determined relative availabilities of O(2)H, O(3)H and O(6)H ($1 \cdot 00 : 0 \cdot 26 : 0 \cdot 72$) for EHC-II agreed acceptably with the theoretical values, $1 \cdot 00 : 0 \cdot 00 : 0 \cdot 75$, estimated for the surfaces of a perfect crystal of cellulose II.

On the basis that the idealized structures shown in Figs 6 and 8 represent a maximum of hydrogen-bonding order on the surfaces of elementary fibrils, data summarized in Table III show that hydrogen bond disorder is substantially higher for the complete fiber, whether of lattice I or II structure, than for the corresponding EHC. The general aspects of these hydrogen bonds in cellulose have been discussed by Nissan (1967).

TABLE III

Relative availabilities of O(2)H, O(3)H, and O(6)H[a]

Cellulose	Lattice I	Lattice II
Idealized, perfectly ordered surface	1·00:0·00:0·50	1·00:0·00:0·75
EHC, experimental	1·00:0·29:0·54	1·00:0·26:0·72
Total fiber, experimental	1·00:0·32:0·80	1·00:0·84:0·85

[a] Rowland and Roberts, 1974; Rowland *et al.*, 1974.

E. HYDROLYTICALLY SENSITIVE SURFACES

Contact of aqueous mineral acid with accessible surfaces of cotton cellulose causes hydrolysis, initially rapid losses in weight and degree of polymerization (DP), formation of hydrocellulose particles, and, subsequently, slower loss of weight with essentially constant DP (Sharples, 1971). Crystallization during hydrolysis was established by Hermans and Weidinger (1949) and confirmed by others (Sharples, 1971). It was suggested that the initial rapid rate of degradation is a physical effect, resulting from strains set up in microstructural units and molecular chains, or from local variations in normal hydrogen bonding (Michie *et al.*, 1961). Frölander *et al.* (1969) found that the rate of acid hydrolysis depends on the number or area of misaligned zones in the cell wall and that this determines the drop in intrinsic viscosity. Results of mild hydrolytic decomposition of differently treated cotton yarns were interpreted by de Boer and Borsten (1969) by relating reactivity to internal stress; they found good agreement between rate or extent of hydrolysis and structural changes introduced into the cotton. Most recently, Shinouda (1974) found that hydrolysis of cotton shifted the OH stretching vibration band to lower frequency, indicating that stronger hydrogen bonds were formed. Interpreting decreases in availability of O(6)H relative to O(2)H as evidence for increasing intermolecular hydrogen bonding of O(6)H, Rowland *et al.* (1971b) concluded that hydrogen bonding order increased through 40 min. hydrolysis of native cotton with 2·5 N hydrochloric acid at reflux, and decreased thereafter. Intramolecular hydrogen bonding of O(3)H remained nearly constant throughout hydrolysis. Completely mercerized cotton behaved similarly, but the hydrogen bonding order of O(3)H increased substantially during the early stage of hydrolysis and remained constant thereafter (Rowland and Roberts, 1974). For both native and completely mercerized cottons, the maxima in O(6)H hydrogen bonding coincided with the minima in moisture regain and in peak width in X-ray diffractograms; these are interpreted as minima in internal pores and maxima in lateral order, respectively.

Consistent with increased accessibility, swollen and decrystallized cottons are more susceptible to acid hydrolysis, having weight losses and reductions in DP larger than native cotton (Zeronian, 1971). Although these features also characterize mercerized cotton, strength retained in mercerized yarns at comparable degradation or hydrolysis is higher than in native cotton yarns. This dichotomy is accentuated when mercerization is conducted under tension or with restretching (Handu et al., 1974). The behavior is rationalized in terms of number and distribution of chain ends in the elementary fibrillar units (Rowland et al., 1976). Alkaline degradation of cellulose is sensitive, complicated, and difficult to interpret in terms of accessible surfaces (Richards, 1971). However, recent studies of hypochlorite oxidation of cotton and this oxidation followed by hot alkaline degradation with sodium bicarbonate (Lewin and Roldan, 1975) disclosed many changes similar to those caused by acid hydrolysis; the same surfaces appear to be involved in both cases. For example, with increased extent of oxidation, packing (i.e. hydrogen bonding) of microstructural units improved and crystallite width increased. Crystallite size increased after oxidation and alkaline extraction, although microfibrils showed a reduction in lateral dimensions and an increase in perfection. The peeling-off reaction, which evidently initiates at carbonyl groups on C(1), C(2), and C(6) of D-glucopyranosyl units on the most readily accessible surfaces, continues along accessible (disordered) surfaces to a point where ordered packing and hydrogen bonding slows down the degradation.

Although it is not evident in the results of all investigators, there are strong similarities between enzymatic and acid hydrolysis of cellulose. The pattern of initially rapid weight loss, followed by slower weight loss, was superimposable for cotton cellulose hydrolysed with cellulase at 50°C and with 2·5 N hydrochloric acid at reflux, when the time scale of the former was in days and the latter in hours (Rowland et al., 1973). Amemura and Terui (1965) observed a similar pattern for enzymatic hydrolysis of refined wood pulp. As hydrolysis proceeded, they observed a minimum in moisture regain and a maximum in crystallinity, conforming closely to the pattern of changes observed by Rowland et al. (1971b) for acid hydrolysis of cotton. The small hydronium ion and the large enzyme molecules (C_1 and C_x) have ultimately the same effects upon cotton in spite of differences in size: about 5 Å maximum for the former and between 13×79 Å to 42×252 Å for the latter (Whitaker et al., 1954).

F. TYPES OF ACCESSIBLE SURFACES

Dimensions of N-methylol reagents and dyes that are commonly applied to cotton are upwards of about $9 \times 6 \times 4$ Å and $30 \times 10 \times 2$ Å, respectively. Sites inside the cotton fiber that serve as surfaces for such solutes may range from lamellae to microfibrils to elementary fibrils, but in any case, the common surface is that of the elementary fibril, individually or in the foregoing composites. To illustrate the nature of surfaces on elementary fibrils, Rowland (1972) proposed a model that is shown in Fig. 9. This

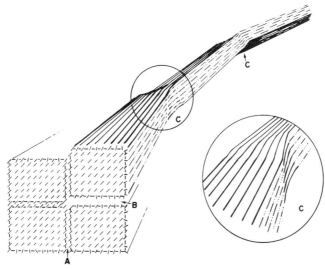

FIG. 9. A schematic representation of the elementary fibril of cotton to illustrate (a) coalesced surfaces of high order, (b) readily accessible disordered surfaces, and (c) readily accessible surfaces on strain-distorted, tilt and twist regions (Rowland, 1972).

model, based on evidence such as that discussed in preceding sections, depicts the following points:

(a) The elementary fibril, the basic microstructural unit from which the fiber is constructed, provides surfaces as walls of pores within the fiber.

(b) The elementary fibril has a width near 50 Å and is highly ordered; portions between C segments become hydrocellulose particles and reach almost complete crystallinity in certain cases.

(c) The elementary fibril is characterized by three types of surfaces, (A) an inaccessible, or less accessible perfectly ordered or highly ordered surface on the exterior of a highly-ordered segment of the elementary fibril, (B) a readily accessible, somewhat disordered surface on the exterior of the same type of segment of elementary fibril, and (C)

a readily accessible, substantially disordered surface on the exterior of a segment of elementary fibril, characterized by internal strain resulting from tilt-twist distortion.

In all of these surfaces, molecular chains of cellulose are envisioned as maintaining a high degree of orientation, and disorder is proposed to be a disruption of intra- and inter-molecular hydrogen bonds by virtue of distortion of the plane of D-glucopyranosyl units to an attitude different from that in the unit cell (refer to Fig. 6). The distortion or separation of adjacent molecular chains, even on C surfaces, is proposed to be insufficient to permit a reagent molecule to penetrate beneath the outer layer of molecular chains.

Confirmation of the existence of two types of accessible surfaces was obtained by Roberts *et al.* (1972). Readily available hydroxyl groups on accessible surfaces of the cotton fiber were tagged by reaction with N,N-diethylaziridinium chloride to introduce 2-(diethylamino)ethyl substituents under mild conditions in an aqueous medium. The partially substituted cotton was hydrolysed progressively into a series of hydrocelluloses and corresponding soluble fractions. This study and a related one (Rowland and Roberts, 1972b) included detailed analysis of (a) the hydrolysis into rapid and slow stages, (b) distribution of 2-(diethylamino)ethyl substituents between the soluble and hydrocellulose fractions, and (c) distribution of substituents at O(2), O(3), and O(6) in each of the products. Relative availabilities of hydroxyl groups in the original cotton fiber were estimated on the basis of the analyses of each of the products. Total readily available hydroxyl groups were distributed equally between B and C surfaces. However, the proportion of B and C accessible surfaces was $0.64 : 0.36$. A unit of accessible surface was defined as one side of a cellobiose unit, i.e. one side of a D-glucopyranosyl unit bearing the O(6)H plus one side of a similar unit rotated through $180°$ and bearing O(2)H and O(3)H. The difference in proportion of accessible surfaces and available hydroxyl groups on B and C is the direct consequence of substantially different levels of ordered hydrogen bonding that restrains O(3)H and O(6)H from being available for reaction. The extent of hydrogen bond disorder on B and C surfaces is best stated in terms of fractions of O(3)H and of O(6)H hydrogen bonds that are disrupted, as compared to the perfectly ordered hydrogen bonding surface, i.e. A of Fig. 9. The perfectly ordered accessible surface has none of the O(2)H, all of the O(3)H, and half of the O(6)H tied up in hydrogen bonds. By comparison with the perfectly ordered surface, B surfaces in the cotton fiber were characterized by 13% disruption of O(3)H hydrogen bonds and 14% disruption of O(6)H hydrogen bonds. C surfaces were characterized by 91% disruption of the former and 58% disruption of the latter hydrogen bonds.

G. SORPTION

Martin *et al.* (1972) studied sorption of solutes from aqueous solution on Sephadex G-15. Results are pertinent to cellulose because elution of a variety of solutes from columns of chopped native cellulose and columns of Sephadex G-15 occurs with parallel displacements (R_g) relative to glucose as a reference material (Rowland and Bertoniere, 1976). Since elution volume is an inverse function of molecular weight of the solute, indications of positive or negative sorption are most clearly evident when solutes having the same molecular volume are compared. For small molecules, molecular weight serves as an acceptable basis for comparison in place of the unknown molecular volumes. Data of Martin *et al.* (1972) and Rowland and Bertoniere (1976), summarized in Fig. 10 show that a carboxyl group in a saccharide causes negative sorption, i.e. such solutes were eluted earlier than glucose, which is the basis for comparison. Cyclic ureas bearing protons on the nitrogen atoms were positively sorbed, but replacement of one or both of the protons with methyl groups reduced sorption. Ethylenediamine, relative to ethylene glycol, showed pronounced sorption. Diethylenetriamine and polyethylenimine were sorbed to the extent that they were not eluted from the column. Evidently, negative sorption of glucuronic acids is caused by repulsion between a negative charge on the carboxyl group of the solute and a similar charge on the substrate due to carboxyl groups and/or acidic hydroxyl groups. Positive sorption of amides is attributable to hydrogen bonding, which is stronger when the nitrogen atom of the amide can donate a proton. Amines are probably sorbed by ionic interactions, as well as by hydrogen bonding.

Baddi *et al.* (1971) studied equilibrium adsorption isotherms of dimethylolethyleneurea on purified native cotton and concluded from affinity enthalpy values that the phenomenon was merely physical adsorption. The sorbed reagent was completely removed by simple washing with distilled water. Dimethyloldihydroxyethyleneurea, another common finishing agent for cotton, exhibited a similar degree of adsorption (see point No. 10 in Fig. 10), consistent with its higher molecular weight (Martin *et al.*, 1972).

Retention of polyethylenimine on cellulose increases with molecular weight (Allan and Reif, 1971; Wade *et al.*, 1972); this behavior is consistent with observations of Martin *et al.* (1972). Alince (1974) showed that both adsorption of polyethylenimine and tenacity of adsorption increased with increasing carboxyl content of cellulose, but that the number of adsorbed molecules exceeded the accessible carboxyl groups. Thus the polyethylenimine–cellulose interaction increases with the number of sites (molecular weight) in the polyethylenimine for interaction, which involves hydrogen bonding and/or other forces in addition to ionic interactions. On a given cellulosic substrate, pickup and adsorption of polyethylenimine are

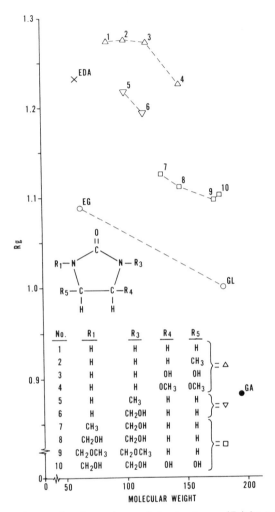

FIG. 10. Elution volumes of various solutes relative to glucose (R_g) from Sephedex G-15 columns, as a function of molecular weight of the solutes. Abbreviations are as follows: GL = glucose, GA = glucuronic acid, EG = ethylene glycol, EDA = ethylenediamine; the numbers refer to 2-imidazolidinones, whose substituents are defined in the insert within the figure (Martin *et al.*, 1972; Rowland and Bertoniere, 1976).

inversely proportional to molecular weight (Kindler and Swanson, 1971; Wade *et al.*, 1972), and for a specific polyethylenimine, adsorption increases as internal surfaces of the cellulose increase (Alince, 1974). The adsorption of polyethylenimine is largely irreversible, water ranging in pH from 4 to 10·5 being no more effective than distilled water. Dilute sodium chloride is somewhat better for desorbing polyethylenimine, but its effectiveness decreases as molecular weight of polyethylenimine increases.

Sorption of poly(vinyl acetate) onto cellulose from non-aqueous solvent is influenced by pore structure of the cellulose and molecular weight of the solute (Alince and Robertson, 1970). Thickness of the layer of sorbed polymer on cellulose surfaces indicates that the fraction of segments of polymer in contact with cellulose is about 0·1 (Alince et al., 1970). Sorption of organic-soluble polymers is reduced in the presence of hydrogen-bonding solvents, which apparently compete with solutes for adsorption sites (Chan et al., 1970).

Zollinger (1960), in studies concerning the mechanism of dye sorption, showed that linearity of structure favored sorption on cellulose and that sorption of α, ω-polyenedicarboxylic acids increased with increasing length of the polyene chain, which provides a flat π-electron bonding sequence. Complexities of the sorption mechanism are well illustrated by Giles and McIntosh (1973) in results of studies of intermolecular bonding between sucrose, as a model for cellulose, and aromatic sulfonates or azo dyes in aqueous solution. These investigators concluded that neither hydrogen bonds nor π bonds directly bind dyes to cellulose. Rattee (1974) stated that the nature of dye binding to cellulose must be complex, involving several kinds of cooperative and inhibiting forces. Within the confines of a pore, hydroxyl-rich surfaces of cellulose limit modes of interaction that would otherwise be possible through more numerous directional approaches of dye and D-glucopyranosyl units in homogeneous solution. Based on the foregoing information on structure of fibrillar surfaces, the predominant opportunity presented to a dye is to associate with hydroxyl groups of D-glycopyranosyl units in the usual type of interaction between ionic, polar, and polarizable molecules (ionic bonds, hydrogen bonds, dispersion forces, etc.), but the importance of one type of interaction over another may vary with changes in dye structure.

Iyer et al. (1968) found that adsorption of Chlorazol Sky Blue FF, in the presence of equimolar concentrations of alkali halide metals, increased as follows: Li < Na ≪ K < Rb < Cs. Since this is the same order as effectiveness of salts for breaking the structure of water, it was proposed that the structure-breaking effect at the surface of cellulose and on the hydrophobic portion of the dye facilitated closer contact between dye and cellulosic surfaces. Dobbins (1970) has discussed this type of effect in detail.

II. Intracrystalline Interactions

A. LATTICE CONVERSION

Some of the confusion in the literature (Warwicker et al., 1966) concerning penetration of swelling agents between crystalline surfaces versus penetration into crystalline units has been resolved (Warwicker, 1971). Warwicker (1969a) showed that mineral acids caused intercrystalline swelling up to the highest concentrations of sulfuric acid that could be tested or at the highest concentrations only of phosphoric, nitric, and perchloric acids. He found that the same degree of swelling, measured by increased fiber width, could be achieved with a single reagent by intercrystalline swelling alone or by a combination of intercrystalline and intracrystalline swelling. Changes in equatorial X-ray scans of cotton show intracrystalline swelling and lattice conversion beginning with $3.0\,N$ sodium hydroxide at 20°C and $2.0\,N$ caustic at 0°C, the same range where fiber width increases markedly (Warwicker, 1967). However, sodium hydroxide affects the fine structure of cotton before it penetrates crystalline regions. Reordering of chains on the surfaces of fibrils is better as a consequence of intercrystalline penetration and swelling with sodium hydroxide up to about $2.0\,N$ at 20°C. Dimick and Atalla (1975) and Vigo et al. (1969) emphasized that decrystallization is distinct from swelling, the effectiveness of alkali metal hydroxides for lattice conversion being $LiOH \cong CsOH < RbOH < KOH < NaOH$, and that for swelling being $CsOH < RbOH < KOH < NaOH < LiOH$ (Heuser and Bartunek, 1925). Various investigators have used different swelling techniques and methods of measurements, so these details must be considered (Zeronian and Cabradilla, 1972).

Addition of ethanol to sodium hydroxide reduces the normality required for intracrystalline penetration and conversion to lattice II (Jeffries and Warwicker, 1969). Combinations of alkali metal hydroxides with benzyltrimethylammonium hydroxide, at concentrations lower than those which individually cause conversion of lattice I to lattice II, give rise to a high degree of conversion (Vigo et al., 1969). This mode of removal of intracrystalline swelling agent also affects the nature and order of surfaces formed in the dried cellulose; the effect may actually occur when the organic solvent contacts the swelling medium in the cellulose. Thus, when acidic ethanol is used in the initial wash to neutralize sodium hydroxide, followed by dimethylformamide and ethanol, there is a much greater conversion to cellulose II. This is accompanied by an increase in hydrogen bond disorder (Jeffries and Warwicker, 1969). When swelling of cotton with benzyltrimethylammonium hydroxide is followed by treatment with polar, water-soluble aprotic solvents, there is a dramatic increase in conversion of lattice I to lattice II and a marked reduction in the crystallinity of the cellulose (Vigo et al., 1970).

Liquid ammonia penetrates cellulose crystallites rapidly and converts lattice I to lattice III when ammonia removal is by evaporation; lattice I is reformed when ammonia is removed by washing with water (Lewin and Roldan, 1971; Warwicker, 1972). Klenkova (1967) reviewed the effective activation and decrystallization of cellulose by monomethylamine and monoethylamine, the lesser effectiveness of higher amines, the very much weaker effects of secondary and tertiary amines, and the greater action of some amines at lowered temperatures. Primary aliphatic diamines penetrate more easily than corresponding monoamines and produce smaller changes in interplanar spacings. Diamines containing one primary and one secondary amine group form stable complexes with cellulose, but similar diamines with tertiary and primary groups or two secondary groups form complexes only if the cellulose is preswollen with liquid ammonia (Creely and Wade, 1975). Synergism in swelling with amines is evident in maximum changes from combinations of morpholine : ethylenediamine at 30 : 70 and morpholine : ethylenediamine : water at 40 : 40 : 20 (Kulkarni and Loikhande, 1975).

In all cases of intracrystalline swelling there is considerable variation in distances between 101 planes, but interchain distances in 101 planes are virtually constant. Warwicker and Wright (1967) pointed out that the true fundamental unit in these reactions is a sheet of molecular chains like those across the tops of Figs 6 and 8. These chains are held together by van der Waals forces between the D-glucopyranosyl units of adjacent molecular chains. Warwicker and Wright (1967) depicted the separation of sheets in 101 planes as shown in Fig. 11, but in this representation the attitude of the D-glucopyranosyl units should not be considered definitive.

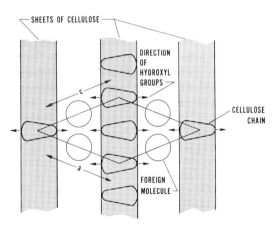

FIG. 11. Diagrammatic representations of sheets of cellulose molecules separated by swelling agent. The shaded portions are stacks of molecular chains lying in 101 planes with molecular chain direction perpendicular to the plane of the projection (Warwicker and Wright, 1967). Reproduced with the permission of John Wiley and Sons, Inc.

Most investigators in the past have interpreted their intracrystalline swelling results in terms of the size of the cation (Vigo *et al.*, 1969; Zeronian and Cabradilla, 1972). On the basis of activity coefficients of concentrated solutions of LiOH, NaOH, and KOH, Dimick and Atalla (1975) related order of activity coefficients for these alkali metal hydroxides to order of lattice conversion. They suggested that these agents begin to penetrate crystallites only when water for complete hydration of the ions is no longer readily available, and that the unhydrated hydroxyl ion is the prime agent responsible for penetration and lattice conversion. This concept, that depletion of water for complete or normal hydration of ions facilitates crystallite penetration, provides a reasonable explanation for observations in the second paragraph of this section.

Dobbins (1970) applied the flickering cluster theory of Frank and Wen (1957) for the structure of water to the swelling of cellulose. Dobbins proposed that the swelling in cellulose by alkali metal hydroxides is not proportional simply to the ionic concentrations or to normal activity coefficients, but is directly related to the concentration of the highly mobile, polar B region water molecules in the solvation shells of the ions. Ions are characterized by A and B regions; the former refers to highly structured immobilized water immediately adjacent to the ion, and the latter refers to the fluid, mobile water farther from the nuclei. Because both the lithium cation and the hydroxide anion are highly solvated, the 4 N solution probably represents the point at which all of the cellulose is in contact with B shell water and crystallite penetration takes place. Sodium hydroxide is less solvated, more base being required to tie up all the free water; thus the swelling threshold is at a higher concentration. Similarly, potassium, rubidium, and cesium, characterized by decreasing charge/size ratio and less solvation, reach their maxima at progressively higher concentrations.

Information concerning specific interactions between alkali metal hydroxide and hydroxyl groups of D-glucopyranosyl units of cellulose chains was obtained over a range of concentrations of sodium hydroxide $(0·5–6·0 N)$ by Rowland et al. (1971b) and Roberts et al. (1971). On the basis of changes in distribution of substituents at O(2), O(3), and O(6) in disordered celluloses and in methyl β-D-glycopyranoside as a function of increasing sodium hydroxide concentration, these investigators proposed that alkoxide formation occurs at all three hydroxyl groups in dilute aqueous base, but that alkoxide formation at O(2)H and O(3)H is reduced with increasing concentration of base by competition with adduct formation between sodium hydroxide and these vicinal hydroxyl groups. At the concentration at which intracrystalline penetration begins, it appears that O(6)H exists in the form of an alkoxide associated with Na^+, and that O(2)H and O(3)H are predominantly in adduct form associated with NaOH.

Liquid ammonia, which is also representative of amines, causes intracrystalline swelling by breaking hydrogen bonds between sheets of cellulose chains (Warwicker, 1972). Both OH . . . N and NH . . . O hydrogen bonds can be formed, the former being stronger than OH . . . O bonds that tie sheets of cellulose chains together in crystallites (Klenkova, 1967).

Hence the sheets are pushed apart, and the ammonia molecule penetrates rapidly between sheets, solvating the hydroxylic edges of cellulose chains. Dissolution is prevented, as in the case of penetration of alkali metal hydroxides, only by cohesive van der Waals forces between chains.

Similarities in changes of fine structure caused by swelling cotton fibers with concentrated sodium hydroxide or liquid ammonia reflect the general disruption of the hydrogen bonding system of natural cellulose. The disordered fraction, measured by X-ray diffraction, was increased from about 25% for native cellulose to 60–70% for swollen celluloses, and accessibilities were increased 35–55% to D_2O and 190–310% to iodine (Rousselle et al., 1976). However, differences have been reported: fibrillar diameter was reduced 3–12% by swelling in 14% sodium hydroxide and 35–47% by swelling in liquid ammonia followed by water washing (Hebert et al., 1975); accessibilities were higher for fibers swollen in liquid ammonia followed by evaporative removal of ammonia than those swollen in 20% sodium hydroxide or liquid ammonia followed by water washing (Rousselle et al., 1976). Swelling under tension, a process that increases fiber, yarn, and fabric strengths and strength retentions after chemical finishing (Murphy et al., 1974; Lawson et al., 1975), increases fibrillar orientation, decreases density, and exerts little effect upon accessibilities in comparison to fibers swollen without tension (Rousselle et al., 1976; Murphy et al., 1974). Improved retention of strength in tension-swollen fibers or in swollen-restretched fibers after chemical finishing (Murphy et al., 1974) or degrading treatments (Handu et al., 1974) has been attributed to orientation of macrofibrils, disorder of hydrogen bonding on surfaces of elementary fibrils, and more uniform distribution of strain and molecular chain ends along elementary fibrils (Rowland et al., 1976).

B. CHEMICAL REACTIONS

Intracrystalline reactions fall into two categories: (1) those in which the chemical reagent is dissolved in an intracrystalline swelling agent, and (2) those in which the chemical reagent and solvent for the modified cellulose penetrate progressively into crystallites with concomitant reaction. In the first category, reactions include etherifications that are commonly conducted in sodium hydroxide of mercerizing strength. Under these conditions, intra- and intermolecular hydrogen bonds of cellulose are disrupted, and the distribution of substituents is defined by relative reactivities of the hydroxyl groups under the specific conditions (Rowland et al., 1969b). These reactions are believed to occur without complete randomness along molecular chains on the surfaces of separated sheets of cellulose molecules (Fig. 11); the degree of randomness varies with the swelling agent. Individual derivatized molecules may be solvated if the product is soluble in the reagent system (e.g. methylcellulose), or the sheets may coalesce, when the reagent system is removed, into crystallites whose structures are disordered to the extent that substituents or crosslinks are present.

Reactions in the secondary category are exemplified by acetylation in pyridine. The rate at which this type of reaction occurs has long been known to be dependent upon accessibility of surfaces in the microstructure of cellulose. Studies of Warwicker (1969b) and Zeronian and Cabradilla (1972) have clarified these dependencies. This type of reaction is envisioned as the unzippering of sheets from one another as reagent and solvent penetrate the structure and react with hydroxyl groups.

A new intracrystalline swelling agent appears to operate by progressive penetration and chemical reaction. This agent, consisting of a combination of dimethyl sulfoxide and paraformaldehyde, was reported very recently by Johnson *et al.* (1975). The key step to the swelling and dissolution of cellulose is the formation of methylolcelluloses by reaction of paraformaldehyde with cellulosic hydroxyl groups. This unstable derivative is solvated and stabilized by dimethyl sulfoxide.

III. Perspective

Although there are some inconsistencies or apparent contradictions among results summarized here, new insights are numerous. Evidence is strong and consistent that water surrounds and plasticizes elementary fibrils of never-dried cellulose to a unique degree; that elementary fibrils of dried or dried and rewet fibers are under stress; that water can reach considerably more hydroxylic surfaces than conventional N-methylol reagents with molecular weights only one order of magnitude higher; and that surfaces readily accessible to conventional reagents are characterized by disorder of hydrogen bonds at O(3)H and O(6)H and, presumably, of attitude of D-glucopyranosyl units. There is now evidence that a substantial fraction of readily accessible surfaces in the cotton fiber is on segments of elementary fibrils that are under stress. These are almost certainly the sites of chain cleavage during chemical finishing. Yet to be measured is the extent to which surfaces of strained segments of elementary fibrils are more readily accessible than surfaces of unstrained segments. Do strained surfaces constitute walls of wider pores, while adjacent, more ordered surfaces form walls of crevices?

Accessibility measurements, except those involving D_2O, are arbitrary, and generally fall short of providing basic information about cellulose. Absolute accessibility methods applicable to compounds of conventional reagent size (mol wt 100–1000) would provide valuable information about surfaces of lamellae, macrofibrils, and elementary fibrils. Hydrogen bonding order on these surfaces might be characterized by an extension of methods based on N,N-diethylaziridinium chloride.

Initial water sorbed by cellulose is more strongly bound than that in subsequent layers, and much additional information is consistent with the concept of bound water on cellulosic surfaces. This is confirmed in NMR spectra. Penetration of solutes in aqueous media into cotton fibers, which is limited according to molecular size of the solute, is evidently further restricted in inverse proportion to the ability of the solute to make use of bound water on pore walls as solvent water.

The beginning of a correlation between sorption of solutes onto cellulose and hydrogen bonding between solute and cellulose is indicated. Sorption is evidently a function of strength, number, and possibly spacing of hydrogen bonding sites in the solute; but we need much more quantitative information to understand this phenomenon. The roles of ionic bonding, hydrogen bonding, and hydrophobic bonding between solutes and cellulose need clarification.

Important aspects of inter- and intra-crystalline swelling and lattice conversion with ammonia, amines, quaternary ammonium hydroxides, and alkali metal hydroxides have been brought to light. Features of the mechanisms of swelling have been elucidated. Lattice penetration by alkali metal hydroxides has been related to concentration and hydration of hydroxide ion or to B shell water surrounding the anion, in contrast to earlier interpretations that emphasized the cation size. Although reasonable theories have been offered to explain improved retention of strength in tension-mercerized or mercerized-restretched fibers after chemical finishing, much additional information is required to verify or refine them.

With all the practical interest in base-catalysed etherifications of cellulose for preparing cellulose gums, it is surprising that the effect of sodium hydroxide concentration upon relative reactivities of $O(2)H$, $O(3)H$, and $O(6)H$ was discovered as recently as 1968.

The combination of dimethyl sulfoxide and paraformaldehyde constitutes a new type of solvent for cellulose, one involving solvation of an unstable chemical derivative of cellulose, and possibly presaging other new solvents and practical developments.

In general, recent work on interaction of liquids with cellulose has clarified concepts and developed mechanisms that appear reasonably representative of reality and that afford a solid foundation for future advances. Complexities of the cotton fiber and other cellulosic substrates are appreciated, but certainly not quantitatively delineated.

References

Aggebrandt, L. G. and Samuelson, O. (1964). *J. Appl. Polymer Sci.* **8**, 2801–2812.
Alince, B. (1974). *Cellul. Chem. Technol.* **8**, 573–590.
Alince, B. and Robertson, A. A. (1970). *J. Appl. Polymer Sci.* **14**, 2581–2593.
Alince, B., Kuniak, L. and Robertson, A. A. (1970). *J. Appl. Polymer Sci.* **14**, 1577–1590.

Allan, G. C. and Reif, W. M. (1971). *Sv. Papperstidn.* **74**, 563–570.

Amemura, A. and Terui, G. (1965). *Proc. 5th Symp. Cellulase and Related Enzymes, Cellulase Assoc.* pp. 33–38, Osaka University, Japan.

Baddi, N. T., Patel, S. B. and Chipalkatti, H. R. (1971). *Text. Res. J.* **41**, 153–158.

Barker, R. H. and Pittman, R. A. (1971). *In*: "Cellulose and Cellulose Derivatives" (N. M. Bikales and L. Segal, eds.) (Part IV), pp. 181–212, Wiley-Interscience, New York.

Bose, J. L., Roberts, E. J. and Rowland, S. P. (1971). *J. Appl. Polym. Sci.* **15**, 2999–3007.

Brunauer, S., Emmett, P. H. and Teller, E. (1938). *J. Amer. Chem. Soc.* **60**, 309–319.

Carles, J. E. and Scallan, A. M. (1973). *J. Appl. Polym. Sci.* **17**, 1855–1865.

Caulfield, D. F. and Steffes, R. A. (1969). *Tappi* **52**, 1361–1366.

Chan, S. S., Minhas, P. S. and Robertson, A. A. (1970). *J. Colloid Interface Sci.* **33**, 586–597.

Child, T. F. (1972). *Polymer* **13**, 259–264.

Child, T. F. and Jones, D. W. (1973). *Cellul. Chem. Technol.* **7**, 525–534.

Cousins, E. R., Bullock, A. L., Mack, C. H. and Rowland, S. P. (1964). *Text. Res. J.* **34**, 953–959.

Cranston, R. W. and Inkley, S. A. (1957). *Advan. Catal.* **9**, 143–154.

Creely, J. J. and Tripp, V. W. (1971). *Text. Res. J.* **41**, 371–373.

Creely, J. J. and Wade, R. H. (1975). *Text. Res. J.* **45**, 240–246.

de Boer, J. J. and Borsten, H. (1969). *Text. Res. J.* **39**, 346–351.

Dechant, J. (1967). *Faserforsch. Textiltech.* **18**, 263–265.

Dimick, B. E. and Atalla, R. H. (1975). Abs. 169 National Meeting Amer. Chem. Soc., Phila., Penn., *CELL* **No. 64**, p. 20.

Dobbins, R. J. (1970). *Tappi* **53**, 2284–2290.

Frank, H. S. and Wen, W. Y. (1957). *Discuss. Faraday Soc.* **24**, 133–140.

Frey-Wyssling, A. (1955). *Biochim. Biophys. Acta* **18**, 166–168.

Frölander, U., Hartler, N. and Nyren, J. (1969). *Cellul. Chem. Technol.* **3**, 499–506.

Giles, C. H. and McIntosh, A. (1973). *Text. Res. J.* **43**, 489–492.

Goring, D. A. I. (1966). *Pulp. Pap. Mag. Can.* **67**, (11) T519–T524.

Grant, J. N. and deGruy, I. V. (1972). *Cotton Grow. Rev.* **49**, 358–368.

Gregg, S. J. and Sing, K. S. W. (1967). "Adsorption, Surface Area and Porosity", pp. 180–182, Academic Press, London and New York.

Guthrie, J. D. and Heinzelman, D. C. (1974). *Text. Res. J.* **44**, 981–985.

Hagamassy, J., Jr., Brunauer, S. and Mikhail, R. S. (1969). *J. Colloid Interface Sci.* **29**, 485–491.

Handu, J. L., Mytheli, J. and Gupta, R. C. (1974). *Text. Res. J.* **44**, 476–478.

Harris, N. R. and Whitaker, G. (1963). *J. Appl. Chem.* **13**, 348–355.

Haworth, S., Jones, D. M., Roberts, J. G. and Sagar, B. F. (1969). *Carbohyd. Res.* **10**, 1–12.

Hebert, J. J., Hensarling, T. P., Jacks, T. J. and Berni, R. J. (1972). *Microscope* **20**, 161–164.

Hebert, J. J., Muller, L. L. and Babin, R. M. (1975). *Microscope* **23**, 139–144.

Hermans, P. H. (1943). *Kolloid Z.* **102**, 169–180.

Hermans, P. H. and Weidinger, A. (1949). *J. Polym. Sci.* **4**, 317–322.

Heuser, E. and Bartunek, R. (1925). *Cellul. Chem.* **6**, 19–26.

Ingram, P., Woods, D., Peterlin, A. and Williams, J. L. (1974). *Text. Res. J.* **44**, 96–105.

Iyer, S. R. S. and Baddi, N. T. (1969). *Cellul. Chem. Technol.* **3**, 561–566.

Iyer, S. R., Srinivasen, G. and Baddi, N. T. (1968). *Text. Res. J.* **38**, 693–700.

Jeffries, R. and Warwicker, J. O. (1969). *Text. Res. J.* **39**, 548–559.

Jeffries, R., Roberts, J. G. and Robinson, R. N. (1968). *Text. Res. J.* **38**, 234–244.

Jeffries, R., Jones, D. M., Roberts, J. G., Selby, K., Simmens, S. C. and Warwicker, J. O. (1969). *Cellul. Chem. Technol.* **3**, 255–274.

Johnson, A., Maheshwari, K. Z. and Miles, L. W. C. (1969). Premier Symp. Internat. de la Recherche Textile Cotonniere, Paris, pp. 557–568.

Johnson, D. C., Nicholson, M. D. and Haigh, F. C. (1975). Abs. 8th Cellulose Conference, State University, College of Forestry, Syracuse, N.Y., May 19–23, p. 58.

Kindler, W. H. and Swanson, J. W. (1971). *J. Polymer Sci. Part A-2* **9**, 853–865.

Klenkova, N. I. (1967). *J. Appl. Chem. USSR* **40**, 2113–2126.

Klenkova, N. I. and Ivaskin, G. P. (1963). *J. Appl. Chem. USSR* **36**, 378–387.

Kulkarni, M. P. and Lokhande, H. T. (1975). *Text. Res. J.* **45**, 108–109.

Lawson, R., Ramey, H. H., Jr. and Elliott, P. W. (1975). *Text. Res. J.* **45**, 510–514.

Lepoutre, P. and Robertson, A. A. (1974). *Tappi* **57** (10), 84–90.

Lewin, M. and Roldan, L. G. (1971). *J. Polymer Sci. Part C* **36**, 213–229.

Lewin, M. and Roldan, L. G. (1975). *Text. Res. J.* **45**, 308–314.

Lin, S. Y. (1972). *Fibre Sci. Technol.* **5**, 303–314.

Mann, J. (1971). *In*: "Cellulose and Cellulose Derivatives" (N. M. Bikales and L. Segal, eds.), Part IV, pp. 89–116. Wiley-Interscience, New York.

Marchessault, R. H. and Sarko, A. (1967). *Adv. Carbohydrate Chem.* **22**, 421–482.

Martin, L. F. and Rowland, S. P. (1967). *J. Chromatogr.* **28**, 139–142.

Martin, L. F., Bertoniere, N. R. and Rowland, S. P. (1972). *J. Chromatogr.* **64**, 263–270.

Michie, R. I. C., Sharples, A. and Walter, A. A. (1961). *J. Polymer Sci.* **51**, 85–98.

Morosoff, N. (1974). *J. Appl. Polymer Sci.* **18**, 1837–1854.

Morrison, J. L. and Dzieciuch, N. A. (1959). *Can. J. Res.* **37**, 1379–1390.

Murphy, A. L., Margavio, M. F. and Welch, C. M. (1974). *Text. Res. J.* **44**, 904–914.

Neal, J. L. and Goring, D. A. I. (1969). *J. Polymer Sci. Part C* **28**, 103–113.

Nelson, R. and Oliver, D. W. (1971). *J. Polymer Sci. Part C* **36**, 305–320.

Nissan, A. H. (1967). *In*: "Surfaces and Coatings Related to Paper and Wood", Symposium (R. Marchessault and C. Skear, eds.), pp. 221–265. Syracuse University Press, Syracuse, N.Y.

Patel, A. R., Kulshreshtha, A. K., Baddi, N. T. and Srivastava, H. C. (1975). *Cellul. Chem. Tech.* **9**, 41–50.

Peterlin, A. and Ingram, P. (1970). *Text. Res. J.* **40**, 345–354.

Pittman, R. A. and Tripp, V. W. (1971). *Appl. Spectrosc.* **25** (2) 235–237.

Porter, B. R. and Rollins, M. L. (1972a). *J. Appl. Polymer Sci.* **16**, 217–236.

Porter, B. R. and Rollins, M. L. (1972b). *J. Appl. Polymer Sci.* **16**, 237–254.

Ramiah, M. B. and Goring, D. A. I. (1965). *J. Polymer Sci. Part C* **11**, 27–48.

Ranby, B. (1969). *In*: "Cellulases and Their Applications" (R. F. Gould, ed.), pp. 139–151. *Advan. Chem. Ser. 95*, Amer. Chem. Soc., Washington, D.C.

Rattee, I. D. (1974). *Text Res. J.* **44**, 728–730.

Richards, G. N. (1971). *In*: "Cellulose and Cellulose Derivatives" (N. M. Bikales and L. Segal, eds.), Part V, pp. 1007–1014, Wiley-Interscience, New York.

Roberts, E. J., Wade, C. P. and Rowland, S. P. (1971). *Carbohyd. Res.* **17**, 393–399.

Roberts, E. J., Bose, J. L. and Rowland, S. P. (1972). *Text. Res. J.* **42**, 217–221.

Rollins, M. L., deGruy, I. V., Cannizzaro, A. M. and Carra, J. H. (1966). *Norelco Rep.* **13**, 119–125, 132.

Rousselle, M. A. and Nelson, M. L. (1971). *Text. Res. J.* **41**, 599–604.

Rousselle, M. A., Nelson, M. L., Hassenboehler, C. B., Jr. and Legendre, D. C. (1976). *Text. Res. J.* **46**, 304–310.

Rowland, S. P. (1972). *Text. Chem. Color.* **4**, 204–211.

Rowland, S. P. and Bertoniere, N. R. (1976). *Text. Res. J.* **46**, 770–775.

Rowland, S. P. and Cousins, E. R. (1966). *J. Polym. Sci.* **4**, 793–799.

Rowland, S. P. and Roberts, E. J. (1972a). *J. Polym. Sci. Part A-1*, **10**, 867–879.

Rowland, S. P. and Roberts, E. J. (1972b). *J. Polym. Sci. Part A-1*, **10**, 2447–2461.

Rowland, S. P. and Roberts, E. J. (1974). *J. Polymer Sci. Polymer Chem. Ed.* **12**, 2099–2103.

Rowland, S. P., Roberts, E. J. and Wade, C. P. (1969a). *Text. Res. J.* **39**, 530–542.

Rowland, S. P., Roberts, E. J., Bullock, A. L., Cirino, V. O., Wade, C. P. and Brannan, M. A. F. (1969b). *Text. Res. J.* **39**, 749–759.

Rowland, S. P., Roberts, E. J. and Bose, J. L. (1971a). *J. Polymer Sci. Part A-1*, **9**, 1431–1440.

Rowland, S. P., Roberts, E. J. and Bose, J. L. (1971b). *J. Polymer Sci. Part A-1*, **9**, 1623–1633.

Rowland, S. P., Wade, C. P. and Roberts, E. J. (1973). *Text. Res. J.* **43**, 351–356.

Rowland, S. P., Roberts, E. J. and French, A. D. (1974). *J. Polymer Sci. Polymer Chem. Ed.* **12**, 445–454.

Rowland, S. P., Nelson, M. L., Welch, C. M. and Hebert, J. J. (1976). *Text. Res. J.* **46**, 194–214.

Segal, L. (1971). *In*: "Cellulose and Cellulose Derivatives" (N. M. Bikales and L. Segal, eds.), Part V, pp. 719–739, Wiley-Interscience, New York.

Sharples, A. (1971). *In*: "Cellulose and Cellulose Derivatives" (N. M. Bikales and L. Segal, eds.), Part V, pp. 991–1006, Wiley-Interscience, New York.

Shinouda, H. G. (1974). *Cellul. Chem. Technol.* **8**, 319–323.

Stone, J. E. and Scallan, A. M. (1966). *In*: "Consolidation of the Paper Web" (F. Bolam, ed.), Brit. Pap. and Board Makers' Ass. Inc., London, Vol. 1, pp. 145–166.

Stone, J. E. and Scallan, A. M. (1967). *Tappi* **60**, 496–501.

Stone, J. E. and Scallan, A. M. (1968). *Cellul. Chem. Technol.* **2**, 343–358.

Stone, J. E., Trieber, E. and Abrahamson, B. (1969a). *Tappi* **52**, (1) 108–110.

Stone, J. E., Scallan, A. M., Doneser, E. and Ahlgreen, E. (1969b). *In*: "Cellulases and Their Applications" (R. F. Gould, ed.), pp. 219–241, *Advan. Chem. Soc. 95*, Amer. Chem. Soc., Washington, D.C.

Vigo, T. L., Wade, R. H., Micham, D. and Welch, C. M. (1969). *Text. Res. J.* **39**, 305–316.

Vigo, T. L., Mitcham, D. and Welch, C. M. (1970). *J. Polymer Sci. Part B* **8**, 385–393.

Wade, C. P., Roberts, E. J. and Rowland, S. P. (1968). *J. Polymer Sci. Part B* **6**, 673–677.

Wade, C. P., Roberts, E. J. and Rowland, S. P. (1972). *Text. Res. J.* **42**, 158–160.

Ward, K. J. and Morak, A. J. (1964). *In*: "Chemical Reactions of Polymers" (E. M. Fettes, eds.), pp. 321–365. Interscience Publishers, New York and London.

Warwicker, J. O. (1967). *J. Appl. Polymer Sci.* **5**, 2579–2593.

Warwicker, J. O. (1969a). *J. Appl. Polymer Sci.* **13**, 41–54.

Warwicker, J. O. (1969b). *J. Appl. Polymer Sci.* **13**, 1037–1048.

Warwicker, J. O. (1971). *In*: "Cellulose and Cellulose Derivatives" (N. M. Bikales and L. Segal, eds.), Part IV, pp. 325–379. Wiley-Interscience, New York.

Warwicker, J. O. (1972). *Cellul. Chem. Technol.* **6**, 85–97.

Warwicker, J. O. and Wright, A. C. (1967). *J. Appl. Polymer Sci.* **11**, 659–671.

Warwicker, J. O., Jeffries, R., Colbran, R. L. and Robinson, R. N. (1966). *Shirley Inst. Pamphlet No. 93*, Manchester, England, pp. 27–28.

Weatherwax, R. C. (1974). *J. Colloid Interface Sci.* **49**, 40–47.

Whitaker, D. R., Colvin, J. R. and Cook, W. H. (1954). *Arch. Bioch. Biophys.* **49**, 257–262.

Zeronian, S. H. (1971). *J. Appl. Polymer Sci.* **15**, 955–965.

Zeronian, S. H. and Cabradilla, K. E. (1972). *J. Appl. Polymer Sci.* **16**, 113–128.

Zimm, B. H. and Lundberg, J. L. (1956). *J. Phys. Chem.* **60**, 425–428.

Zollinger, H. (1960). *Amer. Dyest. Rep.* **49**, 142–149.

8. Chromatography

R. S. ASQUITH and M. S. OTTERBURN

*Department of Industrial Chemistry, The Queen's University of Belfast, Belfast
N. Ireland*

I. Introduction

Forms of chromatography have been known since at least the middle of the 19th century, and it may be claimed that the earliest use of such methods was reported by Pliny (23–79 AD) who described the use of papyrus, impregnated with an extract of gall nuts, to detect iron salts. Alternatively, the work of Runge Schönbein and Goppelsroeder—in the period 1850 to 1910—("Kapillaranalyse"), later adapted by Brown (1939) could be the forerunner of chromatography. Numerous accounts of the early history of chromatography have been given (cf. Weil and Williams, 1950, 1951; Farradane, 1951; Zechmeister, 1951), but there is little doubt that the Russian botanist, Tswett, should be given credit for realizing the great possibilities of separating materials by elution down columns of adsorbents. In his first paper in 1903, published by the Warsaw Society of Natural Sciences, he described more than 100 adsorbents used in conjunction with numerous solvents. A German translation of this paper has been made by Hesse and Weil (1954).

Despite this early work on chromatographic techniques, until the 1940s the average research worker tended to rely almost entirely on classical methods of separation of substances, i.e. sublimation, various forms of distillation, crystallization, selective precipitation with salts or non-solvents, and extraction between two immiscible solvents. From this period onwards, however, numerous methods were refined and became important. These methods can be summarized as "differential migration methods". At the risk of oversimplification it can be said that in all these methods two forces are applied within the system of molecular species which are to be separated. Firstly a general migrating force which is applied to all species present equally, and secondly a force which retards the movement of the molecules. The latter varies in intensity with the nature of each of the substances being separated. This force tends to define the rate of movement therefore of each substance. Some of the more common

TABLE I

Differential migration methods

Method	Force causing migration	Basis of retarding force (differing in intensity for each substance)
Chromatography	Movement of an eluent (liquid or gaseous) over an immobile phase, by pressure or gravity	Adsorption. Partition between phases. Ion exchange. Molecular entrapping in solid phase pores, mass and shape of molecules. (These forces may act together or separately.)
Foam chromatography	Movement of foam by gas flotation	Interfacial tension
Liquid/liquid counter-current extraction	Stepwise or continuous differential movement of two liquid phases	Partition between the liquid phases (cf. chromatography)
Electrophoresis	Electrical potential gradient	Electrical mobility (charge/mass ratio of substances)
Mass spectrometry	Magnetic field	Centrifugal force (ionic charge/mass ratio of substance)
Differential sedimentation	Centrifugal force	Mass and shape of molecules
Diffusion (gases)	Thermal agitation	Mass of molecules
Dialysis (solution)	Thermal agitation	Mass and shape of molecules

separative techniques are summarized in Table I, the major forces applying being indicated.

Possibly one of the reasons why these techniques were not developed more rapidly, despite the fact that the general principles of most were known in the early part of the century, is that they are generally highly efficient for separating small quantities of materials, but become progressively more unwieldy when attempts are made to separate large quantities. Structural studies of organic compounds, in particular, could not be carried out by classical degradation methods unless large quantities were available. Hence until sophisticated techniques of physical chemistry became available, requiring minute amounts, the differential migration methods were of secondary importance.

It can be seen from Table I that numerous different retarding forces can be applied in chromatographic methods, and hence chromatography can be divided according to the major force being used for retardation, i.e., adsorption, partition, molecular seive (or gel permeation), ion exchange chromatography. All techniques are variants based on these concepts with the exception of foam chromatography. Alternatively chromatographic methods can be classified according to the stationary phases used in the technique. Neither system of classification is entirely satisfactory as methods tend to overlap one another to some extent. The second method of division is used here. In it the techniques can be divided into the following groups: liquid/solid, gas/solid, liquid/liquid and gas/liquid. (Thin layer chromatography can be employed in both liquid/solid and liquid/liquid systems.)

II. Liquid/Solid Chromatography

In liquid/solid chromatography all systems rely on the flow of a liquid eluting solvent over, or through, a solid phase. The components of the mixture to be separated are retarded, or held, to differing extents on the solid phase as they are eluted through it by the liquid solvent. The forces which retard the movement of the components of the mixture differ with the type of solid phase employed. Hence liquid/solid chromatography can be subdivided into three types, adsorption, ion exchange and gel filtration.

A. ADSORPTION CHROMATOGRAPHY

The process of adsorption in relation to chromatography has been discussed in numerous standard texts (cf. Lederer and Lederer, 1957; Heftman, 1967; Stock and Rice, 1967) and also in many reviews. Despite the fact that it is the oldest chromatographic technique, the complexity of the adsorptive forces involved make it the least well understood theoretically. Methods, though successful, are generally empirical. The intermolecular forces which are probably primarily responsible for adsorption can be classified as: (1) London forces between surfaces and adsorbed molecules; (2) electrostatic forces between polar surfaces and any absorbed molecules, or between non-polar surfaces and polar adsorbed molecules; (3) charge transfer forces between electron donors and acceptors; (4) hydrogen bonds; (5) covalent bonds between the adsorbent and the adsorbed molecules; (6) ionic bonds between the adsorbent and the adsorbed molecules.

These types of bonding may be roughly divided into physical adsorption (cf. Young and Crowell, 1962) and chemisorption (cf. Hayward and Trapnell, 1964). For chromatographic separations rapid reversibility is essential, so that the substance moves down the adsorbent on progressive elution. Irreversible chemical bonding prevents such movement. Even within the constraint of rapid reversibility, the shape of the band of such a substance eluting down an adsorbing column will be influenced by the nature of the adsorption isotherm of the substance in respect to the solvent and adsorbent used. Giles *et al.* (1960) have classified isotherms for substances in liquid/solid systems and these can be related to the shape of the bands produced by the substances on the chromatographic column. Here it is sufficient to consider two simple types of adsorption isotherm.

(a) Ideal (Langmuir) adsorption. The isotherm is a straight line, the amount of substance adsorbed on the adsorbent being directly proportional to the amount of substance present in the surrounding solution. Such an isotherm is most probable with very low concentrations of substances in solution. With this type of adsorption the substance elutes down a column as a sharp band (the width of the band depending on the number of adsorbing sites per unit surface area of adsorbent).

(b) Normal deviation from ideal adsorption. The isotherm is convex, i.e. the proportion of substance adsorbed on the adsorbent decreases as the concentration in the solution increases. In this case, on a chromatographic column the leading edge of the band of substance is sharp and contains the highest concentration of the substance. It moves forward more rapidly than would be expected if the isotherm were ideal. The tailing edge of the band, where the concentration of substance is lower, approximates more closely to ideality, and tends to move less rapidly. As a result a "tail" of substance follows the main band, and as the concentration of this "tail" becomes more dilute it moves slower, becoming more extended as the band moves down the column.

For sharp separations of bands on an adsorption column it is hence desirable that all the materials to be separated should be present on the column in concentrations in which their adsorption isotherms are linear (i.e. ideal). Within this constraint the migration rates of the bands will be related to the slopes (K) of the respective isotherms. The retention volume R' of a band on a chromatographic column is defined as the total volume of eluent (in ml) required to elute the band centre. R' is related to K by the equation:

$$R' = WK + V_0 \qquad (1)$$

where W is the weight of adsorbent in the column and V_0 is the free volume of the column, i.e. ml of moving phase therein. It is common practice to define the rate of movement of a substance on a column, paper or thin layer chromatogram by the so-called R_F value defined as the distance moved by the substance as a fraction of the distance moved by the eluting solvent front. If all experimental conditions are constant the R_F value for a substance is a constant specific to that substance and is thus related to K by the equation

$$R_F = (V_0/W)/[V_0/W + K] \qquad (2)$$

Unfortunately exactly comparable experimental conditions are difficult to reproduce in different laboratories and hence the R_F value is often replaced by an R value which reports the movement of bands as a fraction of the movement of a band of a common compound of similar structure. Thus glycine (R_{Gly}) is often used as the reference compound for amino acids.

The adsorbent plays an important role in determining the shape of the adsorption isotherms and also the values of K for different adsorbing substances. The more important factors are the adsorbent surface area available and the number of adsorbing sites present thereon. Such factors influence the adsorbent linear capacity, i.e. the maximum amount of sample that can be charged to a column without loss of isotherm linearity (Snyder, 1961). Generally this capacity is high for adsorbents with large surface areas where the adsorption sites do not have widely differing energies. Destruction or selective blocking of the strongest adsorption sites increases linearity and inhibits chemisorption. For this purpose water is often used to deactivate silica or alumina (Snyder, 1964) whilst high molecular weight fatty acids or alcohols are applied to charcoal (Tiselius and Hahn, 1943). Higher temperatures (Hagdhal and Holman, 1950) and stronger eluents generally give straighter isotherms.

The influence of the factors of adsorption sites and surface area of adsorbent are best shown by an example. Bond strengths between the adsorbent and adsorbing material are a measure of the tendency of the material to adhere to the column. Such strengths can be measured by apparent affinity. Thus in Table II the three compounds have widely different affinities for alumina and should easily separate, the order of elution being phenol, nitrobenzene and cellobiose. The spread of these bands is however determined primarily by the K values of their isotherms and the deviation of the isotherms from ideality. These latter factors will be controlled by the number of adsorbing sites available to each compound. If, as in this case, the types of sites required differ, whilst measurement of surface areas will give a measure of all adsorbing sites available (direct proportionality) it will not differentiate between the "adsorbability" of one substance as compared with another. This type of differentiation is often

TABLE II

Apparent affinities for adsorption on Alumina

Substance	Solvent	Apparent affinity (K cals/mole)
Phenol	Water	2.4
Cellobiose	Water	4.5
Nitrobenzene	Water/ethanol	4.0

TABLE III

Comparative availability of adsorbing sites of Alumina

Substance	Solvent	% Adsorbing sites available
Phenol	Water	100% (assumed)
Cellobiose	Water	63.8%
Nitrobenzene	Water/ethanol	0.04%

best determined by an empirical method. If it is assumed that one compound, i.e. phenol, is capable of bonding with all surface sites (if in sufficiently high concentration), the amounts of other substances adsorbed (at the same concentration) can then be expressed as a proportion of the amount of phenol (Table III). There are fewer sites available for nitrobenzene than for cellobiose or phenol, hence at the same concentrations phenol could be expected to elute first (lowest affinity) in a tight band (adsorption isotherm linear), nitrobenzene would elute second, but the band would be more diffuse (non-ideal adsorption) and cellobiose last. The expected elution pattern is shown in Fig. 1.

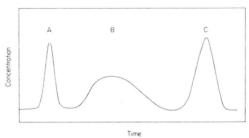

FIG. 1.   Expected elution pattern on alumina column. A = Phenol; B = Nitrobenzene;
C = Cellobiose.

1. Standardization and Types of Adsorbents

The adsorbents used are very susceptible to modification by either deliberate or unintentional contamination, changes in particle size etc. Different samples can, therefore, vary considerably in properties. It is desirable to standardize particle size as far as possible, to improve column flow and also to give a reasonably large surface area for adsorption. With alumina (Jutisz and Teichner, 1947) it has been shown that average particle diameters of 6–9 microns give a surface area of 60–90 sq. metres/gm, but with larger particles (below 6 sq. metres/gm surface area) no chromatographic separations can be achieved.

Adsorbents vary greatly in capacity with the method of formation. Nylon 6 powder, prepared by dissolution of the fibre in formic acid and re-precipitation from water, has been used to separate 2–4 dinitrophenyl derivatives of amino acids (Steurle and Hille, 1959). The original workers found this method very successful but later workers had difficulty with it (Asquith and Jordan, 1961) until it was found that only if the nylon powder was precipitated under strictly controlled conditions, and contained between 0.1% and 0.2% titanium dioxide, could worthwhile separation be achieved.

Adsorbents can be standardized by various test procedures. Thus the Brockmann method (Brockmann and Schodder, 1941) grades alumina samples into five grades (Grade V having the weakest adsorbing power) according to the rates of elution of six azo dyes thereon with different solvents. Similar systems can be used to grade silica (Weiss and Shipman, 1961).

Different adsorbents obviously show different selectivities for different types of compounds. Table IV summarizes a few adsorbents and their uses.

TABLE IV

Adsorbent	Useful for group of compounds which can be separated thereon
Alumina	General
Bauxite	Sugars, Amino-sugars
Magnesium Oxide	General
Magnesium Silicate	Sugar acetates, steroids, acetylated glycosides
Calcium hydroxide	Carotenoids
Silica (anhydrous)	Sterols, fatty acids, terpenes
Charcoal	Branched from . straight chain hydrocarbons. Tyrosine-containing peptides from other peptides
Fullers earth	Pteridines, basic amino acids
Aluminium silicate	Sterols
Calcium carbonate	Xanthophylls, naphthoquinones
Calcium phosphate	Dyes, proteins
Strontium phosphate	Dyes
Sucrose	Chlorophylls, xanthophylls
Cellulose	Dyes, sugars

2. Solvents in Adsorption Chromatography

Obviously the strength of a solvent in elution will depend, to a considerable extent, on the adsorbent used. Relative strengths of different solvents on particular adsorbents have been correlated (Heftman, 1967). Some workers have claimed that sample solubility is the decisive factor (cf. Freundlich and Heller, 1939). Undoubtedly it is essential that all components in a mixture to be separated must be soluble in the solvent used for elution. In general the eluting power of solvents increases with increased polarity (though with non-polar adsorbents, such as charcoal, the eluting power may reverse). It is advisable therefore when changing solvents during elution that they should be changed in ascending order of polarity. With polar adsorbents it has been claimed that the eluting power of solvents is in order of dielectric constant (Jacques and Mathieu, 1946) and various authors (Trappe, 1940; Strain, 1942) have established eluotropic series of solvents for specific purposes. It would appear that the solvents function by being preferentially adsorbed at reactive sites (Jacques and Mathieu, 1946), and with mixtures of solvents that with the highest eluting power (highest dielectric constant) is preferentially adsorbed.

The major drawbacks to liquid/solid adsorption chromatographic methods experimentally can be summarized as:

(1) Variations in adsorbing efficiency, with consequent changes in R values, from one batch of an adsorbent to another.

(2) Recovery of substances from an adsorbent column may be low. Some adsorbents seem to fix a proportion of each substance irreversibly, possibly by chemisorption at a small number of particularly reactive sites.

(3) The adsorbent may act as a catalyst to bring about chemical changes in the substances being separated. Thus, some alcohols can be dehydrated to olefines on alumina columns.

3. Techniques of Liquid/Solid Adsorption Chromatography

The techniques of liquid/solid adsorption chromatography can roughly be divided into three classes (a) liquid flowing, (b) frontal analysis, (c) displacement.

(a) In liquid flowing chromatography, the commonest type, the mixture to be separated is applied directly (in a small quantity of solvent) to the column of adsorbent (already saturated with solvent). The mixture is then eluted with solvent continuously, the components of the mixture separating as discrete bands. A major difficulty of this system is that in a mixture some bands may move rapidly, whilst others are strongly retarded (similar effects are observed in ion exchange chromatography). As a result, these latter bands, even if sharp, will require large volumes of solvent to elute them. (The elution pattern therefore appearing as in Fig. 2a.) To avoid this, step-wise elution (in which the first solvent is changed for a more powerful eluent) can be used. Tiselius (Alm *et al.*, 1952) pointed out that a sudden change of solvent can result in a substance being eluted as apparently more than one band. Thus the second solvent by collecting up the tail of the

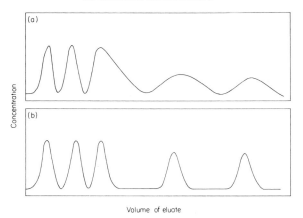

Concentration

Volume of eluate

FIG. 2   (a) Single solvent elution showing tailing (peak 3) and spreading (peaks 4 and 5). (b) Gradient elution with two solvents. Note inhibition of tailing and spreading.

primary band may elute the tailing material as a second band. To prevent this and also to minimize tailing, the system of gradient elution is usually used. In this, by a simple system (Lederer and Lederer, 1957) the proportion of the more powerful eluting solvent is progressively increaseed in the eluting mixture. The tail of any band is therefore always in a higher concentration of the more powerful eluting solvent than is the front. As a result all bands are sharp and elute in a reasonable volume of solvent (Fig. 2b). Instrumentation has now been designed for controlling the mixing of the solvents so that the elution gradient can be varied (*Science Tools*, 1970; Keck, 1971). High molecular weight compounds may take a finite and prolonged time to reach equilibrium adsorption. This time may be sufficiently long for no true reversible adsorption to occur on a column. As a result these substances "smear" down the column without separation, and techniques of gradient elution are unlikely to improve the separation.

(b) The frontal analysis technique, which is not strictly separative but rather analytical, has been well described (Brenner and Niederwieser, 1961). It relies on the fact that if there are slight differences in the affinities of the substances for the adsorbent, then each will be displaced by the next if all the adsorbent sites are saturated. A solution of the mixture is fed continuously to the column and the front of this eluate is monitored. At this front the substances are displaced by one another in order of increasing affinity. Hence each substance appears in the eluate as a step, the final concentration being the total in the original solution (Fig. 3).

(c) Displacement adsorption chromatography. This method is similar to frontal analysis but the mixture to be examined is eluted with a solution of a substance (displacer) which is more strongly adsorbed than any component of the mixture. As a result similar elution patterns are obtained

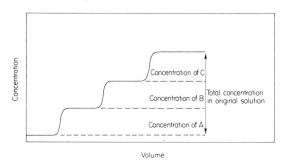

FIG. 3.   Idealized frontal analysis system.

as in frontal analysis (Tiselius, 1942). An improved method, carrier displacement (Tiselius and Hogdhal, 1950), has been described. In this a series of compounds are added to the mixture to be separated. These compounds have intermediate adsorption affinities to the components of the mixture. Ideally each added substance displaces one component of the mixture, and is itself displaced by the next component. Hence the components of the mixture can be separated from each other (Holman, 1951).

Despite the numerous techniques of liquid–solid chromatography devised between 1940 and 1960, with the exclusion of thin-layer-chromatography, this method has, in the writers' opinion, tended to be superseded by other chromatographic systems in many fields.

B. Ion Exchange Chromatography

The definition of an ion exchanger has been given as "an insoluble material containing labile ions that will exchange reversibly with other ions in a surrounding solution without any physical change occurring in the material" (Calmon and Kressman, 1957). For all practical purposes they are, therefore, polymeric materials carrying electrostatic charges neutralized by labile ions of opposite charge. Hence anion exchangers are positively, and cation exchangers negatively, charged. Numerous natural, modified natural, and synthetic polymers, both organic and inorganic, can be used as ion exchangers. The natural and modified natural materials are usually multifunctional, carrying more than one type of exchanging group. Synthetic unifunctional resins are preferable for chromatography as changes in pH of the eluting solution do not result in a change of the relative affinities of the ions undergoing separation. The total capacity of such resins is determined by the total number of active sites per unit weight of material. The active sites may fall into four classes: strongly or weakly acid, strongly or weakly basic. The capacity of the weak exchangers is pH dependent, the total exchange capacities only operating at high and low pH

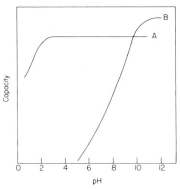

FIG. 4.   Exchange capabilities as function of pH of typical unifunctional strongly (A) and weakly (B) acidic resins.

values respectively. Figure 4 shows the capacities as functions of pH for typical unifunctional strongly and weakly acidic resins. With some natural ion exchangers, exchange occurs on the surface of the particles and hence capacity is a function of particle size. With synthetic resins, ions can enter the pores of the particles, hence the whole of the particle is available for exchange. Ions larger than the pore size (generally 5 to 15 Å) will be excluded. The degree of crosslinking within the resin will determine the swelling in solution and hence the pore size.

The desirable properties of an ion exchange resin for chromatographic purposes may be summarized as: (1) the resin must be unifunctional; (2) the capacity in the operating pH range should preferably be high; (3) the stability of the resin to aqueous solutions should be high, preferably up to temperatures of 60°C; (4) the degree of swelling (and hence pore size) when wet should be controllable by crosslinking; (5) the resin particles should be spherical and the size range small, to pack evenly in columns and allow easy percolation of the eluate. Resins based on polystyrene satisfy most conditions and are commonly used for chromatography. A variety of acidic and basic groups can be introduced into the benzene rings, and the degree of crosslinking can be controlled by co-polymerization with divinyl benzene. The main chain of the polymer is chemically stable and spherical particles can be obtained by the method of polymerization. The particle size range can be controlled between narrow limits by flotation separation. Similarly the degree of crosslinking can be estimated by density in sodium tungstate solutions (Hamilton, 1963).

1. Separation Techniques

The strongly acidic and basic resins find most application because of their constant capacity over a wide pH range. Weakly acidic and basic resins are used for certain separations because of their greater selectivity in specific cases, thus weak acid resins will exchange strong bases but not weak ones.

At low concentrations of exchanging ions the exchange isotherms are linear, though at high concentrations the departure from linearity may result in tailing. Gradient elution, usually with a changing pH gradient, is often used to avoid this.

All the techniques used in adsorption chromatography (i.e. liquid flowing, frontal analysis and displacement) have been used in ion exchange chromatography on columns.

There are some water-immiscible liquids, such as tri-n-octylamine which can be used for liquid–liquid partition solvents. Because of their basicity, or acidity, they are also capable of ion exchange. These liquids can be coated onto a solid support such as cellulose (as a column, thin-layer or paper support material), and thus used to separate ions, combining reversed phase partition (see partition chromatography) with ion exchange effects. The use of such liquid exchangers has been reviewed (Cerrai, 1964). This system has been extended and papers are now available commercially for paper ion-exchange chromatography containing about 50% finely ground ion-exchange resin. Ion-exchange papers have also been produced by impregnation with zirconium phosphate (Ruttenberg et al., 1965) or by modifying the cellulose by introduction of carboxymethyl groups (Miranda et al., 1962) or diethylaminoethyl groups (Marsden, 1965). These latter modified celluloses, as column materials, have been used in the separation of protein fractions from wool (cf. Joubert and Burns, 1967). Separations of compounds on ion-exchange papers, or thin layer plates can usually be carried out on columns of the same exchanger, though the nature of the separation cannot be predicted from one method in relation to another (Sherma and Cline, 1964).

A further use of ion-exchange resins has been the separation of large molecules from small (see also gel filtration chromatography). The pores of the swollen resin particles are of finite size and decrease as the extent of crosslinking increases. Detailed studies of the behaviour of polymers thereon have been made (Deuel et al., 1951). If the pore size of the resin is too small for the polymer to penetrate, exchange only occurs on the surface of particles, the low molecular weight monomers being more retarded. Richardson (1951) used this method to separate direct cotton dyes from inorganic salts, whilst cellulose xanthate can be separated from contaminating inorganic anions similarly (Samuelson and Gärtner, 1951). Even with relatively small ions, highly crosslinked resins show different selectivities. Thus a crosslinked (8% divinyl benzene) sulphonated poly-styrene resin absorbs only 0.79 equivalents/gm dry resin/ml solution of lithium ion whilst the same resin absorbs 1.56 equivalents sodium ion. The high degree of hydration of the lithium ion inhibits penetration (Dean, 1969).

C. Gel Filtration (Permeation) Chromatography

This method of chromatography has been developed for the separation of materials, particularly proteins and polypeptides of high molecular weights. Work in adsorption, and later in ion exchange chromatography, had shown that materials of high molecular weight were excluded, whilst low molecular weight materials were retarded on such columns. Mould and Synge (1954) showed that separations based on this "molecular sieving" occurred with uncharged substances during electro-osmotic migration through gels. Other workers attempted to use sieve effects for estimations of molecular weights (Lathe and Ruthven, 1956). Porath and Flodin (1959) introduced the first crosslinked dextran and poly-(vinyl alcohol) gels and termed the separation by molecular size "gel filtration". Generally the term "gel filtration" is now applied to the use of hydrophillic gels with aqueous solutions, whilst "gel permeation" is applied to the use of hydrophobic gels in organic solvents.

The original dextran gels were co-polymers of epichlorhydrin and a polysaccharide. A wide range of these gels, and newer modifications, are produced under the name of "Sephadex".* These gels exhibit some sorption effects such as interaction of the matrix with aromatic compounds, retaining these to a greater extent than non-aromatic compounds of the same molecular size. They also contain a small number of carboxyl groups which may cause distorting effects (Gelotte, 1960) by ion exchange. Hydrophillic gels of poly-(vinyl ethyl carbitol), poly-(vinyl pyrolidone) and poly-(acrylamide) have also been described (Lea and Schon, 1962). The last is now marketed under the name of "Bio-Gel".† By selection from the different Sephadex and Bio-Gel range of resins it is possible to fractionate and separate molecules from molecular weights, varying from 200 to 400,000, (e.g. Sephadex G25 excludes above 10,000 and fractionates between 500 and 10,000).

Several hydrophobic gels such as vulcanized rubber (Brewer, 1960) and crosslinked poly-(methyl-methacrylate) (Determann et al., 1964) have been described. A series of crosslinked poly-styrene gels have been introduced under the name of "Styragel"‡ for fractionation of solvent-soluble polymers in gel permeation chromatography.

---

* Pharmacia Fine Chemicals, Uppsala, Sweden.
† Bio-Rad Laboratories, Richmond, Calif., U.S.A.
‡ Dow Chemical Company, U.S.A.

1. Basis and Techniques of the Gel Filtration Separation System

The major technique used is column elution chromatography, though the gels can be used in thin-layer chromatography (Morris, 1964). Indeed since 1959 many hundreds of papers have been published on the use of gel filtration, and much of the work on Sephadex gels at least has been correlated and abstracted (such abstracts being available from Pharmacia Fine Chemicals).

In the absence of interfering sorption effects, separation is effected entirely by the ease with which the molecules of different size penetrate the gel matrix. A diagrammatic representation of such separation is shown in Fig. 5.

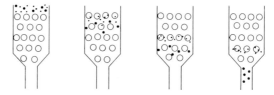

FIG. 5.   Stages in the separation of a mixture of large and small spherical molecules by gel-filtration.

The theory may be briefly summarized as follows. Assuming that a particular resin has a specific pore size, different sized molecules in a mixture, eluted through a column of the resin, must fall into one of the following classes:

(a) Molecules too large to enter the pores at all. These will move down the column at nearly the same rate as the eluent.

(b) Molecules of different sizes within the range of the gel pore size. These will penetrate the gel with increasing difficulty as the molecular size increases. These species will fractionate on the gel, being eluted in order of decreasing molecular weight.

(c) Very small molecules which easily penetrate the gel pores. These will be retarded most and will elute as one band when all the eluent in all the gel particles has been totally charged.

From this it can be seen that for a specific column of a gel two eluent volumes can be defined. The void volume, V_0, is that volume which must elute through the column to change all the solvent between the gel particles. At this volume all molecules of type (a) will be eluted. This volume can be measured practically by applying a suitable high molecular weight material to the column and measuring the volume of solvent required to elute it. A second volume (V_t) can also be defined. This is the volume of eluent required to totally replace the eluent molecules in the gel by fresh molecules. At this volume the very low molecular weight molecules, type (c), will be eluted. The measurement of this volume can be determined using a low molecular weight compound or inorganic salt. Between these two volumes all material applied to the column as a mixture should elute. Andrews (1964) has shown in thin layer gel filtration that the rate of migration is directly proportional to molecular weight (on a logarithmic scale) for a series of globular proteins ranging from cytochrome C to γ-globulin. The method has also been used for determining molecular weight distribution of polymers (cf. Granath and Flodin, 1961; Moore, 1964).

There are two major dangers in the use of gel filtration chromatography. Firstly aggregation of the substances may give apparent molecular weights higher than actual. Secondly, absorption into the gel is influenced by molecular shape as well as size; hence standardization of columns requires materials, not only of known molecular weight, but also of similar shape to those under examination (Fischer, 1969; Ackers, 1970). With proteins this difficulty can sometimes be overcome by reducing the disulphide bonds and then working in a denaturing and disaggregating solvent such as 6M guanidine hydrochloride, thus ensuring that all the molecules adopt similar randomly coiled conformations (Fish *et al.*, 1970; Mann and Fish, 1972).

2. Affinity Chromatography

This method may be used on any solid support, but gel filtration resins are most common. It is used for the separation of specific components of biological fluids by selective adsorption. *In vivo*, "anti-bodies" are produced to counter materials absorbed therein. Specific "anti-bodies" can be isolated and bound covalently to the gel matrix. The resultant gel strongly adsorbs the material with which the "anti-body" reacts. This material can thus be separated from other components of the biological fluid (Cuetracasas and Anfinsen, 1971). The techniques of affinity chromatography have been recently described (Lowe and Dean, 1974).

3. Thin Layer Chromatography

It is sufficient here to include a brief mention of this now well-known technique. Virtually any system of liquid–solid or liquid–liquid chromatography can be converted to the thin-layer technique. It has the advantages of paper chromatography but is more versatile. In this method slurries of adsorbents, gels, ion exchange resins etc. are applied as thin-layers to a support material, (glass or plastic sheets). The mixtures to be separated are applied as spots to the thin-layer and the eluting solvent is allowed to flow up the sheet by capillary action. In general the advantages of the method are:

(a) Very varied types of liquid/solid chromatography can be used.
(b) The separations are faster than column chromatography, and generally faster than paper chromatography.
(c) As with paper chromatography only minute amounts of material are required for separations.
(d) The separated spots can be easily located by use of spray developers, and more reactive developers (which would decompose paper, e.g. strong acids) can be used, providing the thin layer is inert.
(e) By using thicker layers of adsorbent etc. preparative scale separations can be effected.
(f) Better resolution than on paper is often achieved for shorter runs as the thin layer material often has a higher capacity than paper. The shape and size of the spots therefore remains closer to the size of the spot of applied mixture, and tailing is less.

Two minor disadvantages, as compared with paper chromatography, are the difficulty of preserving thin layer chromatograms and the difficulty in obtaining reproducible R_F values.

A number of books (Bobbitt, 1963; Hashimoto, 1963; Truter, 1963; Labler and Schwartz, 1965; Randerath, 1965; Stahl, 1965) are now available, which cover the whole field.

III. Liquid–Liquid Chromatography

This type of chromatography is based on the partition of substances between two immiscible liquid layers. To overcome the low efficiency and prolonged time required for countercurrent separations, Martin and Synge (1945a,b) proposed a system for separating acetyl amino acids, which they termed a partition chromatographic column. It consisted of an inert support on which one solvent (usually water) is tenaciously held and immobilized. A second liquid phase, immiscible with the first, is then used as an eluent for the column. Substances separate down such a column by partition between the support-bound immobilized solvent and the eluting solvent (Martin and Synge, 1945a,b; Synge, 1946). In many cases the support for the immobilized solvent phase is not totally inert, and liquid/liquid partition is not exclusively responsible for retention of substances. Thus some compounds do not migrate in the sequence predicted from their partition coefficients (Moore and Stein, 1948). Kieselguhr, generally accepted as the most inert support material for columns (Moore and Stein, 1952), nevertheless adsorbs alkaloids (Chilton and Partridge, 1950).

The exact state of the "bound" water has been considered, particularly in relation to the fact that water-miscible eluents apparently behave as partitioning solvents on hydrated cellulose. Various suggestions have been made regarding the nature of this phase (Axelrod, 1955; Franek and Mastner, 1959; Ingle and Marshall, 1962).

A. PAPER CHROMATOGRAPHY

In 1944, Consden *et al.* replaced columns of powdered cellulose by sheets of absorbent paper as the solid support in partition chromatography, initiating the technique of paper chromatography. This technique has been fully covered in a number of books (Lederer and Lederer, 1957; Block *et al.*, 1959; Heftman, 1967; Stock and Rice, 1967). Filter paper contains about 14% moisture at normal humidity (65% RH) and absorbs considerably more at high humidities. The materials to be separated are applied as a spot to the paper, which is then enclosed in a suitable cabinet, the atmosphere of which is saturated with water and solvent vapour. The solvent saturated with water is eluted down (descending chromatography) or up the paper (ascending chromatography) by capillary action. The paper is subsequently dried and spots developed by suitable spray-developers. The method now tends to be replaced, for many purposes, by thin-layer-chromatography. The advantages of the method can be summarized as:

(a) Cellulose is not highly adsorptive to most compounds (excluding direct cotton dyes) and therefore partition predominates.
(b) The technique is simple, requiring no tedious column preparation, the paper being self-supporting.
(c) More effective separation can often be obtained than on columns by drying the paper after a first elution (one-dimensional chromatography), rotating it through 90° and eluting with a second solvent prior to development (two-dimensional chromatography).
(d) Very small amounts of material are required for separations.

The disadvantages of paper chromatography are chiefly the following:

(a) Preparative separations are difficult, if not impossible, though various attempts have been made (Fischer and Behrens, 1952; Hagdahl and Damuelson, 1954; Solms, 1955).
(b) Double spots are sometimes given by single components. This can occur when a compound is partially ionized, the ion and the parent compound eluting at different rates, resulting in two spots connected by a "tail".
(c) When adsorption on the paper occurs, "tailing" on the paper results which prevents separation.

(d) R_F values vary and it is difficult to correlate the R_F values of different laboratories apparently using identical systems (Green and Marcinkiewicz, 1963).

(e) Salts (e.g. NaCl) in mixtures of compounds ionize and separate into spots (e.g. NaOH + HCl). These latter spots cause high local concentrations of water which interfere with the desired separations. They may also inhibit colour development.

(f) The time required for elution is governed by the capillary flow of solvent down, or up, the paper. This is relatively slow (12–18 h), unlike thin-layer-chromatography (5 mins to 2 h).

B. Reversed Phase Chromatography

In this type of partition chromatography the organic solvent is bound to the column material, or paper. The mixtures are separated thereon by elution, usually with an aqueous solvent. A number of substances have been used to impregnate paper, in particular, with organic phases such as rubber latex, silicone oils, paraffin oils (Boldingh, 1948; Strain, 1953; Ashley and Westphal, 1955; Michalec, 1955).

IV. Gas Chromatography

Gas chromatography can be divided into two types, gas–solid chromatography based on adsorption from the mobile gaseous phase, and gas–liquid chromatography based on partition between the gaseous eluent (carrier gas) and a non-volatile liquid phase immobilized on an inert support.

Both these systems possess two advantages as compared with liquid–liquid chromatography:

(a) The low viscosity of the carrier gas permits a very rapid elution of the column; separations can be achieved in as little as five minutes. At the higher column temperatures required to volatilize the mixtures of substances equilibrium is rapidly attained and the eluted bands are usually sharp. Due to the ease of elution, very long 30 ft. columns can be used to obtain wide separations of bands. Further such columns do not need to be straight as gravity plays no part in the elution; compact coiled columns are normally used.

(b) The methods of detection of bands of substances eluting in the carrier gas stream are general for all organic compounds, whereas the methods of colour development etc. used in liquid chromatography are specific to each class of substances. Further, the varied detection methods (Section IV, A) are very sensitive, thus ninhydrin will detect 10^{-6} gms of an amino acid on a paper chromatograph but in gas chromatography 10^{-13} gms/ml of eluting gas can be detected by standard systems, and with specialized detector systems as little as 10^{-17} gm/ml of eluting gas have been claimed.

The major limitations of both types of gas chromatography are that the substances to be separated must be volatile (without decomposition) at a reasonable temperature, and preparative separations are not easily possible.

The techniques of both types of gas chromatography are the same in that a carrier gas at a controlled rate of flow is passed continuously through the column of adsorbent, heated in an oven at a suitable temperature, and thence passed through a detector system. Carrier gases commonly employed as eluents are nitrogen, hydrogen, helium, argon and carbon dioxide. Most of these can now be bought sufficiently pure to require no treatment prior to use. The mixture to be separated (usually dissolved in a volatile solvent) is injected into the gas stream at the entry to the column.

For preparative separations, the preferred method is an automated repeating system, which relies on the retention volume of a particular substance remaining constant for repeated injections of mixture onto the same column. Having determined the elution pattern of the mixture, the instrument is pre-set to collect fractions at the appropriate time of elution and to inject samples successively.

To assist in the identification of unknown materials, the chromatographic system may be coupled directly to a mass spectrometer, each eluted band being examined as it leaves the column. In 1957, Homes and Morrel first used such a system to identify constituents of town gas. A molecular separator is necessary to separate carrier gas from the constituents. Two such separations are the Jet Separator (Rhyhage, 1964) and the membrane separator (Llewellyn and Littlejohn, 1968). A palladium separator (Lovelock et al., 1970a) precludes loss of sample as compared with other systems, but is unsatisfactory for sensitive (reducible) components and is poisoned by sulphur compounds. This device was included in an integrated gas chromatography/mass spectra unit, which was part of the Mariner probe to Mars (Lovelock et al., 1970).

Reaction gas chromatography is a modified technique in which specific components of the mixture are removed by irreversible adsorption in an abstractor column, prior to entering the gas chromatograph, or within the chromatograph itself. The methods have been reviewed (Beroza and Coad, 1967; Leathand and Shurlock, 1970). For success this technique relies on a rapid irreversible binding of specific classes of compounds. For example, boric acid coated supports in a pre-column will remove most primary and secondary alcohols by formation of the trialkyl borate esters; dehydration may occur with tertiary alcohols. The technique has been used for distinguishing the different classes of alcohols (Regnier and Huang, 1970).

The use of this system for functional group analysis has recently been described (Ma and Lucas, 1976).

A. DETECTOR SYSTEMS FOR GAS CHROMATOGRAPHY

Most detectors are of the "differential" type, i.e. giving a zero signal to the recorder when carrier gas only is passing through. The bands are thus recorded as a series of peaks, the size of which is proportional to the amount of material in the band. "Integral" detectors give a continuous signal proportional to the total amount of substance eluted. Stepped records are thus obtained, analogous in appearance to "frontal analysis" records. Step height is proportional to the amount of the particular constituent eluted. The integral methods are little used and will not be considered here.

A good detector should have the following properties: (a) sensitivity to a wide range of components and the capability of estimating very low concentration levels; (b) rapid response in order to separate close component peaks when eluted. This is partially related to the size of the detector chamber, which must be small; (c) response signal should be linear with respect to concentration of component passing through the detector; (d) the detector signal should be capable of activating an automatic recorder after amplification; (e) the detector should be stable and simple to operate.

Detectors fall into two classes, those which destroy the components during detection and those which do not. A series of detectors have been developed with widely different sensitivities. All are based on some physical property of compounds such as thermal conductivity, gas density, flame ionization (Giddings, 1965), β-ray ionization, (Lovelock and Lipsky, 1960), photo-ionization glow discharge, flame temperature, dielectric constant. Sensitivity varies from 10^{-6} moles/ml to 10^{-13} moles/ml with different methods. Only three commonly used detectors will be described here.

1. Thermal Conductivity Cells (Katharometer)

The measurement of thermal conductivity has been used for many years to analyse simple gas mixtures (Shakespear, 1921). It relies on the fact that when a steady stream of a pure gas flows over an electrically heated platinum wire the rate of heat loss and hence the resistance of the wire are constant. If the composition of the gas stream changes, the temperature and resistance of the wire change. Obviously fluctuations in gas flow and changes in ambient temperature will affect the resistance of the wire. To cancel the latter effects, two wires in parallel are set up such that the pure carrier gas flows over one and the column eluent over the other at identical rates, both flow channels being mounted in a thermostatted metal block. The change in resistance of one wire when a band of material passes over it, as compared with the control wire, gives a measure of the amount of substance in the band. Sensitivity of the system is enhanced by using carrier gases of high thermal conductivity (e.g. hydrogen, helium), low block temperature and high current through the wires. Thermistors may replace the wires, but this limits the use of hydrogen with which the bead material of the thermistor may react. The disadvantages of the method are over-long response time and insensitivity to small samples. The effective volume is large (unless thermistors are used) hence they are unsuitable for capillary columns, and they are often non-linear in response. The detection limit is about 10^{-6} moles/ml.

2. The Gas Density Balance

First designed by Martin (Martin and James, 1956), this system relies on a sensitive anemometer located between two other channels drilled in a copper block. Pure carrier gas flows through one of these channels and the column eluent through the other. When a band passes through the eluent channel there is a change of gas density and the carrier gas flows through the anemometer. Response is, therefore, a linear function of concentration and molecular weight of the band. No calibration is, therefore, required, and the system can be used for molecular weight determination (Phillips and Timms, 1961).

3. Flame Ionization Detectors

The systems depend on an increase in current passing through a circuit when eluted substances are ionized in a flame. The flow from the chromatographic column enters a fine jet in which hydrogen (which may be the carrier gas or may be added after elution to the carrier) burns in an oxygen atmosphere. This jet forms one electrode (usually negative) of the cell. The other electrode is a platinum grid situated at the tip of the flame. The potential difference across the electrodes is about 200 V. The background ionization current is increased when a band passes through and additional ions are formed. About the only organic substance not detected is formic acid. The detector has the advantage of not indicating the presence of water and hence aqueous solutions can be used. A series of inorganic compounds (i.e. H_2, N_2, O_2, S_iCl_4, S_iHCl_3, S_iF_4, H_2S, SO_2, COS, CS_2, NH_3, NO, NO_2, CO, CO_2 and noble gases) are not detected because of their high ionization potentials (Condon et al., 1960). A further limitation is the need for calibration for each chemical species.

B. Gas–Solid Chromatography

The early work on gas/solid chromatography dealt almost entirely with separations of more permanent gases. The danger of catalytic decomposition of substances on the adsorbent at higher temperatures, required to volatilize mixtures, reduced the use of this technique. The advent of gas/liquid chromatography resulted in the gas/solid systems being re-examined and developed (cf. Ray,1956; McKenna and Idleman, 1960). In particular the use of molecular sieve systems, which retain small molecules, has been investigated. Natural materials, such as zeolites, retain their crystalline porous form on drying. The pore diameters vary between 3.8 nm and 15 nm. With these adsorbent sieves, oxygen and argon can be separated from nitrogen (Lard and Horn, 1960). Palladium black can selectively adsorb and separate hydrogen from deuterium (Glueckauf and Kitt, 1956).

In simple adsorption systems, separations of low molecular weight hydrocarbons are effective on silica gel, charcoal (James and Martin, 1952a) or alumina (James and Martin, 1954).

C. Gas–Liquid Chromatography

The technique was first proposed by James and Martin (1952a). In later papers they described the theory and examined the sequences of separation of materials using different stationary phases (James and Martin, 1952b, 1954).

The system, as with paper chromatography, relies on partition of substances, in this case between the carrier gas and a liquid solvent held in an inert solid support. The solid supports are normally diatomaceous earths or porous materials such as crushed fire brick. A number of commercial supports such as Celite and Chromasorb are now available. The liquid phase is applied to the support, usually by dissolving the required amount in a volatile solvent and mixing this solution with the support material. The volatile solvent is then evaporated prior to packing the column. Packing long (30 ft) coiled columns evenly can present difficulties. Metal columns can be packed when straight and then coiled. An effective method, used in our laboratories, for coiled glass columns, is to attach one end of the column to gentle suction from a water pump, and the other to a tube of similar diameter (glass or rubber). This tube is then used as a gravity feed (about 15–20 ft long). One operator pours the column packing into it in portions through a funnel, whilst at the lower level a second operator gently taps the coil.

The major requirement of the liquid phase is that it should not evaporate at the working temperature of the column. Whilst it is customary to think of liquid phases of very low volatility, such as Apiezon greases, and polyethylene glycol esters, liquids of much higher volatility can be used if the column temperature is low (Porter and Johnson, 1961).

Gas–liquid chromatography possesses certain advantages over gas–solid chromatography and for some time after the discovery of the former, the latter tended to be ignored. The major differences are:

(a) Gas–liquid chromatography is applicable to all vapours and gases (except the permanent gases), whilst gas–solid is useful only for more permanent vapours.

(b) The column life, in gas–liquid chromatography, is almost indefinite, as long as the liquid phase does not evaporate at the working temperature. Due to irreversible adsorption it is preferable to change the column packing for each analysis in gas–solid chromatography.

(c) The reproducibility of the packing material is much better in gas–liquid than in gas–solid columns, hence putting the latter, which require more frequent changing, at a disadvantage.

(d) The gas–liquid system is much less likely to catalyse degradation of the eluting substances than is the gas–solid system.

(e) The large number of liquids, suitable for liquid phases in gas–liquid chromatography, and the possibility of varying the amounts of these present on the support material, makes this method a very versatile separative technique.

(f) Gas–solid chromatography has the advantage that the adsorbent can be used as a molecular sieve system.

V. Applications of Chromatography in Textile Science

There is a very broad use of natural and synthetic polymers, and also chemicals such as dyes and finishes, in the textile industry. The necessity for identification of some materials, the estimation of others and the determination of structures is a fruitful field for the application of a number of chromatographic techniques. Indeed, studies of the structure of keratin in particular have in no small measure contributed to the discovery of many chromatographic techniques. The subsequent development of such techniques for use in other aspects of textile science has been of considerable value both academically and technologically.

The applications of chromatographic techniques may be considered under the broad headings of protein fibres, cellulosic fibres, synthetic fibres, dyes and auxiliaries, and textile finishes.

A. Protein Fibres

The use of chromatographic techniques in determining the structure and reactions of keratin and fibroin have been recently summarized (Asquith,

1977), as has their application to the determination of keratin composition and biosynthesis (Fraser *et al.*, 1972). The proceedings of the Quinquennial Wool Conferences from 1955 to 1975 also provide numerous papers showing the development of the applications of chromatography in this field. Here it is sufficient to highlight some particular aspects of chromatographic systems which have been applied.

The use of automated ion exchange chromatography for amino acid analysis of protein fibres is well estabished. Such methods were first used for determining the amino acid composition of wool in 1962 (O'Donnel and Thompson, 1962), although manual ion exchange columns had been used earlier (Corfield and Robson, 1955).

Ion exchange columns have been used for separation and semi-quantitative estimation of cysteic acid peptides obtained from oxidized fractions of wool after partial hydrolysis (Asquith and Parkinson, 1971; Crewther, 1975) whilst the first isolation of larger protein fractions from oxidized wool was also carried out on ion exchange resins (Corfield *et al.*, 1967). More recently gas–liquid chromatography has been applied to the separation of amino acids in protein hydrolysates, the amino acids being converted to esters of the *N*-trifluoro-acetyl derivatives to render them volatile (Islam and Darbre, 1967; Casagrande, 1970; Jönsson *et al.*, 1973). Numerous trimethylsilylating agents have been investigated (Smith and Shewbart, 1960). The difficulty of complete separation of all amino acids in a protein hydrolysate still delays the acceptance of this general method but undoubtedly its rapidity and accuracy makes it extremely attractive, if satisfactory amino acid derivatives can be found.

The use of nylon 6 powder columns to determine *N*-terminal endgroups in keratin was first reported in 1959 (Steurle and Hill, 1959), and since then the method has been developed to study the extent of reaction of various side-chain groups in keratin fibre with reactive dyes (Corfield *et al.*, 1967; Baumann, 1970), though the acidic groups in the dyes may partially inhibit the reaction of the basic side-chains of lysine with 1-fluoro-2:4 dinitrobenzene (Hille, 1962; Asquith and Chan, 1971; Abbott *et al.*, 1975). The use of polyamide powder to separate 2-4-dinitrophenyl amino acids by thin-layer-chromatography has been reported (Wang and Huang, 1965).

Numerous types of chromatography have been used in isolating summarized fractions from wool and determining their structure. This field has recently been comprehensively summarized (Crewther, 1975). Studies of the primary structure of fibroins has also been summarized (Lucas *et al.*, 1958; Asquith, 1977).

Chromatographic techniques have proved useful in identifying amino acids formed during the modification of wool. Thus cysteic acid content is an indication of oxidative treatments undergone by the fibre and its determination is of some technological importance. Early methods for this estimation involved ion-exchange chromatography, though this has been largely superseded by electrophoresis (Ziegler, 1963). Similarly, lysinoalanine was first identified in alkali-treated wool by Ziegler (1964). The amounts of lysinoalanine and lanthionine present in alkali-treated wool are a useful measure of the extent of alkali damage, and various methods have been described for such estimations (Bauters et al., 1967). Ion-exchange chromatography has been used (Robson et al., 1967; Tasdhomme, 1970) whilst paper chromatography has been particularly investigated for the estimation of lanthionine (Derminot and Tasdhomme, 1965; Miro and Garcia Dominguez, 1966) and accuracies as high as 5 μmoles/gm wool are claimed for one method (Dowling and Crewther, 1964).

Similar chromatographic techniques led to the finding of β-aminoalanine in alkali treated wool (Asquith and Garcia Dominguez, 1968), and ornthino-alanine (Ziegler et al., 1967). Wool treated with amines in alkaline solution yielded further new products, which can be chromatographically separated (Garcia Dominguez et al., 1971; Asquith and Carthew, 1972; Asquith et al., 1974) from hydrolysates. The presence of the isopeptides N^6-γ-glutamyl lysine (Asquith et al., 1970) and N^6-β-aspartyl lysine (Asquith et al., 1971) were first detected in enzymic hydrolysates of wool followed by ion-exchange and paper chromatography. It was shown that these crosslinks increased in quantity when wool is subjected to dry heat (Asquith and Otterburn, 1971). A more efficient method of enzymic hydrolysis has recently been reported (Schmitz et al., 1975) whilst a more effective method of separation of these substances from hydrolysates on ion-exchange resins has been described (Otterburn and Sinclair 1976). It is interesting to note that these crosslinks may considerably influence nutritional values of edible proteins (Hurrell et al., 1976).

The fact that tyrosine and lysine are readily available in the amorphous regions of fibroin enabled Zahn and Zuber (1958) to react the side-chains with difunctionally reactive compounds isolating the products after protein hydrolysis on nylon powder columns. This enabled them, by using different reagents as "molecular calipers" to estimate some distances between tyrosine and lysine side-chains in the fibroin molecules. This work has been extended to the study of wool (Zahn and Hammoudeh, 1973).

Other important findings which have been materially assisted by the use of chromatographic techniques are the detection of cystine in fibroin (Schroeder and Kay, 1955; Zuber et al., 1957), and also its peptides (Robson et al., 1970; Earland and Robins, 1973). A lysine adduct has been isolated from a hydrolysate of wool dyed with a Remazol dye (Zahn and Rouette, 1968) whilst various polymethylene bridged crosslinks have been identified when wool is treated with bis-diazo-hexane (Zahn and Waschka,

1956). The ready addition of cysteine thiol groups to reactive double bonds has been used to convert these groups to more basic derivatives. The latter can then be easily estimated in wool hydrolysates by ion-exchange chromatography (Friedman and Noma, 1970; Friedman et al., 1970).

B. Synthetic Fibres

The contribution of chromatography to the synthetic-fibre field is probably more confined than in the field of natural fibres. Here the overall general structure of the groups of fibres is known, and it is relatively simple to differentiate these groups by more rapid tests. Chromatographic techniques are, therefore, of more importance in such secondary work as: the identification of specific fibre types within a group (e.g. polyamides); the identification of minor constituents within a commercial fibre, which impart particular properties to that fibre; the identification of low molecular weight contaminants (e.g. oligomers) introduced during formation of the polymer; the identification of materials formed during heat decomposition of the fibre (e.g. toxic hazards). Many polymers are used not only as fibres but also as plastics, and many studies have been carried out on the latter.

Early use was made of paper chromatography to identify different polyamide fibres. Zahn and Wolleman (1951) hydrolysed the fibres and separated the constituent monomers by paper chromatography. Acids such as adipic acid were developed on the paper with bromothymol blue, whilst diamines and amino acids were developed with ninhydrin. Estimation of the intensity of the blue ninhydrin spots by densitometry enable the composition of copolymers to be deduced. With the advent of thin-layer-chromatography this method was improved (Mori and Takeuchi, 1970), separations being carried out on silica gel. The R_F values (reproducible to 2%) were found to have a linear relationship to the number of carbon atoms in the chains for each series of hydrolysis products (diamines, dicarboxylic acids, amino acids) and hence complex copolymer constituents could be identified without the necessity of a full series of reference compounds. The identification and determination of difunctional acids in polyamides usually presents more problems than the identification of diamines and amino acids. Methylation of the hydrolysate followed by gas–liquid chromatography can be used to determine these acids (Anton, 1968) and also acids obtained from hydrolysed polyesters. Another more direct approach, but yielding less information, is the separation of polyamides by paper chromatography with 88% formic acid as an eluting solvent. The separated spots can be developed with Solway Fast Blue 2BS. R_F values of nylon 6.6 and nylon 6 respectively were 0.4 and 0.8 whilst copolymers had intermediate values, polyurethanes having an R_F of 0.0.

The concentration of the formic acid used for elution is crucial however (Ayres, 1953).

Whilst amine end-groups in polyamides can be determined by titration, the presence of dye on the fibre makes this impossible. To overcome this difficulty the polyamide can be dissolved in a phenol/ethanol/water mixture and the dye removed by passage down a strongly basic anion exchanger. The eluted nylon solution can then be titrated in the normal manner (Valk *et al.*, 1974, 1975).

A novel use of chromatography in the polyamide field (Byrne *et al.*, 1968) was to extract worn nylon garments with solvents and examine the extracted fats by thin layer chromatography. The yellowing of the fabric on repeated wearing and washing was claimed to be enhanced by deposited body fats.

Thin-layer chromatography has been used to separate low molecular weight oligomers formed during the trans-esterification of dimethyl terephthalate with ethylene glycol (Dorman-Smith, 1967). Dimethyl-terminated dimers and trimers, monohydroxy-ethylene-mono-methyl terephthalate and bis-hydroxyethylene terephthalate were identified. Other workers (Favretto *et al.*, 1970) used thin-layer combined with gas–liquid chromatography to analyse the molecular weight distribution of a series of oligomers of polyethylene glycols $RO(CH_2CH_2O)_nR'$. The oligomer spots were developed with iodine vapour and the intensities measured by reflectance. The gas–liquid and thin-layer techniques gave good agreement and the results tallied with those obtained by classical methods up to $n = 17$. Further characterization studies of glycol ethers have recently been carried out (Halken and Khemangkom, 1971), by gas–liquid chromatography.

The constituents of fibres which can be degraded by hydrolysis have been examined by rendering the hydrolysis products more volatile by conversion to the trimethylsilyl derivatives (Ponder, 1970). By the same system molecular weight distribution of polyethylene glycols has been determined (Fletcher and Persinger, 1967).

Hydrolysis, followed by gas chromatography of derivatized products is limited in application to specific groups of textile fibres. A more general method for the study of non-volatile polymers is the special gas chromatographic technique of pyrolysis. This system can be used to advantage in analysing both synthetic fibres and polymers. The material is heated to a temperature at which decomposition occurs and the volatile products are analysed by gas chromatography. With polymers three types of decomposition may occur (Rybicka, 1963): (a) thermal rupture of the polymer chain. The free radicals so formed yielding the monomers or a variety of other products; (b) thermal rupture between the side-chains and the polymer backbone, the latter crosslinking and remaining in the pyrolysis chamber; (c) crosslinked polymers may tend to carbonize, giving off volatile products difficult to relate back to the constituents of the original polymer.

Obviously the products obtained will be partly determined by the conditions of pyrolysis, but if these are carefully controlled, and the pyrograms of a large number of polymers are catalogued, it is possible to distinguish between them. The technique is of most value when the polymer is in a particular environment, e.g. pigmented.

Various designs for controlled pyrolysis chambers have been reported (Swann and Dux, 1961; Barbour, 1964; Ettre and Varadi, 1964). The pyrolysis products of cellulose esters are difficult to interpret. However gross differences in the materials can be easily seen, whereas subtle differences can be observed when the conditions are carefully controlled (Manka, 1964). Cellulose acetate gives a quantitative recovery of acetic acid when pyrolysed (Groten, 1964). Pyrolysis patterns of poly-(ethylene), poly-(propylene) blends and copolymers thereof are easily distinguishable (Groten, 1964), whilst poly-(vinyl alcohol) gives acetaldehyde and acetic acid (Ettre and Varadi, 1963). Different polyacrylates also give unique pyrograms (Radell and Strutz, 1959) whilst polystyrenes give the monomer up to 700°C (Lehmann and Brauer, 1961). Schemes involving gas chromatography for the characterization of polymers have been summarized (Claver, 1966), and early work (Hasse and Rau, 1966) suggested that textile fibres can be identified by pyrolysis and gas chromatography of the products. Derminot and Rabourdin-Delin (1971) later proposed a potential scheme, based on pyrolysis, which was generally applicable to fibres within a group, e.g. modacrylics, and also showed that some mixtures of fibres could be identified, such as polyester-cotton, nylon 6.6–wool and wool–acrylic blends.

Various techniques, other than pyrolysis, have been used to degrade materials, so that classifiable degradation products can be identified by gas chromatography. Such systems are based on oxidation (Scholz *et al.*, 1966), electrical discharge (Sternberg and Litle, 1966) and mercury-sensitized photolysis (Juvet and Turner, 1965). They do not appear to have been used as yet for fibre identification.

C. CELLULOSE FIBRES

Chromatographic methods have proved immensely useful in carbohydrate chemistry. They often permit identification of sugars by most simple methods, and several classical reviews have been written thereon (e.g. Hirst and Jones, 1949; Partridge, 1950; Binkley, 1955). Techniques such as gel filtration have been used for separating polysaccharides (Churms, 1970) whilst gas–liquid chromatography has been used to a limited extent to separate volatile derivatives of monosaccharides (Aspinall, 1963; Sweeney *et al.*, 1966). Examination of wood pulp by this method has been reported (Laver *et al.*, 1967). The automated determination of monosaccharides by ion-exchange chromatography of their borate complexes has been described recently (Walborg *et al.*, 1975), whilst the whole topic of carbohydrate analysis has been reviewed (Whistler and BeMuller, 1976).

Possibly because the structure of cellulose has been well defined by classical means, less use has been made of chromatography in the specific field of cellulose fibre chemistry. Due to the relatively simple structure, chromatographic techniques have been used mainly to examine modified material. Thus partially methylated celluloses have been examined by hydrolysis, the products being subjected to gas–liquid chromatographic separation (Neeley *et al.*, 1962). Similar work has indicated that furanosides also result from the hydrolysis of methylated cellulose (Kirchner, 1960). Natural cellulosic fibres have been analysed for alkoxy groups by reaction with hydriodic acid followed by chromatography of the resulting iodides on a silicone gum rubber column (Sporek and Danyi, 1962). Uronic acid and aldonic acid groups in cellulose and hemicellulose have been determined by ion exchange chromatography and it has been shown that 90% of the aldonic acid groups in wood pulp can be accounted for as simple aldonic acids (Pettersson and Samelson, 1966). The method was extended to identify uronic acids in spent sulphite liquors (Pettersson and Samuelson, 1967).

Fingerprint pyrograms of various polymers, including cellulose, have been reported (Groten, 1964). Flash pyrolysis of cellulose using a carbon arc source of energy gives a complex mixture of products including acetone acrolein and propionaldehyde, as well as some esters and ethers (Martin and Ramstand, 1961); 2-methyl furan and furan can also be detected (Greenwood *et al.*, 1961).

Generally it would seem reasonable to suggest that studies of cellulose fibres by chromatographic techniques have been incidental to the study of carbohydrates as a general field.

D. DYES AND PIGMENTS

The use of chromatographic methods for the separation of dyes is classical. Long before scientific techniques of chromatography were developed, dyers used elementary frontal analysis techniques (to check whether dyes were "mixed" or "pure") by splashing solutions on adsorbent paper and observing the colour distribution at the edge of the adsorbed spots. The separations of dyes on alumina columns and the comparison of the colours in vintage and cheap port wines by elution on alumina are well known elementary experiments. The purification of synthesized dyes in quantities up to 10 g by column chromatography (either by eluting the dye, or by breaking up the column and extracting the separated band with a suitable solvent) are standard procedures in organic synthesis. Mixtures of dyes are used to assess the effectiveness of different samples of an adsorbent. The application of such techniques as paper, thin-layer and column chromatography to dye separation has been discussed in numerous books (cf. Lederer and Lederer, 1957; Strain, 1942; Truter, 1963).

Despite this, so numerous are the synthetic dyes now available for different purposes, that new chromatographic systems for the identificaïion of dyes within specific classes continue to be reported. Brown (1960) reported the use of paper chromatography in identifying and analysing a number of commercial dyes. He later extended this work to the identification of oxidation bases (used for dyeing hair and furs), reactive dyes, textile finishes and fluorescent brightening agents, demonstrating the interdependence of chromatography and other analytical techniques for this purpose (Brown, 1964).

Rayburn (1969) has reviewed the application of thin-layer chromatography to the identification of dyes and finishes in the textile trade, whilst Freeman (1970) has described a dye analysis scheme using paper and thin-layer chromatography supplemented by infra-red spectra. Comprehensive reviews cover the work on dyestuff identification up to 1969 (Brown, 1969a, b) and the use of thin-layer chromatography in quality control has recently been described (Wood, 1975). Methods for the documentation of thin-layer chromatographs of dyes have also been reported (Jones, 1967).

Direct dyes can be distinguished from basic dyes by pH dependent thin-layer chromatography on cellulose acetate and from acid or metal complex dyes by paper chromatography (Schlegelmilch and Khodadadian, 1973). Food dyes have been a topic of considerable study and separations thereof have been investigated on various thin-layer chromatographic substrates such as cellulose (Wollenweber, 1962), starch (Davidek and Janicek, 1964) and polyamides (Wang, 1967). The detection of the harmful yellow dyes, auramine, picric acid and metanil has also been reported (Chiang and Lin, 1969).

Various attempts have been made to correlate dyeing properties with chromatographic behaviour of dyes. Thus the relationship between R_F values of leuco vat dyes on paper and their substantivity to cellulose indicates that substantivity increases as the R_F value of the leuco base decreases (Artym and Moryganov, 1966). The use of circular cellulose filter paper chromatography enables the influence of pH on the extent and rate of reaction of reactive dyes with cellulose to be studied, and information concerning the kinetics of the interaction can thus be obtained (Iyer and Singh, 1969). Methods for determining the amounts of bound, unfixed and hydrolysed reactive dyes present after printing or dyeing have been described (Kritchevskii and Krainov, 1971), whilst it has been shown that the behavior of mixtures of reactive dyes on thin-layer and paper chromatography is affected not only by differences in the substantivity of the dyes for the support medium, but also by the initial amount of dye applied to the chromatogram (Rattee et al., 1965).

Gas chromatographic systems have little use in dyestuff separation due to the lack of volatility of the materials. The high substantivity of dyes for substrates seems to have inhibited the use of ion exchange resins for purification although acid dyes can be easily converted to the free acids by passage through cation exchanges. Equally, gel permeation does not seem to have been developed for this field, though studies of the purification of dyehouse eluents have been reported recently. In these, natural ion exchange resins are used for dye removal (Poots et al., 1976a,b). Gas chromatography has been used to separate and recover various anthraquinone derivatives which have been rendered volatile by conversion to either trifluoroacetyl or trimethylsilyl derivatives (Terril and Jacobs, 1970).

In the dyeing process it has been shown that considerable savings in dyestuff, water and chemicals can be made when thin-layer chromatography is used to correlate the hydrolysis of reactive dyes with the pH of the dyebath and the dye auxiliary concentration (Szuchy et al., 1974). Silica gel thin-layer-chromatography has also been used to compare the effects of applying reactive disperse dyes to polyamide fibres from aqueous solutions, and from a solvent system (dimethyl formamide/perchlorethylene) respectively. These studies showed that when the dyes were applied from aqueous solution to polyamide the fabric contained covalently-linked, non-reacted and hydrolysed dye. The last was absent in the solvent-dyed material (Zhukov et al., 1975). Gas chromatography has been used to identify carriers, and mixtures of carriers, used in the dyeing of polyesters (Giogas, 1972).

Organic pigments used for the colouration of rubber, plastics, printing inks, lacquers and paints are mainly azo compounds, derivatives of anthraquinone, phthalocyanines or vat dyes. All are practically insoluble in water and some are only soluble in high boiling solvents at higher temperatures. Their low volatility prevents the use of gas–liquid chromatography for separation, and hence it would seem that these classes of coloured compounds would resist chromatographic techniques. Various attempts have been made to separate them on paper and thin-layer chromatography by using solvents such as dimethylformamide or pyridine as eluents, and limited success has been achieved (Stahl, 1965; McClure et al., 1968; Gasparic, 1972). It has been claimed that by careful selection of the adsorption material it is possible to identify organic pigments by thin-layer chromatography without the use of reference pigments Schlegelmilch and Kuss, 1973).

E. Textile Finishes

Textile finishes comprise a large number of different classes of compounds such as fluorescent brightening agents, surface active agents, moth proofing agents and insecticides, resins, and oils. Many published works tend to incorporate details of the identification and analyses of such materials with the assessment of other materials, particularly dyes. A number of the references given subsequently are equally relevant to dye separation techniques.

1. Fluorescent Brightening Agents (FBAs)

Lanter (1966) reported early work on the separation of FBAs by paper and thin-layer chromatography whilst, analogous to his work on organic pigments, Gasparic (1967) studied the separation of insoluble FBAs. He reported that the water soluble FBAs could be separated by thin-layer chromatography on silica gel but reversed phase paper chromatography was more suitable for the insoluble types. Paper impregnated with dimethyl formamide or paraffin was used and amounts as low as 0.1μ gms of FBA could be detected.

In a critical review, Kurz and Schuierer (1967) showed that silica gel was better for FBA separation than a cellulose support. They pointed out that all resolutions of FBAs should be carried out in the absence of light to avoid cis-trans isomerism during analysis.

By using a combination of different supports and solvents Thiedel and Schmitz (1967) separated a series of sixteeen FBAs by thin-layer chromatography. The major supports used were partially siliconized cellulose and polyamide whilst the solvent mixtures were complex. Later workers (Figge, 1968) classified FBAs according to their solubilities in various solvents and described methods for the separation of eighteen FBAs by thin-layer chromatography on Kieselguhr G. or H. Non-ionic and ionic FBAs were first separated as groups by development with basic polar solvents. Within each group, separation was performed by one- or two-dimensional chromatography using solvents selected from nineteen systems.

The R_F values of twenty-four commercial FBAs have been recorded for a paper chromatographic system, which consists of Whatman No. 3 paper impregnated with 50% dimethyl formamide, and uses n-heptane or n-hexane as developing solvents (Gasparic, 1970).

By pH dependent chromatography on cellulose acetate and initial reaction chromatography on cellulose, flavanic acid derivatives of FBAs can be separated out and then specifically identified by thin-layer chromatography on silica gel (Schlegelmilch *et al.*, 1971).

2. Resin Finishes

Paper, thin-layer and gas chromatographic techniques have all been used to identify resin finishes on textiles. In the course of a wide review in 1965 of the use of gas chromatography for research and control in the textile industry, Edel summarized its value not only in such fields as polymer and plasticizer identification, assessment of pyrolysis products of cellulose, and residual odours in solvent degreased materials, but also in assessing the course of crosslinking in cellulose by bis-(methoxymethyl)-amine derivatives. At the same time Kiel (1965) described the use of paper chromatography not only to identify soaps and resin finishes but also to evaluate the fastness to light of the latter. Other work (Hughes, 1966) has shown that resin finish, based on urea, melamine, cyclic ureas and other amino compounds can be identified by stripping from the fabric with dilute acid. The components are separated on thin-layer chromatograms, being visualized with diamino-benzaldehyde. Similar chromatographic methods have also been reported by Buchsbaum and Datyner (1966). Valk (Valk *et al.*, 1968) has described the identification of etherified *N*-methylol crosslinking agents used on cellulose. The finish is hydrolysed and removed from cotton polyester fabrics. Carbamates and alcohols from the etherified *N*-methylol groups are then extracted from the hydrolysate with ether and identified by thin-layer chromatography on silica gel. The scheme of separation was later developed in more detail and the identification of sulphones in alkaline hydrolysates of finish based on divinyl sulphone was included (Valk *et al.*, 1969a). Further work described triazine derivatives (Valk *et al.*, 1969b).

An interesting use of thin-layer chromatography is to cure cellulose-reactive finishing agents on adsorbent cellulose film prior to development (Moore and Babb, 1972). The proportion which has reacted with the adsorbent is immobilized by the curing process, whilst the unreacted material is mobile in the eluting solvent. This would seem to be a useful rapid assay of the reactivity of a new finish. It could also be used for investigating the by-products formed and the mechanism of the curing reaction with cellulose.

Resins based on difunctional acids and amines can be identified and estimated by methanolysis of the resin with methanolic hydrogen chloride, followed by gas chromatography of the methyl esters so formed. Similar identification of polyesters, as fibres and finish, can be performed by trans-esterification of the polymer for one hour under pressure at 175°C in excess methyl acetate. The resultant acid methyl esters in methyl acetate are then determined by gas–liquid chromatography (Stephens, 1969; Swanepoel et al., 1970). The method has been recommended for estima-tion of Hercosett 57 applied to wool fabric as an anti-shrink finish, though recently more direct, less complicated, methods which do not involve chromatography have been proposed (Lewis, 1976).

Apart from the identification and estimation of textile finishes sundry materials allied to the textile auxiliaries have been examined (Rawlinson and Deeley, 1967). Thus lubricants for wool, such as mineral oil, sperm oil, castor oil and wool grease can be identified satisfactorily by their chroma-tographic behaviour on thin-layer silica gel, using alcoholic molyb-dophosphoric acid as a locating agent (Hartley, 1968). Organic additives, coatings and surface treatment materials present in paper have been examined by similar methods (Broniatowski, 1967). Mixtures of plasti-cizers from coated fabrics have been separated; as many as thirty compounds being described (Rau and Haase, 1965). Rusznak et al. (1966) have used paper chromatography to not only determine the relative substantivity for cellulose of a number of dyes and the effect of additives thereon, but also to study the hydrolysis of starch. Their results compare favourably with polarographic and colorimetric methods used to determine starch decomposition.

The examination of surface active agents has been carried out by gas chromatography of volatile derivatives (Suffis et al., 1967; Halken and Khemangkom, 1971), paper chromatography (Borecky, 1966), thin-layer chromatography (Milster et al., 1967) and ion-exchange systems coupled with thin-layer chromatography (Grosse, 1966).

3. Moth Proofing and Insecticides

Early moth proofing agents and insecticides had limited fastness to wool and other fibres. D.D.T., whilst highly potent and water insoluble, requires application from organic solvents as does Dieldrin. Clearly, soluble reagents, comparable in structure to acid dyes but colourless, have the advantages of substantivity and ease of application. From the point of view of chromatographic assay two different systems are necessary. The low volatility of phosphate and sulphonic acid–containing soluble insecticides–precludes separation by gas chromatographic systems, whilst the more volatile insoluble substances can be estimated in very small quantities by this method. Thus the chlorinated moth proofing agents (Dieldrin and D.D.T.) have been estimated by gas–liquid chromatography (Williams, 1966), using silicone oil columns, and their fastness properties estimated, whilst cholinesterase-inhibiting insecticides have been determined on thin-layer-chromatography (Menn and McBain, 1966). In the latter case seven organo-phosphate and seven N-methyl carbamates were identified and the detection limits reported.

Dieldrin is an attractive insecticide finish on the grounds of cost and effectiveness. Unfortunately some doubts have been expressed concerning its toxicity to humans. A number of alternative, less toxic finishes have therefore been investigated (Asquith, 1977), such as O-O-diethyl phosphorothionate-O-ester, phenylglyoxalonitrile, bromosalicylanilides and quaternary ammonium compounds. The identification and estimation of these should prove a further challenge to chromatographic techniques.

VI. Conclusion

The application of chromatographic techniques to textile science have resulted in findings of real value, particularly in relation to wool science and the analysis of many textile materials. The types of techniques used depend on many different fundamental principles and it is to be expected that workers have mainly tended, particularly in analytical work, to report their findings in relation to one technique and its application to various textile problems. As a result no review can cover the whole field of published work on both the techniques and their application to textiles. In this article we have attempted to indicate papers which will inform the reader of the general principles of chromatographic techniques and also other papers which apply these techniques to various aspects of textile science.

References

Abbott, T. J., Asquith, R. S., Chan, D. K. and Otterburn, M. S. (1975). *J. Soc. Dyers Colourists* **91**, 133.

Ackers, G. K. (1970). *Adv Protein Chem.* **24**, 343.

Alm, R. S., Williams, R. J. P. and Tiselius, A. (1952). *Acta Chem. Scand.* **6**, 826.

Andrews, P. (1964). *Biochem. J.* **91**, 222.

Anton, L. (1968). *Anal. Chem.* **40**, 1116.

Aspinall, G. O. (1963). *J. Chem. Soc.* 1676.

Ashley, B. D. and Westphal, U. (1955). *Archs. Biochem. Biophys.* **56**, 1.

Artym, M. I. and Moryganov, P. V. (1966). *J. Textile Inst.* **57**, A20.

Asquith, R. S. (1977). "Chemistry of Natural Protein Fibres" (R. S. Asquith, Ed.). Plenum Press, New York and London.

Asquith, R. S. and Carthew, P. (1972). *Biochim. Biophys. Acta* **278**, 8.

Asquith, R. S., Carthew, P., Hanna, H. D. and Otterburn, M. S. (1974). *J. Soc. Dyers Colourists* **161**, 357.

Asquith, R. S. and Chan, D. K. (1971). *J. Soc. Dyers Colourists* **87**, 181.

Asquith, R. S. and Garcia-Dominguez, J. J. (1968). *J. Soc. Dyers Colourists* **84**, 155.

Asquith, R. S., Gardner, K. L. and Otterburn. M. S. (1971). *Experientia* **27**, 1388.

Asquith, R. S. and Jordan, B. J. (1961). Inst. Textile de France, Colloquium Structure de la Laine 94.

Asquith, R. S. and Parkinson, D. C. (1971). *Makromol. Chem.* **141**, 233.

Asquith, R. S. and Otterburn, M. S. (1971). *Appl. Polymer Symp.* **18**, 277.

Asquith, R. S., Otterburn, M. S., Cole, M., Buchanan, J. H. Fletcher, J. C. and Gardner, K. L. (1970). *Biochim. Biophys. Acta* **221**, 342.

Asquith, R. S. and Shaw, T. (1968). *Makromol. Chem.* **115**, 198.

Axelrod, L. R. (1955). *Anal. Chem.* **27**, 1308.

Ayers, C. W. (1953). *Analyst* **78**, 382.

Barbour, W. M. (1964). *J. Gas Chromatog.* **2**, 1.

Baumann, H. (1970). *Textilveredlung* **5**, 506.

Bauters, M., Lefebure, L. and Van Overbeke, M. (1967). *Bull. Inst. Textile France* **21**, 425.

Beroza, M. and Coad, R. A. (1967). In "The Practice of Gas Chromatography" (L. S. Ettre and A. Zlatkis, Eds.). Wiley Interscience, New York.

Binkley, W. (1955). Adv. Carbohydrate Chem. **10**, 55.

Block, R. J. (1959). "A Manual of Paper Chromatography and Paper Electrophoresis", Academic Press, New York and London.

Bobbitt, T. M. (1963). "Thin Layer Chromatography", Reinhold Publishing Corporation, New York.

Boldingh, J. (1948). *Experientia* **4**, 270

Borecky, J. (1966). *Kolor Ertestto* **8**, 386.

Brenner, M. and Niederwieser, A. (1961). *Experientia* **17**, 237.

Brewer, P. I. (1960). *Nature, Lond.* **188**, 934.

Borckmann, H. and Schodder, H. (1941). *Ber.* **74**, 73.

Broniatowski, A. (1967). *Svensk, Pappierstid.* **70**, 234.

Brown, J. C. (1960). *J. Soc. Dyers Colourists* **76**, 536.

Brown, J. C. (1964). *J. Soc. Dyers Colourists* **80**, 185.

Brown, J. C. (1969a). *J. Soc. Dyers Colourists* **85**, 137.

Brown, J. C. (1969b). *J. Soc. Dyers Colourists* **85**, 150.

Brown, W. G. (1939). *Nature, Lond.* **143**, 377.

Bucksbaum, N. and Datyner, A. (1966). *J. Soc. Dyers Colourists* **82**, 18.

Byrne, G. A., Holmes, F. M. and Lord, J. (1968). *J. Soc. Dyers Colourists* **84**, 20.

Calmon, C. and Kressman, T. R. E. (1957). In "Ion-Exchangers in Organic and Biochemistry", Interscience, New York.

Casagrande, D. J. (1970). *J. Chromatog.* **49**, 537.

Cerrai, E. (1964). *Chromatography Reviews* **6**, 129.

Chiang, H. C. and Lin, S. L. (1969). *J. Chromatog.* **44**, 203.

Chilton, T. and Partridge, W. M. (1950). *J. Pharm. Pharmacol.* **2**, 784.

Churms, S. (1955). *Adv. Carbohydrate Chem.* **10**, 55.

Claver, G. C. (1966). *J. Paint Technol.* **38**, 74.

Condon, R. D., Scholly, P. R. and Averil, W. (1960) In "Gas Chromatography". (R. P. W. Scott, Ed.). Butterworth Press, London.

Consden, R., Gordon, A. H. and Martin, A. J. P. (1944). *Biochem. J.* **38**, 24.

Corfield, M. C., Fletcher, J. C. and Robson, A. (1967). *Biochem. J.* **102**, 801.

Corfield, M. C. and Robson, A. (1955). *Biochem. J.* **59**, 62.

Crewther, W. G. (1975). Proc. 5th Intern. Wool Textile Res. Conf. Aachen **1**, 1.

Cuetracasas, P. and Anfinsen, C. B. (1971). *Ann. Rev. Biochem.* **40**, 259.

Davidek, J. and Janicek, G. (1964). *J. Chromatog.* **15**, 542.

Dean, J. A. (1969). "Chemical Separation Methods", Van Nostrand Inc., London and New York.

Derminot, J. and Rabourdin-Belin, C. (1971). *Bull. Inst. Textile France* **25**, 721.

Derminot, J. and Tasdhomme, M. (1965). *Bull. Inst. Textile France* **19**, 519.

Determann, H., Lueben, G. and Wieland, T. (1964). *Makromol. Chem.* **73**, 168.

Deul, H., Solms, J., Anyas-Weisz, L. and Huber, G. (1951). *Helv. Chim. Acta* **34**, 1849.

Dorman-Smith, V. A. (1967). *J. Chromatog.* **29**, 265.

Dowling, L. M. and Crewther, W. G. (1964). *Anal. Biochem.* **8**, 244.

Earland, C. and Robins, S. P. (1973). *Int. J. Protein Res.* **5**, 327.

Edel, G. (1965). *Bull. Inst. Textile France* **19**, 579.

Eltre, K. and Varadi, P. F. (1963). *Anal. Chem.* **35**, 69.

Eltre, K. and Varadi, P. F. (1964). *Anal. Chem.* **36**, 90.

Farradane, J. (1951). *Nature, Lond.* **167**, 120.

Favretto, L., Marletta, G. P. and Gabrielli, L. F. (1970). *J. Chromatog.* **46**, 255.

Figge, K. (1968). *Fette Seifen Anstri Mittel* **70**, 680.

Fischer, A. and Behrens, M. (1952). *Z. Physiol. Chem.* **291**, 14.

Fischer, L. (1969). "An Introduction to Gel Chromatography", North Holland Publishing Co., Amsterdam.

Fish, W. W., Reynolds, J. A. and Tanford, C. (1970). *J. Biol. Chem.* **245**, 5166.

Fletcher, J. P. and Persinger, H. E. (1967). Pittsburgh Conf. Appl. Chem. Appl. Spectroscopy. Abs. Papers 65.

Franek, F. and Mastner, T. (1959). *Collec. Czech. Chem. Commun.* **24**, 2952.

Fraser, R. D. B., Macrae, T. P. and Rogers, G. E. (1972). "Keratins". American Lecture Series, C. C. Thomas Publishers.

Freeman, J. F. (1970). *Can. Textile. J.* **87**, 83.

Freundlich, H. and Heller, W. (1939). *J. Amer. Chem. Soc.* **61**, 2228.

Friedman, M., Krull, L. H. and Cavins, J. F. (1970). *J. Biol. Chem.* **245**, 3868.

Friedman, M. and Norma, A. T. (1970). *Textile Res. J.* **40**, 1073.

Garcia-Dominguez, J. J., Miro, P., Reig, F. and Anguera, S. (1971). *Appl. Polymer. Symp.* **18**, 269.

Gasparic, J. (1967). *Kolor Ertestio* **8**, 2.

Gasparic, J. (1970). *J. Chromatog.* **49**,

Gasparic, J. (1972). *J. Chromatog.* **66**, 179.

Gelotte, B. (1960). *J. Chromatog.* **3**, 330.

Giddings, F. (1965). "Dynamics of Chromatography", Marcel Dekker, New York.

Giles, C. H., MacEwan, T. H., Nakhwa, S. N. and Smith, D. (1960). *J. Chem. Soc.* 3973.

Giorgas, G. (1972). *Textil. Praxis* **27**, 264.

Glueckauf, E. and Kitt, G. P. (1956). "Vapour Phase Chromatography". Butterworth Press, London.

Granath, K. and Flodin, P. (1961). *Makromol. Chem.* **48**, 160.

Green, J. and Marcinkiewicz, S. (1963). *Chromatog. Rev.* **5**, 58.

Greenwood, G. T., Knox, J. H. and Milne, E. (1961). *Chem. Ind.* 1879.

Grosse, I. (1966). *J. Textile Inst.* **57**, 640.

276   R. S. ASQUITH AND M. S. OTTERBURN

Groten, B. (1964). *Anal. Chem.* **36**, 1206.
Hagdahl, L. and Damelson, C. E. (1954). *Nature, Lond.* **174**, 1062.
Hagdahl, L. and Holman, R. T. (1950). *J. Amer. Chem. Soc.* **72**, 701.
Halken, J. K. and Khemangkom, V. (1971). *J. Oil Colour Chemists Assoc.* **54**, 764.
Hamilton, P. B. (1963). *Anal. Chem.* **35**, 2055.
Hartley, R. S. (1968). *J. Textile Inst.* **59**, 401.
Hashimoto, Y. (1963). "Thin Layer Chromatography", Hirokawa Publishing Co., Tokyo.
Hasse, H. and Raw, J. (1966). *Melliand Textilber.* **47**, 434.
Hayward, D. O. and Trajamell, B. M. W. (1964). "Chemisorption", Butterworth Press, London.
Heftman, E. (1967). "Chromatography" 2nd Edition, Reinhold Publishing Corp., New York.
Hesse, G. and Weil, H. (1954). "Michael Tswett's Erste Chromatographishe Schrift", Woelm Exchange.
Hille, E. (1962). *Textil Praxis* **17**, 171.
Hirst, E. L. and Jones, J. K. N. (1949). *Discussions Faraday Soc.* **7**, 268.
Holman, R. T. (1951). *J. Amer. Chem. Soc.* **73**, 1261.
Holmes, J. C. and Morrell, F. A. (1957). *Appl. Spectrosc.* **11**, 86.
Hughes, C. H. (1966). *Chem. Ind.* **10**, 411.
Hurrell, R., Carpenter, K. J., Sinclair, W. J., Otterburn, M. S. and Asquith, R. S. (1976). *Brit. J. Nutr.* **35**, 383.
Ingle, R. B. and Minshall, E. (1962). *J. Chromatog.* **8**, 386.
Islam, A. and Darbre, A. (1967). *J. Chromatog.* **29**, 49.
Iyer, S. R. S. and Singh, G. S. (1969). *Colourage* loc. cit.
James, A. T. and Martin, A. J. P. (1952a). *Analyst.* **77**, 915.
James, A. T. and Martin, A. J. P. (1952b). *Biochem. J.* **50**, 679.
James, A. T. and Martin, A. J. P. (1954). *Brit. Med. Bull.* **10**, 170.
Jacques, J. and Mathieu, J. P. (1946). *Bull. Soc. Chim. France.* 94.
Jones, R. L. (1967). *J. Soc. Dyers Colourists* **83**, 283.
Jonsson, T., Eyem, J. and Sjoquist, J. (1973). *Anal. Biochem.* **51**, 204.
Joubert, F. J. and Burns, M. A. C. (1967). *J. South African Chem. Inst.* **20**, Jutisz, Jitisz, M. and Teichner, S. (1947). *Bull. Soc. Chim. France* 389.
Juvet, R. S. and Turner, L. P. (1965). *Anal. Chem.* **37**, 1464.
Keck, W. (1971). *Anal. Biochem.* **39**, 288.
Kiel, E. G. (1965). *Textile* **24**, 607.
Kircher, H. W. (1960). *Anal. Chem.* **32**, 1103.
Kritchevskii, G. E. and Krainov, M. M. (1971). *Tekstil. Prom.* **8**, 59.
Kurz, J. and Schuierer, M. (1967). *Fette Seifen Anstri Mittel* **69**, 24.
Labler, L. and Schwarz, V. L. (1965). "Thin Layer Chromatography", Elsevier Publishing Co. Amsterdam.
Lanter, J. (1966). *J. Soc. Dyers Colourists* **82**, 125.
Lard, E. W. and Horn, R. C. (1960). *Anal. Chem.* **32**, 878.
Lathe, G. H. and Ruthven, C. R. J. (1956). *Biochem. J.* **62**, 665.
Laver, M. L. (1967). *Tappi.* **50**, 618.
Lea, J. D. and Sehon, A. H. (1962). *Can. J. Chem.* **40**, 159.
Leathand, D. A. and Shurlock, B. C. (1970). "Identification Techniques in Gas Chromatography", Wiley-Interscience, London.

Lederer, E. and Lederer, M. (1957). "Chromatography", Elsevier Publishing Co. Amsterdam.

Lehmann, F. A. and Brauer, G. M. (1961). *Anal. Chem.* **33**, 673.

Lewis, J. (1976). *J. Soc. Dyers Coloursists* **92**, 440.

Llewellyn, P. and Littlejohn, D. (1968). 16th Ann. Conf. Mass Spectrosc. Allied Topics. ASTM Committee E4 Pittsburgh Pa.

Lovelock, J. E. and Lipsky, S. R. (1960). *J. Amer. Chem. Soc.* **82**, 431.

Lovelock, J. E., Simmonds, P. G. and Shoemake, G. R. (1970). *Anal. Chem.* **42**, 881.

Lowe, C. R. and Dean, P. D. G. (1974). "Affinity Chromatography", Wiley-Interscience, London.

Lucas, F., Shaw, J. T. B. and Smith, S. G. (1958). *Advances in Protein Chem.* **13**, 107.

Ma, T. S. and Lucas, A. S. (1976). "Organic Functional Group Analysis by Gas Chromatography", Academic Press, London.

McClure, A., Thomson, J. and Tannahill, J. (1968). *J. Oil Colour Chemists Assoc.* **51**, 580.

McKenna, T. A. and Idleman, J. A. (1960). *Anal. Chem.* **32**, 1299.

Manka, D. P. (1964). *Anal. Chem.* **36**, 480.

Mann, K. G. and Fish, W. W. (1972). *Methods Enzymol.* **26**, 28.

Marden, N. (1965). *Ann. N.Y. Acad. Sci.* **125**, 428.

Martin, A. J. P. (1948). *Ann. N.Y. Acad. Sci.* **49**, 249.

Martin, A. J. P. and James, A. T. (1956). *Biochem. J.* **63**, 138.

Martin, A. J. P. and Synge, R. L. M. (1945a). *Biochem. J.* **35**, 91.

Martin, A. J. P. and Synge, R. L. M. (1945b). *Biochem. J.* **35**, 1358.

Martin, S. B. and Ramstad, R. W. (1961). *Anal. Chem.* **33**, 982.

Menn, J. J. and McBain, J. C. (1966). *Nature, Lond.* **209**, 1351.

Michalec, C. (1955). *Naturwissenschaften* **42**, 509.

Milster, H., Meckel, L. and Schwimmis, W. (1967). *Z. Ges. Textil-Industrie* **69**, 555.

Miranda, F., Rochat, H. and Lissitzky, S. (1962). *J. Chromatog.* **7**, 142.

Miro, P. and Garcia-Dominguez, J. J. (1966). *Melliand Textilber* **47**, 676.

Moore, D. R. and Babb, R. M. (1972). *Textile Res. J.* **42**, 506.

Moore, J. C. (1964). *J. Polymer Sci.* **2A**, 835.

Moore, S. and Stein, W. H. (1948). *Ann. N.Y. Acad. Sci.* **49**, 265.

Moore, S. and Stein, W. H. (1952). *Ann. Rev. Biochem.* **21**, 521.

Mori, S. and Takeuchi, T. (1970). *J. Chromatog.* **47**, 224.

Morris, C. J. O. R. (1964). *J. Chromatog.* **16**, 167.

Mould, D. L. and Synge, R. L. M. (1954). *Biochem. J.* **58**, 571.

Neeley, W. B., Nott, J. and Roberts, C. B. (1962). *Anal. Chem.* **34**, 1423.

O'Donnell, I. J. and Thompson, E. O. P. (1962). *Aust. J. Biol. Sci.* **15**, 740.

Otterburn, M. S. and Sinclair, W. S. (1976). *J. Sci. Fd. Agric.* **27**, 1071.

Partridge, S. M. (1950). *Biochem. Soc. Symp.* **3**, 52.

Pettersson, S. and Samuelson, O. (1966). *Svensk Pappierstid,* **69**, 729.

Pettersson, S. and Samuelson, O. (1967). *Svensk Pappierstid.* **70**, 462.

Phillips, C. S. G. and Timms, P. L. (1961). *J. Chromatog.* **5**, 131.

Ponder, L. H. (1970). *Textile Chem. Colorist* **2**, 364.

Poots, V. J., McKay, G. and Healey, J. J. (1976a). *Water Res.* **10**, 1061.

Poots, V. J., McKay, G. and Healey, J. J. (1976b). *Water Res.* **10**, 1067.
Porath, J. and Flodin, P. (1959). *Nature, Lond.* **183**, 1657.
Porter, R. S. and Johnson, J. F. (1961). *Anal. Chem.* **33**, 1152.
Radell, E. A. and Strutz, H. C. (1959). *Anal. Chem.* **31**, 1890.
Randerath, K. (1965). "Thin Layer Chromatography", Academic Press, London and New York.
Rattee, I. D., Lewis, D. M. and Stevens, C. B. (1965). *Nature, Lond.* **208**, 269.
Rawlinson, J. and Deeley, E. L. (1967). *J. Oil Col. Chem. Assoc.* **50**, 373.
Ray, L. H. (1956). *Analyst* **80**, 863.
Rayburn, J. A. (1969). *Text. Chem. Col.* **1**, 58.
Rau, J. H. and Haase, H. (1965). *Textilber.* **46**, 1317.
Regnier, F. E. and Huang, J. C. (1970). *J. Chromatog. Sci.* **8**, 267.
Rhyhage, R. (1964). *Anal. Chem.* **36**, 759.
Richardson, R. W. (1951). *J. Chem. Soc.*, p. 910.
Robson, A., Williams, M. J. and Woodhouse, J. M. (1967). *J. Chromatog.* **31**, 284.
Robson, A., Woodhouse, J. M. and Zaida, Z. H. (1970). *Intern. J. Protein Res.* **2**, 181.
Ruttenbug, M. A., King, T. P. and Craig, L. C. (1965). *Biochemistry* **3**, 748.
Rusznak, I., Peter, F., Dolesch, J. and Gergely, A. (1966). *Textil. Praxis* **21**, 432.
Rybica, S. M. (1963). *J. Gas Chromatog.* **1**, 36.
Samuelson, O. and Gartner, F. (1951). *Acta. Chem. Scand.* **5**, 596.
Schlegelmilch, F., Aobelkader, H. and Eckelt, M. (1971). *Textile-Industrie* **23**, 274.
Schlegelmilch, F. and Khodadadian, C. (1973). *Melliand Textilber* **54**, 1098.
Schlegelmilch, F. and Kuss, W. (1973). *Defazet* **27**, 484.
Schmitz, I., Baumann, H. and Zahn, H. (1975). 5th Intern. Wool. Text. Conf. Aachen 2, 313.
Scholz, R. G., Bednarezk, J. and Yarmauchi, T. (1966). *Anal. Chem.* **38**, 331.
Schroeder, W. A. and Kay, L. M. (1955). *J. Amer. Chem. Soc.*, **77**, 3908.
Science Tools (1970), **17**, 47.
Shakespear, G. A. (1921). *Proc. Phys. Soc. London* **33**, 163.
Sherma, J. and Cline, C. W. (1964). *Anal. Chim. Acta* **30**, 139.
Smith, E. D. and Shewbart, K. L. (1969). *J. Chromatog. Sci.* **7**, 704.
Snyder, L. R. (1961). *J. Chromatog. Sci.* **5**, 430.
Snyder, L. R. (1964). *Adv. Anal. Chem. Inst.* **3**, 251.
Solms, J. (1955). *Helv. Chim. Acta* **38**, 1127.
Sporek, K. F. and Danyi, M. D. (1962). *Anal. Chem.* **34**, 1527.
Stahl, E. (Ed.) (1965). "Thin Layer Chromatography. A Laboratory Handbook". Academic Press, London and New York.
Stephens, L. J. (1969). *Text. Res. J.* **39**, 482.
Sternberg, J. C. and Litle, R. L. (1966). *Anal. Chem.* **38**, 321.
Steurle, L. and Hille, H. (1959). *Biochem. Zeitschrift* **331**, 220.
Stock, R. and Rice, C. B. F. (1967). "Chromatographic Methods". Northumberland Press.
Strain, H. H. (1942). "Chromatographic Adsorption Analysis". Interscience, N.Y.
Strain, H. H. (1953). *J. Phys. Chem.* **57**, 638.
Suffis, R., Sullivan, J. J. and Henderson, W. S. (1967). *J. Soc. Cosmetic Chem.* **16**, 783.

Swan, W. B. and Dux, J. B. (1961). *Anal. Chem.* **33**, 654.

Swanepoel, J., van der Merwe, R. and Grabner, T. (1970). *Textilveredlung* **5**, 200.

Sweeney, C. C., Wells, W. W. and Bentley, A. R. (1966). *Methods Enzymology* **VIII**, 96.

Synge, R. L. M. (1946). *Analyst* **71**, 256.

Szuchy, L., Mityko, J. and Gadsacs, M. (1974). *Kolor Ertesito* **16**, 31.

Tasdhomme, C. (1970). *Bull. Inst. Text. France* **24**, 237.

Terril, J. B. and Jacobs, E. S. (1970). *J. Chromatog. Sci.* **8**, 604.

Thiedel, H. and Schmitz, G. (1967). *J. Chromatog.* **27**, 413.

Tiselius, A. (1942). *Chem. Abs.* **36**, 3413.

Tiselius, A. and Hahn, L. (1943). *Kolloid Z.* **105**, 177.

Tiselius, A. and Hagdhal, L. (1950). *Acta Chem. Scand.* **4**, 394.

Trappe, W. (1940). *Biochem. Z.* **350**, 150.

Truter, E. V. (1963). "Thin Film Chromatography", Cleaver-Hulme Press.

Valk, G., Beines, U. and Stein, G. (1974). *Melliand Textilber* **55**, 897.

Valk, G., Beines, U. and Stein, G. (1975). *Melliand Textilber* **56**, 99.

Valk, G., Schleifer, K. and Klippel, F. (1968). *Melliand Textilber* **49**, 92.

Valk, G., Schleifer, K. and Klippel, F. (1969). *Melliand Textilber* **50**, 449.

Valk, G., Schleifer, K. and Klippel, F. (1969). *Melliand Textilber* **50**, 569.

Walbourg, E. F., Kondo, L. E. and Robinson, D. (1975). *Methods Enzymology* **XLI**, 10.

Wang, K. T. (1967). *Nature, Lond.* **213**, 212.

Wang, K. T. and Huang, J. M. K. (1965). *Nature, Lond.* **208**, 281.

Weil, H. and Williams, T. I. (1950). *Nature, Lond.* **166**, 1000.

Weil, H. and Williams, T. I. (1951). *Nature, Lond.* **167**, 906.

Weiss, H. V. and Shipman, W. H. (1961). *Anal. Chem.* **33**, 37.

Whistler, R. L. and BeMuller, J. N. (Ed.) (1976). "Methods in Carbohydrate Chemistry". VII. 257. Academic Press, London and New York.

Williams, V. A. (1966). *Text. Res. J.* **36**, 1.

Wollenweber, P. (1962). *J. Chromatog.* **7**, 557.

Wood, J. H. (1975). "Book of Papers", Nat. Tech. Conf. A.A.T.C.C. 278.

Young, D. M. and Crowell, A. D. (1962) In "Physical Adsorption of Gases", Butterworth Press.

Zahn, H. and Hammoudeh, M. M. (1973). *Kolloid Z. Z. Polymere* **251**, 289.

Zahn, H. and Rouette, P. F. (1968). *Textilveredlung* **3**, 241.

Zahn, H. and Waschka, O. (1956). *Makromol. Chem.* **18**, 201.

Zahn, H. and Wolleman, G. (1951). *High Polymers* **12**, 282.

Zahn, H. and Zuber, H. (1954). *Textil. Rundschau* **9**, 119.

Zechmeister, L. (1951). *Nature, Lond.* **167**, 405.

Zhukov, V. A., Popiko, I. V. and Krichevsky, G. E. (1975). *Tekstil. Prom.* **1**, 64.

Ziegler, K. (1963). *Forschungsber. Landes Nordschein-Westfalen.* 1275.

Ziegler, K. (1964). *J. Biol. Chem.* **239**, P.C. 2713.

Ziegler, K. Melchert, I. and Lurken, C. (1967). *Nature, Lond.* **214**, 404.

Zuber, H., Ziegler, K. and Zahn, H. (1957). *Naturforsch.* **126**, 734.

9. Kinetics of Polymerization

A. D. JENKINS

School of Molecular Sciences, University of Sussex, Brighton, Sussex, England

I. Introduction

The kinetics of a reaction often constitute the most accessible and useful manifestation of the chemical mechanism of the process. Together with spectroscopic observations on the nature and concentrations of intermediates in the reaction, the kinetics enable diagnosis of details to be made which, in turn, facilitates prediction of the extent and means of control which can be exerted to modify the outcome of the reaction. This is particularly true in regard to polymerization because in this case it is not only the rate and direction of reaction which are important but also the average molecular weight and the molecular weight distribution of the resultant polymer: these two quantities are kinetic parameters on a par with rate or order of reaction. In the context of fibre formation they are indeed of much greater importance than reaction velocity because it is just these characteristics of a polymer which determine its suitability for conversion into a useful fibre.

Any initial impression that a survey of polymerization kinetics in relation to fibre-forming polymers encompasses a restricted field is dispelled by a brief consideration of the wide range of polymers which are of commercial importance as fibrous materials. The Encyclopaedia of Polymer Science and Technology (published by Interscience Division of John Wiley and Sons, New York, in 1970) informs us that fibres are produced on the industrial scale from the following types of polymer: acrylic, olefin, vinyl halide, polyester, polyamide, polyurethane. The field thus comprises polymers produced by both step-growth and chain reaction processes, and in the latter case we shall have to take into account several different classes of mechanism. In the space available only a general review is possible and the reader is referred to the texts in the bibliography, of which that by Allen and Patrick (1974) is the most recent and comprehensive, for fuller discussion of individual systems.

A. REACTION STEPS

In a few cases the reaction consists of a single simple process in which two entities react together and combine to give a larger molecule; repetition of this step then generates in time a polymer. For the most part, particularly with chain reactions, several steps are involved, but all polymerizations necessarily include the linking or growth step which is the essence of polymerization and which is known as *propagation.*

Many polymerizations will only start if *initiation* is provoked by an added catalyst or initiator. It should be noted that the use of the word "catalyst", although widespread, is in most cases incorrect since only rarely can the substance be reclaimed from the reaction without change or diminution in quantity. There is a small number of cases where the term "catalyst" is properly applied.

Rarely, there is a distinct initiation step which does not require an additive but which involves reaction between monomer molecules under the influence of heat or light; it is also common to convert monomers to reactive intermediates by reaction with high energy radiation, such as the γ-radiation from ^{60}Co.

Thus far we have mentioned the initiation of reaction and the monomer-addition process which builds up a giant molecule. Since chain reactions usually involve a highly reactive intermediate it is usually (but, as will be seen below, not always) necessary also to include a *termination* step in which the reactive intermediates are destroyed. When this reaction constitutes part of the process, the combination of initiation, propagation and termination provides an example of a classical chain reaction, in fact of a particularly simple kind because the chain-carrying species is a single chemical entity with a spectrum of sizes varying from a single monomer unit to the largest polymer molecule formed in the system.

In general it is necessary also to consider processes which limit the molecular weight of the product without interfering with the rate. In polyesterifications, for example, the phenomenon of ester interchange (p. 286) has exactly this effect while in chain reactions it is represented by chain *transfer.* The relative rates of the various steps can be controlled within limits by changes in the available variables (temperature, reagent concentration) in such a way that the rate and the mean molecular weight are varied. This is of considerable importance commercially where optimization of rate of reaction and product properties (for our purpose molecular weight) is economically essential. Table I indicates very broadly the way in which the increase in rate of one of the reaction steps is likely to affect the overall rate and the mean degree of polymerization $\bar{P}$ in a typical chain reaction polymerization. One should note that, when transfer occurs, the influence on the overall rate is zero only in ideal cases; not infrequently retardation ensues (see p. 296), that is, the rate is decreased.

TABLE I

Influence of increase in rate of reaction steps

Reaction Step	Influence of Increased Rate on	
	Rate of Reaction	$\bar{P}$
Initiation	Increase	Decrease
Propagation	Increase	Increase
Transfer	None (or Decrease)	Decrease
Termination	Decrease	Decrease

B. Classification of Reaction Mechanisms

Although there is a very wide variety of types of polymerization reaction, for kinetic purposes they can be divided between two main groups—"step-growth" and "chain reaction" polymerization. These two categories correspond closely to the older nomenclature of "condensation" and "addition" polymerization but the parallel is not entirely complete because the older system is based on monomer structure while the more modern one is concerned to express the chemical nature of the polymerization process. While it is true as a broad generalization that certain types of monomer tend to react by a certain kind of chemical route there are exceptions to any general rule and indeed some monomers can afford polymers by reactions of both types.

Step-growth reactions proceed by the repetition of a propagation step, frequently a condensation, between molecules of gradually increasing size. Molecular growth is then on an extended time scale. Chain reactions, by contrast, depend upon the agency of a reactive intermediate which itself increases in size by repeated addition of monomer units, often leading to completion of molecular growth in a very short time. Table II gives a list of

TABLE II

Characteristics of step-growth and chain-reaction polymerizations

Step	Chain
Propagation involves species of all molecular sizes	Every propagation step involves monomer
Monomer disappears rapidly in the early stages	Monomer survives until conversion reaches completion
Propagation is mainly between polymer molecules which thereby increase continually in size	Polymer molecules, once formed, do not undergo mutual reaction
High molecular weight material is produced only in late stages	High molecular weight material produced in earliest stages

general characteristics which help to contrast the two types of reaction mechanism.

II. Kinetics of Step-Growth Polymerization

A. POLYESTERIFICATION

In the step-growth field by far the majority of studies have been concerned with polyesterification, with or without added mineral acid. Simple esterification is numbered among the reactions most extensively investigated from a kinetic point of view, yet it cannot be said that it is fully understood, especially in the case of reactions taking place in the absence of an added catalyst (Goldschmidt and Udby, 1910; Rolfe and Hinshelwood, 1934; Flory, 1937, 1939, 1946; Davies, 1949; Davies and Hill, 1953; Au-Chin and Kuo-Sui, 1959; Kemkes, 1967; Hamman et al., 1968). At first sight the kinetics of polyesterification are simple because the acid-catalysed reactions generally obey the simple law

$$\text{Rate} \propto [\text{Alcohol}]\,[\text{Acid}]\,[\text{Acid catalyst}] \tag{1}$$

Since it is usually possible to arrange for the concentration of the added catalyst to be constant, and since it is obviously advantageous to begin with equal concentrations of alcohol and acid functions if a high molecular weight polymer is to be attained, one can formulate the kinetic behaviour as

$$\frac{-dc}{dt} = kc^2, \tag{2}$$

where k is a composite quantity being the product of the true velocity constant and the acid catalyst concentration and c is the momentary concentration of each type of functional group.

The esterification process is, of course, an equilibrium when carried out in a closed system but kinetic studies almost invariably are performed in systems in which the volatile product (water) is removed as it is formed so that the forward reaction can be considered in isolation.

For these reactions it is customary to use as an index of the progress of the reaction a quantity p, the extent of reaction, which represents the fraction of the original functional groups which have undergone reaction, i.e.

$$p = \frac{c_0 - c}{c_0}, \tag{3}$$

the suffix $_0$ denoting the start of reaction.

Carothers (1936) has shown that the number average degree of polymerization $\bar{P}_n$ is related to p through the equation

$$\bar{P}_n = (1 - p)^{-1} \tag{4}$$

Taking together Eqs. (2), (3) and (4) we have

$$(1 - p)^{-1} = 1 + c_0 kt$$

or

$$\bar{P}_n = 1 + c_0 kt \tag{5}$$

which embodies the interesting conclusion that the mean molecular size is a linear function of time.

If no catalyst is added the system relies on the carboxylic acid component to provide the necessary protons, in which case the parallel equation to (1) above is

$$\text{Rate} \propto [\text{Alcohol}] [\text{Acid}]^2 \tag{6}$$

leading to

$$(1 - p)^{-2} = 1 + 2c_0^2 kt \tag{7}$$

It is then natural to test such systems by plotting either $(1 - p)^{-1}$ or $(1 - p)^{-2}$, as appropriate, as a function of time and inspecting for linearity. Figure 1 depicts a typical result which is usually described as representing a satisfactory fit to Eq. (7) beyond an induction phase. This is the point to

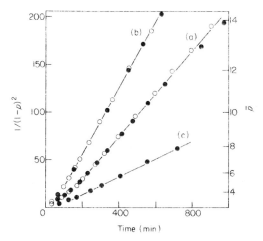

FIG. 1. Polyesterification reactions; (a) diethylene glycol/adipic acid, 166°C, (b) diethylene glycol/adipic acid, 202°C, (c) diethylene glycol/caproic acid, 166°C. [N.B. The time scale for (b) has been doubled.] The linearity of these plots is a test of Eqn. (7). From Flory, (1939); (1940).

bring out the fact that the equations cited above can only be expected to be valid if all OH groups have the same reactivity regardless of the size of the molecule to which they are attached, and the same applies to COOH groups. The initial period of reaction, which corresponds to the curved part of the plot on Fig. 1, therefore is reasonably interpreted as indicating that, at the oligomer state, the reactivity of functional groups is undergoing progressive change, becoming constant for relatively large molecules, whatever their size.

There is a considerable body of evidence that this is a general phenomenon which can be explained as a relative reluctance of small molecules to react because of the loss of translational entropy which results (Hamman *et al.*, 1968; Mareš *et al.*, 1969). The larger the reactant molecule, the less intrusive this effect becomes, hence linearity is approached.

In these terms the plot in Fig. 1 is acceptable as satisfactory by the polymer chemist but it nevertheless usually comes as a surprise to realize that, at the point at which linearity is attained, something like 70% of the functional groups have already reacted, so in straight chemical terms it is only the last 30% of reaction which behaves simply, although it is of course only in the extreme stages of conversion that the system begins to contain high polymer and it is this part of the process which is indeed of real significance for the production of polymers which will be capable of fibre formation.

B. Ester Interchange

In the previous section a simple process of polyesterification was briefly described in its basic essentials and mention must now be made of a step which is usually present during the reaction and which can be turned to good use in a number of ways, including application as an alternative, and indeed preferred, method for preparing polyesters. This step is known as ester exchange, ester interchange or transesterification; it consists of the displacement of one alcohol moiety from an ester by another alcohol, as shown in Eq. (8).

$$R_1COOR_2 + R_3OH \rightleftharpoons R_1COOR_3 + R_2OH \qquad (8)$$

Since acids are among the effective catalysts for this equilibrium and since the rate constants are generally higher than those for ordinary esterification, it must be expected that transesterification will be established during a polyesterification reaction. As a rule of thumb, the equilibrium constants for ester interchange are close to unity (Farkas *et al.*, 1949) and insensitive to temperature but where one alcohol is aliphatic and the other aromatic it is the formation of the aliphatic ester which is preferred (Schildknecht, 1955).

In relation to fibrous polymers the material of outstanding interest is, of course, poly(ethylene terephthalate) which is prepared on the commercial scale by an interchange process. In the first step the dimethyl ester of terephthalic acid is prepared and purified, the purification being far easier than in the case of the free acid. Methanol is then displaced by ethylene glycol in a transesterification reaction and excess ethylene glycol finally driven off to allow the polyesterification equilibrium to favour the formation of polymer.

C. CYCLIZATION

The kinetic description of step growth condensation processes must also take into account the possible formation of cyclic molecules by α, ω reaction of functional groups on a single molecule, e.g.

$$\text{(ring)}\begin{array}{c}\text{OH}\\\text{COOH}\end{array} \xrightarrow{k_1} \text{(ring)}\begin{array}{c}\text{O}\\|\\\text{CO}\end{array} + H_2O$$

which will compete with the polymer-forming step

$$\text{(ring)}\begin{array}{c}\text{OH}\\\text{COOH}\end{array} + \text{(ring)}\begin{array}{c}\text{OH}\\\text{COOH}\end{array} \xrightarrow{k_2} \text{(ring)}\begin{array}{c}\text{OH}\\\text{COOH}\end{array}$$

Obviously, the cyclization step will be first order and the propagation second order with respect to monomer concentration. Any other influences, such as that of a catalyst, will cancel out in any comparative expression because of their equal effects on the two processes. One can therefore write

$$\frac{\text{Rate of cyclization}}{\text{Rate of polymerization}} = \frac{k_1}{k_2 c} \tag{9}$$

From general chemistry, it is well understood that the ease of ring formation is very sensitive to the number of atoms in the resulting ring so that interest centres upon the relationship between ring size and the quantity k_1/k_2, known as the cyclization constant (Lenz, 1967). This relationship is depicted in Fig. 2.

D. POLYAMIDIFICATION

Mention must be made of these systems in view of the pride of place held among fibre-forming polymers by nylon yet surprisingly little can be said

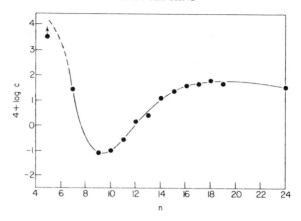

FIG. 2. The cyclization constant as a function of potential ring size for w-hydroxy-acids. Data from Stoll and Rouvé (1935).

regarding the kinetics of the process. The formation of simple amides is itself a reaction that has failed to attract much attention from kineticists and one presumes that commercial nylon production works so well without detailed kinetic analysis that the latter has not been thought worthwhile.

As far as one can see, there is a general parallel with polyesterification, allowing for the fact that the elimination of water leaves an amide rather than an ester link. Both direct and exchange processes are available, the latter having been employed to form block copolymers from units of different nylons. There is some information about the equilibrium in closed systems but interpretation is rendered difficult due to the partition of water between phases and its possible departure from ideal behaviour (Skuratov et al., 1952; van Velden et al., 1955; Fukumoto, 1956; Heikens et al., 1960; Zimmermann, 1964; Sawada, 1969; Ogata, 1970).

E. KINETICS AND MOLECULAR SIZE

Equation (5) shows that the number average molecular weight is a linear function of time in the case of an acid-catalysed polyesterification but, useful as this information may be, a full correlation of the properties of the fibre with the characteristics of the polymer from which it has been drawn calls for a knowledge of the distribution of the molecular sizes present. Kinetics can come to our aid in arriving at this knowledge, as the following derivation indicates.

The general kinetic equation required is

$$\frac{-d}{dt}\left(\sum_{n=1}^{\infty}[N_n]\right) = k\left(\sum_{n=1}^{\infty}[N_n]\right)^2 \tag{10}$$

Here $[N_n]$ is the concentration of molecules containing n monomer residues, the species N_1 being the monomer itself; k, the velocity constant, includes the catalyst concentration.

Integration of Eq. (10) leads to

$$\frac{1}{\sum\limits_{n=1}^{\infty} [N_n]} - \frac{1}{[N_1]_0} = kt \tag{11}$$

Since p is given by

$$p = \frac{[N_1]_0 - \sum\limits_{n=1}^{\infty} [N_n]}{[N_1]_0} \tag{12}$$

we can recast (5) as

$$(1 - p)^{-1} = 1 + k[N_1]_0 t \tag{13}$$

A knowledge of the concentration of individual species requires consideration of differential equations exemplified by

$$\frac{d[N_n]}{dt} = k \sum\limits_{m=1}^{n-1} [N_m][N_{n-m}] - k[N_n] \sum\limits_{n=1}^{\infty} [N_n] \tag{14}$$

This equation recognizes that a molecule of size n is formed by the union of any two molecules whose combined size is equivalent to n and destroyed by reaction of itself with species of any size whatever.

The solution of this equation is (Burnett, 1954)

$$[N_n] = [N_1]_0 p^{n-1}(1 - p)^2 \tag{15}$$

Since the quantity $[N_n]$ is a concentration, this equation effectively provides a count of the numbers of molecules of all sizes present but it is frequently of more interest to work in terms of the relative weights of the species. The weight W_n of $[N_n]$ molecules of size n is obviously

$$W_n = n[N_n]$$

(using the monomer residue as the unit of weight) so that in weight terms, Eq. (15) transforms into

$$W_n = [N_1]_0 n p^{n-1}(1 - p)^2 \tag{16}$$

Typical curves corresponding to Eqs. (15) and (16) are depicted in Fig. 3.

This distribution also arises from other reaction mechanisms, including common examples of radical polymerization, and because of its widespread occurrence it is known as the "most probable" distribution.

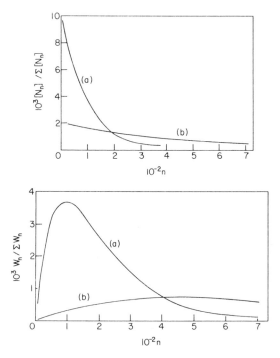

FIG. 3. The "most probable" distribution, corresponding to Eqns. (15) and (16), for (a) $p = 0.990$, $\bar{P}_n = 100$, (b) $p = 0.998$, $\bar{P}_n = 500$.

III. Kinetics of Chain Reaction Polymerization

A chain reaction involves the repetition (a large number of times) of a propagation step in which an active chain carrier is continuously regenerated at ever-increasing size; reactions which generate or destroy chain carriers are comparatively very infrequent and hold each other in balance, thereby ensuring a constant concentration of chain carriers. This is a basic statement of the kinetic approach to the analysis of such reactions under normal conditions in which it is regarded as valid to apply the "steady state approximation", the essential criterion for which is that the rate of change of concentration of chain carriers is very much less than the rate of their formation or destruction.

Usually two other approximations also are implicit in kinetic treatments of chain polymerization; the principle that the reactivity of a chain carrier is independent of the size of the molecule to which it is attached, and the long chain approximation, which regards the consumption of monomer in steps other than propagation as insignificant. Although there are cases where these approximations fail, they are not of interest in the general context of this account.

The constant reactivity principle enables us to treat all chain carriers as equivalent so that a single velocity constant for propagation will suffice, and similarly for transfer and termination processes. Using the subscript n to denote the number of monomer units contained in any entity and the symbol X to denote a chain carrier, it is possible to write for the total concentration of chain carriers

$$[X] = \sum_{n=1}^{\infty} [X_n] \tag{17}$$

and to use simply $[X]$ in all calculations relating to the overall kinetics.

In chain reaction polymerizations there is always an initiation step in which the active intermediates are generated. Rarely, one encounters systems which are devoid of any termination process, and these will be discussed below in connection with certain anionic reactions, but in the general case all three processes have to be taken into account. The termination step may involve one chain carrier or two so that for a great many reactions of this class one can analyse the overall process by means of the scheme below, which serves to define the individual velocity constants. (Monomer is denoted by the symbol M and polymer by P.)

Step	Reaction	Rate
Initiation	$\rightarrow X_1$	R_i
Propagation	$X_n + M \xrightarrow{k_p} X_{n+1}$	$R_p = k_p[X] \times [M]$
Termination		
(i) First Order	$X_n \xrightarrow{k_t} P_n$	$R_t = k_t[X]$
(ii) Second Order		
(a) Dispropor-tionation	$X_n + X_m \xrightarrow{k_t} P_n + P_m$	$R_t = k_t'[X]^2$
(b) Combination	$X_n + X_m \xrightarrow{k_t''} P_{n+m}$	$R_t = k_t''[X]^2$

Since the only point at which the monomer is involved is the propagation step, the rate of this individual process can be equated to the rate of the overall reaction, defined as the rate of disappearance of the monomer,

$$\frac{-d[M]}{dt} = k_p[X][M] \tag{18}$$

from which it is clear that the form of the overall kinetic behaviour in any particular case depends on the manner in which the steady state concentration of the chain carriers $[X]$ is related to the variables of the system. This relationship can be derived from the steady state principle which is embodied in the statement

$$R_i = R_t \tag{19}$$

Before proceeding from this point it is pertinent to note that the long chain approximation enables one to write Eq. (18) in the form given, neglecting any possible consumption of monomer in initiation or transfer processes, while the constant reactivity principle underlies the use of a single velocity constant for each of the reaction steps.

To apply Eq. (19) it is necessary to choose between the first order and second order forms of the termination reaction, indeed the general case would include both, as they can be simultaneously operative, but such expressions are very cumbersome and it suffices in the present context to scrutinize the two extreme cases. As a broad generalization, the second order form is required for radical reactions and the first order form in other cases.

With this in mind, Eqs. (19) and (18) then transform as follows:

$$\text{First order termination } R_i = k_t[X]; \quad \frac{-d[M]}{dt} = \frac{k_p R_i[M]}{k_t} \tag{20}$$

$$\text{Second order termination } R_i = k_t[X]^2; \quad \frac{-d[M]}{dt} = k_p \left(\frac{R_i}{k_t}\right)^{\frac{1}{2}}[M] \tag{21}$$

In these equations a single k_t has been used rather than k_t, k_t' and k_t'' as in the scheme above. This corresponds to the assumption that only one of the three termination steps is operative, in which case there is no need to use discriminating symbols. Sometimes disproportionation and combination occur simultaneously but, even so, the kinetic expression for the overall reaction retains the simple form appropriate to second order termination.

The kinetic formulation is now explicit, except for the specification of the rate of initiation. Since this depends upon the type of mechanism employed, it is convenient at this point to abandon the general discussion of chain reaction polymerization in favour of descriptions of reactions according to the chemical nature of the chain carrier. Before doing so, however, mention should be made of the chain transfer process which has not been represented in the equations presented above.

Chain transfer consists of the transfer of the activity of a chain carrier from a larger entity to a smaller one, for example by the extraction by a large radical of an atom from a simple molecule leaving a fragment which is a small radical

$$X_n^{\bullet} + CCl_4 \rightarrow P_n(X_nCl) + \dot{C}Cl_3 \tag{22}$$

Reaction of the new radical with the monomer then regenerates a chain carrier

$$\dot{C}Cl_c + M \rightarrow X_1 \tag{23}$$

Since X_1 is presumed to have the same reactivity as X_n there is no effect on the rate of reaction, although there are clear implications regarding the molecular weight of the polymer produced. Simple chain transfer thus can be neglected for overall discussion of rate but, as will be seen below (p. 296), the consequences of chain transfer are frequently more complex and do intrude into the kinetic scheme.

A. CHAIN CARRIERS

Apart from reactions involving ring-opening and ring-enlargement, chain reaction polymerization is usually concerned with monomers containing double bonds. These are vinyl or vinylidene compounds and the expression "vinyl polymerization" is used as an umbrella to cover this field of reactions, summarized in the following equation:

$$n\,CH_2 = CHR \rightarrow +CH_2-CHR+_n \tag{24}$$

The agency (chain carrier) by which this reaction is engineered must have a high reactivity but its precise chemical nature is subject to wide variation. Radicals are most commonly employed on the commercial scale and radical reactions are the best understood in fundamental terms but anions and cations can also serve, as well as the more complex species involved in Ziegler–Natta and related systems, in which a more elaborate coordinated species provides the focus of reaction.

Kinetic behaviour is the criterion by which the nature of the reactive intermediate may most readily be identified; in the following section, therefore, the four types of process will be given separate discussion in order to highlight their essentials and their differences.

B. RADICAL CHAIN POLYMERIZATION

It is best to begin by assuming that the termination step is of the second order type. This is reasonable since, by their very nature, radicals can only be generated or destroyed in pairs. Actually, there are examples where radicals of other kinds can be added to the system to react with the chain carrier and so render the termination effectively first order but these are the exception rather than the rule and will be treated separately. It thus remains to consider the various means available for initiation and their kinetic consequences.

(a) Initiation arising from the thermal decomposition of an added substance C, the concentration of which is assumed, in the simplest case, to remain constant.

$$C \xrightarrow{k_i} 2R \cdot \qquad (25)$$

where R denotes a radical derived directly from the initiator. If the only reaction undergone by R is addition to the monomer

$$R \cdot + M \xrightarrow{k_i} X_1 \qquad (26)$$

we may write

$$R_i = k_1[R\cdot][M] = 2k_i[C] \qquad (27)$$

and hence, from Eq. (21),

$$\frac{-d[M]}{dt} = k_p \left(\frac{2k_i[C]}{k_t}\right)^{\frac{1}{2}} [M] \qquad (28)$$

This equation displays the dependence of rate upon the square root of the initiator concentration which arises directly from the second order nature of the termination step and which is so characteristic of radical reactions. (N.B. There are exceptions to this rule; see the discussion of anionic polymerization initiated by potassium amide.)

(b) Initiation by photochemical decomposition of initiator by light of incident intensity I. If only a small percentage of the light is absorbed

$$R_i = kI[C] \qquad (29)$$

and

$$\frac{-d[M]}{dt} = k_p \left(\frac{kI[C]}{k_t}\right)^{\frac{1}{2}} [M] \qquad (30)$$

with half order dependence on both I and $[C]$.

(c) Initiation by photochemical decomposition of monomer by light of intensity I. If only a small percentage of the light is absorbed

$$R_i = kI[M] \qquad (31)$$

and

$$\frac{-d[M]}{dt} = k_p \left(\frac{kI}{k_t}\right)^{\frac{1}{2}} [M]^{\frac{3}{2}} \qquad (32)$$

which also has half order dependence on I but three-halves order dependence on $[M]$.

(d) Initiation by thermal reaction of the monomer. If this is a second order process

$$R_i = k_i[M]^2 \tag{33}$$

and

$$\frac{-d[M]}{dt} = k_p\left(\frac{k_i}{k_t}\right)^{\frac{1}{2}}[M]^2 \tag{34}$$

with second order dependence on the monomer concentration. In fact, these reactions are more complicated, some authors believing that the initiation step is best described as third order with respect to the monomer (Mayo, 1953, 1968; Hiatt and Bartlett, 1959; Kirchner, 1966, 1969; Brown, 1969).

(e) Initiation by high energy radiation (e.g. ^{60}Co γ-radiation).

$$R_i = k_M[M] + k_s[S] \tag{35}$$

where $[S]$ is the solvent concentration and k_M, k_S are velocity constants for initiation by interaction of radiation with M and S respectively

$$\frac{-d[M]}{dt} = k_p\left(\frac{k_M[M] + k_s[S]}{k_t}\right)^{\frac{1}{2}}[M] \tag{36}$$

C. COMPLICATING KINETIC FACTORS IN RADICAL CHAIN POLYMERIZATION

1. First Order Termination

This can occur as a result of a transfer reaction, which generates a new radical of exceptionally long life and low tendency to reinitiate polymerization, by reaction of a growing chain with an organic free radical or by reaction of a growing chain with a suitable transition metal ion.

The first of these possibilities is known as "degradative chain transfer", best exemplified by a reaction which leads to the formation of a delocalized allyl radical (Bartlett and Altschul, 1946), thus:

$$X_n^{\cdot} + CH_2{=}CH{-}CH_2Cl \rightarrow P_n(X_n\,Cl) + \begin{bmatrix} CH_2{=}CH{-}\dot{C}H_2 \\ \updownarrow \\ \dot{C}H_2{-}CH{=}CH_2 \end{bmatrix} \tag{37}$$

This is clearly a case where chain transfer does have a kinetic influence of a detrimental kind.

Reaction with an organic free radical occurs in the presence of stabilized species, such as diphenylpicrylhydrazyl (Barlett and Kwart, 1950; Matheson *et al.*, 1951),

commonly employed as an inhibitor. The third type of reaction, that involving a transition metal ion, is kinetically much more interesting (Bamford *et al.*, 1957, 1962); it is an example of a reaction between an organic free radical (the chain carrier) and an inorganic free radical since the transition metal involved has an unpaired electron in a d or f level.

$$X_n + FeCl_3 \rightarrow P_n + FeCl_2 \tag{38}$$

In all cases the reactions described compete with the normal bimolecular radical termination, hence the overall kinetics may correspond to a rate of termination dependent upon the concentration of chain carriers raised to a power anywhere from $1 \cdot 0$ to $2 \cdot 0$ and to an exponent of initiator concentration (assuming thermal decomposition of initiator to be the effective initiation mechanism) anywhere between $0 \cdot 5$ and $1 \cdot 0$.

2. Transfer with Retardation

This is closely related to the reaction generating allyl radicals outlined in the previous section but it deserves discussion as a general case in its own right as chain transfer may occur by reaction with solvent, monomer or polymer, as well as with an added transfer agent, with a broad range of kinetic consequences available. Denoting the transfer agent by S, the general reaction is

$$X_n^{\cdot} + S \xrightarrow{k_f} P_n + S \cdot \tag{39}$$

and the kinetics will reflect the reactivity of $S \cdot$ as compared to that of $X_n^{\cdot}$. If $S \cdot$ immediately reacts with the monomer to give $X_1^{\cdot}$ or if $S \cdot$ coincidentally resembles $X_n^{\cdot}$ in its reactivity, the kinetic features will be undisturbed by the occurrence of transfer. However, it can hardly be expected that $S \cdot$ and $X_n^{\cdot}$ will behave similarly in all respects, hence the need to take into account the possible reactions

$$S \cdot + X_n^{\cdot} \rightarrow \tag{40}$$

and

$$S\cdot + S\cdot \rightarrow \quad (41)$$

which will obviously have kinetic implications because transfer will have become, in effect, termination.

The detailed kinetic treatment of these cases will be found elsewhere (Jenkins, 1958a,b) but it is useful to record that the ratio λ of the rates of polymerization with and without transfer agent can be employed to determine k_f by means of the equation

$$k_f = \frac{(R_i k_t)^{\frac{1}{2}}}{n[S]}\left(\frac{1}{\lambda} - \lambda\right) \quad (42)$$

where n is a number with the extreme values 1 and 2, depending on the relative importance of Eqs. (40) and (41).

The reduction in rate arising in this way is known as the phenomenon of retardation; it is illustrated in Fig. 4.

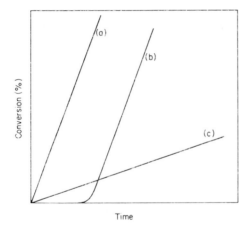

FIG. 4. The effects of retarders and inhibitors on radical polymerization. (a) No additive, (b) in the presence of a good inhibitor, (c) in the presence of a retarder.

3. Inhibition

If the primary radicals (those formed in the first initiation step) or the chain carriers react with very high velocity with an added substance, such as a stabilized radical or a suitable transition metal ion, it can happen that the rate of reaction is so low as to be unobservable. This is the phenomenon of inhibition. No polymerization will be seen until almost all the inhibitor has been consumed when the rate rises rapidly until it attains the normally expected velocity, unless so much initiator has been consumed during the inhibition period that it has become appreciably depleted. Figure 4 shows how retardation and inhibition are manifested as plots of $[M]$ as a function of time.

4. Primary Radical Termination

In the rather unusual conditions of very high initiator concentration or very low monomer concentration (or both), an appreciable number of primary radicals evade reaction (26) and instead terminate chains by combination or disproportionation with chain carriers

$$R\cdot + X_n^{\cdot} \rightarrow P_n \tag{43}$$

or simply destroy one another

$$R\cdot + R\cdot \rightarrow \tag{44}$$

The effect of these reactions is to bring about a departure from the simple kinetics represented by equation (28) by increasing the order of reaction with respect to monomer concentration to a value intermediate between $1 \cdot 0$ and $1 \cdot 5$, at the same time decreasing the order with respect to initiator concentration to a value between $0 \cdot 5$ and $0 \cdot 0$ (Bamford et al., 1959).

5. Diffusion Control and Chain End Inaccessibility

The viscosity of polymer solutions is a well-known characteristic but it is not always appreciated that the macro-radicals intermediate in the polymerization process are just as hindered by mobility problems as the complete polymer molecules. This is a factor to be considered in all cases, but one of its more interesting kinetic facets is the dependence it imposes on the rate of reaction in relation to the viscosity of the medium. Thus the velocity constant for termination seems in general to be inversely related to the viscosity of the system and this effect is probably the source of most of the solvent effects which have been reported as occurring with radical polymerization.

When a reaction is taken to significant conversion, the accumulated dissolved polymer causes the viscosity of the medium to increase very substantially and the first consequence of this trend is a decrease in the rate of termination with concomitant acceleration of overall rate. At sufficiently high polymer concentration diffusion problems hinder transport of the monomer to the radical sites so that the rate reduces again. This type of behaviour is known as the gel-effect, the Trommsdoff effect or the Norrish–Smith effect.

A different complication is encountered if the polymer formed is insoluble in the reaction medium, as in the case of acrylonitrile where the polymer precipitates as a granular solid as it is formed. Individual molecules are very tightly coiled in a hostile medium, and also they tend to coagulate. If a macro-radical finds itself embedded in such a unit it will be quite unable to terminate, moreover at low temperatures and high degrees of occlusion even the monomer will find it difficult to gain access to the radical. The resulting kinetic behaviour has been investigated in some detail: the most spectacular feature of these systems is displayed when a polymer formed photochemically at low temperature (25°C) is heated to 60°C or above at which point the agglomerates disperse liberating the contained radicals. A very fast polymerization can result which can be exploited in favourable cases to prepare block copolymers.

D. COPOLYMERIZATION

When a mixture of two monomers is employed there are four possible propagation processes and three distinct terminations (without distinguishing disproportionation and combination). Using the symbols A and B to denote the monomers and X_A, X_B for macro-radials with terminal A and B groups, respectively, the reaction scheme becomes

Initiation $\rightarrow X_A$ or X_B

Propagation $X_A + A \xrightarrow{k_{AA}} X_A$

$X_A + B \xrightarrow{k_{AB}} X_B$

$X_B + B \xrightarrow{k_{BB}} X_A$

$X_B + B \xrightarrow{k_{BB}} X_B$

Termination $X_A + X_A \xrightarrow{k_A}$

$X_A + X_B \xrightarrow{k_X}$ Polymer

$X_B + X_B \xrightarrow{k_B}$

Using the steady state approximation, it is possible to calculate the rate of polymerization but the expression obtained is cumbersome and of little real use. Of far more value is an equation for the composition of the polymer obtained and this is easily derived because it requires consideration of the propagation steps alone.

Thus, the rates of consumption of the two monomers are obviously given by

$$\frac{-d[A]}{dt} = k_{AA}[X_A][A] + k_{BA}[X_B][A] \tag{45}$$

$$\frac{-d[B]}{dt} = k_{AB}[X_A][B] + k_{BB}[X_B][B] \tag{46}$$

Since a stationary state exists, the rate of formation of X_A must be balanced by the corresponding rate of destruction, hence

$$\frac{d[X_A]}{dt} = k_{BA}[X_B][A] - k_{AB}[X_A][B] = 0 \tag{47}$$

from which

$$\frac{[X_A]}{[X_B]} = \frac{k_{BA}[A]}{k_{AB}[B]} \tag{48}$$

If we divide Eq. (45) by Eq. (46) and substitute Eq. (48) we find

$$\frac{-d[A]}{-d[B]} = \frac{[A]}{[B]} \left\{ \frac{(k_{AA}/k_{AB})[A] + [B]}{(k_{BB}/k_{BA})[B] + [A]} \right\} \tag{49}$$

or

$$\frac{-d[A]}{-d[B]} = \frac{[A]}{[B]} \left\{ \frac{r_A[A] + [B]}{r_B[B] + [A]} \right\} \tag{50}$$

where

$$r_A = k_{AA}/k_{AB} \quad \text{and} \quad r_B = k_{BB}/k_{BA} \tag{51}$$

Clearly the left hand side of Eq. (50) gives the ratio of the concentrations of A and B units in the polymer, hence Eq. (50) provides a simple relationship between the composition of the polymer and that of the monomer mixture which shows that the two composite parameters r_A and r_B control the situation. These quantities are known as the reactivity ratios and the equation is the copolymer composition equation.

Obviously it would be of interest to be able to predict the monomer reactivity ratios in order to anticipate the composition of a copolymer prepared under any chosen conditions. This goal still proves elusive, although for almost thirty years the Q-e scheme has provided an empirical rationale for copolymerization behaviour. Although the newer Patterns approach to radical reactivity has not been widely applied, it is much nearer to affording prediction of the vital parameters; certainly, given some basic knowledge of copolymerization behaviour, it becomes possible to forecast the outcome of other experiments of a similar type with remarkable accuracy in most cases (Jenkins, 1972, 1974).

For fibre-forming polymers the incorporation of a small amount of comonomer provides a delicate method of controlling properties. Thus the inclusion of a modest quantity of a monomer which differs only slightly (say in the size of an alkyl substituent) can reduce the propensity for crystallization and thereby extend the range of properties attainable in the ultimate drawn filaments.

E. EMULSION AND SUSPENSION POLYMERIZATION

The kinetics described above arise from experiments with bulk monomer or in solution in organic media; such conditions are appropriate for fundamental studies but are not suitable on the commercial scale owing to the severe problem of heat transfer (in bulk systems) or the cost of solvents (with organic liquids). Industrially, water offers great advantages as a medium, although only a minority of monomers are soluble in it so that it is necessary to disperse the monomer in some way. Simple dispersion brought about by stirring provides a suspension of small monomer globules in each of which an oil-soluble initiator can cause a bulk polymerization to occur on a very small scale. Such systems require careful control but they are nevertheless attractive and provide a useful product. The kinetics are in no way remarkable.

An alternative procedure with entirely different kinetic character is that of emulsion polymerization (Smith and Ewart, 1948; Haward, 1949; Harkins, 1950; van der Hoff, 1962; Gardon, 1968) where a surface-active agent is employed to solubilize some of the monomer in tiny micelles, perhaps 150 Å in diameter. In fact most of the monomer remains in the form of droplets, about 10^4 Å in size, stabilized by emulsifier on the surface. It is customary to use a water-soluble initiator so that polymerization can only commence when the radicals formed from the initiator enter the micelles, the monomer droplets then constituting a reservoir from which the monomer concentration in the micelles is maintained at a high level.

The general picture of emulsion polymerization then consists of three stages. In the first, polymerization is initiated in the micelles while the monomer diffuses in from neighbouring droplets. As the micellar monomer–polymer systems increase in size they draw emulsifier from their environment. In the second stage, all the emulsifier has been absorbed by the polymer-containing particles; there are no monomer micelles left and no new polymer particles are formed; this period is therefore one of constant rate. In the third step, the monomer supply is tending towards exhaustion so the rate decreases.

In such a system, the kinetic treatment will necessarily be complicated. Pioneering work in this field was published by Smith and Ewart (1948) and by Haward (1949) but many later developments have been reported (van der Hoff, 1962; Gardon, 1968). The salient feature of the Haward–Smith–Ewart theory is that in the steady state the rate of polymerization is proportional to the number of polymer particles but it also carries the implication that each particle can only house one radical at a time. One thus has a picture of a radical entering a particle and initiating a poly-merization which continues until a new radical arrives to terminate the growth of its predecessor. This generally leads to a situation of alternating activity and quiescence but the important point is that conditions are usually conducive to the formation of polymer of high molecular weight. It has become the principal commercial method for the polymerization of dienes for the synthetic rubber industry.

F. Cationic Chain Polymerization

By comparison with radical and anionic reactions, the scope for rigorous kinetic analysis of cationic polymerizations is meagre indeed but, even if little can be offered in the way of useful mathematical equations, it is at least possible, and certainly important, to understand something of the reasons for this state of affairs. Here only a brief summary will be attemp-ted but an excellent full discussion is available (Allen and Patrick, 1974).

In ionic polymerizations the counter-ion plays an important role because it effectively regulates the acts of initiation and propagation. Part of the difficulty with cationic polymerizations is that initiation is brought about by a remarkably wide range of types of substances.

Much of the earlier work in this field employed Lewis acids, such as BF_3 or $AlCl_3$ as the initiator but it was found that an additional substance, the cocatalyst, seemed to be required for polymerization to take place (Olah, 1963; Kennedy and Langer, 1964). Since the cocatalyst was usually a hydroxyl compound, it was believed that its function was to generate protons which then initiate the chain:

$$BF_3 + H_2O \rightleftharpoons H^+ + BF_3OH^- \tag{52}$$

Of course, any excess water present will also act as a terminating agent so complications in the kinetics are readily anticipated.

The necessity for a cocatalyst continued to be assumed for a long time but recently opinion has come to accept that in certain cases direct initia-tion by the halide can indeed occur in highly purified systems. For example, Chmelír et al. (1967) showed that, in completely dry n-heptane, $AlBr_3$ can cause isobutylene to polymerize with overall kinetic behaviour corresponding to

$$\frac{-d[M]}{dt} = k[AlBr_3]^2[M] \tag{53}$$

and other examples of this type of behaviour are known (Matyska et al., 1966; Maślinśka-Solich et al., 1969), although the underlying mechanism is not understood.

Metal halides with organic groups also possess the capacity to initiate, provided that a suitable cocatalyst is present (Kennedy and Gilham, 1972). The latter may be a Brønsted acid or an alkyl halide, in fact in suitable conditions the upshot is the same in either case. Thus, if the monomer is isobutylene, the Brønsted acid route may be formulated as:

$$Et_2AlCl + HCl \rightarrow H^+Et_2Al^-Cl_2 \xrightarrow{\begin{array}{c} CH_3 \\ \diagdown \\ C = CH_2 \\ \diagup \\ CH_3 \end{array}} (CH_3)_3C^+.Et_2Al^-Cl_2 \tag{54}$$

while, using t-butyl chloride, the same ion-pair is produced directly:

$$Et_2AlCl + tBuCl \rightarrow (CH_3)_3C^+.Et_2Al^-Cl_2 \tag{55}$$

Naturally one would expect the kinetics to differ because of the involvement of the monomer in initiation in the former case.

Other types of initiation include the Brønsted–Lowry acids (Hayes and Pepper, 1961; Pepper and Reilly, 1962) (e.g. H_2SO_4, $HClO_4$) and the halogens, notably iodine (Ledwith and Sherrington, 1971). In both cases the kinetics are complicated and, in view of recent work (Ledwith and Sherrington, 1971), it appears that there are important parallels. From the kinetic point of view, the most important and extensively studied case is the polymerization of styrene brought about by perchloric acid in CH_2Cl_2 which is usually described as a "pseudo-cationic" process (Gandini and Plesch, 1968). This system was monitored by spectrometry and conductivity which demonstrated clearly that ionic species do not attain a significant concentration until the polymerization is almost complete. Thus, in what *prima facie* would appear to be an ionic system, the ions have but a peripheral significance and that in phases of reaction when polymerization is *not* taking place, and much the same may be true when the initiator is iodine.

An important contribution to cationic polymerizations was the introduction of stable salts of carbonium ions as initiators. These compounds provide a well-characterized source of pure carbonium ions of known structure. Ledwith (1967, 1969; Bawn et al., 1964), in particular has exploited compounds such as $\emptyset_3C^+SbCl_6^-$ and the tropylium salt

$C_7H_7^+SbCl_6^-$, both of which are largely dissociated in dilute solution in CH_2Cl_2. One virtue of these substances as initiators is that they have enabled absolute values of rate constants to be determined and some measure has been obtained of rate constants for the initiation process.

Mention must also be made of the interesting case of initiation by electron transfer, first clearly exposed by Ellinger (1968). The classical case is N-vinyl carbazole, which is able to donate an electron to a wide variety of acceptors. Here, too, velocity constants have been determined; they are noteworthy for the large discrepancy with corresponding values from reactions initiated by carbonium ions.

Although cationic reactions can also be initiated electrochemically, photochemically, and by high energy radiations, no special relevance to fibre-forming polymers exists to justify further discussion here.

From the foregoing discussion it will be apparent that it is often impossible to identify the initiating species with certainty; similarly the nature of the propagating entity may be a subject for speculation. Thus, leaving aside the problem of pseudo-cationic systems, there may be uncertainty about the degree of freedom enjoyed by the cation in its relationship with the counter-ion. The ion-pairs usually exist in high concentration but exhibit low reactivity; by contrast the free ions are highly reactive but present in very low concentration, except in solvents of the highest polarity. This leads to the dilemma that, if the rate of polymerization is low enough to be followed accurately (with a view to kinetic analysis), it is impossible to measure the concentration of chain carriers whereas, under conditions which permit determination of the free ion concentration, polymerization is uncontrollably rapid. This leads to the sort of compromise in which the half-life of polymerization is only a few seconds so that adiabatic calorimetry is required to enable the course of reaction to be charted (Plesch, 1971). As a broad statement of the general magnitude of the velocity constant for propagation of free cations, one can say that values of about $10^6 \, l \, m^{-1} \, s^{-1}$ seem to be the norm.

Chains can be brought to an end either by formation of a stable carbonium ion (by isomerization, transfer or abnormal addition) or by neutralization with an anion, both processes being first order with respect to the concentration of chain carriers. The monomer will also be involved if transfer or head-to-head addition takes place (Kennedy and Squires, 1967; Plesch, 1968; Gumbs et al., 1969).

IV. Anionic Chain Polymerization

A. INITIATION

Any attempt to study the kinetics of anionic polymerization begins with the problem of the choice of solvent. In early work liquid ammonia or diethyl ether was employed but both are disadvantageous, the former being an active transfer agent and the latter lacking the power to retain the polymer in solution. Most modern studies have concentrated upon solvents of low polarity, hydrocarbons and cyclic ethers, which differ as groups in a striking and useful fashion (Schlenk and Bergmann, 1928; Scott *et al.*, 1936; Fontanille and Sigwalt, 1966; Shatenstein and Petrov, 1967; Moskalenko and Arest-Yakubovich, 1972). In hydrocarbons, the initiation step is relatively slow (for example, the equilibrium in the raction lies well over to the

$$Na + \quad \text{[naphthalene]} \quad \rightleftharpoons \quad Na^+ \quad \text{[naphthalene radical anion]}^{\cdot-} \tag{56}$$

left) and in these conditions one can examine the initiation process in some detail whereas in tetrahydrofuran (THF) initiation is fast enough for reaction (56) to be regarded as a completed reaction, enabling scrutiny to be concentrated on propagation. More flexibility is provided by mixtures of the two solvent classes or by the use of even more strongly solvating compounds, such as hexamethylphosphoramide, which solvates alkaline earth cations sufficiently for their naphthalene derivatives to be employed as initiators (Fontanille and Sigwalt, 1966; Moskalenko and Arest-Yakubovich, 1972).

Although emphasis will fall on initiation by organometallic compounds, it is interesting to note that the first detailed kinetic investigation of an anionic polymerization provided a splendid example of the utility of inorganic bases for this purpose. This was Higginson and Wooding's (1952) study of the polymerization of styrene, in liquid ammonia, initiated by potassium amide.

The kinetics showed the following dependence of rate on reactant concentrations:

$$\frac{-d[M]}{dt} \propto \frac{[KNH_2]^{\frac{1}{2}}[M]^2}{[NH_3]} \tag{57}$$

This equation has the half-order dependence on initiator concentration, usually regarded as diagnostic of radical processes, and a novel inverse dependence on solvent concentration. The half-order with respect to initiator is explained by weak dissociation to the real initiating species NH_2^- while the nature of the dependence on solvent requires the participation of ammonia in what is effectively a termination step. The reaction scheme is then:

$$KNH_2 \rightleftharpoons K^+ + NH_2^- \tag{58}$$

$$NH_2^- + CH_2{=}CH.\emptyset \rightarrow H_2NCH_2{-}C\bar{H}\emptyset \tag{59}$$

$$H_2N(CH_2-CHØ)_{n-1}CH_2-\bar{C}HØ$$

$$+ CH_2{=}CH.Ø \rightarrow H_2N(CH_2-CHØ)_nCH_2\bar{C}HØ \quad (60)$$

$$H_2N(CH_2-CHØ)_{n-1}CH_2\bar{C}HØ$$

$$+ NH_3 \rightarrow H_2N(CH_2-CHØ)_{n-1}CH_2CH_2Ø + NH_2^- \quad (61)$$

Since the equilibrium in reaction (58) lies over to the left, reaction (59) is the rate-determining initiation step and stationary state kinetics then provide the correct formulation of the rate equation. The activation energy for reaction (61) is greater than that for reaction (60) so the overall process has a negative temperature coefficient.

Alkoxides might be expected to initiate in similar fashion; sodium and potassium alkoxides will initiate the polymerization of methyl methacrylate in liquid ammonia but not in hydrocarbon solution, nevertheless, the latter media will support reaction if a lithium t-alkoxide is employed.

This reaction provides an example of initiation by addition of monomer to the free anion; in other cases the monomer becomes inserted into an ion pair or acquires negative charge by an electron transfer process. Typical initiators of the ionic or ion pair type are the alkyl lithiums and Grignard reagents. The Grignard reagents are lower in potency but are adequate where the monomer possesses a really powerful electron-withdrawing group whereas more reluctant monomers require the stimulation of a lithium alkyl.

Many reports are in existence of reactions, initiated by lithium alkyl, conducted in aromatic hydrocarbon solvents. The kinetics of these reactions are complicated by the fact the the lithium alkyl exists in solution mostly in the form of a hexameric complex. Since the reactions sometimes appear to be $\frac{1}{6}$ order with respect to the lithium compound, it has been assumed that the correct kinetic formulation is provided by the equilibrium:

$$(RLi)_6 \rightleftarrows 6(RLi) \quad (62)$$

This is the case with n-butyl lithium and it is supported by the discovery of $\frac{1}{4}$ order for s-butyl lithium which is known to be predominantly tetrameric. Many authorities nevertheless regard this explanation as too facile, preferring a set of equilibria which involve also a tetramer and a dimer (Brown, 1968; Szwarc, 1968).

In aliphatic solvents the rate of initiation is much lower, perhaps by a factor of one thousand, and there is usually an induction period (Worsfold and Bywater, 1960; Hsieh, 1965; Johnson and Worsfold, 1965; Burnett and Young, 1966). The kinetics usually reveal an order of reaction with respect to lithium alkyl of anything from 0·5 to 1·5, a result which is apparently to be attributed partly to slow processes in the association equilibria and the best description in general terms is that provided by Roovers and Bywater (1968).

The rate of initiation, with n-butyl lithium in benzene, is greatly accelerated by as little as 2% of a polar solvent, such as tetrahydrofuran, possibly due to facilitation of the process of dissociation. Bywater (1975) has recently discussed these and other aspects of initiation by lithium alkyls in various solvents.

Lithium alkyls and the other initiators mentioned so far act by direct addition of their daughter anion to the monomer

$$RLi + CH_2 = CHX \rightarrow RCH_2C\bar{H}X.Li^+ \qquad (63)$$

but with other initiators an alternative mechanism is favoured, that of electron transfer, exemplified by the sodium naphthenate compound cited in reaction (56), for which the initiation step is formulated as:

$$Na^+ \quad + CH_2{=}CH\emptyset \rightarrow \dot{C}H_2\bar{C}H\emptyset.Na^+ + \qquad$$

GREEN   RED   (64)

The occurrence of this process is readily appreciated by the observer because the green colour of the sodium naphthenate in THF is replaced by the colour of the monomer radical anion, which, in the case of styrene, is red.

From this point on the kinetics of reactions initiated by sodium naphthenate must recognize that the straightforward propagation by addition of the monomer to $\dot{C}H_2\bar{C}HO.Na^+$ could take place by either radical or anionic mechanisms and that, in any case, there is competition from the radical dimerization of this species which yields a di-anion capable of propagating from two points.

$$2\dot{C}H_2CHX.Na^+ \rightarrow Na^+\bar{C}HXCH_2CH_2C\bar{H}X.Na^+ \qquad (65)$$

It seems that the competition between monomer addition and dimerization has not been kinetically analysed for any polymerizing system but the parallel reactions with 1,1-diphenylethylene exhibit an overwhelming tendency for dimerization (Szwarc, 1968).

The question of the relative propensity for radical and anionic addition also seems to be unresolved quantitatively but, while the general conclusion is that radical reactions cannot be ruled out, there is a total lack of evidence that they play any significant part in the reactions studied to date.

Thus the emergent picture of initiation by sodium naphthanate is the simple sequence of reactions: (56), (64), (65); even though equilibria may well be involved the polymerization process will keep the reactions moving from left to right as written even though, in the absence of the monomer, reaction (56) varies in its stable position from well over to the right with biphenyl to well over to the left with anthracene, with naphthalene in an intermediate position.

B. PROPAGATION: "LIVING" POLYMERS

On the assumption that sodium naphthenate in THF initiates by rapid formation of the monomer di-anion shown in reaction (65), propagation can be pictured as monomer addition at each end but, since the ion pairs are quite stable, there is no termination process to consider. One then has the simplest possible kinetic system in which the rate of reaction is given by

$$\frac{-d[M]}{dt} = k_p[E][M] \tag{65}$$

where $[E]$ is the concentration of anion ends which is equal to the concentration of naphthalene introduced, if no side reactions occur and the equilibria are all over to the right hand side. If this is a true picture, polymerization will continue until the monomer is exhausted but the ion pairs will remain stable, able to resume activity if a further supply of the monomer is provided. Reaction then will clearly continue not only in the sense that the monomer is consumed but also by increasing the molecular weight of the product. For this reason the polymers in the system have been described as "living" or, at the point of zero monomer concentration, "dormant". This is not the place to discuss the obvious extension of such systems to the preparation of block copolymers but accounts are available elsewhere (see review in Noshay and McGrath, 1977).

The kinetics of such systems can be developed by supposing that an initiator In becomes converted by reaction with the monomer into a reactive species X, which by further reaction with the monomer increases in size to $X_2, X_3 \ldots X_n$, the serial number denoting the number of monomer units contained therein. The velocity constant for the first step k_i may be supposed to be different from those of all subsequent steps which are presumed to be identical and denoted by k_p. We thus have:

$$\frac{-d[In]}{dt} = k_i[In][M] \tag{67}$$

$$\frac{d[X_1]}{dt} = k_i[In][M] - k_p[X_1][M] \tag{68}$$

$$\frac{d[X_n]}{dt} = k_p[X_{n-1}][M] - k_p[X_n][M] \tag{69}$$

and the rate of reaction is

$$\frac{-d[M]}{dt} = k_i[In][M] + \sum_{n=1}^{\infty} k_p[X_n][M] \tag{70}$$

It is necessary to recognize that the monomer concentration does not remain constant, indeed it may well decrease to zero, hence a more convenient variable than t is z given by:

$$z = \int_0^t [M] \, . \, dt \quad \text{or} \quad dz = [M] \, . \, dt \tag{71}$$

With this change of variable Eqns. (67)–(70) transform into

$$\frac{-d[In]}{dz} = k_i[In] \tag{72}$$

$$\frac{d[X_1]}{dz} = k_i[In] - k_p[X_1] \tag{73}$$

$$\frac{d[X_n]}{dz} = k_p([X_{n-1}] - [X_n]) \tag{74}$$

$$\frac{-d[M]}{dz} = k_i[In] + \sum_{n=1}^{\infty} k_p[X_n] \tag{75}$$

Eqn. (72) obviously integrates simply to

$$[In] = [In]_0 \exp(-k_i z) \tag{76}$$

Since the total concentration of reactive intermediates $\sum_{n=1}^{\infty} [X_n]$ is equal to the amount of $[In]$ which has reacted with the monomer

$$\sum_{n=1}^{\infty} [X_n] = [In]_0[1 - \exp(-k_i z)] \tag{77}$$

whence the rate of reaction, equation (75), becomes

$$\frac{-d[M]}{dz} = k_i[In]_0 \exp(-k_i z) + k_p[In]_0[1 - \exp(-k_i z)] \tag{78}$$

which can be integrated to

$$\frac{\Delta[M]}{[In]_0} = k_p z - \left(\frac{k_p}{k_i} - 1\right)[1 - \exp(-k_i z)] \tag{79}$$

where $\Delta[M] = [M]_0 - [M]$, that is, the concentration of reacted monomer. If the initiation process is rapid $k_i \gg k_p$ so that $k_p/k_i \ll 1$ and all the initiator can be regarded as having reacted, hence

$$\frac{\Delta[M]}{\Delta[In]} = k_p z + [1 - \exp(-k_i z)] \qquad (80)$$

If reaction is taken to completion and $\exp(-k_i z) \gg 1$, Eqn. (80) becomes

$$\frac{\Delta[M]}{\Delta[In]} = k_p z + 1 \qquad (81)$$

This is an important result because the mean degree of polymerization in these circumstances is the ratio of concentrations of the monomer and initiator consumed, that is:

$$\bar{P} = k_p z + 1 \qquad (82)$$

A particularly interesting feature of these systems is the nature of the molecular weight distribution; Eqns. (67–(69) can be used to deduce the distribution most easily if k_i is set equal to k_p whence

$$[X_n] = [In]_0 \frac{(k_p z)^n}{n!} \exp(-k_p z) \qquad (83)$$

which corresponds to the well-known Poisson distribution. This is a very narrowly disperse type of system and is one of the reasons for the interest in anionic polymerizations since they clearly afford a practical means for the preparation of polymer which approximates to being monodisperse.

More generally in anionic systems the propagation process may involve both ion-pairs and free ions, the two types of reaction occurring simultaneously (Szwarc, 1968). Of necessity the molecular weight distribution will then correspond to a superposition of two separate systems and will be much broader in character than the Poisson. The kinetic analysis of the general case has been presented by Hostalka et al (1964, 1965) and Bhattacharyya et al. (1964, 1965); in favourable cases it is possible to deduce the propagation velocity constant for both types of reactive intermediate and the equilibrium constant for the dissociation of the ion-pairs.

Termination is not a kinetic factor in these systems but it is important to note that it is brought about by quenching the active centres with a reagent that can be used to introduce thereby useful functional groups at the chain ends, for example addition of CO_2 produces carboxyl groups after neutralization. Given sufficient time, termination processes can occur but they are not important on the ordinary reaction time-scale (Sprach et al., 1962; Szwarc, 1968). With monomers containing ester groups the situation may be more complicated (Glusker *et al.*, 1961; Wiles and Bywater, 1965; Allen *et al.*, 1967, 1972).

C. ZIEGLER–NATTA POLYMERIZATION

A new era in polymerization studies began in 1955 when Ziegler (see for example Lenz, 1967; Boor, 1967) announced that he had discovered that certain organometallic systems could cause ethylene to polymerize at ambient temperature and pressure to give a very high molecular weight linear product. Natta (see Lenz, 1967; Boor, 1967) subsequently extended the work to propylene and other α-olefins, and since then the existence of highly crystalline olefin polymers has become an accepted fact.

There are numerous catalyst systems that one can class as Ziegler–Natta initiators; the encompassing definition appears to be that they consist of a combination of (i) an organometallic compound, based on a metal from Group I, II or III of the Periodic Classification and (ii) a halide or an ester of a metal from Group IV, V, VI or VII; a favourite combination is aluminium triethyl and titanium tetrachloride. The true catalyst is undoubtedly a reaction product of the two chosen compounds and while the systems can be homogeneous or heterogeneous, the latter are the more important. The kinetic description of polymerization processes initiated by Ziegler–Natta catalysts thus begins with the handicap that the true catalyst is not in solution, neither is its chemical constitution immediately apparent. The accumulation of polymer in the neighbourhood of the active centres may also constitute a complicating factor.

It cannot be said that there is unanimity even about the nature of the active centre; there are two conflicting views, one of which believes that the chain growth process involves only the transition metal (monometallic mechanism) while the alternative concept also involves the other metal in the transition state (bimetallic mechanism). The majority of chemists now appear to favour the former which is based on the papers of Cossee (1960, 1964, 1967; Cossee *et al.*, 1968) and Arlman (1964) and which has been particularly well discussed by Boor (1967) and by Rodriguez and van Looy (1966).

Early on in kinetic discussion it was pointed out that the heterogeneous nature of the catalysts implied that account might have to be taken of adsorption equilibria in arriving at an expression for the relationship between rate of polymerization and overall reactant concentrations. A number of attempts to derive general kinetic expressions were subsequently made with a variety of assumptions concerning the locus of reaction—whether an adsorbed or non-adsorbed monomer is involved, and so on.

Recently, a new attempt has been made to produce a general kinetic scheme which seems to be more successful than its predecessors and this is outlined below (Burfield *et al.*, 1972). Perhaps it is as well to set out first the general features which any useful kinetic approach must embrace:

(i) overall rate first order with respect to monomer concentration;

(ii) overall rate first order with respect to metal alkyl concentration when the latter is low, becoming zero order when it is high;

(iii) a maximum in rate at sufficiently high metal alkyl concentration;

(iv) the number average degree of polymerization is dependent on reaction time except in the later stages of polymerization, dependent on the monomer concentration but not the transition metal halide concentration and dependent upon both the nature and the concentration of the metal alkyl.

The active centre is assumed to be formed by reaction between the metal alkyl and the transition metal halide. It is important to note that the concentration of actual C_0 or potential C_i active centres is then *not* directly related to the fraction θ_A of surface covered by adsorbed monomeric metal alkyl, as earlier theories held.

It is visualized that a potential active site exists when the two catalyst components have undergone mutual reaction, and that an actual active site is formed by further chemical transformation as a result of interaction with the monomer. This latter step is regarded as chain initiation and subsequent additions of the monomer, of course, are propagations.

Initiation $\quad Cat^+R^- + CH_2{=}\underset{X}{CH} \xrightarrow{k_i} Cat^+CH_2^-{-}\underset{X}{CH}{-}R$ (84)

$\qquad\qquad$ Potential $\qquad\qquad\qquad\qquad$ Actual
$\qquad\qquad$ Active $\qquad\qquad\qquad\qquad\qquad$ Active
$\qquad\qquad$ Centre $\qquad\qquad\qquad\qquad\qquad$ Centre

Propagation $\quad Cat^+CH_2^-{-}\underset{X}{CH}{-}R + CH_2{=}\underset{X}{CH} \xrightarrow{k_p}$

$$Cat^+CH_2^-\underset{X}{CH}CH_2\underset{X}{CH}{-}R \qquad (85)$$

The maximum concentration C_t of active centres is obviously

$$C_t = C_0 + C_i \qquad (86)$$

and the rates of initiation and propagation are respectively

$$R_i = k_i C_i \theta_M \qquad (87)$$

$$R_p = k_p C_0 \theta_M \qquad (88)$$

where reaction is assumed to involve adsorbed monomer occupying a fraction θ_M of the available surface.

If it is further assumed that the molecular weight can be limited by chain transfer steps involving either an adsorbed monomer or an adsorbed metal alkyl, the rate of chain transfer is

$$R_t = k_m C_0 \theta_M + k_a C_0 \theta_A \tag{89}$$

Given stationary state conditions and assuming that chain transfer is effectively termination (see below)

$$R_i = R_t \tag{90}$$

therefore

$$C_0 = \frac{k_i C_t \theta_M}{k_i \theta_M + k_m \theta_M + k_a \theta_A} \tag{91}$$

From what has been said above it is evident that the rate of reaction (the rate of propagation) will be given by

$$R_p = k_p C_0 \theta_M \tag{92}$$

which, in combination with Eqn. (91) yields

$$R_p = \frac{k_i k_p C_t \theta_M^2}{k_i \theta_M + k_m \theta_M + k_a \theta_A} \tag{93}$$

When the concentrations of monomer and metal alkyl are both low the Langmuir isotherm concept enables one to write $\theta_A = K_A[A]$ and $\theta_M = K_M[M]$ so that, in these circumstances, Eqn. (93) reduces to

$$R_p = \frac{k_i k_p C_t K_M^2 [M]^2}{k_i K_M[M] + k_m K_M[M] + k_a K_A[A]} \tag{94}$$

which is more useful when written in the form

$$\frac{[M]}{R_p} = \frac{1}{k_p C_t K_M} \left(1 + \frac{k_m}{k_i} + \frac{k_a K_A[A]}{k_i K_M[M]} \right) \tag{95}$$

which should be compared with expressions derived by other authors. The plot of $[M]/R_p$ versus $[M]^{-1}$ should be linear with

$$\frac{\text{slope}}{\text{intercept}} = \frac{k_a K_A[A]}{(k_i + k_m)K_M} \tag{96}$$

which can be used to evaluate k_i.

It will be noted that the chain transfer has been regarded as a termination process; it is possible to add a term for spontaneous termination but this is regarded as unimportant, except at high temperatures.

The authors of this scheme have examined it with special reference to the polymerization of 4-methylpentene-1 with initiation by VCl_3 and trialkylaluminiums and found it to be satisfactory. The broad pattern of behaviour stems from the fact that the number of active centres first rises to a saturation value with $[AlR_3]$ when they have all become associated with growing chains but that at some point thereafter the rate drops because the alkyl is displacing the monomer from adsorption sites, progressively reducing θ_M. The precise form of the relation between R_p and $[A]$ will in those circumstances depend upon the values of K_A and K_M. Some previous studies did not exhibit the predicted maximum in rate as $[A]$ is increased but this is simply explained by the much lower catalyst concentrations employed in those cases.

Using the same principles the number average mean degree of polymerization at time t after commencement of reaction can be shown to be

$$\frac{1}{\bar{P}_n} = \frac{k_a K_A [A]}{k_p K_M [M]} + \frac{1}{k_p K_M [M] t} + \frac{K_A [A]}{k_p K_M [M] t} + \frac{(k_m + 1/t)}{k_p} \tag{97}$$

This equation is entirely in accord with the observed relation between $\bar{P}$ and reaction conditions detailed above. A study of these relationships led to the conclusion that chain transfer with adsorbed monomer is the principal chain termination process in the particular system investigated.

V. Special Systems

A. INTERFACIAL POLYCONDENSATION

The "Nylon Rope Trick", often performed as a lecture demonstration, typifies an intriguing class of reactions which has only commanded widespread attention since about 1960. The step-growth reactions, discussed at the beginning of this chapter, are essentially high-temperature reactions which are fairly slow, that is they take probably a matter of hours to give a high yield of high molecular weight polymer. Although such reactions provide the basis for industrial methods of preparing fibre-forming polymers, in many cases alternative low temperature syntheses are available which combine high molecular rate with high rate of reaction. These methods rely on the use of highly reactive compounds, separated into two immiscible solutions, so that reaction is possible only at the interface.

The authoritative review of interfacial polycondensations (Morgan, 1965) reveals that interest first became apparent in the early 1950s. No doubt the driving force stemmed from the realization that advantages resided in the simplicity of the procedure, the short reaction times, the self-adjusting stoichiometry and the comparative tolerance of impurities, although against this must be set the solvent recovery problem, the concomitant production of salt, and the high cost of the starting materials.

Not a great deal can be said from the kinetic point of view. In the typical case of reaction between an aliphatic di-acid chloride and a diamine the reaction rate is likely to be greater than the rate of mixing. As Morgan (1965) has pointed out, the rates of these reactions are comparable with the velocity of propagation in addition polymerization and that "if reactants could be brought together at a concentration of 1M, several tons of polymer could be prepared per minute".

One feature affording a measure of control is the use or avoidance of stirring. In the non-stirred systems a film of polymer forms at the static interface in which case the rate of "wind-up" of the polymer is important. The presence of additives, for example monofunctional reagents, would be expected to influence both rate and degree of polymerization but it is necessary to note here that the weight of evidence suggests that polymerization takes place in the organic phase rather than in the aqueous phase or literally at the interface so that due consideration must be given to the solubility of the additive.

The nature of the organic solvent is also an important factor (Morgan, 1965). The degree of polymer–solvent interaction affects the thickness of the polymer film, the overall rate of polymerization and the degree of polymerization. The effects can be surprisingly large; thus in one standard system the substitution of chloroform for carbon tetrachloride causes a ten-fold increase in rate (Morgan, 1965).

In stirred systems the strength of the film is not important, indeed the polymer can be obtained granular, plasticized or even dissolved. Very broadly, in stirred systems there is a wider choice of components (including solvents) available but the results are more sensitive to control of quantities and purity of reagents. One of the chief difficulties in correlating the results from different laboratories is that the precise conditions of stirring are so critical to the outcome that slight variations in technique, even in the shape of stirrer blades, render meaningful comparison impossible.

As a general guide, Morgan (1965) has stated that reactions giving useful interfacial polycondensations must be so rapid that reaction is completed within five minutes.

B. LACTAM POLYMERIZATION

The lactams constitute an important class of fibre-forming polymers which exhibit interesting, if complicated, kinetic patterns of polymerization. Much the most important is the cyclic monomer ϵ-caprolactam and from the commercial point of view it is the anionic polymerization which is of greatest interest.

The overall polymerization reaction can be written as

$$n \overset{CO-NH}{\underset{(CH_2)_5}{\bigcirc}} \longrightarrow \overset{CO-N-[CO(CH_2)_5-NH]_{n-2}-CO-(CH_2)_5NH_2}{\underset{(CH_2)_5}{\bigcirc}}$$

Since the propagation step involves ring-opening at the cyclic unit on one end of the polymer with consequent attachment via the carbonyl to the nitrogen of the monomer, the process can be regarded as essentially *N*-acylation of monomer by polymer, but the variety of equilibria possible in these systems obliges us to take account of the fact that both monomer and polymer may exist in neutral, anionic and cationic forms so that a multiplicity of propagation mechanisms is available. Obviously all possibilities will not co-exist in any individual reaction system but it must be expected that, once the particular type of initiator has been chosen, there will still be a variety of routes by which the polymer is formed.

Reactions are conveniently classed as cationic, anionic and hydrolytic: in the first two cases the initiators are strong acids or bases which do not generate water so that conditions are anhydrous; water or substances which generate water under the selected reaction conditions give rise to hydrolytic polymerization.

Strong bases initiate polymerization by inducing formation of the lactam anion which undergoes addition to the monomer to give the dimer anion. Note that in the equations below a loop represents the ring which, in the specific case of ϵ-caprolactam, represents $-(CH_2)_5-$

$$CO-NH + B \rightleftharpoons CO-N^- + BH$$

$$CO-N^- + CO-NH \rightleftharpoons CO-N-CO-\overset{-}{N}H$$

The dimer anion can then abstract a proton from the monomer, regenerating the monomeric lactam anion, which can then react with the neutral dimer in a propagation step.

$$CO-NH + CO-N-CO-\overset{-}{N}H \rightleftharpoons CO-\overset{-}{N} + CO-N-CONH_2$$

$$CO-N^- + CO-N-CO-NH_2 \rightleftharpoons CO-N-CO \qquad NCONH_2$$

It is interesting to note that, in contrast to more conventional mechanisms, here it is the monomer which is charged and the growing polymer which is neutral.

The kinetic situation is complicated in many ways. To begin with, the reactions shown are all equilibria but, apart from that, there are many side-reactions, including alternative propagation modes, to be taken into account. Moreover, the reaction medium is often the molten monomer, which rapidly becomes viscous as polymer forms, while at a later stage the partially crystalline polymer separates from the liquid state. Nevertheless some useful kinetic studies have related the rate of reaction to reactant concentrations and permitted correlations to be made between the structure and reactivity of a number of initiating species.

The cationic and hydrolytic mechanisms are also far from simple. Polymerization also takes place in the solid state giving highly oriented polymers of high molecular weight.

References

Allen, P. E. M. and Patrick, C. R. (1974). "Kinetics and Mechanisms of Polymerization Reactions", Chapter 7. John Wiley.

Allen, P. E. M., Jordan, D. O. and Naim, M. A. (1967). *Trans. Farad. Soc.* **63**, 234.

Allen, P. E. M., Chaplin, R. P. and Jordan, D. O. (1972). *Europ. Polymer. J.* **2**, 271.

Alter, H. and Jenkins, A. D. (1968). *Encyclopaedia of Polymer Science and Technology* **8**, 84.

Arlman, E. J. (1964). *J. Catalysis* **3**, 89.

Atherton, J. N. and North, A. M. (1962). *Trans. Farad. Soc.* **58**, 2049.

Au-Chin, T. and Kuo-Sui, Y. (1959). *J. Polymer Sci.* **35**, 219.

Bamford, C. H. and Brumby, S. (1967). *Makromol. Chem.* **105**, 122.

Bamford, C. H., Jenkins, A. D. and Johnston, R. (1957). *Proc. Roy. Soc. (A)* **239**, 214.

Bamford, C. H., Barb, W. G., Jenkins, A. D. and Onyon, P. F. (1958). "Kinetics of Vinyl Polymerization by Radical Mechanism", Butterworths.

Bamford, C. H., Jenkins, A. D. and Johnston, R. (1959). *Trans. Farad. Soc.* **55**, 1451.

Bamford, C. H., Jenkins, A. D. and Johnston, R. (1962). *Trans. Farad. Soc.* **58**, 1212.

Bartlett, P. D. and Altschul, R. (1945). *J. Amer. Chem. Soc.* **67**, 812, 816.

Bartlett, P. D. and Kwart, H. (1950). *J. Amer. Chem. Soc.* **72**, 1051.

Bawn, C. E. H., Fitzsimmons, C. and Ledwith, A. (1964). *Proc. Chem. Soc. (London)* 391.

Benson, S. W. and North, A. M. (1959). *J. Amer. Chem. Soc.* **81**, 1339.

Bhattacharyya, D. N., Lee, C. L., Smid, J. and Szwarc, M. (1964). *Polymer* **5**, 54.

Bhattacharyya, D. N., Lee, C. L., Smid, J. and Szwarc, M. (1965). *J. Phys. Chem.* **69**, 612.

Billingham, N. C. and Jenkins, A. D. (1972). *In* "Polymer Science", Chapter I, (Ed. Jenkins, A. D.), North Holland.

Boor, J., Jr. (1967). *Macromolecular Reviews* **2**, 115.

Brown, T. L. (1968). *Accounts Chem. Res.* **1**, 17.

Brown, W. G. (1969). *Makromol. Chem.* **128**, 130.

Burfield, D. R. and Tait, P. J. T. (1972). *Polymer* **13**, 315.

Burfield, D. R., McKenzie, I. D. and Tait, P. J. T. (1972). *Polymer* **13**, 302.

Burnett, G. M. (1954). "Mechanisms of Polymer Reactions", Interscience, N.Y.

Burnett, G. M. and Young, R. N. (1966). *Europ. Polymer J.* **2**, 329.

Bywater, S. (1975). "Progress in Polymer Science" (Ed. A. D. Jenkins), **4**, 17, 27.

Carothers, W. H. (1936). *Trans. Farad. Soc.* **32**, 39.

Chaudhuri, A. K. and Palit, S. R. (1968). *Trans. Farad. Soc.* **64**, 1603.

Chmelíř, M., Marek, M. and Wichterle, O. (1967). *J. Polymer Sci. (C)*, **16**, 833.

Cossee, P. (1960). *Tetrahedron Letters* **17**, 12, 17.

Cossee, P. (1964). *J. Catalysis* **3**, 80.

Cossee, P. (1967). "The Stereochemistry of Macromolecules" (Ed. A. D. Ketley), Vol. 1, Marcel Dekker, New York.

Cossee, P., Ros, R. and Schachtschneider, J. H. (1968). 4th International Symposium on Catalysis, Moscow.

Davies, M. (1949). *Research (London)* **2**, 244.

Davies, M. and Hill, D. R. J. (1953). *Trans. Farad. Soc.* **49**, 395.

Ellinger, L. P. (1968). *Adv. Macromol. Chem.* **1**, 169.

Farkas, L., Schaffer, O. and Vromer, B. H. (1949). *J. Amer. Chem. Soc.* **71**, 1991.

Flory, P. J. (1937). *J. Amer. Chem. Soc.* **59**, 466.

Flory, P. J. (1939). *J. Amer. Chem. Soc.* **61**, 3334.

Flory, P. J. (1940). *J. Amer. Chem. Soc.* **62**, 2261.

Flory, P. J. (1946). *Chem. Rev.* **39**, 137.

Fontanille, M. and Sigwalt, P. (1966). *Comptes Rendus Acad. Sci. Paris (C)* **263**, 316, 1208.

Fukumoto, O. (1956). *J. Polymer Sci.* **22**, 263.

Gandini, A. and Plesch, P. H. (1968). *Europ. Polymer J.* **4**, 55 and earlier papers.

Gardon, J. C. (1968). *J. Polymer Sci. (A)* **1**, 623, 643, 665, 687, 2853, 2859.

Glusker, D. L., Lysloff, I. and Stiles, E. (1961). *J. Polymer Sci.* **49**, 315.

Goldschmidt, H. and Udby, O. (1910). *Z. Phys. Chem.* **70**, 637.

Gumbs, R., Penczek, S., Jaguar-Grodzinski, J. and Szwarc, M. (1969). *Macromol.* **2**, 77.

Hamman, S. D., Solomon, D. H. and Swift, J. D. (1968). *J. Macromol. Sci. (A)* **2**, 153.

Harkins, W. D. (1950). *J. Polymer Sci.* **5**, 217.

Haward, R. N. (1949). *J. Polymer Sci.* **4**, 273.

Hayes, M. J. and Pepper, D. C. (1961). *Proc. Roy. Soc. (A)* **263**, 63.

Heikens, D., Hermans, P. H. and van der Want, G. M. (1960). *J. Polymer Sci.* **44**, 437.

Hiatt, R. R. and Bartlett, P. D. (1959). *J. Amer. Chem. Soc.* **81**, 1149.

Higginson, W. C. E. and Wooding, N. S. (1952). *J. Chem. Soc.* 760.

Hostalka, H., Figini, R. V. and Schulz, G. V. (1964). *Makromol. Chem.* **71**, 198.

Hostalka, H., Figini, R. V. and Schulz, G. V. (1965). *Z. physik. Chem. (Frankfurt)*, **45**, 286.

Hsieh, J. L. (1965). *J. Polymer Sci. (A)* **3**, 163.

Jenkins, A. D. (1958a). *Trans. Farad. Soc.* **54**, 1885.

Jenkins, A. D. (1958b). *Trans. Farad. Soc.* **54**, 1895.

Jenkins, A. D. *In* "Reactivity, Mechanism and Structure in Polymer Chemistry" (Ed. A. D. Jenkins and A. Ledwith), John Wiley, Chapter 4 (1974).

Jenkins, A. D. (1972). *Pure and Appl. Chem.* **30**, 167.

Johnson, A. F. and Worsfold, D. J. (1965). *J. Polymer Sci. (A)* **3**, 449.

Kemkes, J. (1967). *J. Polymer Sci. (C)* **22**, 713.

Kennedy, J. P. and Gilham, J. K. (1972). *Adv. Macromol. Chem.* **10**, 1.

Kennedy, J. P. and Langer, A. V. (1964). *Adv. Macromol. Chem.* **3**, 508.

Kennedy, J. P. and Squires, R. G. (1967). *J. Macromol. Sci. (A)* **1**, 861.

Kennedy, J. P., Pasquon, I. and Porri, I. (1972). *In* "Macromolecular Science" (Ed. C. E. H. Bacon), Butterworths.

Kirchner, K. (1966). *Makromol. Chem.* **96**, 179.

Kirchner, K. (1969). *Makromol. Chem.* **128**, 150.

Ledwith, A. (1967). *J. Appl. Chem.* **17**, 344.

Ledwith, A. (1969). *Adv. Chem.* Series No. 91, *Am. Chem. Soc.* 317.

Ledwith, A. and Sherrington, D. C. (1971). *Polymer* **12**, 344.

Lenz, R. W. (1967). Organic Chemistry of Synthetic High Polymers, 70. Interscience, N.Y.

Mareš, F., Bažant, V. and Krupička, J. (1969). *Coll. Czech. Chem. Comm.* **34**, 2208.

Máslínska-Solich, J., Chmelíř, M. and Marek, M. (1969). *Coll. Czech. Chem. Comm.* **34**, 2611.

Matheson, M. S., Auer, E. E., Bevilacqua, E. B. and Hart, E. J. (1951). *J. Amer. Chem. Soc.* **73**, 1700.

Matyska, M., Svestka, M. and Mach, K. (1966). *Coll. Czech. Chem. Comm.* **31**, 659.

Mayo, F. R. (1953). *J. Amer. Chem. Soc.* **75**, 6133.

Mayo, F. R. (1968). *J. Amer. Chem. Soc.* **90**, 1289, 2733, 6895.

Morgan, P. W. (1965). "Condensation Polymers", Interscience N.Y.

Moskalenko, L. N. and Arest-Yakubovich, A. A. IUPAC Macromolecular Symposium Helsinki, Paper I-54 (1972).

Natta, G., Pasquon, I., Svab, J. and Zambelli, A. (1962). *Chim. Ind. Milan* **44**, 621.

North, A. M. and Reed, G. A. (1963). *J. Polymer Sci. (A)* **1**, 1311.

Noshay, A. and McGrath, J. E. (1977). "Block Copolymers", Academic Press, New York and London.

Novokshonova, L. A., Tsvetkova, V. I. and Chirkov, N. M. (1963). *Izv. Akad. Nauk., Ser. Khim.* **7**, 1176.

Ogata, N. (1970). *J. Polymer Sci. (C)* **31**, 217.

Olah, G. H. (1963). "Friedel-Crafts and Related Reactions", Interscience N.Y. **1**, 201.

Pepper, D. C. and Reilly, P. J. (1962). *J. Polymer Sci.* **58**, 639.

Plesch, P. H. (1968). *Prog. High Polymers* **2**, 139.

Plesch, P. H. (1971). *Adv. Macromol. Chem.* **8**, 137.

Rodriguez, L. A. M. and van Looy, H. M. (1966). *J. Polymer Sci. (A-1)* **4**, 1951, 1971.

Rolfe, A. C. and Hinshelwood, C. N. (1934). *Trans. Farad. Soc.* **20**, 935.

Roovers, J. E. L. and Bywater, S. (1968). *Macromol.* **1**, 328.

Sawada, H. (1969). *J. Macromol. Sci.* (*C*) **3**, 310.

Schildknecht, C. E. (1955). "Polymer Processes", Interscience, N.Y.

Schindler, A. (1963). *J. Polymer Sci.* (*C*) **4**, 81.

Schlenk, W. and Bergmann, E. (1928). *Annalen* **464**, 1.

Scott, N. D., Walker, J. F. and Hansley, V. L. (1936). *J. Amer. Chem. Soc.* **58**, 2442.

Šebenda, J. (1972). *J. Macromol. Sci.* (*A*) **6**, 1145.

Shatenstein, A. and Petrov, E. S. (1967). *Russ. Chem. Rev.* **36**, 100.

Skuratov, S. M., Strepikheev, A. A. and Kanarskaya, E. N. (1952). *Koll. Zh.* **14**, 185.

Smith, W. V. and Ewart, R. H. (1948). *J. Chem. Phys.* **16**, 592.

Sprach, G., Levy, M. and Szwarc, M. (1962). *J. Chem. Soc.* 355.

Stoll, M. and Rouvé, A. (1935). *Helv. Chim. Acta* **18**, 1087.

Szwarc, M. (1968). "Carbanions, Living Polymers and Electron Transfer Processes", Interscience, N.Y., pp. 151, 297, 476, 639.

Van der Hoff, M. B. E. (1962). Advances in Chemistry Series, No. 34, *Am. Chem. Soc.* 6.

Van Velden, P. F., van der Want, G. M., Heikens, D., Kruissink, C. A., Hermans, P. H. and Staverman, A. J. (1955). *Rec. Trav. Chim.* **74**, 1376.

Wiles, D. M. and Bywater, S. (1965). *Trans. Farad. Soc.* **61**, 150.

Worsfold, D. J. and Bywater, S. (1960). *Canad. J. Chem.* **38**, 1891.

Zimmermann, J. (1964). *J. Polymer Sci.* (*B*) **2**, 955.

10. Some Aspects of the Degradation Behaviour of Polymers used in Textile Applications

R. H. PETERS and R. H. STILL

The University of Manchester Institute of Science and Technology, Sackville Street, Manchester, England

I. Introduction

The degradation of both natural and synthetic macromolecules may be brought about by a variety of degradation agencies, such as heat, light, oxygen and ozone, and by chemical and biochemical agents. Degradation may occur by more than one type of agency during the processing or the service life of a polymer. Thus, for example, thermal, oxidative and hydrolytic processes can occur during melt spinning. In addition hydrolytic, oxidative and photochemical degradation may in part be responsible for the deterioration of a finished article.

In order to present a balanced picture of the range of degradation processes involved with textile materials, the degradation behaviour of selected polymers will be discussed. Thus acrylic systems (based on acrylonitrile), cellulose, nylon 6 and 6·6 and poly(ethylene terephthalate) ("polyester") have been chosen as representative textile materials. The general features of polymer degradation are however beyond the scope of this review and are documented elsewhere (Grassie, 1956; Jellinek, 1961; Fettes, 1964; Madorsky, 1964; Frazer, 1968; Jenkins, 1972; Kuryla and Papa, 1973; Plastics Institute Conference, 1973).

II. Poly(acrylonitrile) (PAN)

A. Thermal Degradation

Since 1942 acrylonitrile has been used in acrylic fibres such as Orlon (E. I. Dupont de Nemours), Acrilan (Chemstrand Corporation) and Courtelle (Courtaulds Ltd). These materials generally contain comonomers to improve dyeability, solubility and moisture regain and to permit fabrication. The widespread use of acrylonitrile-containing fibres has prompted interest in two main areas, namely the mechanism of thermal colouration and its inhibition and the overall degradation behaviour leading to carbon fibre.

Initial studies (Kern and Fernhow, 1944) showed that pyrolysis at 400°C gave a carbonaceous residue, a basic distillate and traces of hydrogen cyanide. Houtz (1950) reviewed the properties of "Orlon" acrylic fibre and discussed thermal stability in terms of low temperature repeated ironing and the effects at elevated temperatures ($\simeq 200$°C) where a colour change from white to brown was noted. The rate of colouration was enhanced by heat and oxygen, while the molecular weight remained constant to the point of insolubility. No volatiles were lost during colouration and it was proposed that the residue was formed by cyclization of the nitrile groupings followed by dehydrogenation, yielding a condensed aromatic ring system.

Combustion analysis studies were made in an attempt to substantiate this hypothesis but the data obtained were incomplete and this was ascribed to oxygen incorporation.

La Combe (1957) reported that PAN could be spun to yield white fibres of good heat stability (100 h at 100°C). However such fibres could be discoloured by heating with bases at 100°C and bulk polymerization in the presence of a base also gave coloured products. IR analysis showed changes in absorption at 1667 cm^{-1} indicative of C=N formation.

A base catalysed cyclization reaction with the formation of naphthyridine rings was proposed as the colouration reaction.

followed by regeneration of primary amino grouping etc.

Burlant and Parsons (1956) degraded PAN in air and nitrogen and monitored ammonia, hydrogen cyanide and hydrogen evolution rates. Below 210°C ammonia was the sole volatile product whilst above this temperature hydrogen cyanide was also formed at similar rates in air and N_2. Pyrolysis at 250°C yielded an amber liquid which was not acrylonitrile (AN) but which readily polymerized. Elemental analysis of the residue at temperatures between 200 and 320°C indicated that aromatization occurred. Pyrolysis below 200°C yielded a residue which changed from off-white to blue-black, a change which, from IR studies, was interpreted as a progressive linking together of adjacent nitrile groupings.

Schurz (1958) disputed the mechanism proposed by Houtz (1950) for colour formation: changes in IR and UV spectra and physical properties, he suggested, were not consistent with a naphthyridine structure. Thus, discoloured PAN absorbs at 37,000 cm^{-1} whilst substituted pyridines and naphthyridines absorb at 30–33,000 cm^{-1} and 43–46,000 cm^{-1}, and at 1000–670 cm^{-1} in the IR spectrum, but thermally treated PAN shows no such absorption. Schurz also found that PAN showed decreasing solubility with increasing colouration, which he interpreted as being due to cross-linking by azomethine formation.

Grassie *et al.*, from degradation studies on PAN and poly(methacry-lonitrile (PMAN), suggested that the colour change observed indicated that more conjugation was present than that proposed by the Schurz mechanism. The rigid fused naphthyridine system was considered to account for insolubility and a different interpretation was given to the UV data. It was also considered to be consistent with the formation of naph-thyridine rings, since in the early stages, where the rings were only partially conjugated, the UV absorption would not occur in the same position as for a fully aromatized system.

In order to gain insight into the nature of PAN degradation and colouration, Grassie and McNeil (1958a,b) initially studied the degradation behaviour of PMAN. The essential features of PAN and PMAN degradation are similar but PAN exhibits complicating factors which with its limited solubility makes studies on PMAN experimentally easier. IR spectra of coloured samples of PMAN (Figs 1 and 2) were studied.

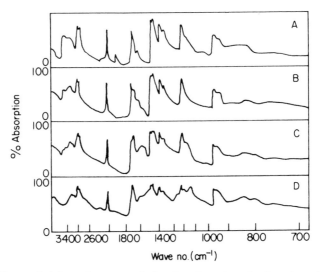

FIG. 1. Changes in infra-red spectrum during the colouration of polymethacrylonitrile. (A) undegraded material, colourless; (B) 3 h at 140°, yellow-orange; (C) 9 h at 140°, orange-red; (D) 23·5 h at 140°, deep red.

A steady decrease in the 2210 cm^{-1} (C≡N) absorption together with an increase in the region of $1693–1490 \text{ cm}^{-1}$ was observed. The 2012 cm^{-1} peak, which disappeared as soon as the sample temperature was raised, is due to ketene-imine groups ($>$C=C=N–) and is not associated with colouration. Changes in the $1660–1490 \text{ cm}^{-1}$ region are shown in more detail in Fig. 2. Initially, at the higher frequency end of this range an increase in absorption was observed. As the sample colour changed from

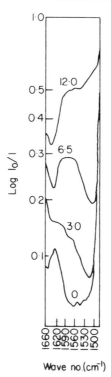

FIG. 2. Changes in the 1660–1490-cm^{-1} region of the infra-red spectrum poly-methacrylonitrile during coloration. Extent of reaction (% decrease in C≡N) is indicated for each spectrum (Grassie and Hay, 1961).

yellow to red a gradual shift to lower frequencies occurred together with increased absorption. This was interpreted as the production of conjugated sequences of C=N bonds which gradually increased in length. The sharp absorption at 1693 cm^{-1} was ascribed to a single unconjugated C=N grouping in accord with that reported by Bircumshaw *et al.* (1954) for paracyanogen which suggested that the structure of thermally coloured PMAN was

IR and colouration studies on PMAN prepared under a variety of conditions (Fig. 3) led to a mechanism for the colouration reaction.

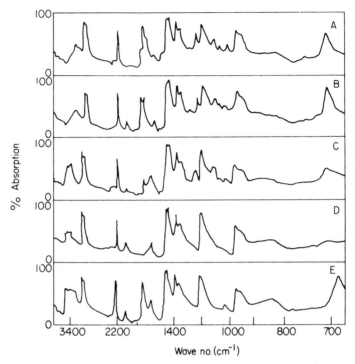

FIG. 3.   Infra-red spectra of various polymethacrylonitrile samples.

	Polymer preparation	Colouration at 160°
A	Monomer after purification by distillation	Rapid
B	Monomer purified by washing with alkali and polymerized in air	Rapid
C	Polymerized in vacuum with benzoyl peroxide as catalyst	Slight
D	Polymerized in vacuum with $\alpha\alpha'$-azoisobutyronitrile as catalyst	None
E	Copolymer with traces of methacrylic acid	Rapid

(Grassie and Hay, 1961).

The ability of PMAN to colour thermally was found to be directly related to the intensity of carbonyl absorption at $1720\,\text{cm}^{-1}$ due to methacrylic acid impurities. When this absorption is absent, as in sample D (Fig. 3), the system was colour stable under the test conditions whilst a copolymer containing methacrylic acid (spectrum E, Fig. 3) showed an enhanced rate of colouration. Pyrolysis of samples free from methacrylic acid gave almost quantitative monomer yields. This behaviour together with the observations of McCartney (1953) on the reaction of PAN and PMAN with alkali led Grassie and McNeil to propose an ionic mechanism for the colouration reaction, similar to that observed at the monomeric level in the intramolecular cyclization of o-cyanobenzoic acid.

Grassie and McNeil showed that organic acids and nucleophiles in general were effective initiators in accord with the proposed mechanism. The ability of a carboxylic acid to initiate colour is related to its pK_a value as shown in Fig. 4.

Madorsky and Straus (1958, 1959; Madorsky, 1964) studied vacuum pyrolysis of high purity PAN in the temperature range 250–800°C. Three fractions were obtained: V_{pyrol}, volatile at pyrolysis temperature, involatile at room temperature; V_{25}, volatile at room temperature, involatile at liquid nitrogen temperature; and a gaseous fraction $V_{-190°C}$. The V_{pyrol}:V_{25} ratio remained constant over the temperature range studied, with $V_{pyrol} = 88\%$ and $V_{25} = 12\%$ based on total volatiles. $V_{-190°C}$ was hydrogen, and was $<0.1\%$ of the volatiles.

V_{25} as first produced was a white liquid, which rapidly changed colour to tan then to brown. This phenomenon was ascribed to a compositional change arising from polymerization, in accord with the observations of

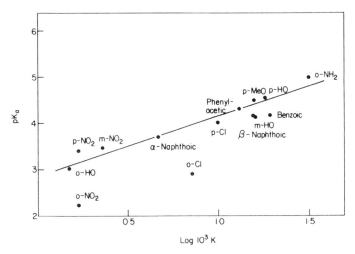

FIG. 4. Effect of variation in acid strength on rate constant for colour initiation by aromatic carboxylic acids. (Grassie and McNeill, 1959).

Burlant and Parsons (1956). Analysis of this fraction was complicated by this change as the composition was time dependent. Hydrogen cyanide, AN, vinyl acetonitrile, pyrrole, acetonitrile, butyronitrile and proprionitrile were however shown to be present. In view of this behaviour, it is not perhaps surprising that previous workers have reported markedly different pyrolysis products. Thus Nagao *et al.* (1958) report considerable evolution of HCN from PAN in air or N_2 at temperatures between 200–350°C whilst Houtz (1950) reports trace quantities at 400°C and Burlant and Parsons (1956) report ammonia evolution. Elemental analysis of the residue indicated that aromatization occurred.

Further studies (Bircumshaw *et al.*, 1954) on relative stabilities of polymers showed that PAN, in contrast to poly(trivinylbenzene) (PTVB), does not become completely carbonized. PTVB forms a highly cross-linked network which loses all of its hydrogen to give hydrocarbon products whilst maintaining most of its carbon (Fig. 5).

Rates of thermal degradation of PAN were studied and rate curves at 218 and 228°C exhibited maxima whilst at higher temperatures the reactions were so fast that neither initial rates nor maximum rates could be established. The shape of curves was interpreted in terms of the degradation proceeding in two steps. Initial rapid evolution of HCN, acrylonitrile and acetonitrile occurs and after the maximum, the rates rapidly reach zero at low total percentage volatilization. An activation energy of 31 kcal/mole was derived from the maximum rate.

Kennedy and Fontana (1959) observed that when PAN was heated to 265°C *in vacuo* the sample underwent "a minute explosion" accompanied

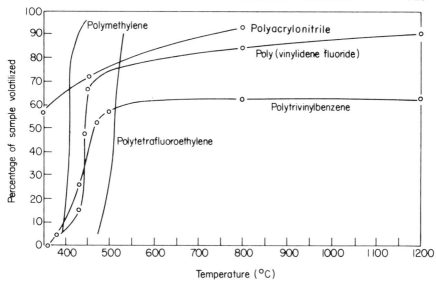

FIG. 5.   Relative thermal stability of high-temperature polymers. (Madorsky, 1964).

by an abrupt colour change to brown. This behaviour was reproducible and was unaffected by preparative conditions. IR absorption at $2237 \, \text{cm}^{-1}$ ($C\equiv N$) abruptly disappeared, which was similar to the behaviour observed in nitrogen by Burlant and Parsons (1956). Solubility in DMF decreased, whilst the *SG* increased and elemental analysis indicated an abrupt loss of hydrogen at the "transformation temperature" while the nitrogen content decreased more gradually with heat treatment. The data is consistent with that obtained by previous workers; no explanation was however advanced for the incomplete analysis figures obtained from the degraded samples. Samples heated to the "transformation temperature" showed 31·6% weight loss and studies by viscometry on heat treated samples indicated that a large degree of chain scission occurred. The authors interpret their results in terms of an auto-catalytic free radical reaction of unspecified nature.

Vosburgh (1960) reports that when Orlon was heated in air above 150°C, an exothermic reaction took place resulting in the formation of a black fireproof textile (AF) which retained up to 40% of the initial fabric tenacity. The important practical variables in the process were found to be (a) the heating time and temperature, (b) the fabric construction, which must allow for rapid free shrinkage and for rapid dissipation of the heat of reaction, (c) an optimum oxygen concentration of 15–20%.

A sample of PAN which had been converted to AF by heating it for 24 hours at 200°C gave analysis figures C 62·44; H 3·45; N 21·80; 0 12·31%, which was explained by assuming that dehydrocyanation occurs in about

15% of the monomer units whilst oxidation occurs in the other 85% followed by dehydration of half the oxidized units. The structure of AF was however thought to be similar to that proposed by Houtz (1950). Ammonia (in trace quantities) was the sole product when PAN was heated below 120°C in air or vacuum. When the heating temperature was increased, the amount of volatile components (NH_3, HCN, CH_4) increased. AF was fireproof and had an unusually high resistance (100 × ordinary fabrics) to destruction upon short-time thermal irradiation. On continued exposure to high temperatures AF fibre gradually lost its tensile properties and at 900°C it was completely consumed in 3 hours. Poor traverse properties suggested that it would be unsuitable for use in fireproof garments where good abrasion resistance was required.

Work on the colouration reaction of PMAN (Grassie and McNeil , 1958a,b; Grassie et al., 1958) was extended to PAN by Grassie and Hay (1961, 1962). Copolymers from acrylonitrile and acrylic acid were studied under vacuum at 175–250°C and the effect of acid concentration on the rate of colouration of PAN is shown in Fig. 6. In PMAN the rate of colouration is proportional to the acid concentration, reducing to zero for acid free polymers. In PAN however there is a large residual rate at zero acid content (Fig. 6). Rigorous purification of both monomer and derived

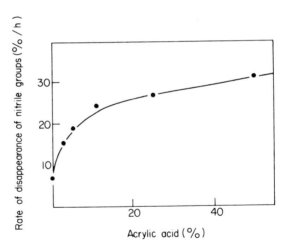

FIG. 6.   Effect of initiator concentration (copolymerized acrylic acid) on the rate of disappearance of nitrile groups in polyacrylonitrile at 200°C. (Grassie and Hay, 1962).

polymer did not affect this residual rate, and it was interpreted as a property of the system rather than the effect of impurities. Mechanisms were proposed to account for both acid catalysis and the self-initiation process. Self initiation was ascribed to initiation by the tertiary hydrogen atom adjacent to a cyano grouping (acidic hydrogen) via a multi-centre intramolecular transfer reaction with the formation of a stable six-membered ring. Thus

Cross-linking occurs with ketimine formation, which structures are capable of initiating cyclization by the mechanism proposed for PMAN (Grassie and McNeil, 1958a,b). The effect of treatment of PMAN with systems containing tertiary C—H centres (Table I), was cited as evidence for self initiation.

TABLE I

Colour initiation in polymethacrylonitrile by C—H centres at 170°C

Initiator	Conc., %	Colour after	
		30 min	60 min
Cumene	10	Orange	Red-Orange
Phenylacetonitrile	10	Yellow	Light red
Benzoic acid	10	Deep red	Deep red
Acrylonitrile (copolymerized)	10	White	Yellow

Distinct solubility differences were also observed, with PMAN remaining soluble even when highly coloured whilst PAN became insoluble before visible colour developed. Attempts were made to establish whether the self-initiation process was solely responsible for insolubility. However, the acid catalysed reaction was found to be much more rapid and was also found to lead to insolubility. It was suggested that in PAN the chain packing resulting from dipolar interaction between adjacent C≡N groupings allowed the colouration process to pass from one chain to another. Such an intermolecular reaction would be more readily achievable in PAN than in PMAN. Thus

In solution cross-linking is inhibited and deeply coloured homogeneous solutions result from alkali attack.

Yoshino and Manabe (1963) investigated the photosensitivity of PAN by heating it in darkness in nitrogen or *in vacuo*. A yellow product which changed to dark orange at room temperature on exposure to light was obtained. This behaviour is similar to that observed by Vosburgh (1960) on samples heated in nitrogen which were then exposed to air. The observations of Yoshino and Manabe suggest that air is not necessary for the process to occur. Visible light absorption of a heat-treated PAN film shifts to longer wavelengths on exposure to light and an ESR absorption is detected. Solubility to acids changes on exposure.

Conley and Bieron (1963) made IR studies on the degradation behaviour of PAN films under oxidative conditions. They showed (Fig. 7)

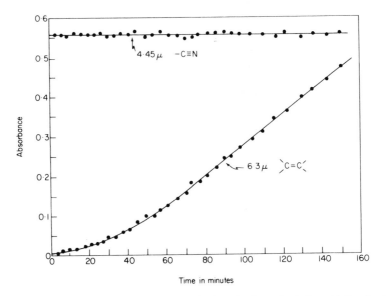

Fig. 7.   Continuous heating and infrared monitoring of polyacrylonitrile oxidation in air at 200°C. (Conley and Bieron, 1963).

that (a) the absorption of the nitrile grouping does not change during the initial oxidation stage of degradation, indicating that it is not involved in the process, (b) the initial degradation product is a 1, 1, 2 trisubstituted olefin, (c) oxygen is required. As a result of their studies they conclude that the mechanisms of Houtz (1950) and Schurz (1958) cannot account for the degradation behaviour in air as these involve the nitrile grouping. Further, the conditions used would not lead to the formation of condensed pyridinoid structures. Reaction is suggested to be initiated at the reactive tertiary hydrogen in the acrylonitrile unit as below

$$
\begin{array}{ccc}
\sim\text{CH} & \text{CH} & \text{CH}\sim \\
| & | & | \\
\text{CN} & \text{CN} & \text{CN}
\end{array}
\xrightarrow{\text{O}_2}
\begin{array}{c}\text{Peroxydic}\\ \text{intermediate}\end{array}
\xrightarrow{\Delta}
\begin{array}{ccc}
\text{CH} & \text{C} & \text{CH}\sim \\
| & | & | \\
\text{CN} & \text{CN} & \text{CN}
\end{array}
$$

Secondary reactions are suggested to occur and the extra absorption bands observed for oxidized PAN at 1685 cm^{-1} and 1710 cm^{-1} were attributed to C=O groups which are present either as acid or amido grouping since absorptions at 3300–3100 cm^{-1} are also present. The tentative reaction scheme below is suggested and it is indicated that these studies are not complete and do not refute in any way the colour forming reactions suggested by other authors.

Lactone structures	Oxidative cleavage	Lactam structures

Complex multifunctional Species

From a n.m.r. study of residues from PAN degraded *in vacuo* Maklakov and Pimenov (1964) describe the behaviour in terms of transformation to structure I and its conversion to II.

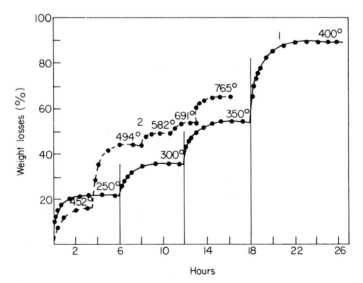

(I)　　　　　　　　　**(II)**

Structure II was formed after 3 h at 210°C and the fraction of II present was temperature dependent.

Geiderikh and his colleagues (Drabkin *et al.*, 1964; Davydov *et al.*, 1965) monitored weight loss and UV and IR spectra changes on PAN degraded *in vacuo*. Below 220°C the C≡N absorption decreased whilst the conjugated C=N intensity increased without weight loss. At 230–250°C gaseous products were liberated and decomposition was autocatalytic. PAN treated below 200°C showed small weight losses and first order kinetics. The behaviour recorded on programming the temperature 50°C every 6 h shows a high rate of decomposition at the beginning of each time interval which is zero at the end (Fig. 8). Raising the temperature caused degradation to begin again and the activation energy is observed to decrease with increasing time of thermal pre-treatment (61 → 4·3 kcal/mole after 200 h at 200°C). A catalytic effect of the conjugated regions on the thermal degradation of non-conjugated sections was

FIG. 8.　Curves of the decomposition of PAN(1) and polyoxyphenylene (2) at successively increasing temperatures. (Davydov *et al.*, 1965).

suggested and shown to be intramolecular by addition of thermally pre-treated PAN which did not affect the rate of decomposition. Catalytic action thus diminishes with increasing distance of the degrading unit from the conjugated system.

Thompson (1966) made observations by DTA in air and nitrogen on PAN samples of different molecular weight and observed an exothermic peak which was molecular weight dependent (Figs. 9 and 10). TG studies indicated that the exotherms were accompanied by substantial weight losses which in nitrogen amounted to 40–50% in the range 200–500°C, after which the residue was stable to 700–800°C. In air the samples degraded completely. No interpretation was given for the thermal analytical behaviour of the system.

Monahan (1966) in an extensive study of PAN degradation *in vacuo* at 280–450°C showed that $(CN)_2$, HCN, AN, CH_3CN and vinyl acetonitrile were obtained as degradation products. Initial rate studies led to activation energies of 15 and 23 kcal/mole for HCN and $(CN)_2$ formation. Studies on the residue by IR, UV and elemental analysis suggested that it was poly 1,4,4-trihydronaphthyridine and 3,6-dimethyl-1,8 naphthyridine was isolated from the degradation products. In addition it was suggested that the formation of vinyl acetonitrile may be linked mechanistically with the production of the black residue, the weight percentage of which increased with increasing rate of formation of vinyl acetonitrile.

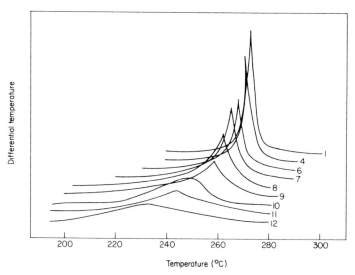

FIG. 9. High-temperature DTA curves for polyacrylonitrile in nitrogen. Samples 1–12 are of decreasing molecular weight. (Thompson, 1966).

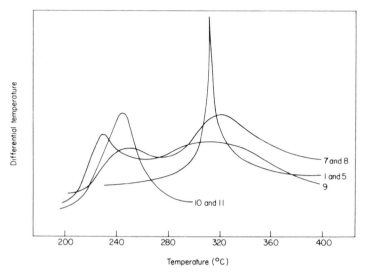

Temperature (°C)

FIG. 10.   High-temperature DTA curves for polyacrylonitrile in air. (Thompson, 1966).

Noh (1967) made studies on the structural changes occurring in PAN by following the disappearance of the nitrile group absorption in the IR spectrum. First order kinetics were shown to be obeyed which are not consistent with long sequences of cyclized C=N groups produced by intra or intermolecular reaction which had previously been assumed to be responsible for the C≡N group disappearance. Some C≡N groupings (19–22%) were left isolated which did not contribute to the partially hydrogenated 1,8-naphthyridine rings. Random chain scission was observed to take place at temperatures in excess of 250°C together with ring formation. Weight loss from the system was related to the rates of the two competitive reactions.

Bonvicini and Caldo (1969) degraded PAN in air and nitrogen and evaluated degradation behaviour in terms of changes in viscosity, solubility and gel formation in DMF, discolouration and HCN evolution. Molecular weight was found to influence the HCN evolution rate and activation energies of 25 kcal/mole in nitrogen and 11 kcal/mole in air were evaluated.

The discovery and manufacture of carbon fibres, by pyrolysing acrylic fibre to high temperature led to a considerable interest in the degradation behaviour of PAN under programmed conditions and wider temperature ranges than previously studied. Studies by Thompson (1966), Kaesche-Krisher (1965) and Gillham and Schwenker (1966) by DTA have been reviewed by Reich (1968) who suggests that the exotherm in the range 250–300°C is due to nitrile group polymerization, and this is supported by

Turner and Johnson (1969) but disputed by Hay (1968) who concluded that it was due to ammonia formation.

Grassie and McGuchan (1970) studied PAN by thermal analytical methods utilizing DTA, TG and TVA (Thermal Volatilization Analysis). Under dynamic conditions in an inert atmosphere at a heating rate of 10°C/min PAN remains unchanged to 245°C when it turns to pale yellow. As the polymer is heated it changes from yellow to tan-brown during an exothermic process. Colour remained until approximately 350°C when the residue became black. Various thermal treatments were applied to the system to assess their effect upon the exothermic process in an attempt to elucidate its nature. Thus samples were preheated to the onset of weight loss and rapidly quenched to prevent further degradation. The effect of this pretreatment on the magnitude of the exotherm (ΔT) is shown in Table II.

TABLE II

Colour and DTA of PAN preheated to onset of weight loss

TG weight loss %	Colour	DTA(ΔT)°C
0	white	32
0·1	pale yellow	27
0·2	yellow	24
0·3	yellow	15
0·4	yellow	11
1·0	tan	5
6·0	tan	0·4

where ΔT is the maximum temperature difference between the sample and the reference.

Previous studies had shown that colouration could occur without weight loss. Accordingly, studies were made after isothermally ageing samples at 180–190°C (Table III).

Exotherm removal and colouration thus involve very small weight losses, a part of which appears to be due to removal of contaminants. When the pre-heated samples were heated to 300°C their colour deepened but did not reach that obtained by direct heating to the same temperature. This was interpreted as an effect of reducing endothermicity, and the temperature of colouration, on the average length of conjugated sequence produced.

IR studies (Fig. II) show that the C≡N intensity decreases relative to methylene absorption at 2940 and 1450 cm^{-1} on isothermal ageing, and absorptions also appear at 1640–1610 and 1580 cm^{-1}. When PAN is

TABLE III

Colour and DTA of PAN after isothermal ageing at 180°C

Duration of heating (hr)	Wt. loss %	Colour	Exotherm $\Delta T°C$
1	0·4	pale yellow	27·5
2	0·6	pale yellow	24
5	0·6	yellow	24
7·5	0·8	gold	17·5
10	0·5	gold	10
16	0·8	gold-tan	3
65	2·0	tan	0·5

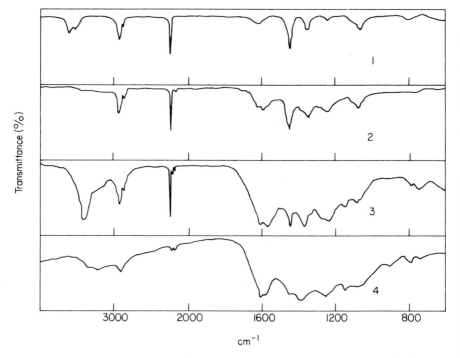

FIG. 11. Infra-red spectra of (1) PAN, (2) PAN heated at 180° for 4 h, (3) PAN heated at 180° for 16 h and (4) PAN heated at 10°/min to 300°. (Grassie and McGuchan, 1970).

heated through the exotherm the C≡N absorption almost disappears and the small band left is accompanied by a band at 2190 cm^{-1} of approximately the same intensity. Beevers (1968) has reviewed IR studies on PAN and the above results are in general agreement with previous workers.

The tan residues obtained on heating to 300°C in nitrogen were observed to change to red-brown on exposure to air at room temperature. This occurred at the surface and gradually penetrated the material with a weight gain of 2% in a few hours, suggesting oxidation involving nitrone formation as described by Friedlander *et al.* (1968). However the weight gained suggests that approximately 10% of the nitrogen atoms would be converted. As an alternative possibility oxidation of terminal imino groups was suggested. Since such systems are also hygroscopic it is possible that the photosensitivity reported by Yoshino and Manabe (1963) is responsible.

Pyrolysis in air indicates colouration to occur at 250°C prior to the exotherm and weight loss, followed by a colour change (orange-tan to black) at temperatures in excess of 300°C. On thermal pretreatment in air similar DTA behaviour to that observed in nitrogen is obtained. The colour is different, and IR studies suggest that carbonyl groupings are present. The $C\equiv N$ absorption is strong in the temperature range 250–300°C but above 300°C the polymer turns black and the absorption decreases rapidly. Grassie and McGuchan interpret the exotherm as being the polymerization of the nitrile grouping which results in colouration. This however is not in accord with the observations by Hay (1968) on PMAN which does not give an exotherm whilst colouring by a similar mechanism. Grassie and McGuchan consider that with PMAN the endothermic pyrolytic reactions could mask the exotherm and suggest that the cyclic structure obtained from PMAN would be more highly strained, leading to reduction in the exothermicity of the polymerization process.

In the case of PAN the exotherm promotes weight loss which suggests that chain scission is competitive with propagation of conjugated sequences. Grassie and McGuchan suggest a model in which initiation occurs at frequent intervals along the chain, with conjugation being terminated when it reaches an adjacent conjugated segment. Scission of the non conjugated connecting units would produce short conjugated chains in accord with the observations of Noh (1967).

Self initiation as proposed by Grassie and Hay (1962) has been disputed by Peebles *et al.* (1968) who concluded that abnormal structures present in PAN are responsible for colour initiation. Such structures arise during free radical polymerization and involve propagation via the nitrile grouping leading to ketonitrile and/or enamine structures.

Peebles concludes that such structures are present in all free radical polymerized AN, and in addition showed that the rate of yellowing in air at low temperatures was proportional to the ketonitrile concentration. Grassie and McGuchan (1971a) made studies on the effect of sample preparation on the TG and DTA behaviour of PAN in order to establish if either the self initiation theory or the Peebles approach were tenable. DTA traces for systems polymerized in bulk and in a slurry were different. Bulk systems gave exotherms at lower temperatures which was interpreted in terms of enamine structures being present, with ketonitrile structures being present in the slurry polymer. The effect of hydrochloric acid treatment as shown in Table IV was presented as evidence for these structures.

TABLE IV

Effect of 0·1 N HCL treatment on DTA of PAN

| | | DTA | | | |
| | | Original | | After treatment | |
Polymer	Treatment*	ΔT	T_R	ΔT	T_R
T3 (bulk)	A	34	264	22	288
	B			24	275
S3 (redox)	A	44	281	13	298
	B			21	289
S4 (redox)	A	18	279	1	232
				2	295
	B			1	239
				3	295
S6 (redox)	B	28	269	17	268
O5 (solution)	B	31	256	38	260
A1 (anionic)	B	10	295	11	285

* A sample heated to boiling
B sample washed with cold HCl
$T_R = (T_S - \Delta T)$ where T_R is reference temperature, ΔT is the maximum temperature difference between the sample and reference, and T_S is the sample temperature.

In the redox polymers (S3, S4, S6) the acidification of the end groups makes them more efficient in initiating the reaction, especially for low molecular weight samples. Bulk systems were more consistent with slurry polymers after treatment. The authors conclude that in bulk systems initiation occurs at enamine structures and this provides radical activity capable of being transferred along the chain. In contrast the ketonitrile structures were judged to be less prone to transfer.

Polymerization of the nitrile groupings should not involve weight loss but the exotherm has been shown to be associated with loss of HCN, NH₃ and chain fragments, particularly under vacuum where large weight losses due to chain fragments are observed. Such data is not consistent with the concept of a small concentration of initiating centres and a long zip length. Thus Grassie and McGuchan concluded that the zip length is short but the kinetic chain length and radical activity is maintained by transfer. Such a pathway would produce cyclized segments linked by segments of unchanged monomer units. Scission of the latter segments can occur to give a low molecular weight wax fraction which is the major component of volatiles produced on weight loss. The radical nature of the reaction was shown for some but not all samples by inhibition by diphenyl picryl hydrazyl (DPPH). Ammonia and HCN production was accounted for on the basis of the terminal imine structures and the unchanged AN segments respectively. A scheme for PAN degradation was outlined as

In air the behaviour of bulk and slurry systems was similar, suggesting that oxidation of the enamine to the ketonitrile grouping occurs.

With anionic PAN, DPPH had no inhibition effect and its behaviour was ascribed to the presence of ionic impurities, together with structural abnormalities resulting from reaction of the strongly basic catalyst and the nitrile grouping.

In order to assess whether their pyrolysis model applied to pre-heated PAN, Grassie and McGuchan (1971b) performed further thermoanalytical studies. The work stemmed from the use of acrylonitrile-containing fibres in the manufacture of carbon fibres. Low temperature reactions are important in this context since the preferred manufacturing process involved "pre-heating" in air in the temperature range 150–250°C (Turner and Johnson, 1969; Bailey and Clarke, 1970). Whilst oxidative processes complicate the situation, the cyclization reaction is of fundamental importance to the conversion of PAN to carbon fibre in high yield (McNulty, 1968; Watt, 1970; Muller et al., 1971; Fitzer and Muller, 1971).

The degradation model (p. 341) was applied to pre-heated PAN omitting chain scission and fragmentation. The model satisfies the experimental observations and the re-initiation step is of particular importance since it explains why small amounts of initiation can cause almost complete CN removal without implying long conjugated sequences. This behaviour is not in accord however with the Peebles model (Peebles et al., 1968). Muller (Muller et al., 1971; Fitzer and Muller, 1971) has however explained shrinkage of PAN during pre-heating and pyrolysis in terms of short conjugated sequence lengths with cross-linking. Such a model is not compatible with the observations of Watt (1970) on mechanical properties which indicated that cross-linking did not occur.

Grassie and McGuchan conclude that cross-linking is not consistent with the volatile fragments produced on pyrolysis. They explain shrinkage in terms of the re-initiation step which would involve cyclization of three monomer units. The model also accounts for the production of appreciable amounts of NH_3 and HCN and subsequent aromatization with hydrogen evolution.

$$+ 2HCN + NH_3 + H_2$$

The products depend on the average length of conjugated sequence. HCN evolution is competitive with main chain scission which would leave an aromatic nitrile grouping in the residue. In the IR spectrum of the pre-heated polymer strong NH bands arise from both the imino end groupings and their isomeric amino counterparts. When cyclization is blocked at an initiation step a β-iminonitrile can be obtained which is responsible for the splitting of the nitrile absorption and the appearance of a new band at $2190\,cm^{-1}$ similar to that observed for 1-cyano-guanidine (Long and George, 1964).

When PAN was pre-oxidized the basic scheme was found to be tenable but the rate of oxidation was found to vary with polymer structure. The cyclization reaction was dependent of the initiating centres already present in the polymer rather than initiation caused by oxygen as proposed by Peebles (Peebles et al., 1968). Thus in one case Grassie and McGuchan found that oxygen inhibited the reaction, probably by reacting with, and thus deactivating, the colour initiating centres in the macromolecules. IR analysis showed that oxygen is incorporated into the system and that carbonyl containing structures are produced. Watt (1970) favours the formation of ketonic structures whilst Danner and Meybeck (1971) favour complete aromatization in carbon fibre formation. Grassie and McGuchan found evidence for both types of behaviour and suggested a mechanism in which the ketonic groupings catalyse dehydrogenation,

resulting in increased conjugation which explains the chromophoric shift and the formation of a black residue. Aromatization is only partial and continues when the temperature is raised. This concept of the pre-oxidized structure led Grassie and McGuchan to suggest that the pre-oxidized material resists fragmentation on pyrolysis as a result of stabilization of the methylene bridges by dehydrogenation prior to chain scission.

In addition conversion of the methylene groupings to carbonyl groupings would also reduce scission by conjugation with adjacent aromatic structures. Such ketonic groupings could isomerize, and cross linking with water evolution could follow. Since water is found amongst the pyrolysis products the route is considered to have experimental justification.

The exothermicity associated with PAN pyrolysis is in the manufacture of carbon fibres, an undesirable feature which can cause excessive shrinkage, melting and disintegration of the fibres. In the manufacturing process a low temperature isothermal treatment (pre-oxidation) is used to overcome this problem. Grassie and McGuchan (1971c) used thermal analysis to investigate the use of additives to accelerate pre-oxidation or to remove the need for such a step, and to study the effect of additives on carbon yield at high temperatures. Organic additives (Grassie and McNeil, 1958a,b) and inorganic systems, such as KCN, NaSCN and Na_2S, which behaved most efficiently were studied. TG showed that in general the organic materials were volatile prior to effectively catalysing the reaction.

In pure PAN, initiation appears to be the slow step and propagation is very rapid. In the presence of the additives much broader exotherms are observed and the propagation reactions appear to be slower, suggesting an ionic or concerted mechanism. Ionic mechanisms were suggested for system such as KCN and NaOH in which propagation involves separation

of charge with carbonium ion migration which would of necessity be slow. The IR spectra of PAN degraded in presence of NaOH and KCN show new nitrile absorptions at 2150 cm^{-1} suggesting the presence of iminonitrile species analogous to that described previously for pure PAN.

Previous work (Grassie and McNeil, 1958a,b; Grassie and Hay, 1961, 1962) led to an ionic initiation mechanism, followed by a concerted stepwise propagation reaction for colouration initiated by organic acids and phenols (p.   ). the acid used becomes an ester which is not conjugated with the —C=N— system and the ester carbonyl group should absorb at 1750 cm^{-1} similar to vinyl esters. For PAN degraded in the presence of sebacic acid the carbonyl absorption disappeared, suggesting that the acid was incorporated into the conjugated structure. Grassie and McGuchan suggest that the intermediate iminoester isomerizes to the imide, which is the normal product of reaction between acids and nitriles (Polya and Spotswood, 1948), and that reaction proceeds as shown below.

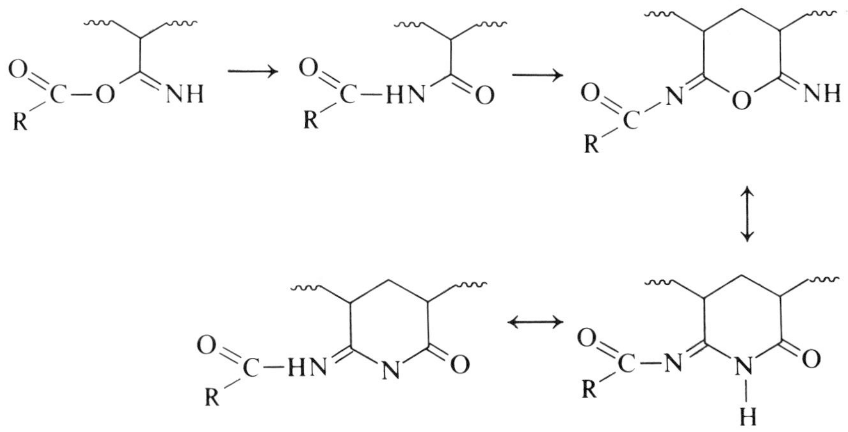

$$
\begin{array}{ccc}
\underset{R}{\overset{O}{\parallel}}C-O \quad NH & \longrightarrow & \underset{R}{\overset{O}{\parallel}}C-HN \quad O & \longrightarrow & \underset{\underset{R}{\overset{O}{\parallel}}C}{O} \quad N \quad O \quad NH
\end{array}
$$

$$
\begin{array}{ccc}
\underset{R}{\overset{O}{\parallel}}C-HN \quad N \quad O & \longleftrightarrow & \underset{R}{\overset{O}{\parallel}}C-N \quad \underset{H}{N} \quad O
\end{array}
$$

Colour changes, IR spectra and the heat liberated in additive initiated pyrolysis, which is not reduced, suggest that the polymer structure is little changed. In some cases the heat liberated is increased, because the heat of reaction is the summation of various processes, namely exothermic polymerization and endothermic processes associated with chain scission and evolution of volatiles. In the case of pure PAN *in vacuo* the weight loss between 275–450°C is caused by distillation of chain fragments; however in nitrogen the relative involatility of these fragments results in lower weight loss. TG studies show that the weight loss associated with the exotherm is reduced in the presence of initiating additives (Fig. 12) indicating that the endothermic contribution to the net ΔH is reduced. Further to this, weight retention studies on PAN at 500°C in the presence of additives suggest that fragmentation is reduced. Thus whilst carbon yield can be increased by pyrolysis in the presence of additives fragmentation is not completely eliminated. Thus additives of the type studied could not therefore be used as a substitute for pre-oxidation whose major function is to prevent fragmentation. Additives could however accelerate pre-oxidation and reduce differential temperatures during processing of carbon fibre.

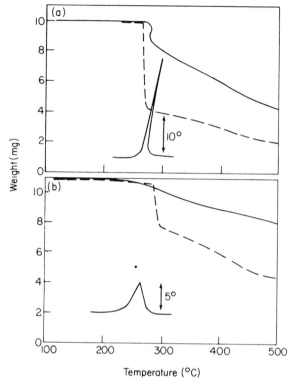

FIG. 12. Comparison of N_2 (——) and vacuum (---) TG curves of (a) PAN and (b) PAN–tannic acid. Samples containing 10 mg PAN heated at 10°/min. DTA curve included. (From Grassie and McGuchan, 1971).

Standage and Matkowsky (1971) studied the controlled thermal oxidation of Dralon T® acrylic fibre (Baytex Inc.), a copolymer of 99·6% AN and 0·4% methyl acrylate in the temperature range 200–325°C. Samples produced were analysed by combustion analysis and IR.

Oxidation proceeded by a continuous exchange of one oxygen atom for two hydrogen atoms with a final oxygen incorporation of 23·9%. From the experimental evidence cited the authors suggest that of the two possible structures suggested for the fully oxidized material the epoxy structure was more acceptable.

Bell *et al.*, (1971) applied micropyrolysis-glc to the study of the thermal degradation kinetics of thin films of PAN (thickness ~750 Å) between

200–850°C. Ammonia, HCN, AN and MAN were characterized over the whole temperature range. Three independent processes were suggested to occur:

(1) ammonia evolution via a stepwise process yielding a ladder polymer containing five-membered rings

+ 4NH$_3$

(2) HCN evolution from head to head links or via a chain reaction leaving double bonds at random in the main chain

(3) Monomer evolution by a depropagation process, involving inter-molecular and intramolecular transfer steps leading to AN and MAN production.

At low temperatures (200°C) the only process observed was the formation of a trihydronaphthyridine ladder system which stabilized the polymer to subsequent degradation at higher temperatures: kinetic studies at high temperature showed the limitations of this stabilization and that fragmentation of uncyclized sequences occurs. An activation energy of 14 kcal/mole was obtained.

Further thermal analytical studies by Grassie and McGuchan (1972a) compared the behaviour of poly(methacrylonitrile), poly(α-chloro-acrylonitrile) and poly(vinylidene cyanide) with PAN. There was a reduced tendency to oligomerization of the nitrile groups in the substituted polymers and the preferred process in the methyl, phenyl and cyano substituted systems was depolymerization to monomer. Cyclization could, however, be induced using suitable initiating systems. The behaviour of poly(α-chloro-acrylonitrile) is complicated by the elimination of HCl but some oligomerization is suggested by the exothermic behaviour recorded on DTA. The data is interpreted as being consistent with the concept that the α-hydrogen atoms in PAN play a major role in the chain process involved in nitrile group oligomerization.

Grassie and McGuchan (1972b) then proceeded to study the pyrolysis of AN copolymers containing carboxylic acid and amide comonomers. One commercial pre-cursor to carbon fibres, Courtelle (Courtaulds Ltd), contains methyl acrylate as the principal comonomer and this prompted a desire to produce materials having superior processing properties. Studies on the ease of pre-oxidation, the exothermicity of direct pyrolysis, and carbon yield, were made on a series of copolymers. These contained the comonomers acrylic acid, methacrylic acid, itaconic acid, acrylamide or acrylamidoxime. Such comonomers would be expected to accelerate the reaction and shift the exotherm to lower temperatures. Pyrolysis of the acid copolymers involved negligible weight loss during the exotherm, confirming that nitrile oligomerization was involved. It was concluded that an ionic mechanism operates similar to that proposed for "monomeric organic acids" (Grassie and McNeil, 1958a,b; Grassie and Hay, 1961, 1962), replacing the free radical mechanism operative for pure PAN. The different initiating efficiency of the acids studied are discussed by Grassie and McGuchan in terms of the stereochemistry of the carboxyl grouping in relation to the nitrile grouping.

With acrylamide a free radical mechanism was suggested

Propagation occurs via

Amideoxime containing polymers were suggested to colour via a concerted mechanism involving migration of the hydroxyl group.

Acids were thought to show most promise in carbon fibre applications with itaconic acid-containing copolymers showing increased carbon yields and lower fragmentation than pure PAN.

These studies were extended to copolymers containing acrylate, methacrylate and styrene as comonomers (Grassie and McGuchan, 1972c). Nitrile reactions occur in the acrylate and methacrylate copolymers since the reaction is able to bypass the comonomer unit. The reaction is slowed down, resulting in a broadening of the exotherm observed on DTA. Hydrogen transfer reactions important in PAN (Grassie and McGuchan, 1970, 1971a, 1972c) were suggested as being involved, but the unusual degradation behaviour of these comonomers in the AN environment was considered to offer a more realistic solution to the problem. When poly(methylacrylate) is pyrolysed, methanol is a major product, but this is not generally produced when the methyl acrylate units are isolated between other monomer units (Madorsky, 1953; Grassie and Torrance, 1968). However, in the AN copolymers methanol arose from single methyl acrylate units. In addition copolymers containing methyl methacrylate also yielded methanol, in contrast to PMMA and its copolymers which normally give high yields of monomer rather than ester decomposition products. These results suggested that the comonomers participated in the nitrile reaction in the following manner

$$CH_3OH + \text{Polymer radical}$$
$$\downarrow$$
$$\text{Cyclization}$$

The sequence is thus terminated by the comonomer, but the kinetic chain length is continued by hydrogen abstraction which reinitiates the cyclization reaction via the polymer radical. Since CO and CH_4 are also produced, chain scission at the comonomer unit also occurs.

Systems containing benzyl acetate should yield benzyl alcohol on pyrolysis (Cameron and Kane, 1968). However, when incorporated into an AN copolymer system, toluene is found to be a major product which could arise from a benzyl radical split out from the copolymer as a result of steric interaction. The benzyl acetate unit could thus act as an initiator rather than a terminator for nitrile oligomerization.

All the acrylate and methacrylate systems studied showed comonomer participation with reasonable weight retention and less chain scission than would have been expected from the degradation behaviour of the parent homopolymer system.

Styrene incorporation has a major effect on the degradation behaviour of PAN with the retardation of the nitrile reaction and restriction of the amount of reaction. That this is not purely a steric effect is suggested from the nature of polystyrene degradation products in whose formation hydrogen transfer reactions play a major role (Madorsky, 1954; Still and Whitehead, 1972).

However since α-methyl styrene shows an identical blocking action it was concluded that the α-hydrogen in the styrene unit is not involved in the process. Grassie and Bain (1970) have shown that the AN-styrene bonds in copolymer systems are easily broken and Grassie and McGuchan suggest that the radical activity of the nitrile reaction may be deactivated by transfer of the radical activity to the comonomer in a chain breaking process. The comonomer radical will be relatively stable and thus the formation of polymer radicals necessary for further cyclization reactions is inhibited or retarded.

Whilst colouration is closely related to the nitrile reaction the colour is not a true measure of nitrile reaction. In the acrylate systems studied, colouration is of the same order as in PAN which may be explained on the basis of the small zip length in the homopolymer. In styrene systems colouration is more intense even though the amount of nitrile reaction is reduced. Since all PAN residues are black when dehydrogenation has occurred, the authors suggested that in the styrene systems cyclization and dehydrogenation reactions may overlap. However, this was not confirmed by TVA since little hydrogen evolution occurred during the exotherm. As an alternative possibility the enhancement of colouration by formation of quinonoid structures was suggested. In terms of carbon fibre production these systems are inferior to acid copolymers.

The work was extended to copolymers containing comonomer units capable of dehydrohalogenation (Grassie and McGuchan, 1973a) in order to assess whether the nitrile reaction was blocked by unsaturation in the main chain and whether these materials were superior to Courtelle. Vinyl chloride, vinylidene chloride and α-chloroacrylonitrile were used as comonomers and dehydrohalogenation was the dominant reaction, yielding AN sequences interrupted by double bonds. This reaction occurs at lower temperatures than nitrile oligomerization and hence the effect on the latter reaction could be studied. Slower reactions with broader DTA exotherms occur and the structure exerts a blocking effect which is compensated for by the effect of radical activity induced into the AN chain as a result of the dehydrohalogenation reaction. The overall effect is extensive nitrile reaction with little retardation of the exotherm. There is however, no evidence of a comonomer participation in the nitrile reaction.

Copolymerization with vinyl chloride improves the DTA characteristics but TG behaviour is not greatly improved giving weight loss via fragmentation at 300–450°C. Vinylidene chloride and α-chloroacrylonitrile incorporation improves carbon yields obtained on direct pyrolysis and the results are similar to those obtained from pre-oxidized Courtelle (Watt, 1970; Grassie and McGuchan, 1971b). It was suggested that this behaviour results from cross-linking reactions involving intermolecular HCl elimination, and, for α-chloroacrylonitrile, HCN elimination from the unsaturated nitrile initially produced.

Grassie and McGuchan suggested that vinylidene chloride copolymers could be of use in carbon fibre production without pre-oxidation, but it was indicated that further work was required to evaluate the properties of the basic fibre and the fibre derived by direct pyrolysis.

Comonomers containing carbonyl groupings have also been studied by DTA to assess their effect on nitrile oligomerization (Grassie and McGuchan, 1973b). Vinyl acetate, vinyl formate, acrolein and methyl vinyl ketone were used. Vinyl acetate exerted a blocking effect which was less than for styrene systems (Grassie and McGuchan, 1972c). this was ascribed to concurrent decomposition yielding backbone unsaturation which dies not prevent nitrile reactions.

However since chain scission is important in the degradation of vinyl acetate polymers, it is possible that chain scission and blocking are interrelated. Grassie and McGuchan suggest that from steric considerations the acetate structure should be *trans* in the propagating nitrile species. This would imply that hydrogen transfer would occur in preference to interaction with the acetate structure. Such a process should result in the formation of the vinyl ester (III). No such species was detected on IR analysis of the polymer, presumably because such a structure would be

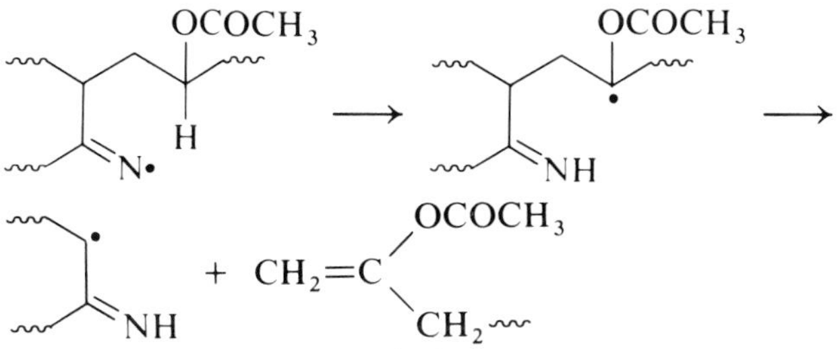

(III)

unstable, leading to rearrangement.

Copolymers containing vinyl formate gave CO as a major product, suggesting a different mechanism to be operative. If this system behaved as vinyl acetate, then formic acid or its degradation products, H_2, CO_2, H_2O and CO are to be expected. CO_2, H_2O and HCOOH were not detected and the following mechanisms were proposed for this system.

(a)

(b)

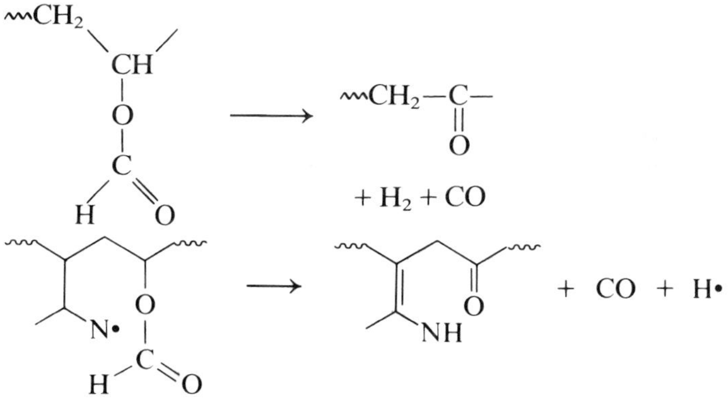

Acrolein accelerates nitrile oligomerization, but to a smaller extent than acids and amides. This may result from increased radical activity in the system as a result of acrolein decomposition.

IR analysis indicates the disappearance of the carbonyl absorption in the residue, suggesting participation in the cyclization reaction. Thus both initiation and termination of oligomerization are possible.

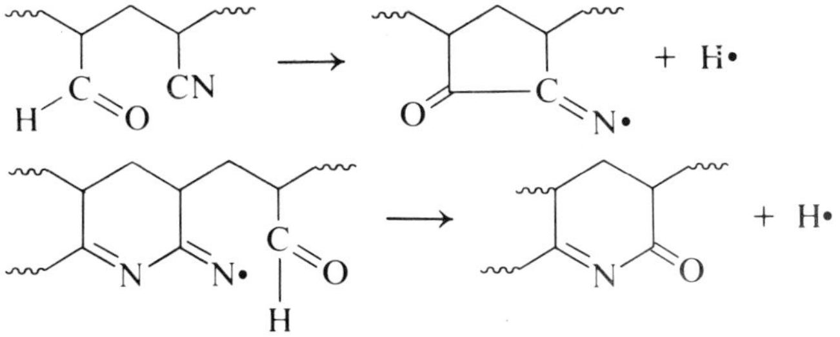

In contrast to the behaviour reported by Grassie and Hay (1961, 1962) no evidence was found for initiation by methyl vinyl ketone. The group can however take part in the cyclization and termination reaction.

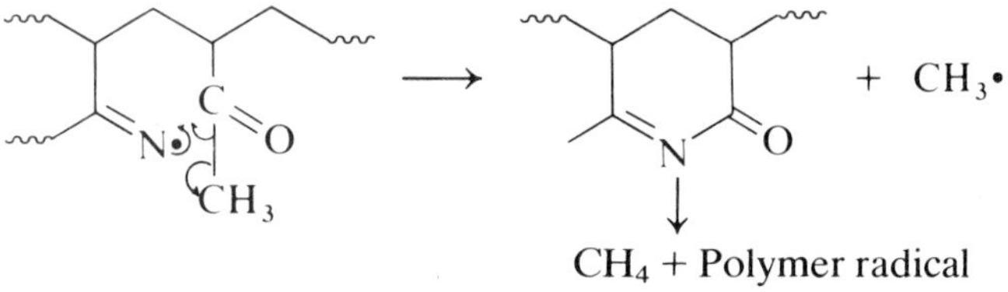

$+ CH_3\bullet$

$CH_4 +$ Polymer radical

Since the steric requirements of such a reaction are more stringent than with acrolein, the alternative reaction involving hydrogen abstraction and chain scission becomes more probable.

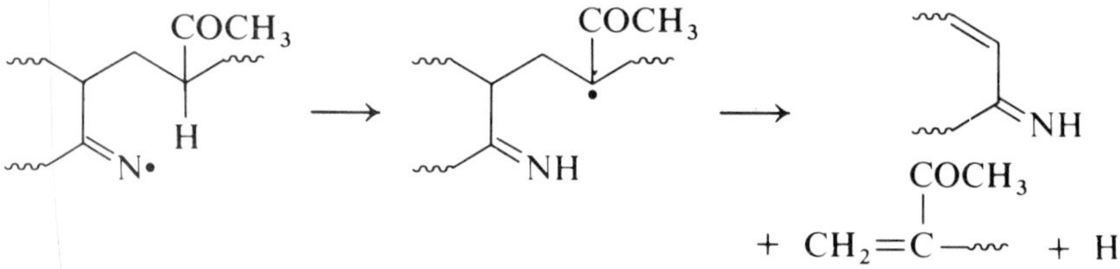

$+ CH_2{=}C \text{—}\text{ⵡ} + H\bullet$

Idol (1974) reviewed the prospects for all AN containing systems, and for carbon fibres including sources of fibre material, and gave a schematic interpretation of the method used for production involving wet or dry spinning and steam orientation followed by pre-oxidation and carbonization in hydrogen. If graphite fibres are required, further heat treatment at 2500–3000°C is used. The properties of composites derived from such systems are discussed and the opportunities open for this system if costs could be reduced are stressed.

The inherent instability of AN containing systems at temperatures below their softening point is emphasized, and the search for suitable thermal stabilizers/processing aids is described as continuing unabated. Stabilizers including halide salts, stearates, organic phosphites and carbonates are discussed and maleic anhydride inhibition of PAN oligomerization, as described by Runge (1971; Runge and Nelles, 1970),

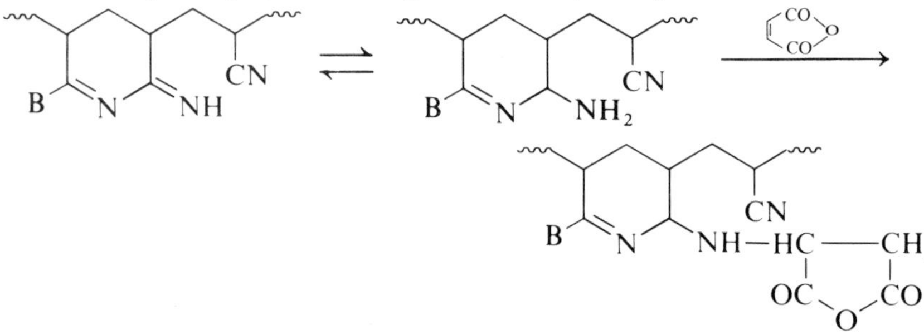

involving blocking via a Michael Addition reaction is described.

Patron and Bastianelli (1974) have studied the side reactions occurring during the production of PAN by free radical initiation. They studied the effect of monomer and initiator concentrations on structural defects and have evaluated conditions for the preparation of high quality polymers of significance for colouration studies.

Recently Peebles (1975) reviewed the degradation behaviour of polymers containing AN from the standpoint of acrylic textile degradation and conversion to carbon fibres. Degradation by a variety of degradation agencies (including oxidative, nucleophilic induced, mechanical, photochemical, radio-chemical and thermal degradation agencies) is discussed.

Flash pyrolysis of radical and anionic PAN between 400–800°C has been studied by Galin and Le Roy (1976) and the degradation products were characterized as HCN, $(CN)_2$, AN and the dimers α-methylene glutaronitrile and a cyanopicoline. The effect of molecular weight and branching density on product yields are described. The experimental results obtained are interpreted in terms of random chain scission, hydrogen transfer and nitrile group cyclization.

III. Poly(ethylene terephthalate)

A. THERMAL DEGRADATION

When poly(ethylene terephthalate) (PET) is heated under nitrogen in the temperature range 280–325°C, the melt degrades and becomes cream coloured and finally black (Held, 1954; Goodings, 1961). In addition the possibility of branching and cross-linking reactions have been suggested (Sobue and Kajiura, 1959) and Hoffrichter (1968) has reported the presence of a gel-like material in the knob found in commerical polyester filaments. This gel-like material, it is suggested, is formed by thermal or thermo-oxidative degradation during melt spinning (Yoda *et al.*, 1970).

A complex mixture of volatile and non-volatile products are formed on degradation as shown in Tables V and VI. Acetaldehyde is the major product of pyrolysis at 288°C (Table VI) whilst the non-gaseous products shown in Table VI were isolated from pyrolysed PET by Straus and Wall (1958). When PET was heated for 65 h *in vacuo* at 310°C, in addition to the products listed in Tables V and VI a sublimate was obtained. This material contained naphthalene and low molecular weight oligomers, and an unsaturated ester whose structure was not identified but was suggested to contain a vinyl grouping (Held, 1954). Such a system may be the precursor to acetophenone and *p*-acetyl benzoic acid, also found on decomposition of PET between 340 and 475°C (Allan *et al.*, 1957a).

At 320°C under nitrogen some oligomer is formed as a sublimate (Pohl, 1951) to which the structure HOOC $C_6H_4COOC_2H_4$ OOC C_2H_4OH has been assigned (Marshall and Todd, 1953). This is in accord with the chain/cyclic structure equilibrium known to occur for PET to the extent of 1·5% (Goodman, 1960).

TABLE V

Products isolated from extensively pyrolysed poly(ethylene terephthalate)

Product	Concentration in degraded polymer, wt %
(a) Dimethyl biphenyl-4-4'-dicarboxylate, $CH_3O.OC.C_6H_4.C_6H_4.CO.OCH_3$	0·058
(b) Dimethyl dibenzyl-4-4'-dicarboxylate, $CH_3O.OC.C_6H_4.CH_2.C_6H_4.OC.OCH_3$	0·028
(c) Methyl p-benzoylbenzoate, $C_6H_5.CO.C_6H_4.CO.OCH_3$	0·007
(d) Methyl p-toluate, $CH_3.C_6H_4.CO.OCH_3$	0·085
(e) Methyl p-ethylbenzoate, $CH_3.CH_2.C_6H_4.CO.OCH_3$	0·085
(f) Methyl biphenyl-4-carboxylate, $C_6H_5.C_6H_4.CO.OCH_3$	0·005

From Madorsky (1964) and Straus and Wall (1958).

In general during degradation the functional group concentration changes, with a decrease in hydroxyl group concentration accompanied by an increase in carboxyl group concentration. Further to this it has been shown that when all the hydroxyl end groups have been removed anhydride formation occurs (Pohl, 1951). A coke obtained on treatment of PET at 306°C in nitrogen for 65 h was shown to have carboxyl and anhydride groups present in the ratio 1 : 2·6. (Goodings, 1961). The results of Goodings' (1961) comparison of the increase in carboxyl group concentration and the decrease in hydroxyl group concentration during polycondensation and degradation are shown in Table VII together with the associated changes in molecular weight. The mechanism of pyrolytic degradation has been elucidated (Allan et al., 1957a,b; Ritchie, 1961) using the model compounds shown in Table VIII.

TABLE VI

Mass-spectrometer analysis of gaseous products from pyrolysis of poly(ethylene terephthalate) at 288°C.

(*In mole per cent of total gases*)

Constituent	%
CO	8·0
CO_2	8·7
H_2O	0·8
CH_3CHO	80·0
C_2H_4	2·0
2-Methyldioxolan	0·4
CH_4	0·4
C_6H_6	0·4

From Madorsky (1964) and Straus and Wall (1958).

TABLE VII

End-group concentrations in poly(ethylene terephthalate) during polycondensation and degradation at 285°C

Stage	Time		Intrinsic viscosity (dl./g.)	End-group concentration (g.-equiv./10^6 g.)			Apparent molecular weight $\bar{M}_n$
				Hydroxyl	Carboxyl		
					Infra-red	Titration	
	h.	min.					
Poly-	2	00	0·424	162	2		12,200
conden-	3	00	0·600	82	8		22,200
sation	4	05	0·747	49	14		31,800
	5	10	0·855	24	27		39,200
	6	20	0·832	7	38	34	44,400
Degra-	7	20	0·731	4	52	51	35,700
dation	8	35	0·605	4	66	72	28,600
	9	43	0·492	4	90	100	21,300
	10	46	0·431	3	99	128	19,600
	11	55	0·386	5	122	153	15,700
	13	05	0·370	6	125	177	15,300

From Goodings (1961).

TABLE VIII

Model compounds used for PET in thermal degradation studies

	Compound	Comment
IV	⬡—$CO_2CH_2CH_2O_2C$—⬡	Represents simple repeat segments.
V	⬡—$CO_2CH_2CH_2O_2C$—⬡—$CO_2-CH_2CH_2-O_2C$—⬡	Represents the repeat unit in the chain.
VI	⬡—$CO_2CH_2CH_2OCH_2CH_2O_2C$—⬡	Represents the unit which contains an ether link.
VII	⬡—$CO_2CH_2CH_2OH$	Represents an hydroxyl ended chain.
VIII	⬡—$CO_2CH.=CH_2$	Represents a vinylic end grouping arising from primary scission of the main chain.

On decomposition IV yields benzoic acid and vinyl benzoate

At low temperatures recombination occurs to yield

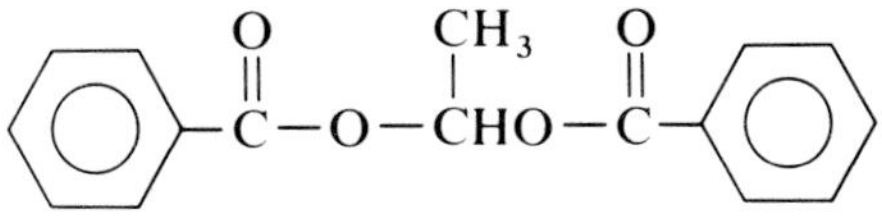

$$\langle\bigcirc\rangle-\overset{\overset{O}{\|}}{C}-O-\overset{\overset{CH_3}{|}}{CHO}-\overset{\overset{O}{\|}}{C}-\langle\bigcirc\rangle$$

which can disproportionate to

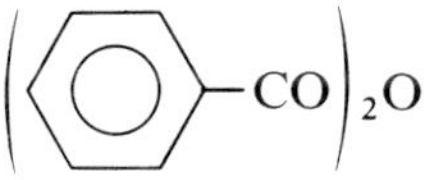

$$\left(\langle\bigcirc\rangle-CO\right)_2O$$

and CH_3CHO. In addition vinyl benzoate may decompose further to yield the products shown in Table IX.

TABLE IX

Decomposition Products of Vinyl Benzoate (Allan et al., 1955)

Compounds	Yield %
 $\langle\bigcirc\rangle-CO_2H + CH{\equiv}CH$	6
$CO_2 + CH_2{=}CH-\langle\bigcirc\rangle$ 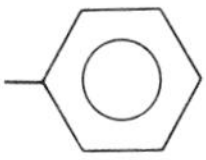	15
 $\left[\langle\bigcirc\rangle-COCH_2CHO\right]$ $\downarrow$ $CO + PhCOCH_3$	76

If these results are applied to PET it suggests that degradation occurs primarily via scission of the main chain in a random manner to yield carboxyl end groups and a vinyl product which can combine to form an ethylidine product which decomposes further. On vacuum pyrolysis carbon monoxide is a major product which arises from the breakdown of vinyl ester units.

Compound V (2-benzoyloxyethyl terephthalate)

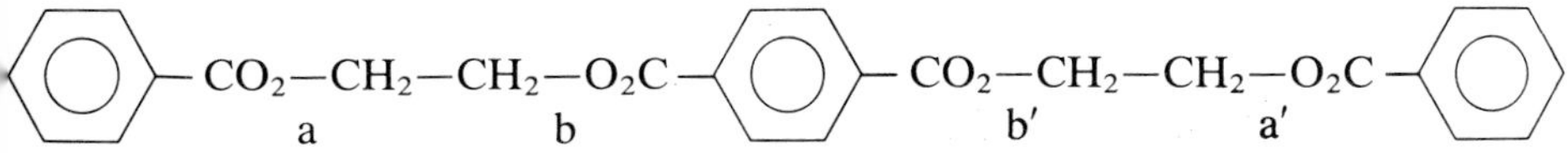

$$\underset{}{\langle\bigcirc\rangle}-\underset{a}{CO_2}-CH_2-CH_2-\underset{b}{O_2C}-\langle\bigcirc\rangle-\underset{b'}{CO_2}-CH_2-CH_2-\underset{a'}{O_2C}-\langle\bigcirc\rangle$$

(V)

contains two structurally similar weak links at a and b and a′ and b′ where alkyl oxygen scission may occur.

Pyrolysis of V yields benzoic and terephthalic acids, acetophenone, p-acetylbenzoic acid, acetaldehyde, acid anhydrides, carbon monoxide and dioxide, methane, ethylene, acetylene, ethylene dibenzoate, vinyl benzoate and benzene. Allan *et al.* (1957b) have provided an explanation for all these products. Thus

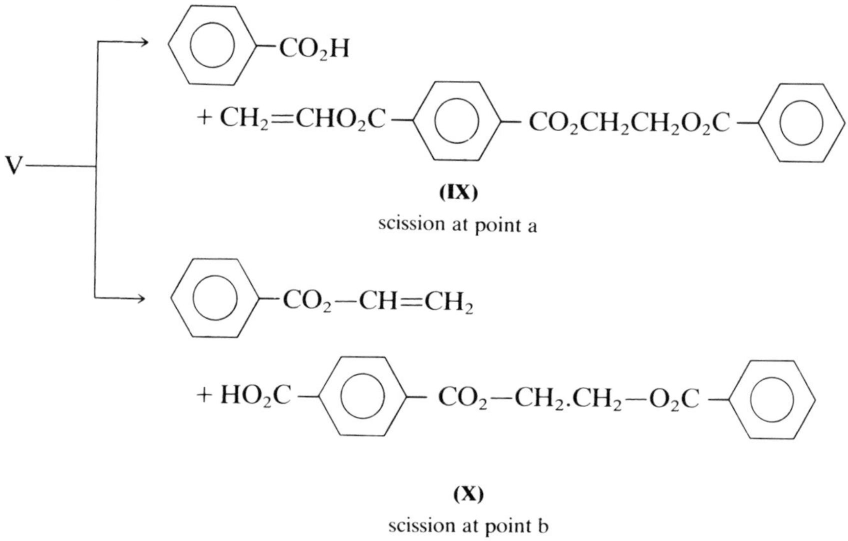

(IX)

scission at point a

(X)

scission at point b

Thereafter further alternative alkyl oxygen scissions of esters (IX) and (X) can occur.

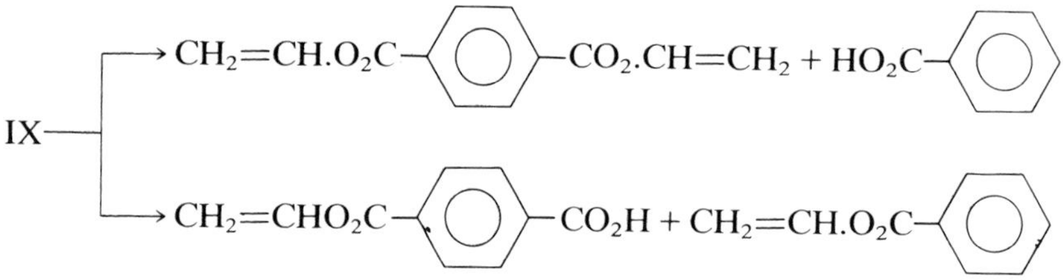

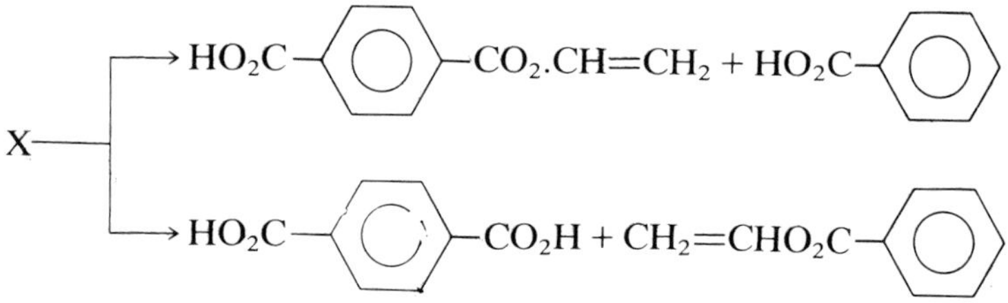

As each vinyl carboxylate species is formed it will break down by the route previously established (Allan *et al.*, 1955). Thus

$$C_6H_5-CO_2.CH=CH_2 \longrightarrow C_6H_5-COCH_3$$

$$HO_2C-C_6H_4-CO_2.CH=CH_2 \longrightarrow HO_2C-C_6H_4-COCH_3$$

$$CH_2=CH.O_2C-C_6H_4-CO_2-CH=CH_2 \longrightarrow CH_3OC-C_6H_4-COCH_3$$

With the exception of diacetyl benzene all the above compounds are found in the pyrolysis of PET and the ester V. On pyrolysis compound VI yields

$$C_6H_5-\overset{O}{\overset{\|}{C}}-OCH_2CH_2-OCH_2CH_2O-\overset{O}{\overset{\|}{C}}-C_6H_5 \longrightarrow$$

(VI)

$$C_6H_5-\overset{O}{\overset{\|}{C}}-O-CH_2CH_2OCH=CH_2$$

$$+$$

$$C_6H_5-\overset{O}{\overset{\|}{C}}-OH$$

which undergoes a series of secondary reactions

$$C_6H_5-\overset{O}{\overset{\|}{C}}-OCH_2CH_2OCH=CH_2 \longrightarrow$$

$$C_6H_5-\overset{O}{\overset{\|}{C}}-O-CH=CH_2 + CH_3CHO$$

$$\begin{matrix} CH_2-O-\overset{O}{\overset{\|}{C}}-C_6H_5 \\ | \\ CH_2-O-\underset{O}{\underset{\|}{C}}-C_6H_5 \end{matrix} \quad + \quad \begin{bmatrix} CH_2OCH=CH_2 \\ | \\ CH_2OCH=CH_2 \end{bmatrix}$$

$$C_6H_5-\overset{O}{\overset{\|}{C}}-OH + CH_2=CH-O-CH=CH_2$$

$$\begin{bmatrix} CH_2-OCH=CH_2 \\ CH_2-OCH=CH_2 \end{bmatrix} \longrightarrow CH_3CHO + CH_2=CHOCH=CH_2$$

$$CH_2=CH-O-CH=CH_2 \longrightarrow CH_3CHO + CH\equiv CH$$

These latter reactions are important since PET contains occasional random ether linkages which constitute weak links in the main chain of the polymer (Pohl, 1951).

Compound (VII) represents one of the chain end groupings and can decompose to yield a variety of products

In PET the decomposition of such end groups will not be important in the early stages of degradation but as the number of hydroxyethyl groups increases with extent of degradation the importance of this type of reaction will grow.

In addition to these studies Goodings (1961) has used glycol dibenzoate IV as the most appropriate model for PET. Degradation of IV at temperatures between 282–323°C under nitrogen gave similar products to those from the polymer, namely benzoic acid, vinyl benzoate and smaller quantities of acetaldehyde and carbon monoxide, with vinyl benzoate and benzoic acid accounting for 94% of the products. PET under similar conditions yields carbon monoxide, acetaldehyde and species containing carboxyl groupings. It appears therefore, that there is a strong analogy between the decomposition reactions of model esters and the polymer. The products from glycol dibenzoate suggest that the initial step in degradation is a β-olefin elimination via a cyclic transition state, and alkylation of the carbon atom β to the ester group improves stability (Pohl, 1951; Buxbaum, 1968) in accord with the above mechanism. The following reaction scheme has been suggested for the initiation and subsequent propagation stages of thermal degradation.

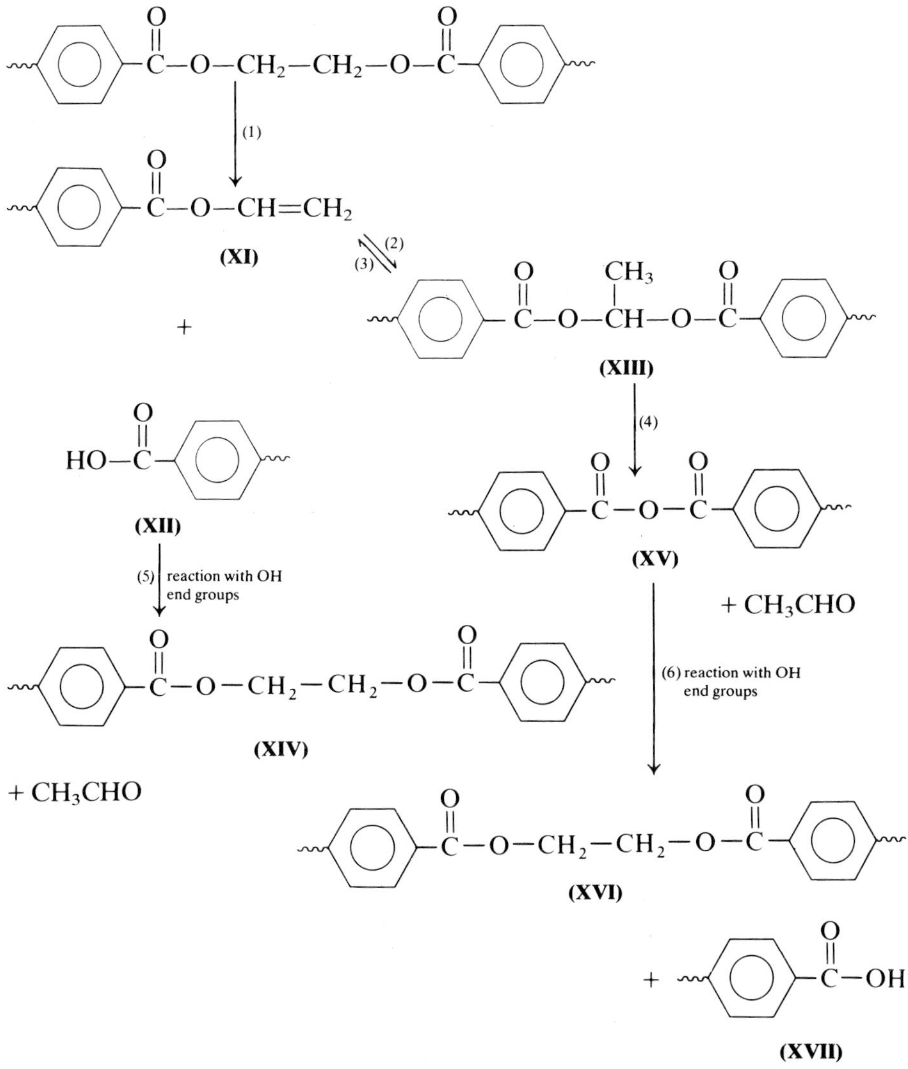

In this scheme the chain is broken to yield a compound (XI) containing a vinyl grouping and a compound bearing a carboxyl chain end (reaction 1). It is possible for these two compounds to react to yield the ester (XIII) (reaction 2). The hydroxyl ended chains which are present may react with the vinyl grouping to reform the chain (XIV) and thereby eliminate acetaldehyde (reaction 5). The ester (XIII) may also decompose further yielding acetaldehyde and produce the anhydride (XV) which can react with hydroxyl ended chains to reform a chain (XVI) and to yield a carboxyl ended chain (reaction 6). This reaction scheme thus offers an explanation of the observed production of carboxyl chain ends and reduction of hydroxyl chain ends. Stirring of the melt causes an increase in the evolution of acetaldehyde which is accompanied by a decrease in the rate at which the carboxyl groups are formed (Goodings (1961)). It is possible that acetaldehyde retained in an unstirred melt condenses to yield a polyene and water, which in turn is capable of hydrolysing the chain to give carboxyl and more hydroxyl ended chains

$$n\ CH_3CHO \longrightarrow CH_3(CH=CH)_{n-1}CHO + (n-1)H_2O$$

Such a polyene would be responsible for colour development in the unstirred melt.

Hydroxyethyl end groupings may also decompose in a similar manner

In addition further reactions can occur in the melt between the hydroxyethyl group ended chains and a variety of systems

Thus as long as free hydroxyl groupings are present each broken polymer linkage may be reformed and it is only when hydroxyl groupings are consumed that the degree of polymerization shows any significant decrease.

In addition to such reactions, Sobue and Kajiura (1959), who investigated the vacuum pyrolysis of PET, suggest that branching or cross-linking occurs during degradation. They proposed the following structures for the branches or cross-links formed

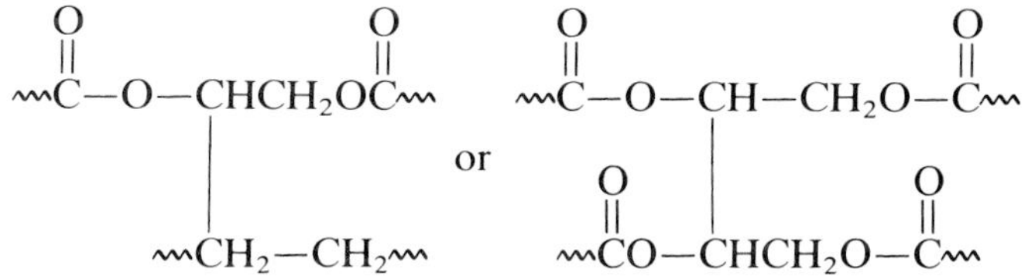

$$
\begin{array}{cc}
\underset{\overset{O}{\parallel}}{\sim\sim C}-O-CHCH_2O\overset{O}{\overset{\parallel}{C}}\sim\sim & \underset{\overset{O}{\parallel}}{\sim\sim C}-O-CH-CH_2O-\overset{O}{\overset{\parallel}{C}}\sim\sim \\
| & | \\
\sim\sim CH_2-CH_2\sim\sim & \sim\sim CO-CHCH_2O-C\sim\sim \\
\end{array}
$$

or

Zimmermann (1966; Zimmermann and Leibnitz, 1965) and co-workers have summarized the main reactions occurring during thermal degradation of PET in a similar manner to that described previously (p. 363) with the inclusion of reactions involving the polymerization of the vinyl ester produced in the primary degradation process (see next page).

As evidence for vinyl ester formation during PET degradation *in vacuo* the mass-spectrometric thermal analysis and thermogravimetric analysis studies of Ozawa (Ozawa *et al.*, 1969) may be cited. Ozawa reports that the following vinyl esters are formed

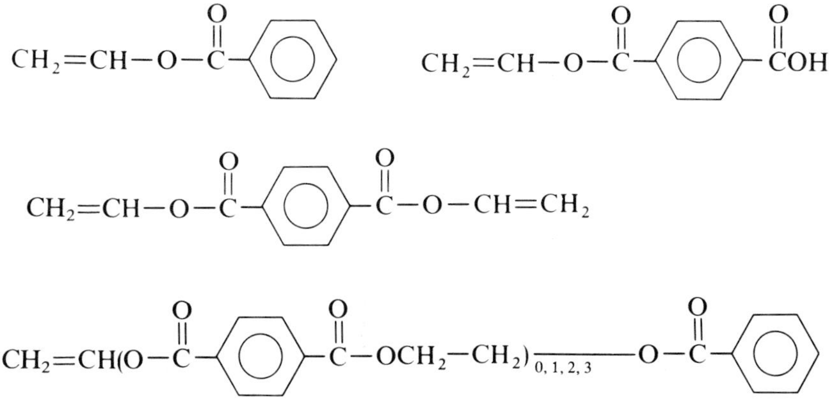

$$CH_2{=}CH{-}O{-}\overset{O}{\overset{\parallel}{C}}{-}\langle\bigcirc\rangle \qquad CH_2{=}CH{-}O{-}\overset{O}{\overset{\parallel}{C}}{-}\langle\bigcirc\rangle{-}\overset{O}{\overset{\parallel}{C}}OH$$

$$CH_2{=}CH{-}O{-}\overset{O}{\overset{\parallel}{C}}{-}\langle\bigcirc\rangle{-}\overset{O}{\overset{\parallel}{C}}{-}O{-}CH{=}CH_2$$

$$CH_2{=}CH(O{-}\overset{O}{\overset{\parallel}{C}}{-}\langle\bigcirc\rangle{-}\overset{O}{\overset{\parallel}{C}}{-}OCH_2{-}CH_2)_{\overline{0,1,2,3}}{-}O{-}\overset{O}{\overset{\parallel}{C}}{-}\langle\bigcirc\rangle$$

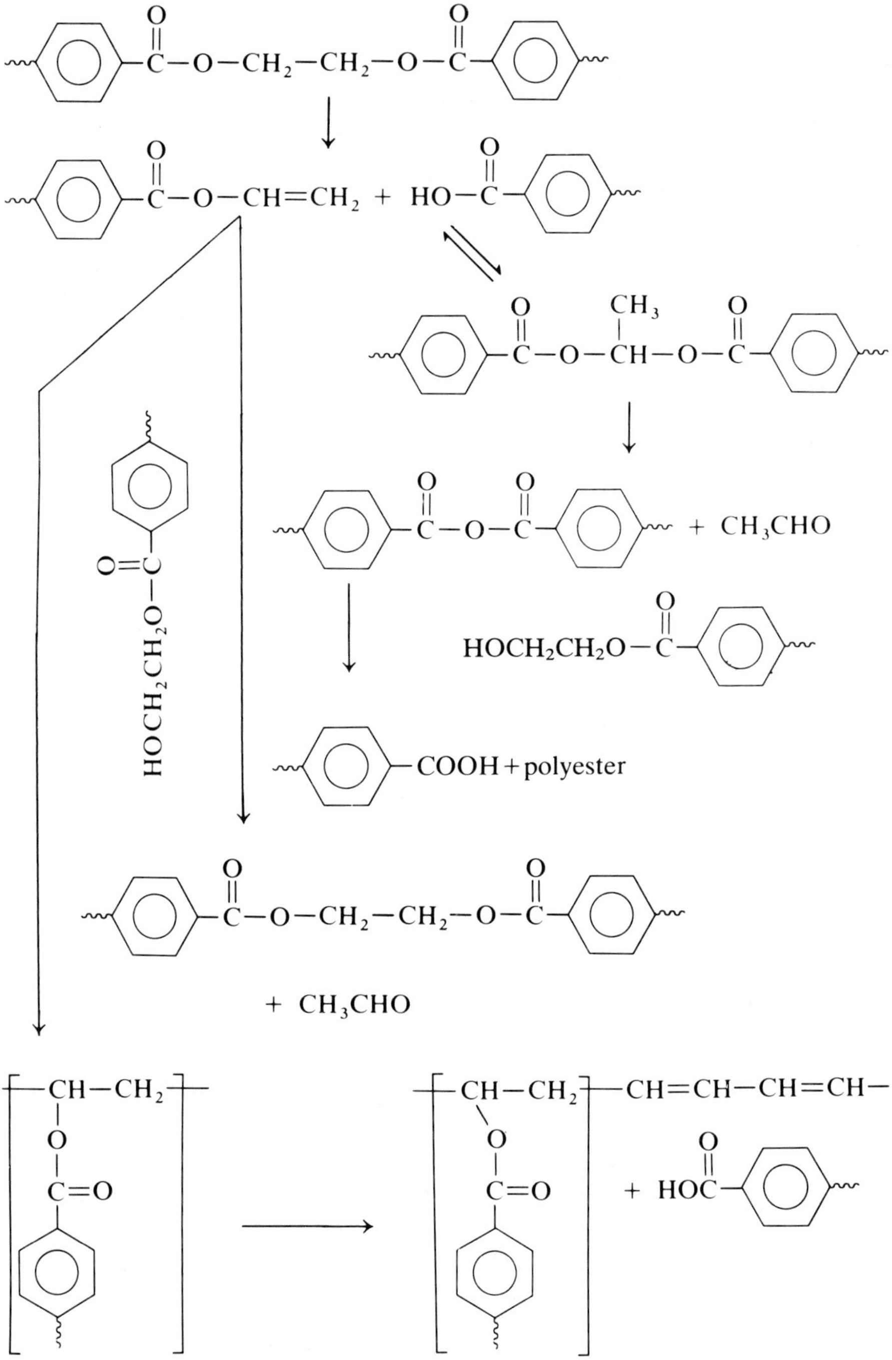

HOCH₂CH₂O—C=O
+ CH₃CHO
HOCH₂CH₂O—C
COOH + polyester
+ CH₃CHO
CH=CH—CH=CH—
+ HOC

B. THERMOXIDATIVE DEGRADATION

The thermooxidative degradation of PET has been studied by a number of workers (Mikhailov *et al.*, 1962; Buxbaum, 1967; Kovarskaya *et al.*, 1968; Voda *et al.*, 1970; Wiesener, 1970, 1971; Zimmermann and Schaaf, 1970, 1971a, b, c; Nealy and Adams, 1971; Spanninger, 1974; Krzeminski and Kedziora, 1975; Birladeanu *et al.*, 1976). Buxbaum (1967), Mikhailov (Mikhailov *et al.*, 1962) and Kovarskaya (Kovarskaya *et al.*, 1968) have studied the degradation of PET in the presence of oxygen. This work was concerned with the increase in colour and carboxyl group content, decrease in inherent viscosity, and product analysis when oxygen was present. As a result of his studies, Buxbaum proposed the following scheme for the thermo-oxidative degradation of PET involving radicals generated from the thermal destruction of the hydroperoxide formed by reaction of the methylenic grouping with oxygen.

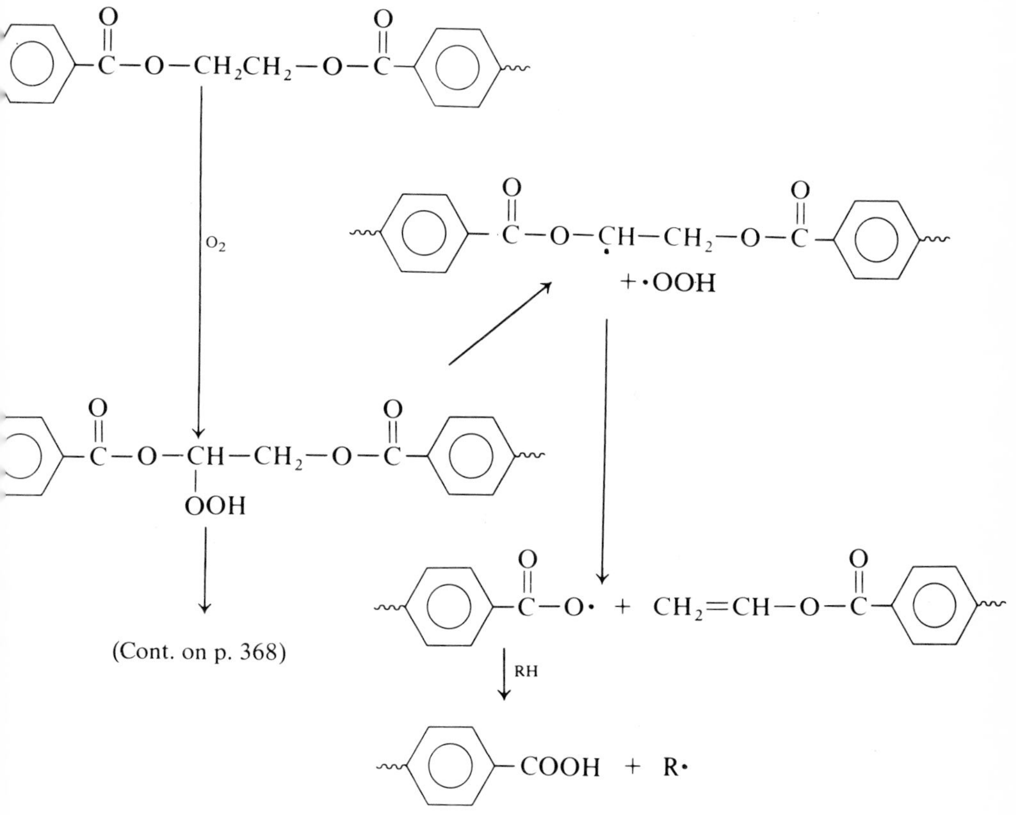

(Cont. on p. 368)

RH = PET

Such a scheme does not account for cross-linking observed by other authors (Allan *et al,* 1955; Buxbaum, 1968). Thermo-oxidative cross-linking was first studied in detail by Yoda *et al.* (1970) and by Nealy and Adams (1971). These authors suggest that cross-linking at 300°C in air occurs via butane triol (Yoda *et al.,* 1970) and biphenyl tricarboxylic acid (Nealy and Adams, 1971) respectively.

Nealy and Adams (1971) found that PET heated in air at 300°C cross-linked as evidenced by the formation of gels on attempted solution in hexafluorisopropanol. The cross-linked samples were depolymerized in excess methanol at 200°C and analysed by GLC which indicated that the biphenyl trimethylcarboxylate (XVIII) was present.

(XVIII)

As a result of these studies a free radical degradation reaction involving cross-linking by two possible routes which generate substituted phenyl radicals was proposed as shown below.

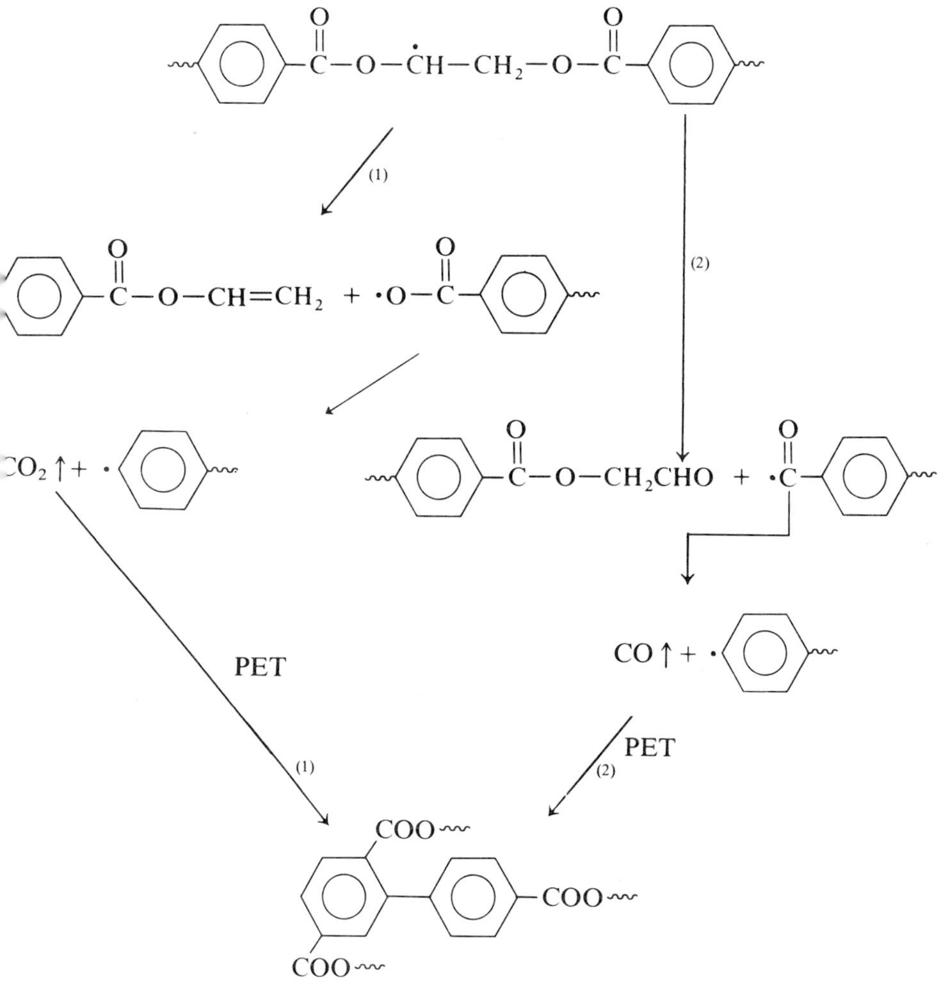

Mechanistic pathway (2) was suggested to predominate on the basis of the product analysis studies of Kovarskaya *et al.* (1968) who report that more carbon monoxide than carbon dioxide is formed on degradation in air.

Yoda *et al.* (1970), on the basis of similar studies in which ethylene glycol, dimethylterephthalate and 1,2,4-butane triol were isolated after methanolysis concluded that the mechanism was

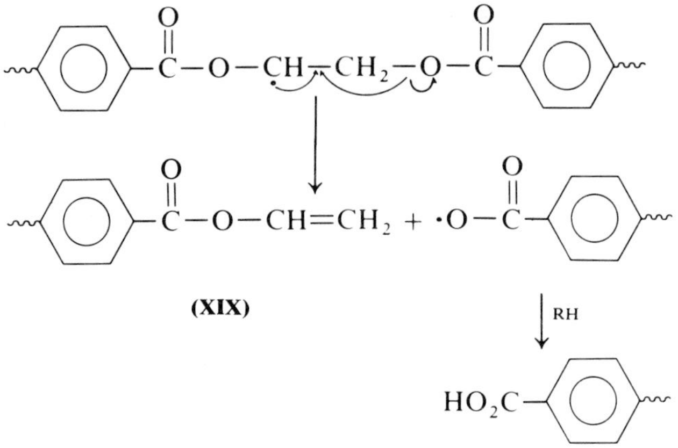

(XIX)

The vinyl ester (XIX) was then considered to decompose in the following manner

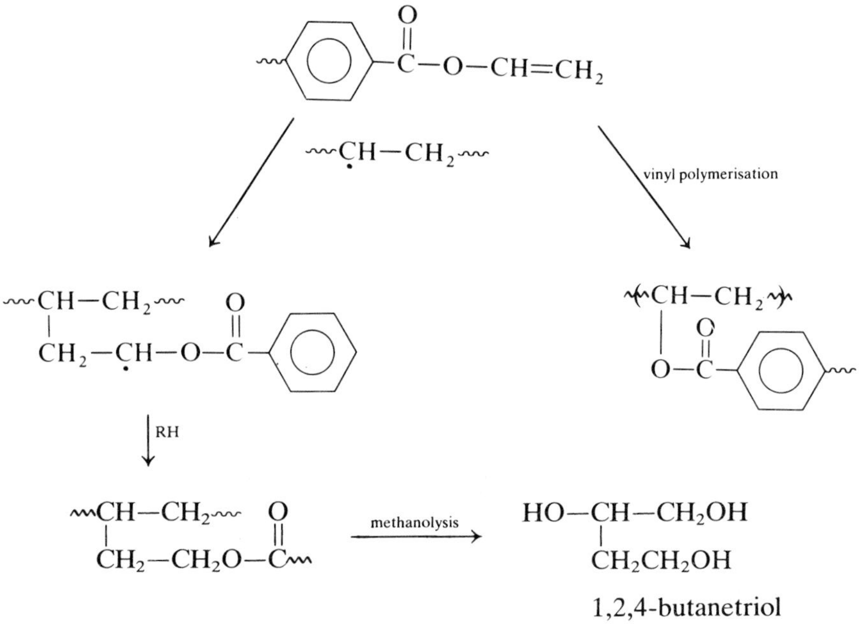

vinyl polymerisation

methanolysis

1,2,4-butanetriol

Poly(vinyl methyl terephthalate) (PVMT) was used as a model compound for the gel and subjected to thermoxidative degradation. The main reaction in the degradation of PVMT should be β-elimination to yield olefinic double bonds and monomethyl terephthalate. The degradation product from PVMT was monomethyl terephthalate and a brown residue which had a similar infra-red spectrum to that obtained from the gel from PET.

Spanninger (1974) repeated studies on PET in air at 300°C and found that the cross-linking and colouration reaction occurred as previously reported (Yoda et al., 1970; Nealy and Adams, 1971). The gelled polymer was separated and methanolysed.

A small amount of methanol insoluble infusible vinyl polymer was isolated which compared well with that described by Yoda (Yoda et al., 1970). On GLC analysis ethylene glycol, dimethyl terephthalate and diethelene glycol (small quantity) and numerous other species were observed. The following esters and acid esters were positively identified.

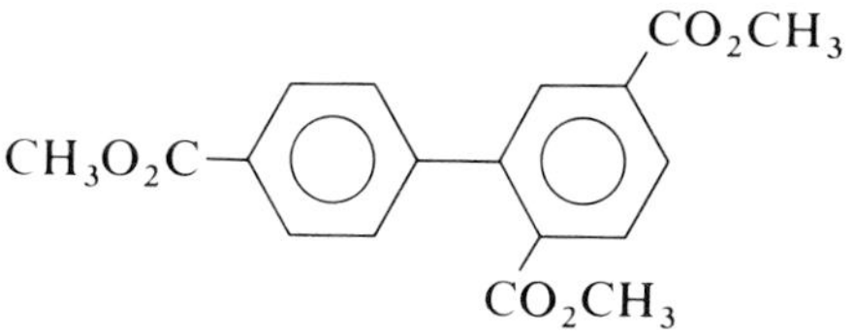

(XX)

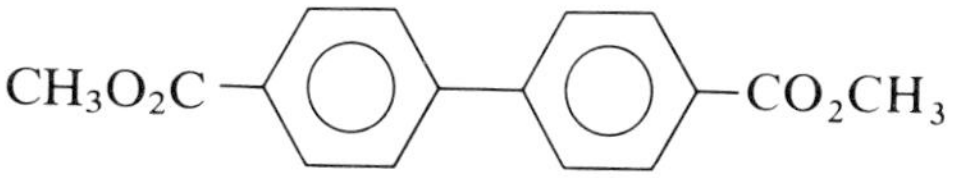

(XXI) 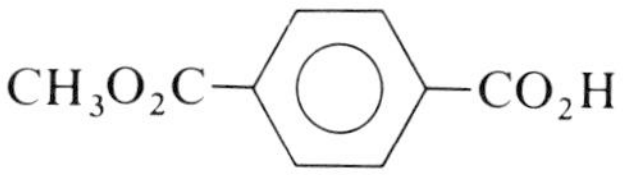   (XXII)

Many other systems which remained unidentified were shown to contain carboxyl and carbomethoxy groupings by mass spectrometric analysis.

These results confirm and amplify the results of Nealy and Adams and support their proposed mechanism. The mechanism relies on the formation of phenyl radicals via initial oxidation at a methylenic hydrogen as described by Buxbaum (1967). After chain scission and decarboxylation the phenyl radical arylates at an adjacent terephthalate nucleus to form a cross-link.

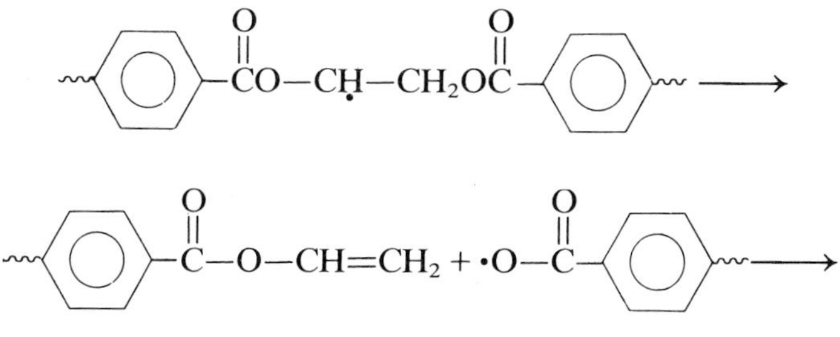

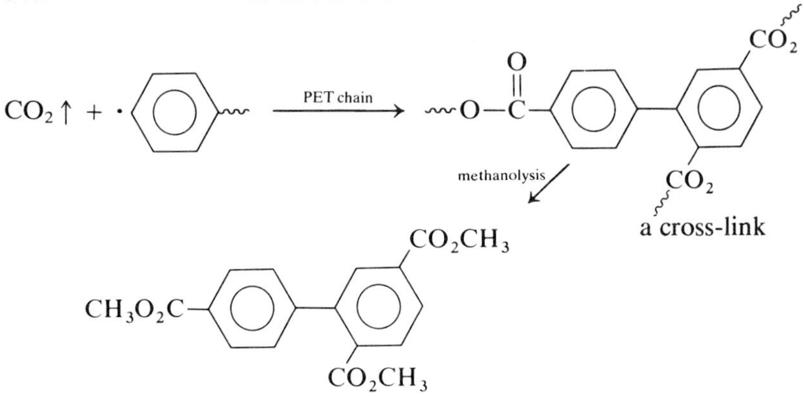

$$CO_2 \uparrow + \cdot \langle \bigcirc \rangle \sim \xrightarrow{\text{PET chain}} \sim O - \overset{O}{\overset{\|}{C}} - \langle \bigcirc \rangle - \langle \bigcirc \rangle - CO_2$$

methanolysis

a cross-link

$$CH_3O_2C - \langle \bigcirc \rangle - \langle \bigcirc \rangle \overset{CO_2CH_3}{\underset{CO_2CH_3}{}}$$

The presence of

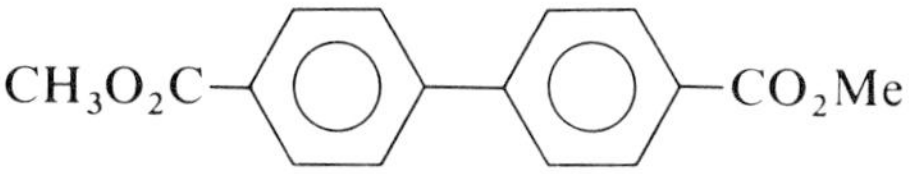

$$CH_3O_2C - \langle \bigcirc \rangle - \langle \bigcirc \rangle - CO_2Me$$

(XXIII)

is also in accord with the mechanistic scheme as is the compound tentatively assigned as

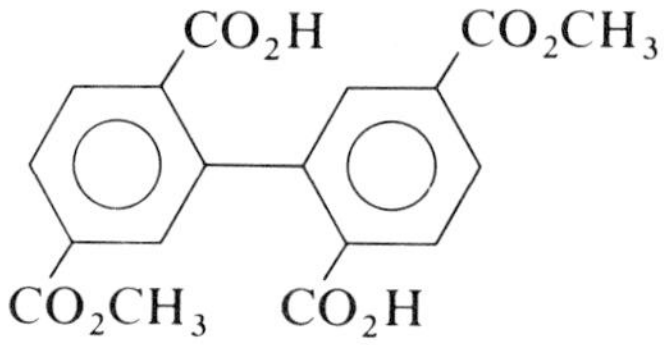

$$\overset{CO_2H}{\underset{CO_2CH_3}{\langle \bigcirc \rangle}} \overset{CO_2CH_3}{\underset{CO_2H}{\langle \bigcirc \rangle}}$$

(XXIV)

both of which could arise in the following manner

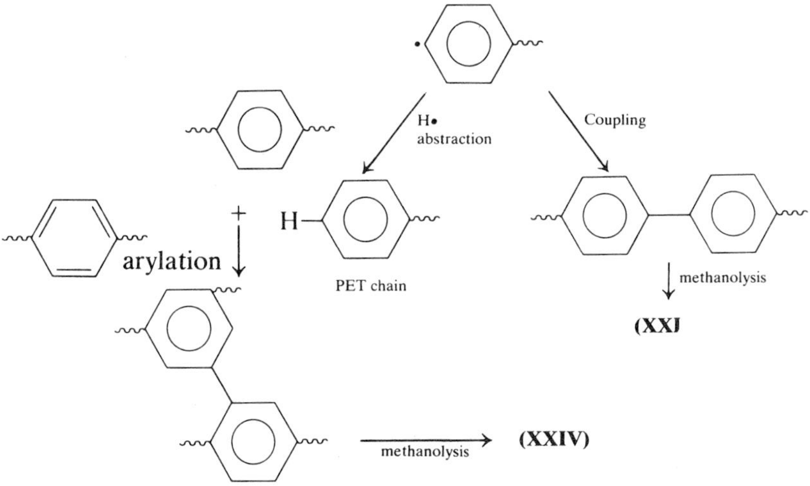

arylation

H• abstraction

Coupling

PET chain

methanolysis

(XXI

$$\xrightarrow{\text{methanolysis}} \textbf{(XXIV)}$$

Spanninger did not isolate any 1,2,4-butane triol in his experiments but the Yoda mechanism is in part substantiated by isolation of a vinyl polymer of the vinyl benzoate type. Spanninger also comments that in a system as complex as PET degradation, cross-linking may occur by more than one mechanism and the different products may well arise as a result of differences in experimental techniques, particularly the gas flow rate used.

Wiesener (1970, 1971) has used isothermal DTA at 220–260°C to determine the oxidative stability of PET. The induction period to the onset of oxidative degradation exponentially decreases with oxidation temperature. The area of the exothermic process is molecular weight dependent. Also, addition of pre-oxidized PET to virgin PET reduces the induction period as does pre-treatment at 240°C.

Similarly Zimmermann (Zimmermann and Schaaf, 1970, 1971a, b, c) has used thermal analysis to study the thermal oxidative stability of PET with particular reference to the effect of the catalysts used in the preparative stages.

Krzeminski (Krzeminski and Kedziora, 1975) has studied the effect of residence time on thermal oxidative degradation and cross-linking of PET during extrusion of film.

Birladeanu (Birladeanu et al., 1976) has used dynamic thermogravimetry in air and nitrogen to investigate the effect of molecular weight, heating rate and degree of conversion on the activation energy for degradation. It was concluded that degradation occurred by a short chain mechanism whereas the DTA behaviour in air suggests overlapping of chain degradation and oxidative processes.

1. Kinetic Studies

These have been made by a number of workers who have utilized measurements of acetaldehyde evolution; changes in melt and intrinsic viscosity, and carboxyl and hydroxyl group content (Goodings, 1961); DTA studies on the temperature dependence of the induction period in oxidatively initiated degradation (Spanninger, 1974); dynamic thermogravimetry (Krzeminski and Kedziora, 1975); and isothermal weight loss studies (Straus and Wall, 1958).

Difficulties arise when acetaldehyde evolution is measured for when the melt is stirred a value of 1.7×10^{-3} moles/repeat unit/h at 282°C is obtained for the evolution rate, whereas the rate was one tenth of this in an unstirred system (Goodings, 1961).

As may be expected the rate of acetaldehyde evolution and the rate of change in viscosity did not correspond when the assumption was made that each acetaldehyde molecule was formed as a result of cleavage of the PET main chain. Both retention and addition of acetaldehyde to the melt results in a fall in intrinsic viscosity.

The rate of degradation has been determined (Goodings, 1961) from the fall in melt viscosity in the range 280–320°C. The first order rate constant $ks_{(MV)}$ for bond scission was calculated according to the equation.

$$ks_{(MV)} = \frac{1 \cdot 68 \times 10^{-4} \times 192}{(\log (M.V.)_0 - 1 \cdot 14)^2} \frac{d \log (M.V.)}{dt}$$

In the same study use was also made of the change of intrinsic viscosity, and the first order chain scission constant $ks_{(IV)}$ was determined using the equation.

$$ks_{(IV)} = \frac{192 \times 4 \cdot 8 \times 10^{-5}}{t} \left[\frac{1}{(IV)_t} - \frac{1}{(IV)_0} \right]$$

Values for $ks_{(IV)}$ ranged from $0 \cdot 6 - 1 \cdot 4 \times 10^{-3} h^{-1}$ for unstirred systems and zero for stirred systems.

The change of carboxyl or hydroxyl groups content has also been monitored. In the non-stirred melt it is interesting to note that carboxyl groupings are formed at a rate which is $1 \cdot 4$ to $2 \cdot 0$ times as fast as that in the stirred system.

The total concentration of end groups N was used to evaluate the rate constant k_N for the change of N from the equation

$$k_N = \frac{192 \times 10^{-6}}{2t} (N_t - N_0)$$

and was found to be $1 \cdot 1 \times 10^{-3}$ in an unstirred system but zero or even negative in a stirred system. All rate constants have the same order of magnitude as shown in Tables X and XI. In this study activation energies

TABLE X

Degradation rate constants for polyethylene terephthalate and glycol dibenzoate
$(h^{-1} \times 10^3)$

	k_s				k_{gas}	
	I.V.		M.V.			
Temp. (°C)	282	306	282	306	282	306
Normal polymer	1·3	4·8	1·7	—	—	5·5
Polymer of reference[a]	2·3 (290°)	4·65 (300°)	2·2	5·4	2·4	
Glycol dibenzoate						5·3

[a] Marshall and Todd (1953).

TABLE X—(*continued*)

	k_{COOH}	k_{OH}	k_{ald}	B.E.R.		k_N	
Temp. (°C)	282	306	282	282	282	306	282
Normal polymer	3·5	4·9	2·4	0·2	—	—	1·2
Polymer of reference[a]							
Glycol dibenzoate					0·8	4·8	

I.V. = intrinsic viscosity. M.V. = melt viscosity. B.E.R. = benzoic acid elimination rate constant (h^{-1}). From Goodings (1961). [a] Marshall and Todd (1953).

TABLE XI

Effect of agitation on polymer degradation rate constants at 282° ($h^{-1} \times 10^3$)

Agitation	k_{COOH}	k_{OH}	k_N	k_{ald}	$k_{s(i.v.)}$
Unstirred	3·5	2·4	1·1	0·2	1·3
Stirred (26 r.p.m.)	1·8	2·3	negative	1·7	negative

From Goodings (1961).

were found to vary according to the method chosen for the measurement of rate constant and range from 22·5 to 62·3 Kcal/mole (94·1–260·7 KJ/mole) as shown in Table XII.

Strauss and Wall made studies utilizing isothermal weight loss techniques *in vacuo* and evaluated an activation energy of 38 kcal/mole

TABLE XII

Activation energies of thermal degradation of poly(ethylene terephthalate)

Method	Activation energy, Kcal/mole
Intrinsic viscosity	62·3
COOH end point	
By titration	37·4
By infrared	33·9
OH end point	22·5
Aldehyde	37·7
Total end points	58·7

From Madorsky (1964) and Straus and Wall (1958).

(159 KJ/mol^{-1}) which they interpret as initial homolytic scission of the C—O bond in the system.

Birladeanu and co-workers have used dynamic thermogravimetry and have evaluated activation energies between 72·4 and 248·6 KJmol^{-1} in air and 227·8 and 272·2 KJmol^{-1} in nitrogen, dependent on heating rate and the initial intrinsic viscosity of PET used.

IV. Polyamides

A. THERMAL DEGRADATION

Both Nylon 6 and 6.6 are susceptible to degradation at the temperatures used in melt spinning in the region of 280°C. If the exposure time is long, as may happen in "dead pockets" in the spinning system, the polymer may form minute particles of gel whose presence in the final filament gives rise to irregularities. The decomposition of the two common polyamides, namely nylon 6.6 and 6 will be discussed.

1. Nylon 6.6

When heated for extended periods of time at 270°C under a gentle stream of nitrogen, no changes in the molecular weight occurred whereas at 300°C a considerable decrease occurred within 2 h (Korshak et al., 1958). At the higher temperature of 330°C, gelation occurs. The degradation is very dependent on whether the volatile material which is formed is removed. Heating at 305°C under a vacuum in a sealed tube gave a product which remained white and was still soluble at the end of the heating period; in contrast to this, by heating in nitrogen with removal of the volatile substances, the polymer rapidly became black and insoluble (Peebles and Huffman, 1971). The rate of gelation thus depends on the rate of distillation of volatile material.

In experiments in which the volatiles were retained, the molecular weight of the sample increased over the first hour, presumably because of some further polymerization; this was followed by a considerable decrease until ultimately it rose when gelation set in (Fig. 13) (Peebles and Huffman, 1971). During this degradation, the number of amine ends increases rapidly but finally drops to a low value as the gelation occurs (Fig. 14) (Peebles and Huffman, 1971). The content of carboxyl groups drops rapidly to a negligible value.

The products of thermal degradation fall into three broad classes namely small molecules (CO_2, NH_3 and H_2O), less volatile higher molecular weight material and the polymeric residue.

Heating in a nitrogen stream at 305°C showed that NH_3 and CO_2 were continuously evolved together with H_2O (Achhamer et al., 1951a,b;

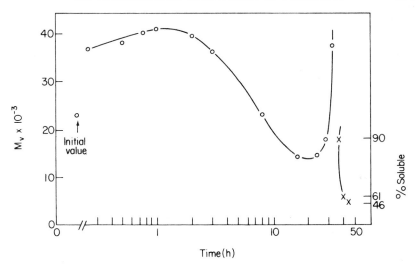

FIG. 13.   Molecular weight as a function of log time, for nylon 6.6 at 282°C in a sealed tube.
O, M_v determined on total sample; ×, on soluble portion only. (Peebles and Huffman, 1971).

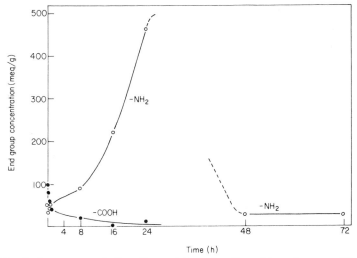

FIG. 14.   Endgroup concentrations (in μeq/g) as a function of time. Same conditions as in
Fig. 13. (Peebles and Huffman, 1971).

Kamerbeck *et al.*, 1961). Changes in elemental composition over different
periods of time are given in Table XIII (Kamerbeck *et al.*, 1961) and these
have been shown to correspond approximately to the amounts of CO_2 and
NH_3 evolved. Ammonia arises from deamination and hydrolysis of amide
groups and carbon dioxide from decarboxylation (Meacock, 1952; Hill,
1954).

TABLE XIII

Elemental composition of 6.6 nylon before and after heating at 305°C

Heating time, h	% C	% H	% N
0	63·72	9·73	12·39
44	70·5	10·26	9·63
100	78·28	10·38	7·02

From Madorsky (1964) and Kamerbeck (1961).

$$RCONH(CH_2)_6NH_2 + H_2N(CH_2)_6NHCOR'$$

$$\rightarrow RCONH(CH_2)_6NH(CH_2)_6NHCOR' + NH_3 \qquad (1)$$

$$RNHCO(CH_2)_4COOH + HOOC(CH_2)_4CONHR'$$

$$\rightarrow RNHCO(CH_2)_4CO(CH_2)_4CONHR' + CO_2 + H_2O \qquad (2)$$

Reactions 1 and 2 offer an explanation of the formation of NH_3 and CO_2 and result in molecular weight increases. Further reactions which lead to branching and cross-linking are possible (Achhamer *et al.*, 1951a,b; Kamerbeck *et al.*, 1961).

$$\diagdown\!\!NH + HOOCR' \rightarrow \diagdown\!\!NCOR' + H_2O \qquad (3)$$

$$\diagdown\!\!CO + H_2NR' \rightarrow \diagdown\!\!C = NR' + H_2O \qquad (4)$$

However hydrolysis of gelled samples did not reveal the presence of 5-oxononane-1,9 dicarboxylic acid expected to result from reaction 2. On the other hand, appreciable amounts of di(ω-amino hexyl) amine expected to arise from reaction 1 were found in the hydrolysate. The importance of the latter was also demonstrated by the gelation of a low molecular weight nylon by the addition of the diamine.

The initial increase in molecular weight may be regarded as polymerization via unreacted chain ends. However since deamination takes place, the NH_3 produced is capable of undergoing aminolysis reactions with reduction in molecular weight and the formation of new amine chain ends (Peebles and Huffman, 1971),

$$RCONHR' + NH_3 \rightarrow RCONH_2 + NH_2R' \qquad (5)$$

It has been suggested that the amide end groups are capable of reaction with the secondary amine formed in reaction 1 to form a cross-linkage thereby liberating more NH_3,

$$R_1CONH_2 + R_2-\overset{\overset{\displaystyle H}{|}}{N}-R_3 \rightarrow R_2-\overset{\overset{\displaystyle R_1}{|}\overset{\displaystyle CO}{|}}{N}-R_3 + NH_3 \qquad (6)$$

Thus chain scission and cross-linking occur simultaneously but eventually sufficient cross-linking sites are created so that the molecular weight can rise again.

The reaction scheme above does not however present the full picture for two reasons, namely appreciable amounts of water are formed and other products are evolved, e.g. hexylamine, cyclopentanone (Peebles and Huffman, 1971). In addition, the residue after hydrolysis contains a variety of products such as succinic acid, adipic acid, and hexamethylene diamine.

Water has been suggested to arise in some measure from reaction 2 but in part is likely to come from thermal cracking followed by dehydration (Kamerbeck et al., 1961).

Examination of the structure of the polymer suggests that main chain cleavage can occur at bonds which are located in a position β to the carbonyl group (Straus and Wall, 1958, 1959) as indicated by dotted lines below

$$-CH_2\!\mid\!CH_2CONH\!\mid\!CH_2-$$

Of the two linkages, the NC bond is the weaker and its breakage could occur by homolytic scission to yield free radicals. The primary dissociation yields two radicals which may stabilize themselves by hydrogen transfer (East et al., 1973),

$$-CO(CH_2)_4CONH(CH_2)_6- \rightarrow -CO(CH_2)_4CO\dot{N}H + \dot{C}H_2(CH_2)_5NH-$$

$$\rightarrow -CO(CH_2)_4CONH_2 + CH_2=CH(CH_2)_4NH- \qquad (7)$$

Hydrogen transfer from the amino radical to the alkyl radical could lead to a saturated end and an imide diradical which rearranges to form an isocyanate

$$-CO(CH_2)_3CH_2CN + CH_3(CH_2)_5NH-$$

$$\rightarrow \text{\tiny WWW}CO(CH_2)_3CH_2NCO$$

Isocyanate groups are not detectable since they will react further.

The amide group which is formed (Eqn. 7) may then dehydrate to form a nitrile group

$$R_1CONH_2 \rightarrow R_1CN + H_2O$$

Nitrile groups have been detected in the volatile fraction and in the residue. One possible reaction product which is to be expected but not found in the degradation products in large quantities is cyclopentanone (Twilley, 1961; Wiloth, 1973). This substance is expected to be formed from reactions of chain ends or from the chain itself

However cyclopentanone is only found in the degradation products in small quantities. It is likely that it undergoes further condensation reactions with itself or with amine end groups to form more complicated structures and eliminate water (Peebles and Huffman, 1971). Substituted cyclopentanone derivatives have been observed in the decomposition products of nylon 6.6 (Peebles and Huffman, 1971) and model compounds (Goodman, 1954, 1955).

2. Nylon 6

When heated under a stream of nitrogen Nylon 6 yields NH_3, CO_2 and H_2O, the quantities depending on the temperature (Table XIV) (Kamerbeck et al., 1961). The rates of formation of these gases do not change in the same way with temperature, indicating that each may be the product of a separate type of reaction. For short retention times, an increase in viscosity was observed (cf. nylon 6.6) due to after-condensation; however, heating at 281°C in a stream of nitrogen reduced the viscosity of the sample until after about 10 days dramatic increases occurred when gelation set in (Fig. 15) (Kamerbeck et al., 1961).

The amount of water evolved is larger than can be ascribed to the reaction of amino and carboxyl end groups. The water however may play a

TABLE XIV

Evolution of NH_3, CO_2, and H_2O after heating nylon for 120 h at various temperatures (millimoles per 113 g of polymer)

Temp., °C	NH_3	CO_2	H_2O
305	156	114	134
	165	117	154
270	10·9	18·6	40
	8·3	19·0	37
257	1·8	5·5	31

From Madorsky (1964) and Kamerbeck *et al.* (1961).

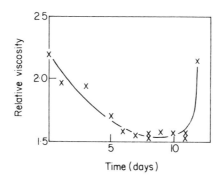

FIG. 15. Relative viscosity (1% solution in 90% formic acid) of nylon 6 after times of heating at 281°. (Kamerbeck *et al.*, 1961).

rôle in hydrolysis of the chains, leading to a reduction in the molecular weight, and in the increase in low molecular weight material which is extractable as the reaction proceeds.

As with nylon 6.6, the formation of secondary amines with the elimination of NH_3 (reaction 1) and of carboxyl groups (reaction 2), are possible. Hydrolysis of the degraded product revealed only a small amount of 1,11-diamino-6-oxo-undecane, indicating that the decarboxylation reaction only occurs to a small extent,

$$RCONH(CH_2)_5COOH + HOOC(CH_2)_5NHCOR$$

$$\rightarrow RCONH(CH_2)_5CO(CH_2)_5NHCOR + CO_2 + H_2O \qquad (8)$$

On the other hand, substantial quantities of di(ω-carboxypentyl) amine were obtained (Kamerbeck *et al.*, 1961) as is expected from elimination of NH_3 from amine ends. However the quantities of this substance were inadequate to account for the quantity of NH_3 released. The former of these substances when added to the melt caused gelation whereas additions of the latter did not, suggesting that although the former was present in small amounts, the quantity was adequate to effect gelation.

The reactions postulated above do not account for the quantity of water evolved nor for the large increase in basic groups which develop in the polymer (Reimschuessel and Dege, 1970) particularly at the higher temperatures.

As with nylon 6.6, main chain cleavage can occur leading to unsaturated chain ends and nitrile groups by dehydration of the amide, or

$$-CONH(CH_2)_5CONH(CH_2)_5CO- \rightarrow -CONH(CH_2)_5CONH_2$$
$$+ CH_2=CH(CH_2)_3CONH-$$

to an isocyanate group and a saturated end,

$$-CONH(CH_2)_5CONH(CH_2)_5CONH- \rightarrow -CONH(CH_2)_5NCO$$
$$+ CH_3(CH_2)_4CONH-$$

Nitrile groups to an extent of about 5% of the original amide groups have been detected in degraded nylon 6 by infra-red techniques (Kamerbeck *et al.*, 1961) but the unsaturated hexenoic acid has not been found amongst the products after hydrolysis, presumably due to further polymerization.

Caproic acid has been found in the hydrolysate and infra-red analysis indicated some isocyanate groups to be present. However, the latter groups would be expected to undergo further reaction (Kamerbeck *et al.*, 1961).

The possibility of splitting the chain at the C—C bond in the β position adjacent to the carbonyl group would lead to the formation of 1-amino-3 butene and acetic acid as products after hydrolysis of

$$-CONHCH_2CH_2CH=CH_2 + CH_3CONH(CH_2)_5CONH-$$

Acetic acid has been found in small quantities but not 1-amino-3 butene. This observation suggests that breakage of this bond occurs probably to only a minor extent. An examination of the probabilities of the splitting of the various bonds in nylon samples has further confirmed this point (East *et al.*, 1973).

In the case of nylon 6, there is an increase in basic groups in the polymer which is greater than expected from the liberated NH_3 (Reimschuessel and Dege, 1970). To account for this increase, the presence of cyclic amidine groups has been suggested (Schlack, 1968). In nylon 6, particularly after some heating, some cyclic monomer is likely to be present (Reardon and

Barker, 1974): this monomer can, in its enol form, condense with an amine end,

$$(CH_2)_5 \overset{\overset{N}{\parallel}}{-}C-OH + H_2NR \rightarrow (CH_2)_5 \overset{\overset{N}{\parallel}}{-}C-OR + NH_3 \tag{9}$$

or the amino end of a chain may cyclize

$$RNH\overset{\overset{O}{\parallel}}{C} \underset{(CH_2)_5}{\diagdown\diagup} NH_2 \rightarrow RN=C\overset{\diagup}{\underset{(CH_2)_5}{\diagdown}} \bigg| + NH_3 \tag{10}$$

The imino ether end groups are basic and hence ammonia is liberated without loss of basic groups. The end groups so formed would be titrated as amino functions and hence the loss of an end group would not be detected by titration of the polymer. Moreover these groups may to a small extent undergo a Chapman rearrangement (Chapman, 1925, 1927, 1928).

$$-CO(CH_2)_5-O-\overset{\overset{N}{\parallel}\diagdown}{C}-(CH_2)_5$$
$$-CO(CH_2)_5N=C\overset{\diagup}{\underset{(CH_2)_5}{\diagdown}}\bigg|\underset{O}{} \Biggr\} \rightarrow -CO(CH_2)_5N\underset{(CH_2)_5}{\diagdown\diagup}C=O \tag{11}$$

When hydrolysed these end groups will yield the di(ω-carboxy pentyl)amine found in the hydrolysis products of previous workers.

However the replacement of one amino group by another basic group does not explain the observed increase in the concentration of basic function. An alternative reaction has been suggested. The caprolactam present in the system because of polymer/monomer equilibrium can react with the carboxyl end groups as follows

$$(CH_2)_5\overset{\overset{NH}{\diagup|}}{-}C=O \;+\; HO\overset{\diagup}{\underset{}{}}\overset{\overset{O}{\parallel}}{C}-R \longrightarrow \underset{(CH_2)_5-C}{\overset{N=C\diagup R}{\bigg|}}\overset{\diagdown}{\underset{\diagdown O}{}}O + H_2O \tag{12}$$

$$(\textbf{XXV})$$

The intermediate ring (compound XXV) so formed may then rearrange and decarboxylate,

$$
\begin{array}{ccc}
\underset{|}{N}{=}\underset{\diagdown}{C}{-}R & \underset{\diagup}{N}{-}CO{-}R & N \\
(CH_2)_5{-}\underset{\|}{C} \to (CH_2)_5{-}CO & \text{or} & (CH_2)_5{-}C{-}R + CO_2 \quad (13) \\
O & & \\
\textbf{(XXVI)} & \textbf{(XXVII)} &
\end{array}
$$

Thus the reaction of the caprolactam with an acid end group liberates one mole of water and the ring compound XXV, which may then isomerize to the imide XXVI or decarboxylate to form the cyclic Shiff's base XXVII. The latter is a titratable basic group. The water so formed may either leave the system or hydrolyse an amide group, the latter yielding more amino and acid end groups. It may be noted that the postulated ring intermediate XXV can be formed without the participation of caprolactam if reaction occurs between the terminal carboxyl group and the adjacent amide group

$$
\underset{|}{-CONH(CH_2)_5\overset{OH}{\underset{|}{C}}O} \to -\overset{OH}{\underset{|}{C}}{=}N(CH_2)_5\overset{OH}{\underset{|}{C}}O \to XXV + H_2O. \quad (14)
$$

This latter reaction could result in branched structures if it occurs between an acid end group and an amide linkage of a second polymer molecule.

3. Kinetics of Reaction of Nylon 6

The rate of liberation of NH_3 and CO_2 have been determined from the polymer in a stream of inert gas (He) over a range of temperatures from 250°C to 290°C. For the liberation of NH_3, rate equations have been derived on the basis of reactions 1, 9 and 10. For the rate of liberation of CO_2, reactions 12 and 13 have been used together with the assumption that the water formed is capable of hydrolysing amide groups (Reimschuessel and Dege, 1970). The kinetic equations offer a satisfactory explanation of the data and hence add credence to the reaction schemes proposed.

B. EFFECTS OF OXYGEN

In the presence of oxygen, nylon is susceptible to attack. Exposure of the hot polymer in the threadline as it leaves the spinneret can cause the formation of hydroperoxides (Waller et al., 1948; Rieche et al., 1963). It is likely that these peroxides however decompose at temperatures greater than 90°C so that their presence is difficult to detect. Work on the thermal oxidation of nylon 6 has suggested that oxygen abstracts a hydrogen atom adjacent to the amide nitrogen leaving a radical which forms a hydroperoxide (Valk *et al.*, 1967). The hydroperoxide may decompose forming a substituted imide and a further radical which in turn can abstract hydrogen,

$$\underset{\underset{-CONH-CH-CH_2-}{|}}{\overset{\overset{O\!+\!OH}{|}}{}} \rightarrow \underset{\underset{-CONH-CH\!+\!CH_2-}{}}{\overset{\overset{O\cdot}{\|}}{}}$$

$$+ \cdot OH$$
$$\downarrow$$
$$\overset{O}{\overset{\|}{-CONH-CH}} + \cdot CH_2-$$

By such a mechanism the chain is broken and on hydrolysis (Tokareva and Mikharlow, 1966; Valk *et al.*, 1967) a new set of products can be detected which agree with the reaction scheme.

Nylon 6 and 6.6 when exposed in air lose strength to an extent which increases with increase in temperature (Valko and Chaklis, 1965) (Fig. 16), a change which is accompanied by a loss in breaking elongation. The fibre also loses some of its basic groups but the carboxyl content is unaltered

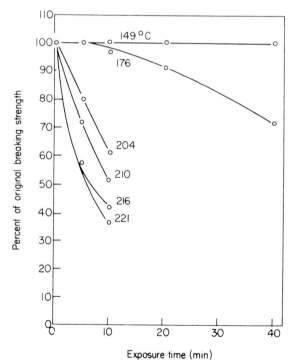

FIG. 16.   Effect of time and temperature of exposure in air on the breaking strength of nylon 6.6 yarn. (Valko and Chiklis, 1965).

TABLE XV

Analytical results on nylon 6.6R and nylon 6.6HS before and after thermal exposure in air

Yarn	Exposure	Amine content equiv./kg	Carboxyl content equiv./kg	Viscosity-average molecular weight M_v
Nylon 6.6R	Control	0·045	0·086	13,340
Nylon 6.6R	176°C/10 min	0·040	0·082	13,490
Nylon 6.6R	216°C/10 min	0·028	0·082	8,892
Nylon 6.6HS	Control	0·046	0·083	12,940
Nylon 6.6HS	216°C/10 min	0·029	0·082	16,180

Valko and Chiklis (1965).

(Table XV). In air, the molecular weight decreases in contradistinction to changes in nitrogen which increase the molecular weight (Fig. 17).

At temperatures greater than 120°C in oxygen or air, polyamides over longer periods of time discolour and eventually become dark brown (East *et al.*, 1975). Discolouration is accompanied by a fall in the breaking strength of the fibre and in the elongation at break. Using nylon 6, an analysis of the dark substances produced when the sample was heated in

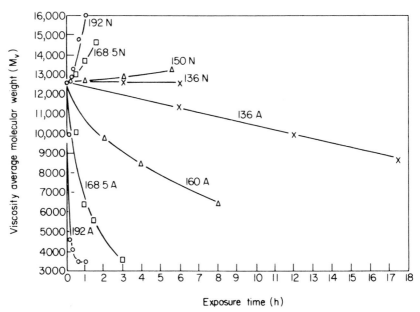

FIG. 17.   Effect of exposure to various temperatures in air (A) and in nitrogen (N) on the viscosity-average molecular weight of nylon 6.6E. (Valko and Chiklis, 1965).

oxygen to 160 or 170°C has been made. The reactions which are suggested are that the secondary amine formed by elimination of NH_3 is able to oxidize by a route similar to that for pyrrolidine; the latter, together with piperidine, has been detected among the products of oxidized nylons (East *et al.*, 1973). The methylene groups in a position β to the nitrogen atom are considered most likely to be oxidized to carbonyl groups, i.e.

$$
\begin{array}{cc}
R & R_1 \\
| & | \\
CO & CO \\
| & | \\
(CH_2)_3 & (CH_2)_3 \\
| & | \\
CH_2 & CH_2 \\
| & | \\
CH_2 & CH_2 \\
\searrow & \swarrow \\
& NH
\end{array}
\quad
\xrightarrow[\text{Heat}]{\text{Air}}
\quad
\begin{array}{cc}
R & R_1 \\
| & | \\
CO & CO \\
| & | \\
(CH_2)_3 & (CH_2)_3 \\
| & | \\
CO & CO \\
| & | \\
CH_2 & CH_2 \\
\searrow & \swarrow \\
& NH
\end{array}
$$

The compound so formed in its enol form may then condense with another secondary amine to give a branched substance which is capable of further reaction to form a polymeric material.

$$
\begin{array}{cc}
R & R_1 \\
| & | \\
CO & CO \\
| & | \\
(CH_2)_3 & (CH_2)_3 \qquad \overset{OH}{\underset{|}{CH=C}}-(CH_2)_3COR \\
| & | \qquad\quad \nearrow \\
HO-C & C-N \\
\| & \| \qquad \searrow \\
CH & CH \qquad CH=C-(CH_2)_3COR \\
\searrow & \swarrow \qquad\qquad\quad | \\
NH & \qquad\qquad\quad OH
\end{array}
$$

The introduction of the first carbonyl will make the methylene group in the β position more susceptible to oxidation to a second keto group to give unit XXVIII

$$
\begin{array}{cc}
R & R_1 \\
| & | \\
CO & CO \\
| & | \\
CH_2 & CH_2 \\
| & | \\
CH_2 & C-OH \\
| & \| \\
CH_2 & CH \\
| & | \\
HO-C & C-OH \\
\| & \| \\
CH & CH \\
\searrow & \swarrow \\
& NH
\end{array}
$$

(**XXVIII**)

Many permutations of the number and position of the active groups are possible so that the brown material is not a homogeneous product. The chromophore is presumably developed from the increasing degree of conjugation in the product.

C. Degradation under Aqueous Conditions

Nylon 6.6 is oxidatively degraded in water at temperatures as low as 100°C. Immersion in distilled water over periods of up to 60 days caused considerable loss of strength (Fig. 18) (Mikolazcuski *et al.*, 1964); little loss was observed at the boiling temperature, presumably because of the

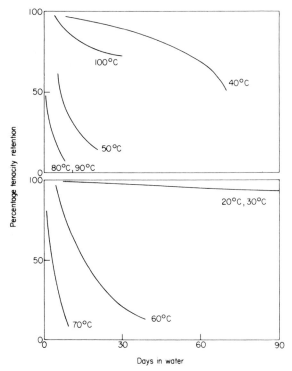

Fig. 18. Effect of distilled water at various temperatures on tenacity of undrawn nylon. (Mikolazcuski *et al.*, 1964)

absence of oxygen. The molecular weight of the material decreases to an extent which depends on pH. Below pH 5·3, little change occurs but above this degradation is rapid and reaches a maximum at a pH of approximately 8 (Vachon *et al.*, 1968) (Fig. 19). The degradation is not hydrolytic, since in the absence of oxygen degradation did not occur. The effects are related in

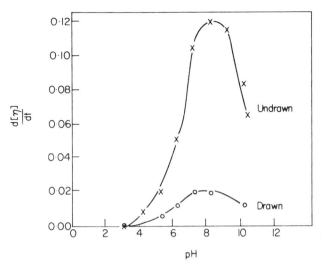

FIG. 19.   Rates of oxidative degradation, determined from intrinsic viscosity measurements at 80°C, of undrawn and drawn nylon 6.6 filaments as a function of pH. (Vachon *et al.*, 1968).

part to the diffusion of oxygen since undrawn fibre degrades more rapidly than the drawn. During treatment acidic compounds are formed as the pH of the liquor drops.

Exposure to aqueous conditions caused the development of cracks and fissures in the undrawn fibre after 6 h at 80°C (Vachon *et al.*, 1965). In the drawn fibre this development takes longer. Although the cause of crack formation is uncertain, its presence seems to be related to the oxidative degradation since degradation of the fibre by hydrolysis did not cause the formation of cracks.

Soxhlet extraction by water of undrawn nylon over a long period yielded a polymer of low molecular weight which is acidic in character (Mikolazcuski *et al.*, 1964) having a molecular weight of about 600 and carrying two carboxyl groups. Using O_2 in the system instead of air, hydrogen peroxide was detected in the liquor after 1 h at 60°C and hydroperoxides have been found in the nylon itself (Vachon *et al.*, 1965). The liquor also showed the presence of ammonia, formic acid, and aldehydes. In the solid polymer, carboxyl and aldehyde end groups increased with time of treatment. CO_2 was evolved. Measurements of the oxygen sorption show a continuous increase with time (Vachon *et al.*, 1965). As with changes in carboxyl or amine groups, the oxygen sorption is pH dependent and parallels the changes in viscosity of the solutions (Fig. 20) (Vachon *et al.*, 1965).

Oxidation proceeds by a free radical mechanism since radical scavengers such as quinone and hydroquinone inhibit the process. The reaction mechanism is likely to arise from the initial formation of hydroperoxides

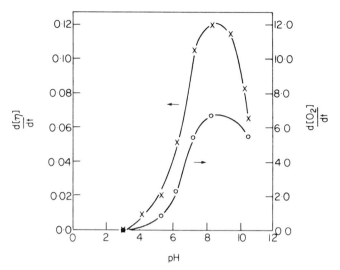

FIG. 20. Rates of oxidative degradation, determined from intrinsic viscosity measurements, and rates of oxygen absorption of undrawn nylon 6.6 filaments at 80°C as a function of pH. (Vachon *et al.*, 1968).

which act as initiators for the autoxidation. These peroxides could be formed on the nylon or in the water in the presence of trace amounts of heavy metals. The lack of oxidation at the low pH values suggests that the initiation step is suppressed under acid conditions, possibly due to stabilization of the intermediate hydroperoxides. At pH 7 and 8, these hydroperoxides are capable of homolytic cleavage to give free radicals, whereas at lower and at higher pH values, they probably decompose through homolytic or heterolytic means. An alternative explanation of the pH effect lies in the possibility that since the attack is likely to occur at the carbon atom adjacent to the nitrogen in the amide group, the protonation of the amide group at low pH values could reduce the electron density at the methylene group and eliminate the centre of oxidation. This explanation is however unlikely since amides are weakly basic.

V. Cellulose

A. THERMAL DEGRADATION

Thermal degradation of cellulose proceeds essentially through two types of reaction. At the lower temperatures, say between 120°C and 250°C, there is a gradual degradation which includes depolymerization, hydrolysis, oxidation, dehydration and decarboxylation. At higher temperatures, a rapid volatilization occurs, accompanied by the formation of levoglucosan leaving charred material (Bikales and Segal, 1971). In general, decomposition

is accompanied by a loss of fibre strength and a marked reduction in the
D.P. (Waller *et al.*, 1948; Findel'shtein *et al.*, 1950).

1. Initial Degradation Reactions

It is difficult to differentiate the thermal degradation at low temperatures
from the normal ageing of cellulose. Thus, rag paper heated for 6 months
at 38°C loses 19% of its folding strength (Richter, 1934); tyre cord held at
150 and 170°C loses strength (Waller *et al.*, 1948); and at higher temperatures (75 to 220°C), the degradation of cotton over periods of 4 to 24 hours
is greater in air than in nitrogen and is enhanced by the presence of
moisture (Farquhar *et al.*, 1956). Heating cotton linters at 170°C has shown
that the D.P. drops significantly and in the presence of oxygen reaches
about 200, a value which is associated with the size of the crystalline
regions (Major, 1958; Shafizadeh, 1968). During degradation, water,
carbon dioxide and monoxide are evolved and carbonyl and carboxyl
groups are formed; examples of the behaviour are shown in Figs. 21 and
22. The development of carbonyl and carboxyl groups as well as the
evolution of CO_2, CO and H_2O is much influenced by the presence of
oxygen: such observations suggest that at these temperatures, the reaction
is mainly an oxidative one probably non-specific in character.

Weight losses are relatively small. Heating cotton for 4 hours at
temperatures from 100°C to 175°C in nitrogen gave losses in the range 1 to

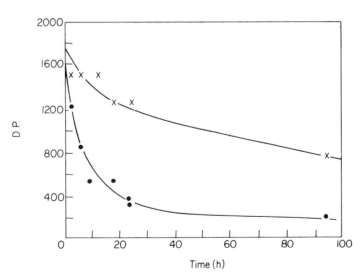

FIG. 21.   Depolymerization on heating at 170°C. [(×) in nitrogen; (●) in oxygen.]
(Shafizadeh, 1968).

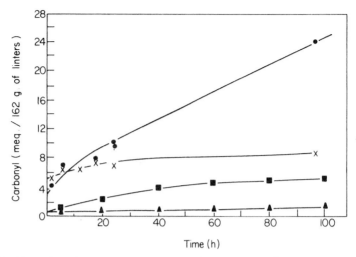

FIG. 22. Formation of carbonyl groups on heating at 170°C. [(●) in oxygen; (×) in nitrogen; (■) in oxygen, calculated from D.P.; (▲) in nitrogen, calculated from D.P.] (Shafizadeh, 1968)

2·0% and in air up to 2·5%. Over the same range (Farquhar *et al.*, 1956) however the fluidity increased from 0·15 to about 11 poise^{-1}.

More recently (Rusznak and Zimmer, 1970), because of interest in dyeing by the Thermosol process, cotton has been heated to 190°C and substantial decreases in the D.P. have been shown to occur in the first 30 sec, the extent of damage being greater if alkali is present (Fig. 23). On the other hand, the aldehyde content decreases rapidly to a plateau whereas the carboxyl groups rapidly increase to a maximum after about 60 sec and then slowly decrease. It is suggested that three parallel reactions take place, namely

(1) hydrolysis at certain of the glucoside linkages and at bonds which have become sensitive to hydrolysis due to oxidation reactions at carbon atoms 2 or 3;

(2) oxidation of the C_1 atom leading to the formation of a carboxyl group in addition to the C_6 atom. Oxidation may also convert secondary alcoholic groups into ketone groups resulting in higher sensitivity to hydrolysis;

(3) decarboxylation of the carboxyl groups in the various positions.

The position is made more complicated if carboxyl and carbonyl groups are initially present in the cellulose (Philipp *et al.*, 1972). The presence of such groups accelerates the decomposition.

In contrast to cotton, presumably because of its lower molecular weight, the D.P. of viscose shows little change on heating at 190°C for periods

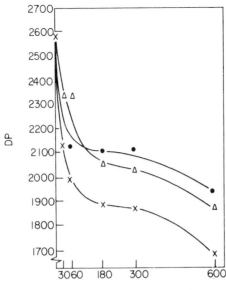

Time of thermal treatment (sec)

FIG. 23.   The effect of convective thermolysis on the average degree of polymerization of cotton treated with alkali. Initial moisture content: ●—5%; △—12%; ×—90%. (Rusznǎk and Zimmer, 1970).

ranging from 30 to 600 secs. It has been suggested that for morphological and accessibility reasons, fissure occurs at the ends of the chains. However, the reducing power of the aqueous extract of thermolysed viscoses increases with time of heat treatment, presumably due to the oligomers which are extracted (Rusznak and Zimmer, 1970).

One minor reaction has been suggested which the end groups are likely to undergo (Browne, 1958; Schwenker and Pacsu, 1958). During ageing, the C_1 aldehyde groups may form anhydrides with the hydroxyls on the C_2 and C_6 atoms of the same anhydroglucose unit. The formation of the 1,2 anhydro linkages is considered to be the first step because it is sterically favoured to be followed by a rearrangement to the more stable 1,6 linkage at temperatures above 110°C (Browne, 1958).

2. Improvement of Thermal Degradation

It has been shown that the degradation of cotton is arrested when the sample is immersed in ethanolamine (B.P. 172°) (Moore, 1948). Improvements have been brought about by using other alkaline substances, e.g. urea, biuret or other amines. Reduction of the carbonyl groups with $NaBH_4$ reduces colour formation (Albreck et al., 1975).

3. Reactions at Higher Temperatures

Pyrolysis gives the gases CH_4, CO, CO_2, a spate of volatile products usually obtained as an aqueous distillate laced with intractable tars and a carbonaceous residue. An example of the products is given in Table XVI (Madorsky et al., 1956). The major product in the tar is levoglucosan (1,6 anhydro-β-D-glucopyranose), which may be present in yields of 40% or more (Shafizadeh, 1968), together with polymeric materials which are

TABLE XVI

Pyrolysis of cotton cellulose under vacuum at 280°C

Period	Duration, min.	Weight loss, %	Volatile fractions			
			Tar, %	H_2O, %	CO_2, %	CO, %
1	38	3·3	33·9	50·7	10·5	4·9
2	58	7·0	55·9	36·1	6·0	2·0
3	53	9·1	63·8	29·6	5·4	1·2
4	100	15·9	—	—	—	—
5	77	9·8	70·2	23·6	4·7	1·5
6	180	16·7	—	—	—	—
7	155	7·8	70·8	21·9	5·5	1·8

From Shafizadeh (1968).

similar to condensation products of levoglucosan and apparently randomly linked oligosaccharides (Ballon, 1954; Shafizadeh, 1975). Many other products of the decomposition have been detected in small quantities, e.g. aldehydes, ketones, organic acids, 5-hydroxymethyl furfuraldehyde, polynuclear hydrocarbons etc. (Golova et al., 1957a; Gelbert and Lindsey, 1957; Schwenker and Pacsu, 1957; Golova and Krylova, 1960; Holmes and Shaw, 1961; Schwenker and Beck, 1963; Gardiner, 1966; Lipska and Wodley, 1969; Wodley, 1971). The yields of the various products depends on the source of cellulose and the rate of heating (Lipska and Parker, 1960; Baker, 1975; Shafizadeh, 1975): the yield of levoglucosan from different fibres is shown in Table XVII (Golova et al., 1957); preheating at or below 200°C for several hours gave larger yields of char and smaller quantities of volatile substances when the cellulose was subsequently pyrolysed between temperatures of 250 to 400°C (Lipska and Parker, 1966).

However, similar volatile products have been found to arise when the decomposition is carried out in different atmospheres and hence it has been suggested that the degradation mechanism at the higher temperatures is not an oxidative one but is essentially thermal in nature (Schwenker and Pacsu, 1957; Schwenker and Beck, 1963).

TABLE XVII

Formation of levoglucosan from different types of cellulose

Cellulosic material	D.P.	Hydrolyzability	Levoglucosan formed, %
Cotton hydrocellulose	1000	5·21	60–63
Cuprammonium cellulose	1000	10·76	14–15
Mercerized cotton cellulose	1200	9·55	36–37
Viscose fibre (without orientation)	380	23·50	4·0–4·5
Viscose fibre (maximum orientation)	400	23·20	4·8–5·0

From Shafizadeh (1968).

Unfortunately, a description of the reactions is made more difficult by the sensitivity of the decomposition to the presence of impurities particularly inorganic materials whose presence decreases the quantity of tar but increases the amount of residue. For example, the yield of levoglucosan rose to 70% for ash free material (Golova and Krylova, 1960).

4. Structure of the Char

Controlled heating up to 1500°C has shown that when a temperature of 250°C is reached, corresponding to a weight loss of 40%, the crystalline structure is destroyed (Davidson and Losty, 1965). Other workers have shown the breakdown of crystal structure at 300°C (Cobb, 1944). On heating to higher temperatures, the number of hydroxyl groups decreases and the carbonyl groups increase (Otani, 1959; Otani *et al.*, 1960), and stable unsaturated rings containing carboxyl groups are formed (Higgins, 1957, 1958). Increasing the temperature up to 370°C results in further weight losses, 40–60%, and the formation of aromatic structures which coalesce at the higher temperatures to give a graphitic structure.

5. Changes in D.P.

The D.P. of the cotton or filter paper falls rapidly, as a result of pyrolysis, to about 200, the value expected if only the crystalline regions remained (Golova and Krylova, 1957; Golova *et al.*, 1959a; Halpern and Patai, 1969a); even though the temperature was in the range of 250°C, this dramatic drop is accompanied by very small weight losses (2–4% from cotton and 1·5% from paper). At higher temperatures, the weight loss is greater but the behaviour is similar. This primary degradation of the cellulose chains is virtually complete before appreciable quantities of volatile products are formed; in fact in the earlier stages, most of the weight loss can be accounted for by evolution of water.

The degradation is thus associated in the initial stages with the disordered regions of the cellulose and indeed less crystalline samples are more rapidly degraded by heat (Weinstein and Broido, 1970; Patai and Halpern, 1970; Ramiah, 1970). Moreover most of the levoglucosan was formed (at 300°C) once the D.P. had dropped to about 200, the yield depending on the type of cellulose used (Golova and Krylova, 1957; Golova et al., 1957b).

6. Examination of Pyrolysis using DTA

Regardless of the source of cellulose, a slight endotherm is observed just above 100°C, presumably due to volatilization of residual moisture (Bikales and Segal, 1971). This is followed by a strong endotherm in a range of temperature between 260 to 350°C which appears to be related to the formation of levoglucosan; this may be followed at higher temperatures by a considerable exotherm. The magnitudes of the endo- and exotherms vary from worker to worker, being dependent *inter alia* on the geometry of the apparatus, in particular on how easily the gases may be removed, the presence of oxygen (Schwenker et al., 1964) and the heating cycle; as a result of the last of these, complications may arise from an overlap of the endotherm by the exotherm (Mack and Donaldson, 1967). On the other hand, some workers have obtained only an endotherm in the presence of an inert gas (Tang and Neill, 1964). The results obtained are also dependent on the manner in which the sample is presented. Thus a thick sample gave an exotherm at 350°C which was absent when a thin sample was used, suggesting that the exotherm may be due to the breakdown of decomposition products trapped in the cellulose (Arsenau, 1971).

A comparison of a series of DTA traces with the corresponding weight loss curves is given in Fig. 24 (Perkins et al., 1966). However interpretation of the endotherms is not simple; the presence of acidic, neutral or basic materials reduces the endotherm, with the very basic materials (NaOH, LiOH, Na_2CO_3 etc.) transforming it into an exotherm (Fig. 25 and Table XVIII). In general the presence of the inorganic materials lowers the temperature at which the weight loss begins. At the same temperature, the addition of $KHCO_3$ markedly increases the rate of decomposition with only small changes in the decomposition products but the quantity of water and CO_2 is relatively increased (Lipska and Wodley, 1969). The presence of the base reduces the temperature at which weight losses occur and since it catalyses the dehydration reaction, the subsequent exothermic reaction becomes more prominent. The point is confirmed by the fact that the

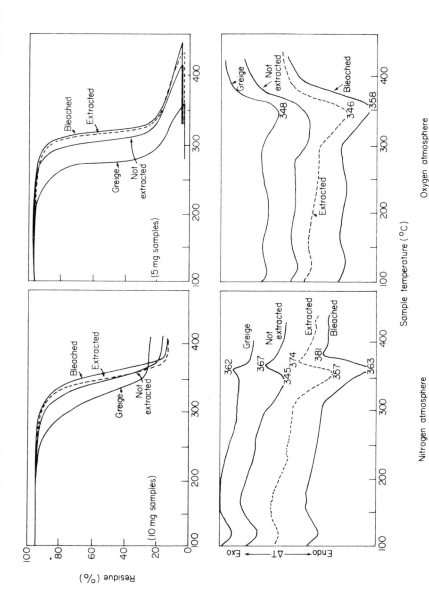

FIG. 24. TG and DTA (gas diffusion, GD) curves of untreated cotton controls in nitrogen and in oxygen atmospheres. Heating rate = 15°C/min. (Perkins et al., 1966).

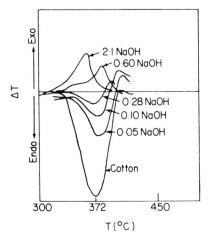

FIG. 25.   DTA curves for cotton and NaOH-impregnated cotton. Numbers are percent concentration. (Mack and Donaldson, 1967).

TABLE XVIII

Results of DTA–TG determinations performed simultaneously on the same sample. Heating rate 10°C/min. 200 mg samples

	DTA			TG	
Additive 10^{-4} mole/g	Beginning of peak °C ± 3°C	Peak maximum °C ± 5°C	Peak area* arbitrary units ± 2	Beginning of weight loss °C ± 5°C	Residue at 400°C % ± 2%
None	310	380	− 150	312	13
NaHSO₄	280	350	− 75	280	16
Na₂SO₄	260	390	− 76	290	21
0·1 × Na₂CO₃	330	387	− 23	290	16
0·2 × Na₂CO₃	330	382	− 19	283	20
NaH₂PO₄	320	358	− 13	287	31
NaOAc	343	370	− 10	260	23
NaCl I	345	370	− 7	290	23
II	375	385	+ 6		
Na₂CO₃ I	315	337	+ 5	240	27
II	355	375	+ 5		
Borax	265	350	+ 55	280	43

* − = endotherm   + = exotherm.
Halpern and Patai (1969c).

extent of the heat change is concentration dependent (Mack and Donald-son, 1967). The endotherm is related to the tar formation (levoglucosan) (Kilzer and Broido, 1965; Broido, 1966; Tang and Eickner, 1968) and its magnitude is correlated with the degree of crystallinity of the cellulose (Basch and Lewin, 1973) whereas the exotherm arises from a combination of several competitive reactions in which compounds of small molecular weight are formed (H_2O, CO_2 etc.) together with the char (Kilzer and Broido, 1965; Broido, 1966; Tang and Eickner, 1968). Again, the thermal history of the sample is important; for example, a sample held at 250°C for a day forms about three times as much char at 400°C and loses the characteristic endotherm (Berkowitz 1957; Kilzer and Broido, 1965).

7. Mechanism of Degradation

The elucidation of the mechanism of pyrolysis is made difficult by the development of new chemical and physical structures in the solid together with a bewildering array of volatile products and only general charac-teristics have been established (Bikales and Segal, 1971; McCarter, 1972).

The first noticeable effect of heating is dehydration which occurs at random along the chain of glucose units (Browne, 1958; Madorsky, 1964) in the amorphous regions. As shown earlier, thermal degradation in the lower temperature range is affected by oxygen whereas at the higher temperatures oxygen does not play a major role (Rutherford, 1947).

In the temperature range in which the main decomposition occurs (say 250°C and greater), water is eliminated and the D.P. drops considerably; at the same time, the sample becomes to some extent cross-linked, as indicated by increases in the wet strength of paper (Back and Klinga, 1963) or difficulties in dissolution of the nitrated derivatives (Golova *et al.*, 1959a).

The production of water at the lower temperatures suggests therefore that intra- or inter-ring dehydration occurs. The latter could give rise to stable etheric linkages whereas intra-ring dehydration would result in a keto-enol tautomeric system (Halpern and Patai, 1969a; Hofman *et al.*, 1960), as indicated below,

The carbonyl groups so formed could participate in a variety of reactions leading to carboxyl groups which may fragment to give volatile acids and oxides of carbon. In the later stages (at 250°C), water and levoglucosan are formed simultaneously at rapid and comparable rates but when the weight loss reaches about 50%, the rate of water formation drops off, the ratio of levoglucosan/water increasing with time or with temperature (Halpern and Patai, 1969a). This picture has been approximately confirmed by mass spectrophotometric results on pyrolysing cellulose at 300°C (Lipska and Wodley, 1969).

The initial drop in D.P., presumably resulting from the dehydration reaction, occurs in the less ordered regions; for example, in a vacuum pyrolysis experiment the tars produced after 150 minutes at 251°C, whilst constituting 10–20% of the weight loss, did not contain measurable quantities of levoglucosan (Basch and Lewin, 1973). Alternatively, dried cellulose heated at 300°C for different time intervals gave very similar yields of levoglucosan (between 15 and 20%) until the D.P. reached 200: once reached, the yield increased to 44%. Such observations are in agreement with the suggestion that levoglucosan production is a function of the degree of crystallinity of the sample (Golova *et al.*, 1957c; 1959b).

The two reactions—dehydration and levoglucosan formation— are considered to be distinct, that of dehydration being only slightly endothermic and that of levoglucosan formation being strongly so. The exotherm is considered to occur when the dehydrocellulose is further decomposed. Schematically the picture is given below,

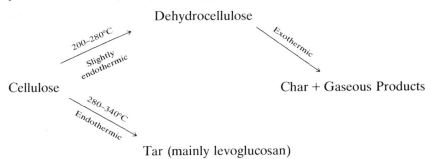

The formation of the tar is considered to give rise to the major weight loss and, being strongly endothermic, accounts for the fact that the temperature at which the endotherm develops is not markedly different from that at which the weight losses become evident (Kilzer and Broido, 1965; Mack and Donaldson, 1967). However even such a simple scheme as that given above suggests that, since the various reactions occur at different rates, the products produced will be markedly dependent on the heating programme to which the cellulose has been subjected. Thus heating at lower temperatures favour dehydration and subsequent char formation; heating at higher temperatures gives higher yields of inflammable volatile products. More likely is the possibility that both reactions compete with each other at all temperatures, one becoming prominent over the other depending on the temperature and the condition of the substrate.

More recent work has suggested that the pyrolysis of glucosidically linked compounds takes place through transglycosylation to provide a mixture of anhydro sugars and randomly linked oligosaccharides (Shafizadeh, 1975). It is possible that such a heterolytic scission (by nucleophilic displacement of glucosidic groups by the free hydroxyl groups present) is one of the important steps in the degradation rather than the homolytic (free radical) cleavage mentioned later.

In the case of cellulose degradation, the anhydro sugar levoglucosan is the most important volatile product of the high temperature degradation ($> 250°C$). Levoglucosan has been put forward as a key intermediate which may undergo further degradation to volatile low molecular weight compounds and tarry residues, the latter being formed by repolymerization or as a result of degradation of the oligomers. The polymerized material is considered to aromatize into carbon char (Parko et al., 1955; Pernikis et al., 1967; Hinojosa et al., 1973). The quantity of this substance varies considerably with the source of cellulose and the additives present (Halpern and Patai, 1970; Hendrix et al., 1970) (Table XIX).

TABLE XIX

Tarry fraction in mg (total volatiles less aqueous fraction), with weight of levoglu-cosan (in mg) given in brackets

	None		Na$_2$SO$_4$		NaCl		NaHSO$_4$		Na$_3$PO$_4$	
	2 h	4 h	2 h	4 h	2 h	4 h	2 h	4 h	2 h	4 h
544	1·3(1·3)	28(7)		4(3)[a]		40(0)[a]		85(35)[a]		50(0)[a]
544P	33(2)	40(12)	4(2)	14(3)	15(1)	29(2)	110(50)	113(58)	54(0)	67(0)
MC	64(4)	65(10)	42(0·8)	42(1·2)	30(tr)	39(1)	80(31)	104(58)	b	b
544BM	115(36)	267(75)	11(tr)	18(7)	24(tr)	27(4)	106(40)	123(50)	b	b

[a] Data at five hours. [b] Sample frothed. tr = trace. From Halpern and Patai (1970).

Since levoglucosan production is low in the initial stages and the initial breakdown occurs in the amorphous regions, it would seem that its formation is likely to arise to a large extent from breakdown of the crystalline regions (Golova et al., 1959).

Addition of glucose to the pyrolysing cellulose decreases rather than increases the yield of levoglucosan suggesting that glucose is not an intermediate in its production (Golova et al., 1959b) and does not yield

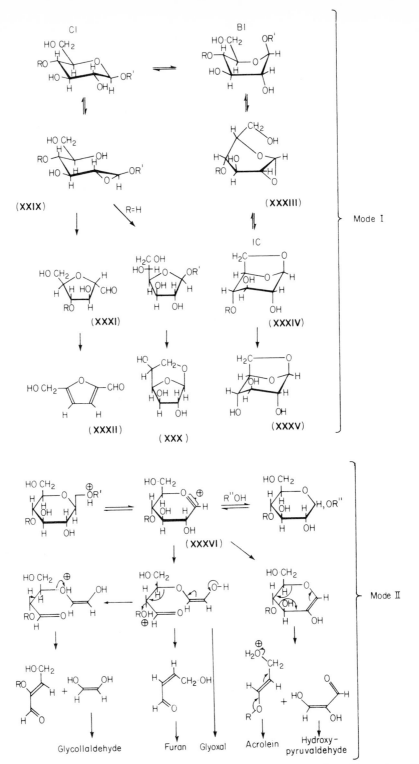

FIG. 26. Two modes of cellulose pyrolysis reactions. Mode 1: mechanism of formation of anhydroglycose and furfurals. Mode 2: mechanism of formation of carbonyl compounds. (Chatterjee and Schwenker, 1972).

levoglucosan by dehydration. Many workers have proposed mechanisms for the formation of levoglucosan and a variety of intermediates which lead to this substance have been suggested (Shafizadeh, 1968).

It seems reasonable to suppose that cleavage occurs at the C—O bond (Madorsky, 1964); the hexose units so formed can then rearrange to form levoglucosan. Cleavages of other C—O bonds (at C_5 or the ring C_1) would break the ring and lead to smaller molecules as degradation products and carbonaceous residues.

Two modes of decomposition have been suggested (Gardiner, 1966). In one of them, the boat conformation of the glucose residue is considered to be in equilibrium with the chair (C1 and B1 in Fig. 26). It is feasible for the boat form then to change through a series of reactions which ultimately result in levoglucosan XXXV (Madorsky et al., 1956; Gardiner, 1966). It may be noted that the formation initially of structure XXXIII splits the levoglucosan molecule from the end of the chain so that an unzipping process can occur. An alternative reaction with ring opening is the sequence from C1 to the furfural XXXII, a reaction which it is considered could go through a carbonium ion intermediate (Byrne et al., 1966). Ionic mechanisms have also been put forward to account for the formation of carbonyl compounds (Byrne et al., 1966); this is shown in Fig. 26.

Other workers have found that free radicals are formed during the decomposition (Ermolenko et al., 1967; Arthur and Hinojosa, 1966) to an extent which depends on heating conditions and the presence of additives (Hinojosa et al., 1973). It has been suggested that the formation of free radicals is associated with char formation and hence their number is increased when chemicals (e.g. borax) which accentuate char formation are added. Mechanisms involving cleavage of chain glucosidic units to give free radicals have been proposed (Pakhamov et al., 1957, 1958; Bull, 1943; Kato, 1967); these may subsequently react to yield levoglucosan or volatile compounds (Fig. 27). Two possible modes of reaction have been suggested, namely the random scission of the cellulose chain followed by rearrangement of the anhydroglucose end which splits off as an anhydroglucose unit leaving a new radical end (Bull, 1943). The alternative is to split the radicals from the chain followed by interaction with each other to form levoglucosan.

There are clearly many features of the reactions not understood. One in particular is the base catalysis of the dehydration and subsequent reactions (Madorsky et al., 1956; Mack and Donaldson, 1967; Halpern and Patai, 1969b). The tar formation on the other hand increases markedly in the absence of base: the formation of levoglucosan and 1,6 anhydrofuranose is reduced and the residue contains a large quantity of ethylenic double bonds.

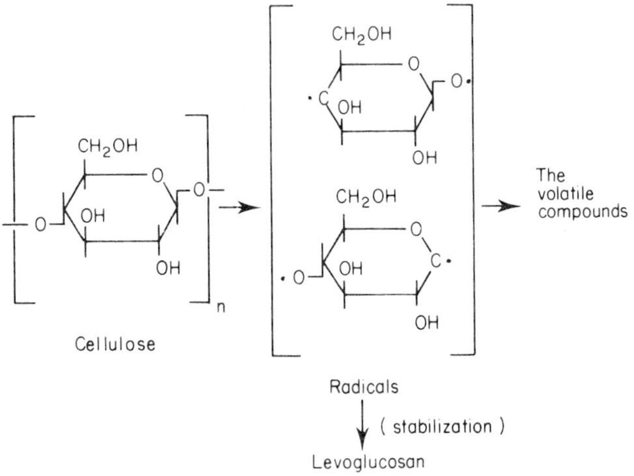

FIG. 27. Intermediate free radical mechanism for cellulose pyrolysis. (Chatterjee and Schwenker, 1972).

8. Kinetics

Determination of rates of decomposition have been carried out in more than one way, e.g. by weight loss, determination of the volatiles or levoglucosan, each method yielding a different feature of the pyrolysis. Moreover since there are competitive reactions occurring, the results will depend on the thermal programme (Baker, 1975). The results are also dependent on the chemical and physical states of the samples (Andraev, 1959, 1961).

Some workers have postulated first order kinetics (Parks *et al.*, 1950; Stamm, 1956; Murphy, 1962), others zero order (Golova and Krylova, 1957). Some data indicate a pseudo zero order decomposition in the initial stages to be followed by a pseudo first order reaction (Tang and Neill, 1964). An example of the results from two kinds of measurement is shown in Fig. 28; both show zero order behaviour until about 70% weight loss has occurred (Lipska and Parker, 1966). However whereas the curve based on the residual glucosan residues decays to zero, that based on weight loss levels out at 85%. Zero order kinetics were also obtained from measurements of the quantities of gas evolved. Corresponding data over a range of temperatures up to 360°C have shown that the initial weight loss becomes less important as the temperature is increased (Lipska and Wodley, 1969).

Thus there would appear to be two modes of decomposition after a rapid initial weight loss. It is reasonable to picture the initial step as a chain scission followed by further decomposition from the end of the fragmented chains to yield levoglucosan molecules (Chatterjee and Conrad, 1966).

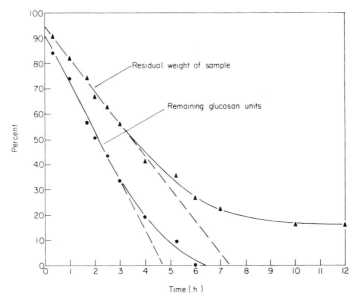

FIG. 28.   Pyrolysis of α-cellulose in nitrogen at 288°C. (Lipska and Parker, 1966).

"Unzipping" of the chains continues until other processes—repolymerization or carbonization—become rate controlling. In the early stages, both chain scission and unzipping occur but as the reaction proceeds the latter becomes more dominant. The zero order behaviour is compatible with unzipping since in the initial stages the number of chains will only decrease slowly; as the reaction proceeds, the number of reactive species corresponds more closely with the number of molecules and the reaction shifts to first order kinetics (Chatterjee and Conrad, 1966; Chatterjee, 1968a,b).

Other workers, on the basis of chain scission and unzipping, have shown that the weight loss turns out to be proportional to $(time)^2$; this plot has been shown more recently to be linear up to about 40% weight loss. The lines however did not pass through the origin, suggesting that there was in addition a rapid initial weight loss (Chatterjee and Conrad, 1966; Basch and Lewin, 1973). However this relation is only applicable as long as chain scission is important. For the later stages where the chain scission has eased, the kinetics become first order.

The kinetics as suggested do not encompass the complete situation; for example, the D.P. drops to 200 even though the weight losses are small (Halpern and Patai, 1970). It is not possible to define a completely satisfactory kinetic scheme because of the wide range of products produced and because of the ill-defined nature of the reactions involved. The position is reviewed elsewhere (Okamoto, 1973).

9. *Importance of Fibre Structure*

Many observers have shown that the rate of pyrolysis is different for different samples of cellulose. In general the more amorphous the fibres, the more rapid the decomposition (Major, 1958; Weinstein and Broido, 1970; Ramiah, 1970; Halpern and Patai, 1970; Madorsky *et al.*, 1970). Examination of cellulose degraded in oxygen showed no change in the X-ray pattern but a rapid diminution of the amorphous content as observed by X-ray spectroscopy (Hurdoc and Schneider, 1970). More recently using a set of carefully purified fibres, the (weight loss)$^{\frac{1}{2}}$ when plotted against time has been shown to give linear plots (Fig. 29) (Basch

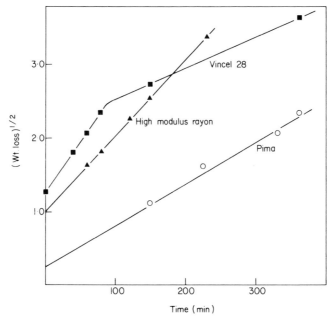

FIG. 29. Kinetics of vacuum pyrolysis at 251°C: (weight loss)$^{\frac{1}{2}}$ vs. time. (Basch and Lewin, 1973).

and Lewin, 1973). The rate constants (Table XX) show considerable differences between samples, the constants being larger the greater the quantity of amorphous material present. Correlations of rate constants which bring in other properties have been made. In fact, a good correlation has been obtained with the parameter

$$\frac{\%\text{ amorphous region} \times f_0}{(\text{D.P.})^{\frac{1}{2}}}$$

where f_0 is the orientation factor as measured from birefringence (Fig. 30) (Basch and Lewin, 1973). One or two exceptions were noted namely those materials which had been chemically treated, e.g. mercerized or hydrolysed cotton, and it may be assumed that some modifications which are not detectable by X-ray methods, used to determine the percentage of amorphous regions, have occurred. It was interesting to note that the extent of the initial reaction (intercept on the rate plot) increases with the quantity of amorphous material, suggesting that there was an initial rapid reaction in these regions. Moreover, in contradistinction to other workers, the dependence on D.P. suggests that some degradation occurs via the chain ends.

TABLE XX

Vacuum pyrolysis at 251°C

No.	Sample	Kinetics		Residue after 150 min	
		Initial weight loss, %	Rate . (wt. loss)$^{\frac{1}{2}}$ $\times (10^3$, min$^{-1})$	Percent crystallinity	Degree of polymerization
1	Cotton, Deltapine	0·25	4·7	63·7	213
2	Cotton, Deltapine, hydrolysed	1·10	7·8	72·5	231
3	Cotton, Pima	0·25	4·7	70·8	235
4	Cotton, Pima, mercerized	0·25	10·2	65·3	185
5	Ramie	1·10	3·2	70·3	565
6	Ramie, hydrolysed	0·08	4·9	69·5	238
7	Tire yarn	3·96	14·0	57·1	—
8	High modulus yarn	1·00	10·4	67·5	—
9	Vincel 64	3·24	12·4	54·9	13
10	Vincel 28	1·56	11·7	66·0	27
11	Textile rayon	1·00	13·6	51·2	24
12	Evlan	3·60	10·4	49·8	—

Basch and Lewin (1973).

The importance of crystallinity is also demonstrated by the increase in activation energy with this factor (Basch and Lewin, 1973; Chatterjee and Conrad, 1966). This is expected since the reaction may be regarded as arising from decomposition of both the amorphous and crystalline regions, the latter requiring a greater activation energy for decomposition because of their higher stability.

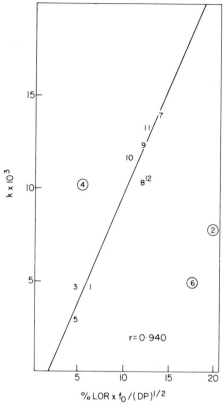

FIG. 30.   Vacuum pyrolysis rate constant vs. %LOR/f_0/(DP)$^{\frac{1}{2}}$. Sample numbers are given in Table XX. (Basch and Lewin, 1973).

10. Reactions in Air

In general, reaction in air occurs faster than in an inert atmosphere (Millet and Goedkin, 1965). As indicated earlier, at the lower temperatures oxidation reactions are very marked. Oxygen consumption is different during the thermal oxidation of different cellulose samples and the rate of reaction decreases with increase in crystallinity. Moreover the degradation arising from oxidation is shown by the oxygen consumption being proportional to the loss in D.P. at 160°C (Kosik and Reiser, 1973). However, at higher temperatures, the products of pyrolysis in air and in nitrogen are the same (Schwenker and Beck, 1963; Wodley, 1971) and it would seem that oxygen acts more as a catalyst (Basch and Lewin, 1973). The rates are dependent on the same factors in air as in a vacuum but increasing orientation decreases the rate, presumably because this factor decreases the rate of diffusion of oxygen in the substrate.

B. ALKALINE DEGRADATION

Degradation of cellulose by alkali can occur in two ways. The first of these is a simple nucleophilic attack at the C_1 atom, either intramolecularly by an ionized hydroxyl group (Fig. 31, route A) or intermolecularly by direct attack by the hydroxyl ion (route B). This reaction becomes more important under more severe alkali treatments, e.g. at 170°C and diminishes with reduction of temperature (Lewin, 1965). The second is a more rapid reaction which has been referred to as a peeling reaction which brings in its train the formation of a yellow colour.

1. Yellowing of Cellulose

Cotton and other celluloses yellow on storage to an extent depending on the pretreatments given to the fabric. In the laboratory, standard tests have been set up for determining the ability of a sample to develop this colour. Yellowing is determined by the measurement of the colour reversion of post colour (Albeck et al., 1975). Essentially, the reflectance is measured and, using the Kubelka–Monk formula, the ratio of the scattering coefficient S to the absorption coefficient K is calculated from the equation

$$\frac{S}{K} = \frac{(1 - R_\infty)^2}{2R_\infty}$$

where R_∞ is the reflectance of an infinite layer (Giertz, 1945). The degree of yellowing (p.c. number) is taken as the difference between the values of

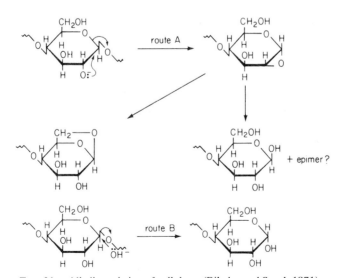

FIG. 31.   Alkaline scission of cellulose. (Bikales and Segal, 1971).

S/K of the treated and untreated material. The p.c. numbers have been shown to increase with the copper number of and the carboxyl groups in cellulose pulp. Although the experiments on the yellowing are not well standardized, most workers associate the formation of colour with the presence of functional groups, presumably carbonyl. In fact, linear relationships between the copper number and the increase in p.c. number have been obtained by several workers (McMillan, 1958; Virkola et al., 1958; Jappe and Kaustmen, 1959); moreover, reduction of the carbonyl groups with sodium borohydride treatment decreases the p.c. number (Giertz and McPherson, 1956). The role of carboxyl groups therefore appears to be uncertain since the evidence is somewhat contradictory (Albeck et al., 1975).

An examination of the yellowing of cotton and modified cottons has been made by measuring the yellow colour extracted from the sample after the latter had been refluxed for an hour in a solution containing 5% $NaHCO_3$ (Albeck et al., 1975). The degree of yellowing, as determined by the optical density, D, of the extract, increases as the aldehyde content of the cotton increases. Changes in the aldehyde content have been made by hydrolysing or oxidizing the sample. An example is given in Figs 32 and 33 where for cotton oxidized with sodium hypochlorite, the yellowing runs parallel with the quantity of aldehyde groups present (Albeck et al., 1975). No such correlation was observed between D and the number of keto or carboxyl groups, a fact confirmed from examination of cotton containing keto groups when only a small degree of yellowing was observed. Alternatively, conversion of the aldehyde groups to carboxyl or alcohol reduces

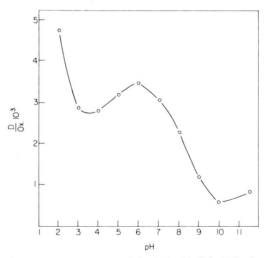

Fig. 32.   D/Ox vs. pH for cellulose oxidized with NaClO. (Albeck et al., 1965).

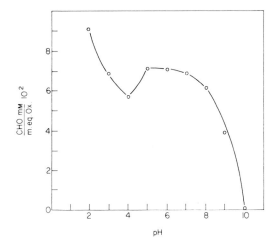

FIG. 33.   Aldehyde groups of cotton oxidized with NaClO vs. pH. (Albeck *et al.*, 1965).

the colour (Rapson and Hakim, 1957; Rapson *et al.*, 1958; Virkola and Lehtikosi, 1958, 1960; Albeck *et al.*, 1975). Unfortunately although the development of a yellow colour is a common feature of sugars (e.g. glucose, galactose etc.) as well as cellulose, identification of the chromophore has not yet been achieved.

2. Peeling Reaction

The alkaline degradation of cellulose and modified cellulose is important in the textile field, e.g. in the ageing of soda cellulose, scouring of cotton etc. (Entwhistle *et al.*, 1949). Cellulose is slowly attacked by hot alkali and the loss in weight is related to the aldehyde content. Stability to alkali may be achieved by oxidation of the aldehyde groups to carboxyl (Meller, 1951), by reduction to alcohol (Meller, 1953c) or blocking by forming the methyl glucoside (Reeves *et al.*, 1946). On the other hand, chemical attack leading to an increase in carboxyl groups, e.g. by oxidation or hydrolysis, makes cellulose more susceptible.

Studies of the degradation of hydrocellulose, oxycellulose etc. have shown that the degradation occurs by removal of an alkoxyl group via a β-alkoxyl elimination reaction. For a simple derivative XXXVII, the reaction proceeds through the formation of the enediol XXXVIII from which the alkoxyl group is eliminated to give XXXIX

$$CH_3CH-CH_2-CHO \rightleftharpoons CH_3-CH-CH=CH \rightleftharpoons CH_3CH=CH-CHO$$
$$\quad\;\; OR \qquad\qquad\qquad OR \quad\; O \qquad\qquad\qquad + OR^-$$

(XXXVII)   **(XXXVIII)**   **(XXXIX)**

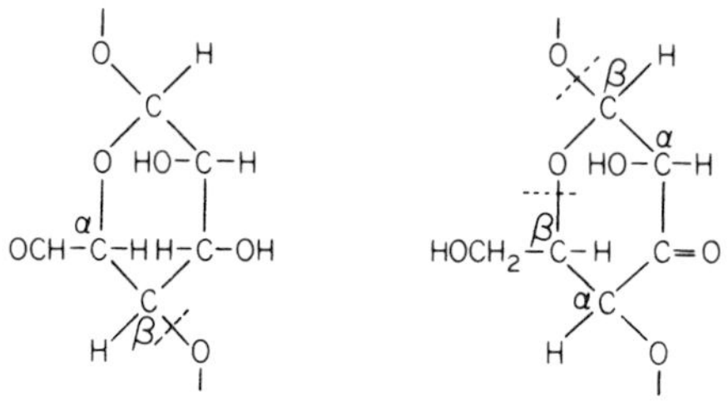

(XXXX)   (XXXXI)

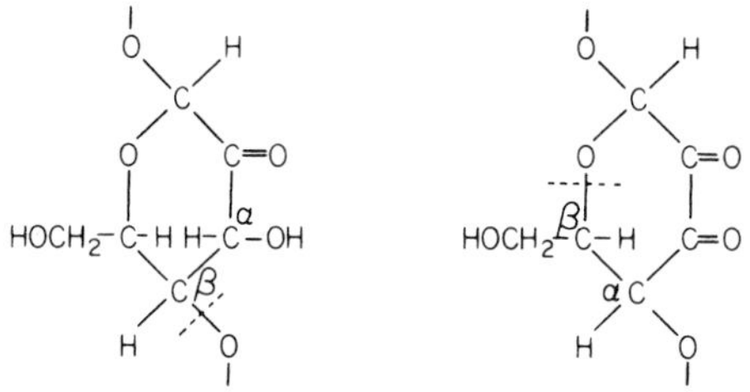

(XXXXII)   (XXXXIII)

Fig. 34.

For carbonyl groups in the anhydroglucose ring, the action of alkali causes ring opening and chain scission is indicated in Fig. 34 (Peters, 1967). However in the case of glucose or the reducing anhydroglucose end of cellulose, an intervening step occurs in which the ketose is formed (Fig. 35). Elimination at the β position (C in the chain) occurs, splitting the unit from the end of the chain, leaving a new reducing end for further reaction. The diketone XXXXVI which is eliminated then undergoes a benzilic rearrangement to yield isosaccharinic acid XXXXVIII. An alternative but less dominant reaction is the direct elimination of the hydroxyl group from the C_3 carbon atom to yield the di-carbonyl end group XXXXIX which undergoes the benzilic acid rearrangement to yield metasaccharinic acid L. In this latter process the end group is stabilized (Corbett and Kenner, 1955; Corbett and Richards, 1957; Machell *et al.*, 1957; Machell and Richards, 1957).

The former of these reactions is a peeling process. The progressive stepwise removal of the units from the chain end is more dominant than the latter stabilization reaction. The number of units peeled off has been shown to be in the range 15–65 according to the conditions and the sample of cellulose or hydrocellulose examined (Lewin, 1965; Corbett and Kenner, 1955; Corbett and Richards, 1957; Machell and Richards, 1957; Machell *et al.*, 1957; Richzenhaim *et al.*, 1954; Immergut and Rånby, 1956; Samuelson and Wemnerbloom, 1954; Richtzenhaim and Abrahamson,

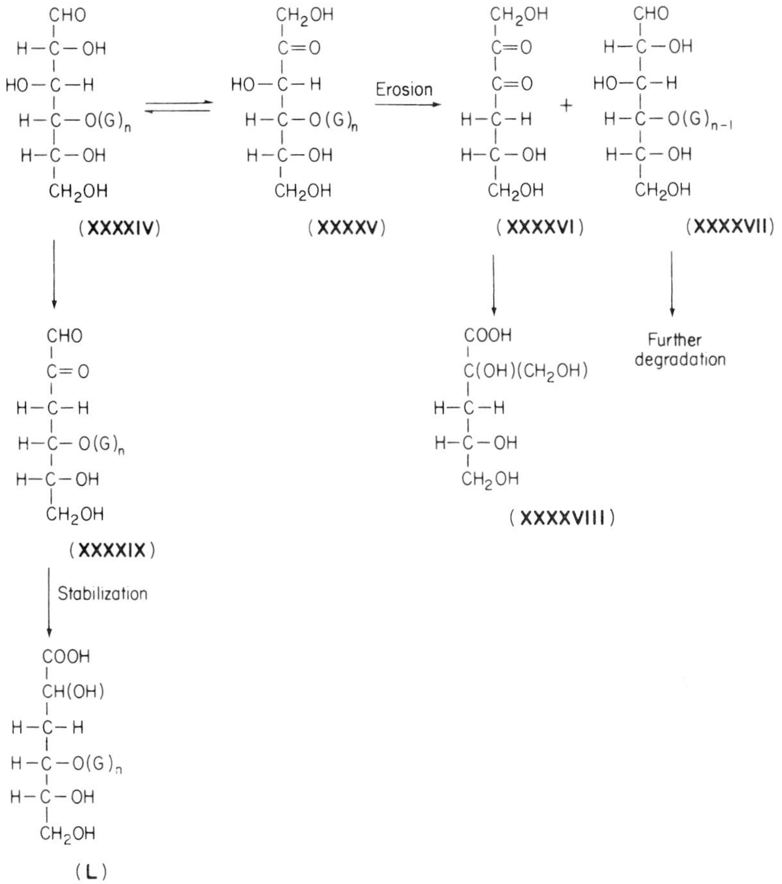

```
   CHO              CH2OH              CH2OH            CHO
    |                 |                  |               |
H—C—OH             C=O                 C=O           H—C—OH
    |                 |                  |               |
HO—C—H            HO—C—H    Erosion    C=O           HO—C—H
    |          ⇌      |         ⟶        |        +      |
H—C—O(G)n         H—C—O(G)n           H—C—H         H—C—O(G)n-1
    |                 |                  |               |
H—C—OH            H—C—OH             H—C—OH         H—C—OH
    |                 |                  |               |
  CH2OH             CH2OH              CH2OH          CH2OH

 (XXXXIV)          (XXXXV)           (XXXXVI)        (XXXXVII)
```

```
   CHO                              COOH              Further
    |                                |             degradation
   C=O                           C(OH)(CH2OH)
    |                                |
H—C—H                            H—C—H
    |                                |
H—C—O(G)n                        H—C—OH
    |                                |
H—C—OH                            CH2OH
    |
  CH2OH                           (XXXXVIII)

 (XXXXIX)
```

Stabilization

```
  COOH
    |
  CH(OH)
    |
H—C—H
    |
H—C—O(G)n
    |
H—C—OH
    |
  CH2OH

  (L)
```

FIG. 35. Alkaline erosion and stabilization of cellulose. (Bikales and Segal, 1971).

1954) and is not dependent on the D.P. of the latter (Albeck *et al.*, 1975). Peeling occurs in the amorphous regions and will stop when a metasaccharinic acid end is formed or slow down at the point where the chain enters the crystalline regions. It must be mentioned that the presence of 1·6 anhydrides have been suggested as acting as a stop for the peeling reaction although some doubt has been thrown on this point (Meller, 1953; 1968; Franzen and Samuelson, 1957).

More recently, the end groups arising from alkali treated hydrocellulose have been determined (Johansson and Samuelson, 1975). The hydrocellulose was treated with borohydride and hydrolysed. The number of carboxylic acid groups (Table XXI) increases and the hexose moieties decrease with time of treatment in conformity with the proposed reaction scheme. The table shows that glucose is the predominant reducing end

TABLE XXI

Yield and analyses of cellulose samples after borohydride reduction

Alkali treatment, h	Yield, %	Glucitol, mmole/ 100 g	Mannitol, mmole/ 100 g	Xylitol, mmole/ 100 g	Carboxyl groups mmole/ 100 g	Intrinsic viscosity dm³/kg	$(DP)_n$
0	100	3·15	0·11	0·009	0·40	161	189
3	79·3	1·59	0·16	0·000	0·96	160	228
6ᵃ	73·1	1·12	0·14	0·000	1·34	160	237
10	68·9	0·84	0·10	0·000	1·59	159	244

ᵃ The following results were obtained with the sample used for characterisation of the carboxylic acid groups: glucitol, 1·10 moles; mannitol, 0·14 mmole; carboxyl groups (before reduction), 1·36 mmoles.
Johansson and Samuelson (1975).

group although the presence of a small amount of mannitol shows that isomerized moieties are formed. Calculations from the yields of the various products indicate that both the peeling and stopping reactions are proportional to the number of reducing ends. In addition, the D.P. increases as the reaction proceeds, indicating that the degradation is accompanied by loss of short cellulose molecules.

TABLE XXII

Nonvolatile organic acids isolated from the hydrolysate of 100 g of hot-alkali-treated hydrocellulose

Acids	μmole
3-Deoxy-*ribo*-hexonic	205
3-Deoxy-*arabino*-hexonic	163
2-Deoxy-*threo*-pentonic	30
2-Deoxy-*erythro*-pentonic	52
2-C-Methylribonic	8
2-C-Methylarabinonic	5
Gluconic	41
Mannonic	7
Ribonic	3
Arabinonic	67
2-C-Methylglyceric	207
Erythronic	39
Threonic	3

Johansson and Samuelson (1975).

The reaction rates decreased rapidly with time; this decrease is not due to changes in alkali concentration since this has a small effect (Colbran and Davidson, 1961) but rather to a decrease in accessibility (Richtzenhain *et al.*, 1954).

Analysis of the hydrolysate from hydrocellulose is shown in Table XXII (Johansson and Samuelson, 1975). The main components are the 3 dioxy-hexanoic acids (as expected), together with 2-C-methylglyceric acid. It has been suggested that this arises from a rearrangement of the D glucose end which gives rise to terminal moieties having a keto group at the C_3 atom LI (Johansson and Samuelson, 1974).

$$
\begin{array}{llll}
CH_2OH & CH_2OH & CH_3 & COOH \\
\ |\ OH & \ |\ OH & CO & \ |\ CH_3 \\
\ |\diagup & \ |\diagup & \ | & \ |\diagup \\
\ C & \ C & CO & \ C \\
\ |\diagdown & \ |\diagdown & \ | & \ |\diagdown \\
\ |\quad H & \ |\quad H & CH_2OR & \ |\quad OH \\
\ | & \ | & (LIV) & \ | \\
C{=}O & C{=}O & & CH_2OR \\
\ | & \ | & & (LV) \\
H{-}C{-}OR & CH_2OR & & \\
\ | & (LIII) & & \\
H{-}C{-}OH & & & \\
\ | & CH_3 & & \\
CH_2OH & \ | & & \\
(LI) & CH_2OH & & \\
& (LII) & &
\end{array}
$$

This compound may undergo a reverse aldol reaction splitting off glyco-aldehyde LII and a residue LIII which first forms a diketone via a β-alkoxy elimination at C_1. The diketone then undergoes a benzilic acid rearrangement to give 2-C-methylglyceric acid LV. Analogous results are obtained from treatments at 170°C, conditions akin to those of pulping.

The main peeling reaction occurs from the reducing end of the chain. Carbonyl groups in positions other than at the end of the chain contribute less to the yellowing than to the peeling reaction. The products formed from carbonyl groups in the chain are discussed elsewhere (Lewin, 1965) but some of the reactions yield new reducing ends. As an example of this, an aldehyde at the C_6 atom will cleave the chain at the C_4 atom and hence produce a new reducing end capable of undergoing peeling. It therefore follows that the amount of anhydroglucose units eroded per carbonyl groups is less when they are located along the chain.

The kinetics of the reactions have been examined and it has been estimated that the rate of peeling was about 65 times that of the stopping reaction (Franzen and Samuelson, 1957). From measurements of the rates of degradation of hydrocellulose, rate constants have been determined for the propagation and termination reactions as well as an estimate of the termination due to inaccessibility (Hass *et al.*, 1967). The results show clearly that the propagation rate is many times that of termination. Moreover the activation energies of the propagation reaction are substantially lower than that of total termination (24·6 as against 32·2 kcal/mole), showing that the stopping reaction becomes more important as the temperature is raised. This is reflected also in the fact that the average degradable chain length decreases as the temperature is raised.

C. DEGRADATION BY ACID

Cellulose hydrolyses under acid conditions to yield degradation products (the hydrocelluloses); the progressive attack causes the cellulose to lose strength and ultimately its fibrous structure, yielding a friable powder. The first step in the hydrolysis is the rapid formation of an intermediate complex between the glucosidic oxygen atom of the cellulose and a proton

$$-O- + H_3\overset{+}{O} \underset{\text{Fast}}{\rightleftharpoons} \underset{H}{-\overset{+}{O}-} + H_2O \xrightarrow{\text{slow}} \text{hydrocellulose} \qquad (15)$$

The complex undergoes a slow rate determining step which is the scission of the glucosidic bond adjacent to the C_1 carbon atom (Bunton *et al.*, 1955). Hydrolysis, besides breaking the chain, produces both reducing and non-reducing groups at the newly formed chain ends.

The rate of reaction in the initial stages is best followed by the number of glucosidic bonds which are hydrolysed, all of which are assumed to possess equal reactivity. The rate of loss of bonds may be related to the number n, through a first order constant k, i.e.

$$\frac{dn}{dt} = -kn \qquad (16)$$

The number of bonds may be calculated from the changes which occur in the D.P. since the latter, P, is related to n by the equation

$$n = n_0 - \frac{n_0}{P} \qquad (17)$$

where n_0 is the number of hydrolysable bonds initially present.

Equation (17) may be substituted into Eqn. (16) and integrated to give

$$\ln\left(1 - \frac{1}{P_0}\right) - \ln\left(1 - \frac{1}{P_t}\right) = kt \qquad (18)$$

where P_0 and P_t are the values of P at the commencement of the experiment and after time t.

Equation (18) may be approximated to Eqn. (19), i.e.

$$\frac{1}{P_t} - \frac{1}{P_0} = kt \tag{19}$$

Homogeneous systems, e.g. methylated cellulose or cellulose in phosphoric acid, obey Eqns. (18) or (19) (Wolfrom *et al.*, 1939a, b; Immergut *et al.*, 1955). An example of the data for cotton and wood pulp is shown in Fig. 36. It is noticeable that different samples give different results but in particular wood fibres hydrolyse faster than cotton; this observation is considered to arise from the presence of some oxidized groups present in the cellulose, such groups enhancing the susceptibility of the glucosidic oxygen to hydrolysis (Sharples, 1971; Lvetzow and Theandor, 1974). With wood pulp, there is a correlation between the hydrolysis rate and the content of oxidized groups; by the same token, reduction of oxidized groups yields a cotton less susceptible to hydrolysis (Rånby and Marchessault, 1959; Marchessault and Rånby, 1959a, b). Such groups are carboxylic acid or aldehyde groups in the C_6, C_2 or C_3 positions. However there appears to be no evidence to suggest that in the native state, wood or cotton contain anything other than glucose linked in the $(1 \rightarrow 4)\beta$ position (Richter, 1934).

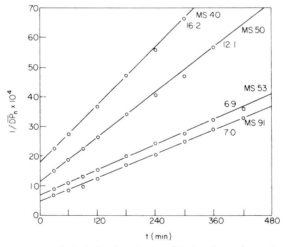

FIG. 36. Homogeneous hydrolysis in phosphoric acid, showing enhanced rate for woodpulp samples (MS40 and MS50) compared with that for cotton samples (MS53 and MS91). (Bikales and Segal, 1971; Immergut *et al.*, 1955).

However, measurement of the rate constant of cellulose of chain length about 100 in dissolving strength sulphuric acid shows it to be consistent with values obtained from cellobiose, cellotriose and cellotetrose, indicating that the reactivity of the majority of bonds may be taken to be identical

(Freundenberg *et al.*, 1930). This does not exclude the possibility that there may be a small number of links which are more readily attacked.

Hydrolysis of solid cellulose by dilute acids is complicated by the heterogeneous nature of the system. Moreover it is the amorphous regions which are most susceptible to attack; hydrolysis is initially rapid but soon slows down (Davidson, 1943). The readily degraded fraction has been equated to the amorphous content (Nickerson and Habrl, 1947), and the method has been used for determination of it (Nelson and Conrad, 1948; Meller, 1953b; Modi *et al.*, 1963; Krejci, 1961; Sharkev and Nenanova, 1963; Kudlacek and Ruzicba, 1964; Kuniak and Alince, 1965).

In this context, the rate of reaction may be found from the rate of loss of weight. The initial weight loss is rapid and followed by slower reaction (about 100 times as slow) which is taken to be hydrolysis of the crystalline residues (Meller, 1953b; Philipp *et al.*, 1947; Daruwalla and Shet, 1962); the rate of weight loss is semi-logarithmic (Fig. 37) (Millett *et al.*, 1954) and hence the amorphous content may be obtained from the extrapolated value at zero time.

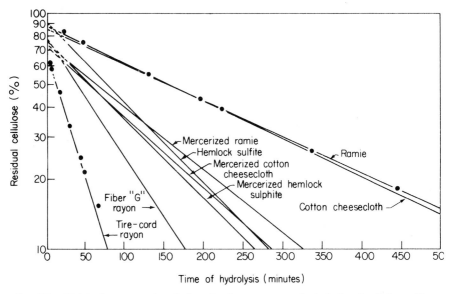

FIG. 37. Weight loss occurring during the heterogeneous hydrolysis of cellulose. The semilogarithmic relation to time is illustrated for a range of natural and regenerated celluloses. (Millet *et al.*, 1954; Bikales and Segal, 1971).

The degradation leads to rod-shaped particles which for cotton are about 400 Å long and 100 Å wide (Immergut and Rånby, 1956); these sizes agree well with examinations of the residues by the electron microscope.

Attack by acid of these residues occurs at the ends of the particles (Sharples, 1957, 1958) rather than at the resistant surfaces of the crystallites.

However the weight-loss method for determining crystallinity yields high values because the increased mobility of the broken chains enables them to crystallize (Sharples, 1975; Hermans and Werdinger, 1949; Lindrst, 1956; Daruwalla and Nabar, 1956; Yu *et al.*, 1966; Sharkov and Levanova, 1960). A preferable method for the determination of crystallinity is the determination of the rate of bond scission; the rate of hydrolysis of the solid cellulose is less than that for completely accessible $(1 \rightarrow 4)$-β-glucosidic linkages by a factor α, the fraction of bonds available for hydrolysis (Sharples, 1954). For this purpose equation (19) may be written as

$$\frac{1}{P_t} - \frac{1}{P_0} = k \propto t \tag{20}$$

k is measured from the hydrolysis of soluble oligosaccharides in which all the bonds are available. The results for the crystallinity are lower than those derived from weight loss. The rate of hydrolysis of samples of cellulose from different sources are clearly different and reflect their different morphology (Sharples, 1957, 1958; Nelson and Tripp, 1953; Nelson, 1960).

D. MODELS FOR THE DEGRADATION

Although it is thought that degradation of cellulose occurs at chain linkages in a more or less random fashion, other possible models must be considered. This is particularly important for cotton where the fibrillar structure may be expected to play an important role. One model for this fibre is that of continuous fibrils, each of which contain 30 cellulose molecules per cross section. The molecules of D.P. 14,000 have their chain ends located in a regular way with a separation of 500 Å. Each chain end causes a local dislocation in the regularity of the arrangement producing regions which are more reactive and susceptible to attack. The model is exemplified in Fig. 38 (Elema, 1973). Attack of a reagent on such a model yields a different molecular weight distribution for the degraded product as

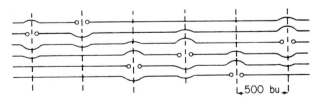

FIG. 38.   Situation of chain ends in the elementary fibril of undegraded cotton. bu—base unit. (Elema, 1973).

compared with that expected from a random attack. Comparison of the ratio of the weight average D.P. to the number average D.P. as a function of the weight average D.P. shows that this model gives better agreement than the assumption of random attack.

However more detailed examination of the hydrolysis products by G.P.C. analysis has shown that the single distribution curve of the cellulose obtained after mild hydrolysis changes to a bimodal one on more severe treatment, suggestive of a two phase structure (Van Lancker, 1974).

This more recent work shows that the understanding of the degradation processes of fibres must take account of the morphology of the substrate and leaves many questions unanswered.

References

Achhamer, B. G., Reinhardt, F. W. and Kline, G. M. (1951a). *J. Appl. Chem.* **1**, 301.

Achhamer, B. G., Reinhardt, F. W. and Kline, G. M. (1951b). *J. Res. Nat. Bur. Stds.* **46**, 391.

Albeck, M., Ben-Bassat, A. and Lewin, M. (1975). *Text. Res. J.* **35**, 836, 935.

Allan, R. J. P., Forman, R. L. and Ritchie, P. D. (1955). *J. Chem. Soc.* 2717.

Allan, R. J. P. Iengar, H. V. R. and Ritchie, P. D. (1957a). *J. Chem. Soc.* 2107.

Allan, R. J. P., Jones, E. and Ritchie, P. D. (1957b). *J. Chem. Soc.* 524.

Andraev, K. K. (1959). *Ind. Chim. Belge Suppl.* **2**, 272.

Andraev, K. K. (1961). *Chemical Abstracts*, **55**, 22824.

Arseneau, D. T. (1971). *Can. J. Chem.* **49**, 632.

Arthur, J. C. and Hinojosa, O. (1966). *Text. Res. J.* **36**, 385.

Back, E. L. and Klinga, L. O. (1963). *Svensk Papperstidn* **66**, 745.

Bailey, J. E. and Clarke, A. J. (1970). *Chem. Brit.* **6**, 484.

Baker, R. R. (1975). *J. Therm. Anal.* **8**, 163.

Ballon, C. E. (1954). *In* "Advances in Carbohydrate Chemistry", Vol. 9, Academic Press, New York and London.

Basch, A. and Lewin, M. (1973). *J. Polym. Sci., Chem. Ed.* **11**, 3071, 3095.

Beevers, R. B. (1968). *Macromolek. Rev.* **3**, 195, 236.

Bell, F. A., Lehrle, R. S. and Robb, J. C. (1971). *Polymer*, **12**, 579.

Berkowitz, W. (1957). *Fuel*, **36**, 355.

Bikales, N. M. and Segal, L. (Eds.) (1971). "High Polymers", Vol. V, Part V. Interscience.

Bircumshaw, L. L., Taylor, F. M. and Whiffen, D. H. (1954). *J. Chem. Soc.* 931.

Birladeanu, C. Vasile, C. and Schneider, I. A. (1976). *Makromol. Chem.* **177**(1), 121.

Bonvicini, A. and Caldo, C. (1969). *Mater. Plast. Elastomeri*, **35**(9), 1243.

Broido, A. (1966). *W.S.C.I.* Paper 66–20, U.S. Dept. of Agriculture.

Browne, F. L. (1958). "Theories of Combustion of Wood and its Control", Forest Products Lab. Rept. No. 2136, Madison, Wisconsin.

Bull, H. B. (1943). "Physical Biochemistry", Wiley, New York.

Bunton, C. A., Lewis, T. A., Llewellyn, D. R. and Vernon, C. A. (1955). *J. Chem. Soc.* 4419.

Burlant, W. J. and Parsons, J. L. (1956). *J. Polym. Sci.* **22**, 249.

Buxbaum, L. H. (1967). *ACS Polymer Preprints*, **8**, 552.

Buxbaum, L. H. (1968). *Angew. Chem. (Int. Ed.)*, **7**, 182.

Byrne, G. A., Gardiner, D. and Holmes, F. H. (1966). *J. Appl. Chem.* **16**, 81.

Cameron, G. G. and Kane, D. R. (1968). *Polymer*, **9**, 461.

Chapman, A. W. (1925). *J. Chem. Soc.* 1992, 2296.

Chapman, A. W. (1927). *J. Chem. Soc.* 1743.

Chapman, A. W. (1930). *J. Chem. Soc.* 2463.

Chatterjee, P. K. and Conrad, C. M. (1966). *Text. Res. J.* **36**, 487.

Chatterjee, P. K. (1968a). *J. Polym Sci.* Part A-1, 3217.

Chatterjee, P. K. (1968b). *J. Appl. Polym. Sci.* **12**, 1859.

Chatterjee, P. K. and Schwenker, R. F. (1972). *In* "Instrumental Analysis of Cotton Cellulose and Modified Cotton Cellulose" (Ed. R. T. O'Connor), Dekker, N.Y.

Cobb, J. W. (1944). *Fuel,* **23**, 121.

Colbran, R. L. and Davidson, G. L. (1961). *J. Text. Inst.* **52**, T73.

Conley, R. T. and Bieron, J. F. (1963). *J. Appl. Polym. Sci.* **7**, 1757.

Corbett, W. M. and Kenner, J. (1955). *J. Chem. Soc.* 1431.

Corbett, W. M. and Richards, G. N. (1957). *Svensk Papperstidn.* **60**, 791.

Danner, B. and Meybeck, J. (1971). Int. Conf. "Carbon Fibres, their Composites and Applications", London.

Daruwalla, E. H. and Nabar, G. M. (1956). *J. Polym. Sci.* **20**, 205.

Daruwalla, E. H. and Shet, R. T. (1962). *Text. Res. J.* **32**, 942.

Davidson, G. F. (1943). *J. Text. Inst.* **34**, T87.

Davidson, H. W. and Losty, H. H. W. (1965). Conf. Ind. Carbon Graphite Papers, London, 20.

Davydov, B. E., Geiderikh, M. A. and Krentsel, A. (1965). *Izvestia, Akad. Nauk SSSR Khim.* **4**, 636.

Drabkin, I. A., Rozenshtein, L. D., Geiderikh, M. A. and Davydov, B. E. (1964). *DAN SSSR* **154**, 197.

East, G. C., Kavan, S. K., Lupton, C. J. and Truter, E. V. (1973). *Proc. Soc. Analyt. Chem.* **10**, 93.

East, G. C., Lupton, C. J. and Truter, E. V. (1975). *Text. Res. J.,* **45**, 863.

Elema, R. J. (1973). *J. Polym. Sci. Symposium*, **42**, 1545.

Entwistle, D., Cole, E. H. and Wooding, N. S. (1949). *Text. Res. J.* **19**, 527.

Ermolenko, I. N., Sviridova, R. N. and Potapovich, A. K. (1967). *Chem. Abstracts* 66-24303.

Farquhar, R. L. W., Pesant, D. and McLaren, B. A. (1956). *Can. Text. J.* **73**, 51.

Fettes, E. M. (1964). "Chemical Reactions of Polymers", High Polymer series XIX, 1964, Interscience.

Findel'shtein, T. A., Rogouin Z. A. and Karkin, V. A. (1950). *Textile Prom.* **9**, 9.

Fitzer, E. and Muller, D. J. (1971). *Die Makromol. Chemie*, **144**, 117.

Franzen, O. and Samuelson, O. (1957). *Svensk Papperstidn*, **60**, 720, 872.

Frazer, A. H. (1968). "High Temperature Resistant Polymers", Interscience.

Freundenberg, K., Kuhn, W., Dürr, N., Bolz, F. and Steinbrunn, G. (1930). *Ber.* **63**, 1510.

Friedlander, H. N., Peebles, L. H., Brandrup, J. and Kirby, J. R. (1968). *Macromolecules*, **1**, 79.

Galin, M. and LeRoy, M. (1976). *Europ. Poly. J.* **12**, 25.

Gardiner, D. (1966). *J. Chem. Soc.* Part C1, 1473.

Gelbert, J. A. G. and Lindsey, A. J. (1957). *Brit. J. Cancer,* **11**, 398.

Giertz, H. W. (1945). *Svensk Papperstidn.* **48**, 317.

Giertz, H. W. and McPherson, J. (1956). *Svensk Papperstidn.* **59**, 93.

Gillham, J. K. and Schwenker, R. F. (1966). *Appl. Polym. Symp.* **2**, 59.

Golova, O. P. and Krylova, R. G. (1957). *Dokl. Akad. Nauk SSSR,* **116**, 419.

Golova, O. P., Krylova, R. G. (1960). *Dokl. Akad. Nauk SSSR,* **135**, 1391.

Golova, O. P., Krylova, R. G. and Nikolaeva, I. I. (1959a). *Vysokomol Soedin,* **1**, 1295.

Golova, O. P., Pakhamov, A. M. and Andrievskaya, E. A. (1957a). *Dokl. Akad. Nauk SSSR,* **112**, 430.

Golova, O. P., Pakhamov, A. M. and Andrievskaya, E. A. (1957b). *Proc. Acad. Sci. USSR,* **112**, 3.

Golova, O. P., Pakhamov, A. M. and Nikolaeva, I. (1957c). *Proc. Acad. Sci. USSR,* **112**, 533.

Golova, O. P., Pakhamov, A. M., Andrievskaya, E. A. and Krylova, R. G. (1959b). *Dokl. Acad. Nauk SSSR,* **115**, 1122.

Goodings, E. P. (1961). "Thermal Degradation of Polymers", S.C.I. Monograph No. 13, 211.

Goodman, I. (1954). *J. Polymer Sci.* **13**, 175.

Goodman, I. (1955). *J. Polymer Sci.* **17**, 587.

Goodman, I. (1960). *Polymer,* **1**, 384.

Grassie, N. (1956). "Chemistry of High Polymer Degradation Processes", Butterworths, London.

Grassie, N. and Bain, D. R. (1970). *J. Polym. Sci.* A1, **8**, 2653.

Grassie, N. and Hay, J. N. (1961). S.C.I. Monograph No. 13, 184.

Grassie, N. and Hay, J. N. (1962). *J. Polym Sci.* **56**, 189.

Grassie, N. and McGuchan, R. (1970). *Europ. Polym. J.* **6**, 1277.

Grassie, N. and McGuchan, R. (1971a). *Europ. Polym J.* **7**, 1091.

Grassie, N. and McGuchan, R. (1971b). *Europ. Polym. J.* **7**, 1357.

Grassie, N. and McGuchan, R. (1971c). *Europ. Polym. J.* **7**, 1503.

Grassie, N. and McGuchan, R. (1972a). *Europ. Polym. J.* **8**, 243.

Grassie, N. and McGuchan, R. (1972b). *Europ. Polym. J.* **8**, 257.

Grassie, N. and McGuchan, R. (1972c). *Europ. Polym. J.* **8**, 865.

Grassie, N. and McGuchan, R. (1973a). *Europ. Polym. J.* **9**, 509.

Grassie, N. and McGuchan, R. (1973b). *Europ. Polym. J.* **9**, 113.

Grassie, N. and McNeil, I. C. (1958a). *J. Polym. Sci.* **27**, 207.

Grassie, N. and McNeil, I. C. (1958b). *J. Polym. Sci.* **33**, 171.

Grassie, N. and Torrance, B. J. D. (1968). *J. Polym. Sci.* A1, **6**, 3303.

Grassie, N., Hay, J. N. and McNeil, I. C. (1958). *J. Polym. Sci.* **31**, 205.

Halpern, Y. and Patai, S. (1969a). *Israel J. Chem.* **7**, 673.

Halpern, Y. and Patai, S. (1969b). *Israel J. Chem.* **7**, 685.

Halpern, Y. and Patai, S. (1969c). *Israel J. Chem.* **7**, 691.

Halpern, Y. and Patai, S. (1970). *Israel J. Chem.* **8**, 655.

Hass, D. W., Hrutfroid, B. F. and Sarkanen, K. V. (1967). *J. Appl. Polym. Sci.* **11**, 587.

Hay, J. N. (1968). *J. Polym. Sci.* A1, **6**, 2127.

Held, F. (1954). *Melliand Textilber*, **35**, 483.

Hendrix, J. E., Anderson, T. K., Clayton, T. J., Olsen, E. S. and Barker, R. H. (1970). *J. Fire and Flammability*, **1**, 107.

Hermans, P. H. and Werdinger, A. (1949). *J. Polym Sci.* **4**, 317.

Higgins, H. G. (1957). *Aust. J. Chem.* **10**, 496.

Higgins, H. G. (1958). *J. Polym. Sci.* **28**, 645.

Hill, R. (1954). *Chem. and Ind.* 1083.

Hinojosa, O., Arthur, J. C. and Mares, T. (1973). *Text. Res. J.*, **43**, 609.

Hoffrichter, S. (1968). *Faserforsch Textiltech.* **19**, 304.

Hofman, W., Ostrowski, T., Urbanski, T. and Witanowski, M. (1960). *Chem. Ind.* 95.

Holmes, F. H. and Shaw, C. J. G. (1961). *J. Appl. Chem.* **11**, 210.

Houtz, R. C. (1950). *Text. Res. J.* **20**, 786.

Hurduc, N. and Schneider, A. (1970). *Bull. Inst. Politechn.* **14**, 357.

Icksam, Noh. (1967). *Kungnip Kongop Yonguso Pogo*, **17**, 1 *Chem. Abstracts* 72 101154.

Idol, J. D., Jnr. (1974). "Acrylonitrile in Prospect and Retrospect", Applied Polymer Symposium No. 25, 1.

Immergut, E. H. and Rånby, B. G. (1956). *Ind. Eng. Chem.* **48**, 1183.

Immergut, E. H., Rånby, B. G. and Mark, H. (1955). *Ricera Sci.* **28**, 308.

Jappe, H. A. and Kaustmen, O. A. (1959). *Tappi*, **42**, 206.

Jellinek, H. H. G. (1961). "Photodegradation and High Temperature Degradation of Polymers", International Symposium on Macromolecular Chemistry.

Jenkins, A. D. (Ed.) (1972). "Polymer Science" (see chapter 22 by Grassie and chapter 23 by Charlesby). North-Holland Publ. Co.

Johansson, M. H. and Samuelson, O. (1974). *Carbohydrate Res.* **34**, 33.

Johansson, M. H. and Samuelson, O. (1975). *J. Appl. Polym. Sci.* **19**, 3007.

Kaesche-Krischer, B. (1965). *Chem. Ing. Techn.* **37**, 944.

Kamerbeck, B., Kroes, G. H. and Grolle, W. (1961). "Thermal Degradation of Polymers", S.C.I. Monograph No. 13, 357.

Kato, K. (1967). *Agr. Biol. Chem. (Tokyo)*, **31**, 657.

Kennedy, J. P. and Fontana, C. M. (1959). *J. Polym. Sci.* **39**, 501.

Kern, L. W. and Fernhow, H. (1944). *Rubber Chem. Technol.* **17**, 356.

Kilzer, F. J. and Broido, A. (1965). *Pyrodynamics*, **2**, 151.

Korshak, V. V., Slonunski, G. L. and Kronganz, E. S. (1958). *Izvest. Akad. Nauk SSSR Otdel. Khim. Nauk*, 221.

Kosik, M. and Reiser, V. (1973). *Zb. Vyzt. Pr. Odbour Pap. Cell.* **18**, 29.

Kovarskaya, B. M., Levantorskaya, I. I., Blyumenfel'd, A. B. and Dralyuk, G. V. (May 1968). *Sov. Plast.* **34**.

Krejci, F. (1961). *Sb. Ved. Praci, Vysoka Skola Chem-Technol, Pardubice*, **2**, 131.

Krzeminski, J. and Kedziora, W. (1975). *Polimery (Warsaw)*, **20**(8), 401.

Kudlacek, L. and Ruzicba, J. (1964). *Chem. Abstracts* **61**, 2043.

Kuniak, L. and Alince, B. (1965). *Faserforsch u. Textiltech.* **16**(3), 153.

Kuryla, W. C. and Papa, A. J. (1973). "Flame Retardancy of Polymeric Materials", Vols. 1 and 2, Dekker.

La Combe, E. M. (1957). *J. Polym. Sci.* **24**, 152.

Lewin, M. (1965). *Text. Res. J.* **35**, 979.

Lindrst, J. (1956). *Svensk Papperstidn.* **59**, 37.

Lipska, A. E. and Parker, W. J. (1966). *J. Appl. Polym. Sci.* **10**, 1439.

Lipska, A. E. and Wodley, F. (1969). *J. Appl. Polym. Sci.* **13**, 851.

Long, D. A. and George, W. O. (1964). *Spectrochim. Acta,* **20**, 1799.

Lvetzow, A. E. and Theander, O. (1974). *Svensk Papperstidn.* **77**, 312.

McCarter, R. J. (1972). *Text. Res. J.* **42**, 709.

McCartney, J. R. (1953). *Nat. Bur. Stds., Circ.* **525**, 123.

Machell, G. and Richards, G. N. (1957). *J. Chem. Soc.* 4500.

Machell, G., Richards, G. N. and Shepton, H. H. (1957). *J. Chem. and Ind.* 467.

Mack, C. H. and Donaldson, D. J. (1967). *Text. Res. J.* **37**, 1063.

McMillan, W. R. (1958). M.A. Sc. Thesis, Toronto.

McNulty, B. J. (1968). *Explos. Res. & Develop Est.* Rep. 1966 ERDE-9/M/66 available CFST1 from *Sci. Tech. Aerosp. Rep.* **6**(11), 1720.

Madorsky, S. L. (1953). *J. Polym. Sci.* **11**, 491.

Madorsky, S. L. (1954). *Poly. Rev.* No. 7, 34.

Madorsky, S. L. (1964). "Thermal Degradation of Organic Polymers", Interscience Publishers, New York.

Madorsky, S. L. and Straus, S. (1958). *J. Res. Nat. Bur. Stds.* **61**, 77.

Madorsky, S. L. and Straus, S. (1959). *J. Res. Nat. Bur. Stds.* **63A**, 261.

Madorsky, S. L., Hart, V. E. and Straus, S. (1956). *J. Res. Nat. Bur. Stds.* **56**, 343.

Madorsky, S. L., Hart, V. E. and Straus, S. (1970). *J. Res. Nat. Bur. Stds.* **1**, 207.

Major, W. D. (1958). *Tappi,* **41**, 530.

Maklakov, A. I. and Pimenov, G. G. (1964). *Dokl. Akad. Nauk SSSR,* **157**(6), 1413.

Marchessault, R. H. and Rånby, B. G. (1959a). *Proc. 2nd Cell. Conf., Syracúse,* 34.

Marchessault, R. H. and Rånby, B. G. (1959b). *Svensk Papperstidn.* **52**, 230.

Marshall, I. and Todd, A. (1953). *Trans. Farad. Soc.* **49**, 67.

Meacock, G. (1952). *J. Appl. Chem.* **4**, 172.

Meller, A. (1951). *Tappi,* **34**, 171.

Meller, A. (1953a). *Appita Proc.* **7**, 263.

Meller, A. (1953b). *J. Polym. Sci.* **10**, 213.

Meller, A. (1953c). *Tappi,* **36**, 366.

Meller, A. (1968). *Holzforschung,* **14**, 78, 129.

Mikhailov, N. V., Tokareva, L. G., Buravchenko, K. K., Terekhova, G. M. and Kirpichinkov, P. A. (1962). *Vysokomol. Soedin,* **4**, 1186.

Mikolazcuski, E., Swallow, J. E. and Webb, M. E. (1964). *J. Appl. Polym. Sci.* **8**, 2067.

Millet, M. A. and Goedkin, V. L. (1965). *Tappi,* **48**, 367.

Millet, M. A., Moore, W. E. and Saeman, J. F. (1954). *Ind. Eng. Chem.* **46**, 585, 1493.

Modi, J. R., Trivedi, S. S. and Mehta, P. C. (1963). *J. Appl. Polym. Sci.* **7**, 15.

Monahan, A. R. (1966). *J. Polym. Sci.* A1 **4**, 2391.

Moore, H. B. (1948). *Text. Res. J.* **18**, 749.

Muller, D. J., Fitzer, E. and Fielder, A. K. (1971). Int. Conf. on "Carbon Fibres, their Composites and Applications", London.

Murphy, E. J. (1962). *J. Polym. Sci.* **58**, 649.

Nagao, H., Uchida, M. and Yamaguchi, T. (1958). *Kogyo Kagaku Zasshi,* **59**, 698.

Nealy, D. L. and Adams, L. J. (1971). *J. Polym. Sci.* **9** (A1), 2063.

Nelson, M. L. (1960). *J. Polym. Sci.* **43**, 351.

Nelson, M. L. and Conrad, C. M. (1948). *Text. Res. J.* **18**, 149.

Nelson, M. L. and Tripp, W. V. (1953). *J. Polym. Sci.* **19**, 577.

Nickerson, R. C. and Habrl, J. A. (1947). *Ind. Eng. Chem.* **39**, 1507.

Okamoto, H. (1973). *Mokazai Gakkaishi*, **19**, 353.

Otani, S. (1959). *Kogyo Kagaku Zasshi*, **61**, 871.

Otani, S., Ogura, A. and Hurudatsu, (1960). *Kogyo Kagaku Zasshi*, **73**, 1448.

Ozawa, T., Kanasashi, M., Sakamoto, R. and Ohama, M. (1969). Preprints for the 18th Conference on Polymers, Japan, 321.

Pakhamov, A. M. (1957). *Izv. Akad. Nauk SSSR, Otd Khim Nauk*, 1497.

Pakhamov, A. M. (1958). *Chemical Abstracts*, **52**, 5811e.

Parks, G. W., Erhardt, J. R. and Roberts, D. R. (1950). *Amer. Dyes. Rep.* **39**, 294.

Parks, G. W., Esteve, R., Collis, M., Guercia, R. and Petrareer, A. (1955). Abstracts of Papers, 127th Meeting A.C.S., Cincinatti, Ohio.

Patai, S. and Halpern, Y. (1970). *Israel J. Chem.* **8**, 605.

Patron, L. and Bastianelli, V. (1974). *Appl. Polym. Symp. No.* 25, 105.

Peebles, L. H. (1975). *Govt. Rep. Announce. Index (U.S.)*, **75**, (11), 55.

Peebles, L. H. and Huffman, M. W. (1971). *J. Polym. Sci.* A1, **9**, 1807.

Peebles, L. H., Brandrup, J., Friedlander, H. N. and Kirby, J. R. (1968). *Macromolecules*, **1**, 79.

Perkins, R. M., Drake, G. L., Jnr. and Reeves, W. A. (1966). *J. Appl. Polym. Sci.* **10**, 1041.

Pernikis, R. Y., Kiselis, O. V., Sergev, V. A., Surna, Y. A. and Stina, U. K. (1967). *Izv. Akad. Nauk Letv SSSR Ser Khim* No. 3, 373.

Peters, R. H. (1967). Textile Chemistry, Elsevier.

Philipp, B., Baudisch, J. and Stöhr, W. (1972). *Cellul. Chem. Technol.* **6**, 379.

Philipp, H. J., Nelson, M. L. and Ziifle, H. M. (1947). *Text. Res. J.* **17**, 585.

Plastics Institute Conference (1973). Degradability of Polymers and Plastics.

Pohl, H. A. (1951). *J. Amer. Chem. Soc.* **73**, 5660.

Polya, J. M. and Spotswood, T. M. (1948). *Rec. Trav. Chim.* **67**, 927.

Ramiah, M. V. (1970). *J. Appl. Polym. Sci.* **14**, 1323.

Rånby, B. G. and Marchessault, R. H. (1959). *J. Polym. Sci.* **36**, 561,

Rapson, W. H. and Hakim, K. H. (1957). *Pulp and Paper Mag. Canada*, **88**, 7, 151.

Rapson, W. H., Anderson, C. B. and King, G. F. (1958). *Tappi*, **41**, 442.

Reardon, T. J. and Barker, R. H. (1974). *J. Appl. Polym. Sci.* **18**, 1903.

Reeves, R. E., Schwartz, W. M. and Giddens, J. E. (1946). *J. Amer. Chem. Soc.* **68**, 1383.

Reich, L. (1968). *Macromolek Rev.* **3**, 49.

Reimschuessel, H. K. and Dege, G. J. (1970). *J. Polym. Sci.* A1 **8**, 3265.

Richter, G. A. (1934). *Ind. Eng. Chem.* **26**, 1154.

Richtzenhain, H. and Abrahamsson, B. (1954). *Svensk Papperstidn.* **57**, 538.

Richtzenhain, H., Lindgren, B. O., Abrahamsson, B. and Holberg, K. (1954). *Svensk Papperstidn.* **57**, 363.

Rieche, A., Schmitz, E. and Schultz, M. (1963). *Z. Chemie*, **3**, 443.

Ritchie, P. D. (1961). Thermal Degradation of Polymers, S.C.I. Monograph No. 13, 106.

426   R. H. PETERS AND R. H. STILL

Runge, J. (1971). *Faserforsch Textiltech.* **22**(1), 1.

Runge, J. and Nelles, W. (1970). *Faserforsch Textiltech.* **21**(3), 105.

Rusznak, I. and Zimmer, K. (1970). *Proc. 18th Hungarian Text. Conf.*, 119.

Rutherford, H. A. (1947). "Flame Proofing of Textiles (Ed. R. W. Little), Reinhold, N.Y.

Samuelson, O. and Wemnerbloom, A. (1954). *Svensk Papperstidn.* **57**, 827.

Schlack, P. (1968). *Chemiefasern,* **12**, 911.

Schurz, J. (1958). *J. Polym. Sci.* **29**, 438.

Schuyten, H. A., Weaver, J. W. and Reid, J. K. (1954). *Adv. Chem. Ser.*, **9**, 7.

Schwenker, R. F. and Beck, L. R. (1963). *J. Polym. Sci.* Part C, 331.

Schwenker, R. F. and Pascu, E. (1957). *Chem. Eng. Data,* **2**, 83.

Schwenker, R. F. and Pacsu, E. (1958). *Ind. Eng. Chem.* **59**, 91.

Schwenker, R. F., Beck, L. R. and Zucearello, R. H. (1964). *Amer. Dyest. Rep.* **53**, 817.

Shafizadeh, F. (1968). "Advances in Carbohydrate Chemistry", Academic Press, N.Y.

Shafizadeh, F. (1975). *In* Proc. Eighth Cellulose Conf., Interscience, **1**, 153.

Sharkov, V. I. and Levanova, V. P. (1960). *Chem. Abstracts* **54**, 10312.

Sharkov, V. I. and Levanova, V. P. (1963). *Chem. Abstracts* **59**, 4157.

Sharples, A. (1954). *J. Polym. Sci.* **13**, 393.

Sharples, A. (1971). *In* "High Polymers", Vol. V (Eds. N. M. Bikales and L. Segal), Interscience .

Sharples, R. (1957). *Trans. Farad. Soc.* **53**, 1003.

Sharples, R. (1958). *Trans. Farad. Soc.* **54**, 913.

Sobue, H. and Kajiura, J. (1969). *Kogyo Kagaku Zasshi,* (*J. Chem. Soc. Japan, Ind. Chem. Sect.*), **62**, 1766.

Spanninger, P. A. (1974). *J. Polym. Sci.* (*Chem. Ed*), **12**(4), 709.

Stamm, A. J. (1956). *Ind. Eng. Chem.* **49**, 1.

Standage, A. E. and Matkowsky, R. D. (1971). *Europ. Polym. J.* **7**, 775.

Still, R. H. and Whitehead, A. (1972). *J. Appl. Polym. Sci.* **16**, 3216 and 3223.

Straus, S. and Wall, L. A. (1958). *J. Res. Nat. Bur. Stds.* **60**, 39.

Straus, S. and Wall, L. A. (1959). *J. Res. Nat. Bur. Stds.* **63A**, 269.

Tang, W. K. and Eickner, H. W. (1968). *U.S. Forest Service* Paper FPL, 82.

Tang, W. K. and Neill, W. K. (1964). *J. Polym. Sci.* Part C, 65.

Tashulatov, Yu. T., Saidaliev, T. and Usmanov, Kl. U. (1966). *Chem. Abstracts* **64**, 19865.

Thompson, E. V. (1966). *Polymer Letters,* **4**, 361.

Tokareva, L. G. and Mikhailov, N. V. (1966). *Mendelev Chem. J.* **11**, 226.

Turner, W. N. and Johnson, F. C. (1969). *J. Appl. Polym. Sci.* **13**(10), 2073.

Twilley, I. C. (1961). "Thermal Degradation of Polymers", S.C.I. Monograph No. 13, 388.

Vachon, R. N., Taylor, H. S., Coe, A. B. and Rebenfeld, L. (1965). *Text. Res. J.* **35**, 473.

Vachon, R. N., Rebenfeld, L. and Taylor, H. S. (1968). *Text. Res. J.* **38**, 716.

Valk, G., Kruessmann, H. and Diehl, P. (1967). *Makromol Chem.* **107**, 158.

Valko, E. I. and Chiklis, C. K. (1965). *J. Appl. Polym. Sci.* **9**, 2855.

Van Lancker, M. (1974). *Ann. Scientific Text. Belge,* **22**, 169.

Venn, H. J. P. (1924). *J. Text. Inst.* **15**, T414.
Virkola, N. E. and Lehtikosi, O. (1958). *Norsk Skogindustrie*, **12**, 87.
Virkola, N. E. and Lehtikosi, O. (1960). *Paper and Timber*, **42**, 589.
Virkola, N. E., Hentola, Y. and Sihtola, H. (1958). *Papperi ja Puri*, **40**, 635.
Vosburgh, W. G. (1960). *Text. Res. J.* **30**, 882.
Waller, R. C., Bass, K. C. and Roseveare, W. E. (1948). *Ind. Eng. Chem.* **40**, 138.
Watt, W. (1970). *Proc. Roy. Soc.* A319, 5.
Weinstein, M. and Broido, A. (1970). *Comb. Science & Technology*, **1**, 207.
Wiesener, E. (1970). *Faserforsch. Textiltech.* **21**(12), 514.
Wiesener, E. (1971). *Chem. Abst.* **74**, 64571.
Wiloth, F. (1973). *Die Makromol. Chemie*, **111**, 283.
Wodley, F. A. (1971). *J. Appl. Polym. Sci.* **15**, 835.
Wolfrom, M. L., Bowden, J. C. and Lassette, E. N. (1939a). *J. Amer. Chem. Soc.* **61**, 1072.
Wolfrom. M. L., Myers; D. R. and Lassette, E. N. (1939b). *J. Amer. Chem. Soc.* **61**, 2172.
Yoda, K., Tsuboi, A., Wada, M. and Yamadera, R. (1970). *J. Appl. Polym. Sci.* **14**, 2357.
Yoshino, T. and Manabe, Y. (1963). *J. Polym. Sci.* A1 **1**, 2135.
Zimmermann, H. (1966). *Textiltech.* **17**, 228.
Zimmermann, H. and Leibnitz, E. (1965). *Faserforsch. Textiltech.* **16**, 282.
Zimmermann, H. and Schaaf, E. (1970). *Proc. Anal. Chem. Conf.* 3rd, **2**, 329.
Zimmerman, H. and Schaaf, E. (1971a). *Chemical Abstracts*, **74**, 3969.
Zimmermann, H. and Schaaf, E. (1971b). *Faserforsch. Textiltech.* **22**(5), 255.
Zimmermann, H. and Schaaf, E. (1971c). *Chemical Abstracts*, **75**, 36923.

11. Molecular Weight and Molecular Weight Distribution in Fibre Forming Polymers and their Consequences to Fibre Properties

BURKART PHILIPP and PETER FRITZSCHE

Institut für Polymerenchemie der Akademie der Wissenschaften der DDR, Teltow–Seehof, GDR

I. Introductory Remarks

Besides the chemical composition the most important criterion of an organic polymer material is its molecular weight (MW) or its degree of polymerization (DP), as these data are relevant to:

(i) the elucidation and control of polymerization and polycondensation processes;

(ii) the processing of polymers, for example to fibres, due to the strong influence of MW on the rheological properties of polymer melts and solutions;

(iii) the product properties, for example the tensile strength of fibres; and it is well known that the DP of a fibre-forming polymer is required to exceed a certain minimum value to permit fibre formation at all (Mark, 1956).

In talking about MW and DP we have to remember the fact that the chain molecules of nearly all linear and branched polymers are not of uniform length, i.e. by no means all polymers contain the same number of monomeric units, but instead show a more or less broad distribution in MW. Usually this distribution is not irregular or confined to a few discrete molecule species, but follows a continuous distribution function. A generalized expression of this function:

$$w_P = W(P) \text{ resp. } dw = W(P)\, dP$$

correlates for each single molecule species the weight fraction w_P with the corresponding degree of polymerization P. Integration according to

$$I(P) = \int_1^P W(P)\, dP$$

gives the so called "integral weight distribution function", from which we derive, by differentiation, the so-called "differential weight distribution function" $W(P)$ as the most conspicuous mode of presenting a distribution function, which is usually "unimodal" (linear polyesters, polyamides, polyacrylonitrile) or may, in some cases, be "bi- or trimodal" (cellulose, polyacrylonitrile). By division of each ordinate value of $W(P)$ by the corresponding P we get the so-called "differential number distribution

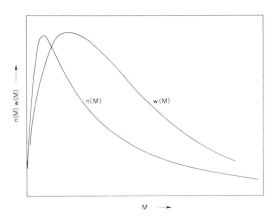

FIG. 1.   Differential number $n(M)$ and weight $w(M)$ distribution function/different abscissae for $n(M)$ and $w(M)$.

function" (see Fig. 1). The existence of these distribution functions implies two important consequences for our problem of correlating MW and fibre properties:

(1) All experimental methods used for MW-determination give average values only. The kind of average is determined by the type of method and, of course, offers no information about the broadness of the MW-distribution if only one kind of average is considered.

(2) Although one well-defined average of MW or DP already provides sufficient information in many cases, especially in process control, this information is sometimes supplemented by an experimental determination of the distribution function, usually of $I(P)$, especially for product characterization. This concerns the cellulosics especially, where no prediction of $W(P)$ can be made from polymerization kinetics. But with synthetic linear polymers also, where $W(P)$ principally can be calculated from an "idealized model" of the chain-building process, an experimental determination is often required to determine deviations from the idealized course of chain-building, or to evaluate the effects of processing and after-treatment on the polymer.

In the following sections we will first deal with the experimental methods for MW and MWD determination, giving a general review and stressing in detail some points not covered in other chapters of this book but considered of some importance to fibre manufacture. Subsequent to this, some examples of application to the different fibre forming polymers are considered.

II. Methods for Determination of Molecular Weight of Polymers

As already explained in the Introduction, molecular weight (MW) and degree of polymerization (DP) will be used in the following to denote an average value for a polydisperse sample. The basis of an experimental determination of such an average value can be any physical or physicochemical effect, X_i, which for each individual molecule species depends in a defined manner on the molecular weight, M_i, of this species and on its weight concentration, c_i, according to $X_i = K \cdot c_i \cdot M_i^\alpha$. Summing up over the whole polydisperse sample gives

$$X_{total} = \sum X_i = \sum K c_i M_i^\alpha \text{ and with } c_i = n_i M_i$$

$$\bar{M}_x = \left[\frac{X_{total}}{K c_i} \right]^{1/\alpha} = \left[\frac{\sum c_i M_i^\alpha}{\sum c_i} \right]^{1/\alpha} = \left[\frac{\sum n_i M_i^{\alpha+1}}{\sum n_i M_i} \right]^{1/\alpha}$$

The average $\bar{M}_x$, of course, increases with an increasing exponent α, and thus depends on the kind of physical effect used for measurement; high values of α imply a strong influence, especially of the longest chains, on the total effect, X_{total}, measured. MW averages thus defined and relevant to practical use are: number average $M_n = \sum n_i M_i / \sum n_i$ with $\alpha = -1$, found for example by osmometry or by end-group titration; viscosity average

$$M_v = \left[\frac{\sum n_i M_i^{\alpha+1}}{\sum n_i M_i} \right]^{1/\alpha} \text{ with } \alpha = 0, 5 \ldots 1, 0,$$

found by viscometry of dilute solutions, and approaching M_w for $\alpha \to 1$; weight average

$$M_w = \frac{\sum n_i M_i^2}{\sum n_i M_i} \text{ with } \alpha = 1,$$

found in rather good approximation by light scattering and by combination of sedimentation measurements and viscometry.

The numerous methods proposed for the determination of MW, of which about ten are in practical use, are generally based on measurements in dilute solutions, within a concentration range of about $0 \cdot 5$ to $5 \text{ g}/l$, and can be classified according to: the kind of average found (see above); the type of effect used for the measurement; the self-consistency of the method, distinguishing between "absolute methods", with X_i depending on M_i and c_i only (osmometry for example), and "relative methods" (such as viscometry), with X_i additionally depending on the shape of the molecule and thus requiring calibration by an absolute method.

In the following, a classification according to the type of the effect measured will be used, thus discerning between chemical end-group methods, methods based on thermodynamic properties of polymer solutions, methods based on transport phenomena and optical methods. A summary of each group will be given considering the theoretical background, experimental procedure including interpretation of data and information available from the method.

A. CHEMICAL END-GROUP METHODS

Assuming an unbranched polymer of known chemical structure and a pre-set weight of the sample, the titrimetric or gravimetric determination of chain end-groups gives directly the number average, M_n, without any extrapolation or further assumption. Unfortunately, the use of this straight-forward method is rather limited due to the following restrictions:

(1) each chain has to carry the same kind and the same number of end-groups;

(2) the end-groups have to be composed of atoms, like S or N, or atom groups, like $-COOH$, $-NH_2$ or $-OH$, which are easily discernible from the "inner links" of the polymer chain and which can be determined analytically with the accuracy and precision required;

(3) M_n should not exceed a value of about 50,000, if conventional analytical methods are to be used. This limit, however, may be shifted to higher M_n by application of special techniques such as, for example, radiochemical methods.

These restrictions are complied with linear polycondensates especially, i.e. by fibre-forming polyesters and polyamides. The NH_2- and COOH-groups of polyamides, for example, can be titrated conductometrically or, by using a suitable indicator, in a solution of the polymer with a binary solvent of phenol and methanol, or with benzylic alcohol as a solvent, using $0·02$ N HCl in methanol and $0·02$ N KOH in benzylic alcohol as a titrant, respectively (van der Want and Kruissink, 1959; Turska and Wolfram, 1958). Hydroxylic end-groups of polyesters can be determined subsequent to a quantitative polymer analogous esterification, for example with 3,5-dinitrobenzoylchloride or o-sulphobenzoic acid anhydride (Zimmermann and Tryonadt, 1967; Zimmermann and Kolbig, 1967).

With cellulose and with polyacrylonitrile, end-group analysis is. of no practical importance for routine determination of M_n, although numerous publications have dealt with quantitative chemical transformation of the aldehyde (hemiacetale) end-groups of cellulose, mostly using the reagent consumption as a measure of the amount of end-groups in the sample. An exception is a method in which HCN is added to the hemiacetale groups, thus transforming them quantitatively to a cyanhydrine suitable for further analysis (Frampton, 1972). With polyacrylonitrile, especially with samples polymerized by a redox system in an aqueous medium, useful information on the steps of initiation, transfer and termination in this polymerization was obtained by conductometry, potentiometry, polarography, and by dye absorption (Philipp et al., 1964; Kiuchi and Nishio, 1964; Shibukawa et al., 1967).

B. METHODS BASED ON THERMODYNAMIC SOLUTION PORPERTIES

All methods cited here give a number average M_n, and, except for "vapour pressure osmometry", are "absolute methods". The physical effect measured (X_{total}) varies with $1/M_n$, thus setting a limit of application in the high MW range, and rendering these methods very sensitive to impurities of low MW as, like all "physical" methods and in contrast to chemical end-group methods, they are unspecific with regard to the chemical structure. Another common feature of the physical methods covered in sections II, B to II, D is that due to the non-ideality of polymer solutions (c.f. Berry and Casassa, 1970) an extrapolation of the effect measured per unit concentration to "infinite dilution" $\lim_{c \to 0} (X_{total}/c)$ is always necessary for the calculation of $\bar{M}_x$, this extrapolation being accomplished either graphically, subsequent to measuring at three different concentrations at least, or per calculation after a "single point measurement" at one specified concentration, if the concentration dependence $X_{total} = f(c)$ is known. Cryometry and ebulliometry, well known from MW determination of low molecular weight compounds, are of very limited importance to polymers due to their insensitivity at higher DP, although by use of thermistors the borderline at about $M_n = 10^4$ was shifted to higher values. Thus, successful M_n-determinations by ebulliometry with polyethylene and polyethylene oxide up to $M_n = 90,000$ were reported some ten years ago (Takeuchi and Amanuma, 1966), but certainly this limit cannot be set for routine analysis.

In spite of being limited to M_n values of about 10^4 too, the so-called 'Vapour Pressure Osmometry" has found widespread use in recent years due to its easy and time-saving procedure and the commercially available automated equipment. Thus this method proved very suitable for analysis of prepolymers and auxiliaries in the textile field. The physical effect underlying this method is the vapour pressure difference between an atmosphere of saturated vapour of the pure solvent and a drop of dilute solution of the sample in this solvent, with the resulting temperature difference, ΔT, being registered as ΔT arises from the heat of condensation of solvent vapour being condensed on the drop of solution. ΔT increases nearly linearly with this vapour pressure difference, and the calculation of M_n proceeds analogous to an ebulliometric measurement according to

$$\frac{1}{M_n} = \lim_{c \to 0} \frac{\Delta T}{c} \cdot \frac{1}{K_D}$$

with K_D being found by calibration with a substance of known MW. As some dependence of K_D on the substance used for calibration has been reported recently (Brzezinski et al., 1973), the proper choice of substance is important for precise measurements. The principle of Membrane Osmometry is treated in detail in Chapter 3 of this volume, including the evaluation of M_n according to the van't Hoff law by the formula

$$\lim_{c \to 0} \frac{\pi}{c} = \frac{RT}{M_n}$$

Though the effect measured (X_{total}) again varies with $1/M_n$, the method is applicable up to $M_n = 500,000$, thus permitting an M_n-determination not only with fibre-forming polycondensates, but also with polyacrylonitrile of the commerical M_n-range, and cellulose derivatives. A lower limit is set, at $M_n \sim 10,000$, by the permeability of the membrane to smaller molecules. While the time-consuming procedure of waiting for osmotic equilibrium (classical "static method") is rendered unnecessary by modern equipment—either achieving a very fast equilibrium by minimizing the volume flux through the membrane, or by using a feed-back control by rate of pressure change ("dynamic method") for a fast, stepwise approach to equilibrium by a servo-mechanism—thus cutting the time for a single measurement down to about 5 to 20 minutes, there still remains the problem of selecting the optimum membrane for the system in question. The pore size of the membrane structure should be rather uniform and large enough to avoid deviations caused by low molecular weight compounds and to ensure a rather high rate of solvent permeation; on the other hand, it should be small enough to avoid permeation of low DP macromolecules (especially of unfractionated samples) as far as possible and, finally, the membrane should not be corroded by the solvent used, a serious problem for example in the osmometry of polyamides. The most widely used membrane materials are regenerated cellulose for organic solvents and cellulose acetate or cellulose nitrate for aqueous solutions; synthetic polymers were not widely used until recently. Though being treated in detail in Section III, C, it should be mentioned in this survey of thermodynamic methods that stepwise addition of non-solvent to a polymer solution, i.e. decreasing the solvent power of the system until the polymer precipitates, can be used with advantage for the estimation of MW, especially of rather sharp fractions. As an experimental procedure the turbidimetric titration is used, in which the amount of titrant (non-solvent) consumed up to the beginning of turbidity increases with decreasing MW (Glöckner, 1966, Molyneux, 1968).

C. METHODS USING TRANSPORT PHENOMENA

All physical effects covered by this category of methods depend not only on MW, but also on the shape of the chain molecules ("hydrodynamic volume", "coil density"), thus either classifying these methods as "relative methods" affording calibration with samples of known MW of the appropriate polymer, or requiring the combination with another independent method of this category to acquire definite information on MW and the shape of the molecules. The type of average found varies widely with the kind of method, or combination of methods, used and, especially in viscometry, with the interaction between polymer and solvent. Usually the larger chains receive a "higher weight" in averaging, thus leading to averages higher than M_n and in some cases coming close to M_w. In contrast to the thermodynamic methods based on static equilibria, the transport phenomena covered here, i.e. free or constrained diffusion of polymer molecules across a boundary between solution and pure solvent; motion of macromolecules relative to the solvent component of the solution in a centrifugal field (sedimentation rate method); and flow of polymer solution by action of a shear gradient (viscometry), not only depend on average MW, but also to some extent on the MW distribution. This "second order effect" is discussed in some recent publications (Fritzsche, 1968) and is already utilized for determining MWD by the sedimentation rate (see Section III, A.).

The correlation between the coefficient of diffusion D and MW according to

$$\lim_{c \to 0} D = D_0 = k \cdot (MW)^{-\alpha}$$

with α varying between 0·5 and 1·0 for linear polymers was sometimes made the basis of a "relative method", but so far is of no practical importance as a method *per se* for fibre-forming polymers. Performance and application of sedimentation measurements with an ultracentrifuge are covered in Chapter 1 of this Volume, but some additional remarks may be useful here. Measurements of the sedimentation rate can be combined either with diffusion measurements, the well established procedure commonly used, or with viscometry (Mandelkern and Flory, 1952; Linow and Philipp, 1968), a procedure which proved suitable for polyacrylonitrile and cellulose derivatives but which is still somewhat open to question with regard to its theoretical foundation and generalization. The MW average calculated according to the following formulae comes rather close to a weight average.

Formulae for determination of MW by sedimentation rate measurements

$$M_{s,D} = \frac{RT_{s_0}}{(1 - v\rho)D_0} \quad \text{(Svedberg)} \tag{I}$$

$$M_{s,[\eta]}^{\frac{2}{3}} = \frac{s_0 \cdot \eta_0 \cdot [\eta]^{\frac{1}{3}} \cdot N_L}{10^{\frac{2}{3}} \cdot \Phi^{\frac{1}{3}} \cdot P^{-1}(1 - \bar{v}\rho)} \tag{II}$$

(Mandelkern and Flory, 1952)

$$M_{s,[\eta]} = 2{,}406 \cdot 10^{25} \left(\frac{s_0 \cdot \eta_0}{1 - v\rho}\right)^{\frac{3}{2}} ([\eta] \cdot k_{SB})^{\frac{1}{2}} \tag{III}$$

(Linow and Philipp, 1968)

where s_0 = constant of sedimentation at "infinite dilution"; D_0 = coefficient of diffusion; η_0 = viscosity of solvent; $[\eta]$ = intrinsic viscosity; N_L = Avogadro's number; $\bar{v}$ = partial specific volume of solute; ρ = density of solvent; k_{SB} = system-specific constant from Schulz–Blaschke formula; R = molar gas constant; T = temperature in °K; Φ, P = constants.

As the sedimentation rate increases with MW there is no upper limit of application of this method, but there is a lower one, at about $M_w = 15{,}000$ to 20,000, due to the increase in the sedimentation time and/or centrifugal force required. Some limitations are set with regard to the polymer–solvent system, too, as the bouyancy term $(1 - \rho v)$ should not be too small. Also the increment dn/dc, decisive for accuracy and precision of the optical measurement of concentration change in sedimentation, should exceed a minimum value.

Viscometry is the most inexpensive, universal and convenient "relative method" for determining MW and is covered in Chapter 2 of this Volume. According to

$$\lim_{c \to 0} \frac{\eta_{\text{solution}} - \eta_{\text{solvent}}}{\eta_{\text{solvent}} \, c} = \lim_{c \to 0} \frac{\eta_{\text{spec}}}{c} = [\eta] = K \cdot M_v^{\alpha} \qquad 0 \cdot 5 \leqslant \alpha \leqslant 1$$

a calibration with samples of known MW is usually evaluated by means of a linear log $[\eta]$-log M_v plot, which should be used for interpolation only and not for extrapolation, especially down to low MW. At low MW the shape of the chain may deviate from that of a statistical coil, thus changing the slope of the plot in this range of low MW. A time-saving procedure in routine viscometry of well-known polymer–solvent systems is the use of "single-point measurements" at some pre-set concentration or specific viscosity with a subsequent calculation of $[\eta]$ by means of a known relation between η_{spec} and the polymer concentration. Of the more than 100 formulae proposed for this purpose, only a few are in practical use, most of these containing one system-specific constant to be determined experimentally for the system in question:

$$\eta_{spec}/c = [\eta](1 + k_{SB} \cdot \eta_{spec}) \quad \text{(Schulz and Blaschke, 1941)}$$

$$\eta_{spec}/c = [\eta](1 + k_H \cdot [\eta] \cdot c) \quad \text{(Huggins, 1942)}$$

$$\ln \eta_{rel}/c = [\eta](1 - k_K \cdot [\eta] \cdot c) \quad \text{(Kraemer, 1938)}$$

$$\ln \eta_{spec}/c = \ln [\eta] + k_M \cdot [\eta] \cdot c \quad \text{(Martin, 1962)}$$

A critical review of the publications relevant to this problem was given by Linow and Philipp (1971b).

D. OPTICAL METHODS

The only method of practical importance in this category is the measurement of light scattering by the polymer coils in a dilute solution, giving after suitable extrapolation to zero concentration and zero scattering angle an M_w-average (for details see chapter 10 of Volume 1). As the physical effect measured, i.e. the scattered light intensity, increases with MW there is no upper limit of application, but the method is very sensitive to contaminating particles of microscopic size (dust particles for example). The lower limit of the MW-range covered by this method is set at about $M_w = 5000$.

III. Determination of Molecular Weight Distribution (MWD)

Although some information on the non-uniformity of molecular weight of a polymer sample is already available from a comparison of M_w and M_n ["non-uniformity parameter" $U = (M_w/M_n) - 1$ (G. V. Schulz)]—an increasing M_w/M_n indicating a more pronounced non-uniformity—there are still many problems of research and process-control as well as product control, where the whole distribution function $w_p = W(P)$ needs to be known. For an experimental determination of this function three types of method are now available for practical use:

(1) evaluation of "second order effects" of MW-determination by physical methods, especially by sedimentation;
(2) evaluation of the coil-size controlled diffusion of dissolved macromolecules in a porous gel (gel permeation chromatography);

(3) phase separation by fractional precipitation or dissolution, using the MW-dependent free energy of phase transition of a single molecule as a basis of fractionation on an analytical or preparative scale.

The following discussion will be centred on methods of phase separation, including the problem of superposition in fractionation with regard to MW as well as to chemical composition in dealing with copolymers, for example acrylics or partially substituted cellulose derivatives. The first two types will be surveyed rather briefly, referring to relevant publications for further details.

A. CALCULATION OF MWD FROM SEDIMENTATION DATA

From the broadening of the concentration gradient curve in sedimentation rate measurements as well as from the concentration profile in sedimentation equilibrium, information is available with regard to MWD, as shown in a review by Benoit and Jacob (1967) which also covers some more recent methods involving the presence of a second non-macromolecular solute. Application of these methods is still rare, but is of increasing interest, partly due to the use of computers simplifying the evaluation of experimental data. Good results were obtained by the sedimentation rate method by Pavlova *et al.* (1971), with fibre-forming polycondensates, and by Linow (1978), with copolymers of acrylonitrile.

B. DETERMINATION OF MWD BY GEL PERMEATION CHROMOTOGRAPHY ("GPC")

GPC is a time-saving procedure for obtaining information about the MWD and averages of MW of polymers. The cost of commercially available equipment is still rather high, but nevertheless the method has found wide application in recent years as a routine method, especially with the "big tonnage polymers" such as polyethylene, where large series of similar samples have to be analysed. Quite satisfactory results are reported also with acrylics and with cellulose derivatives (see Section IV in this context). A very interesting field of application, now rapidly progressing is the quantitative separation of oligomers up to a DP of about 20.

The procedure can be summarized as follows: A solution of the polymer to be fractionated (concentration about 0·5%) is put on top of a column filled with a cross-linked macroporous gel, and eluted through this coloumn by the same solvent. Hereby the shortest time of retention is attached to the largest chains, as these coils, due to their large volume, cannot or can only to a small extent penetrate into the pores of the gel, while smaller molecules are retained for a longer time within this pore structure. Materials widely used for these gels are cross-linked dextrane, polystyrene, poly-acrylonitrile, polyvinylacetate, polyglycolmethacrylate as well as silica gel and porous glass beads. Thus, via the coil size of the macromolecules there exists, at constant elution rate within a limited range, a definite relation between the MW of a type of macromolecule and the "elution volume" v_E, i.e. the volume of eluate washed through the column until the molecule species in question appears in the eluate. With different fractions of the eluate, a separation of the macromolecules according to a molecular volume is achieved, regardless of whether the initial sample is a homo-polymer or a copolymer. Thus, with copolymers the separation is barely superposed by a separation according to DS, in contrast to the phase separation methods treated in Section III, C. There have been many attempts to obtain a universal calibration curve for quite a large number of polymers. Rather widely used is a correlation between $\log[\eta] \cdot M$ and v_E, as proposed by Benoit et al. (1966).

The different fractions of the eluate are usually analysed for weight of polymer per volume. With these weight fractions and the corresponding MW obtained from the calibration curve the MWD can be obtained. MWD curves constructed in this manner are valid in the case of polymers having a rather broad MWD. With samples of narrow distribution a correction has to be applied which takes into account a longitudinal diffusion process.

The analytical processing of the eluate can be automated by using appropriate detectors. The choice of this detector depends on the problem to be solved. Useful equipment for registering the polymer concentration in the eluate is an UV-detector or a refractometric detector of high sensitivity. Additionally, there is a trend for on-line determination of further polymer parameters by applying a second or a third detector. Cantow et al. (1966) adapted to a column an automatic registration of polymer concentration in the eluate by differential refractometry and, simultaneously, a continuous MW-determination by the light-scattering method. Ouano et al. (1974) proposed an instrumental design with computer interface for a GPC molecular weight detector system relying on viscometry and refractometry.

Of course, the outline given here on GPC is somewhat idealized and oversimplified to make clear the principle of the method. In practical use, several additional factors need to be considered, as for example additional effects of diffusion and convection influencing the separation process. Furthermore, as shown by Heitz et al. (1967), the separation is not completely controlled by pore size and the pore size distribution of the gel, but also depends to some extent on chemical interaction between gel structure and polymer coils. On the basis of this effect, the so-called "affinity chromatography" was developed especially for some types of biopolymers.

In practical analytical work, the relationship between log MW and v_E is usually still determined empirically by calibration with rather narrow fractions of known MW, in spite of the many interesting and successful studies already mentioned on general correlations between v_E and physical parameters of the appropriate molecular species. Though rather self-evident, it is to be mentioned finally that the average pore size and the pore size distribution of the gel must be adapted to each special problem of separation; if a sample of rather wide MWD is to be fractionated, several columns with gels of different pore size are to be employed, each of these giving a useful log MW–v_E relationship within a definitive range of MW of about two to three powers of ten. Reviews are given by Heitz (1970) and Johnson and Porter (1970).

C. DETERMINATION OF MWD BY PHASE SEPARATION

The methods for determining MWD by phase separation include: fractional precipitation; fractional solution; column fractionation by solvent–gradient elution with or without an additional temperature gradient; turbidimetric titration.

This section will deal with the use of fractional precipitation in characterizing fibre-forming polymers. The determination of the MWD of a polymer by fractional precipitation is a time-consuming process. Nevertheless, fractional precipitation has become an important method in recent years because, besides the theories of polymer solutions, no other theories or premises are needed for the theoretical explanation of experimental data. Also, vice versa phase separation experiments are a tool for proving and developing the theories of polymer solutions, and no expensive apparatus is needed for the experimental work.

1. Theories of Phase Separation

The basis of modern developments of phase separation theories is mainly the Flory–Huggins theory of polymer solution (see Flory, 1953). A good review had been given by Huggins and Okamoto (1967). In a solution of a polydisperse polymer in one solvent the excess chemical potential of the solvent is:

$$\mu_1 - \mu_1^0 = RT[\ln(1 - \varphi) + (1 - P_n^{-1})\varphi + \chi_1\varphi^2]$$

and of the polymer of a polymerization degree P is:

$$\mu_P - \mu_P^0 = RT[\ln \varphi_P - (P - 1) + \varphi P(1 - P_n^{-1}) + \chi_1 P(1 - \varphi)^2]$$

where φ is the volume fraction and χ_1 the Flory–Huggins interaction parameter. Setting equal the chemical potentials of the polymer P in both phases, we obtain the quotient of the volume fractions of the polymer with the degree of polymerization P in the two phases

$$\frac{\varphi_{P,I}}{\varphi_{P,II}} = \exp -\sigma_s^0 P$$

where

$$-\sigma_s^0 = (1 - \bar{P}_{n,II}^{-1})\varphi_{II} - (1 - \bar{P}_{n,I}^{-1})\varphi_I + \chi_1[(1 - \varphi_{II})^2 - (1 - \varphi_I)^2]$$

In the case of a polydisperse polymer in a mixture of two solvents the expression of σ will be more complicated.

The quantitative agreement between theory and experimental data is not satisfactory. The behavior of phase separation is described by this theory only qualitatively and thus there are efforts in the literature to fit the theory and the experimental data somewhat better. For instance, the interaction parameter was set as a function of temperature and polymer concentration (Koningsveld, 1967)

$$g = \sum_{k=0}^{n} g_k^k; k = 0, 1, 2, \ldots, n$$

n – number needed for adequate description

$$g_k = g_{k_1} + g_{k_2}/T + g_{k_3}T + g_{k_4} \ln T$$

Another possibility for overcoming the noncorrespondence of theories and experimental results is to take into account that the polymer–solvent or the polymer–solvent–nonsolvent systems are not binary or tertiary systems but "quasibinary" and "quasitertiary" systems (Koningsveld, 1969) which means that for quantitative calculations of phase diagrams the MWD of the polymer has to be considered. On the other hand in contemporary investigations, the shape of the phase diagram is considered to be the sign of highest sensitivity for the broadness of MWD. Recently, the Flory–Huggins theory has been extended further to cover in a general way copolymeric and branched systems (Kennedy et al., 1972). Another method of theoretical treatment uses the principle of "corresponding states" (see for example Flory, 1970). these theories aim at modelling the general behavior of a given polymer–solvent system over a wide range of experimental conditions including the range from an upper to a lower critical solution temperature.

Based on these theories of polymer solution it is possible to predict by calculation the effect of fractionation. Many years ago Schulz (1940) calculated a phase separating fraction of a polymer on the basis of his theory of phase separation. In the last few years Koningsveld (1969) has calculated the influence of the MWD, the initial polymer concentration and the size of the fraction on the MWD of the polymer fraction in the polymer-lean and polymer-rich phase. Kamide *et al.* (1973a, b) published some contributions about systematic "mathematical fractionation". They calculated the dependence of MW and MWD (Schulz–Zimm distribution and five Wesslau distribution), as well as the operation conditions, such as initial polymer concentration, fraction size, and the polymer–solvent parameter, including its concentration dependence on the partition coefficient, the volume ratio of the polymer-devoid to the polymer-rich phase, and the MWD of the fraction.

2. Some Rules for the Experimental Procedure

From the mathematical fractionations, the authors mentioned above derived rules for the selection of optimum operative conditions. Utilizing these rules, we have to take into account the fact that these calculations often start from theories of polymer solutions which do not exactly describe the real properties of the special system in question, or the properties of a special case. Actual fractionations are mostly rather far from the idealized behavior. Therefore the rules and published diagrams of the correlation are useful for beginning experimental work in this field and for solving more simple problems of separation, like the fractionation of polydisperse polymer in a single solvent with a constant polymer–solvent interaction parameter. For separation problems of a polymer in a solvent–nonsolvent system, Koningsveld (1969) gave some first calculations. But this problem is still open for discussion.

Some important practical rules of fractionation can be summarized as follows:

(1) Decreasing the initial polymer concentration: the MWD of the polymer in the polymer-rich and polymer-lean phase decreases.

(2) Increasing the number of fractions (decreasing the fraction size in the fractional precipitation): the MWD of the polymer in the polymer-rich phase passes a minimum; the MWD of the polymer in the polymer-lean phase increases.

(3) The MWD of the polymer in the polymer-rich phase is broader than in the polymer-lean phase, regardless of the type and broadness of the MWD of the original polymer, and the order of fractionation, provided the fraction size is small.

(4) The initial concentration dependence of the MWD in the fraction is higher for the fractions having smaller fraction size and for the fractions separated from a whole polymer of wider MWD.

(5) Klein (1970) showed the possibility to estimate also the influence of the solvent–nonsolvent system upon the MWD of the fraction.

(6) For a successful experimental fractional precipitation the following conditions are usually adopted: the number of fractions should be about 10 and the initial concentration is adjusted to less than 1% (Kotera, 1967; but see also Reinisch et al., 1969). This rule is not valid for multimodal distributions, where a larger number of fractions is necessary, of course

(7) To achieve the thermodynamical equilibrium between the phases and thus to arrive at a good reproducibility the polymer has to be precipitated as a liquid phase by selecting a suitable solvent–nonsolvent system or by changing the temperature.

3. Calculation of MWD from Fractionation Data

The fractions obtained by a fractionation experiment again always have a more or less broad MWD. Therefore the MWD of neighbouring fractions overlap each other. Schulz (1940) assumed that half of the weight of the fraction consists of molecules having MW's below the average MW determined for that fraction, and the other half is above the average MW. He calculated a cumulative weight fraction $C(M_i)$, for the ith fraction, according to

$$C(M_i) = \frac{1}{2}w_i + \sum_{j=1}^{i-1} w_j$$

where $w_k (k = i, j)$ is the weight fraction of the kth fraction. The integral distribution curve is obtained by passing a smooth curve through the points on this plot of $C(M_i)$ versus M_i. The differential distribution curve is obtained by graphical differentiation or by calculation (see for example Teichgräber, 1968). Schulz recommended the calculation of the contribution of the highest fraction by

$$C(M\lambda) = \sum_{j=1}^{\lambda-1} w_j$$

because there is no overlapping on one side.

A more recent analytical treatment of fractionation data is given by Kamide and co-workers (1968). Approximating the weight distribution of the fraction by an equilateral triangle shape distribution having its peak at M_w of the fraction with its weight being identical to that of the fraction, the differential distribution curve was calculated directly from the fractionation data.

Both methods are accurate for polymers having rather wide MWD. With rather monodisperse polymers ($M_w/M_n < 1\cdot10$) the MWD so evaluated from fractionation data deviates significantly from the true MWD of the original sample. The method of Kamide *et al.* is more advantageous if the fractions are separated from less dilute solution. The Schulz method becomes more preferable as the initial polymer concentration is lowered.

Another method of analytical treatment of fractionation data is to fit the experimental points with two-parameter mathematical distribution functions. The most important functions are: the "most probable" distribution function according to Flory

$$w_p = Pp^{P-1}(1 - p)^2$$

the Schulz–Zimm distribution function:

$$I(P) = \frac{k}{k! \, P_n} \int_0^P \left(\frac{kP}{P_n}\right)^k \exp\left[-\frac{kP}{P_n}\right] dP$$

the Tung distribution function:

$$I(P) = 1 - \exp(-aP^b]$$

the Wesslau (logarithmic normal) distribution function:

$$I(P) = \frac{1}{\beta\sqrt{\pi}}\frac{1}{P}\exp\left[-\left(\frac{\ln P/P_0}{\beta}\right)^2\right].$$

A review of these two parameter functions suitable for practical use was given by Berger (1969). Kubin (1969) recommended a three-parameter function permitting a still better adaptation to experimental data. After estimating the parameters by drawing the experimental points in suitable coordinates it is easy to arrive at the mean values of the distribution and the integral or differential distribution curve. In Fig. 2 some distribution functions were plotted possessing all the same P_w- and P_n-values.

The plot in Fig. 3 shows the Tung treatment of a fractional precipitation of a polyacrylonitrile sample with 9 and 19 fractions (Fritzsche, 1971b). Both fractionations yielded the same linear Tung plot. That means, with

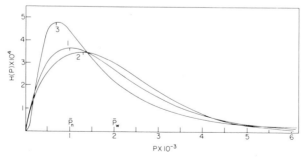

FIG. 2. Polymerization degree distribution functions according to Schulz (1), Tung (2), and Wesslau (3).

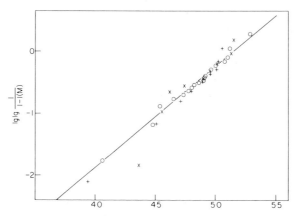

FIG. 3. Tung-treatment of a fractional precipitation of a polyacrylonitrile sample. ×, + —9 fractions; 0—19 fractions (Fritzsche, 1971b).

the premise valid that the MWD of the polymer type can be described by a pre-set mathematical function, only a few fractions are needed to estimate the whole MWD. Another conclusion can be drawn from Fig. 3 and Table I.

TABLE I

Fractional precipitation of polyacrylonitrile

Number of fractions	M_w	M_n	M_w/M_n
9	68,400	49,800	1·38
9	71,200	51,500	1·38
19	73,000	44,600	1·64
Tung function	72,800	36,400	2·00

Increasing the number of fractions and evaluating the data without use of a pre-set function, M_w remains nearly constant, but M_n decreases and the polydispersity increases. Thus, it is difficult to discuss polydispersities calculated from a few experimental points of a fractionation only, and it seems to be more correct to fit these data into a pre-set function. Finally, there is the problem, already touched upon, of how to be sure that the MWD obtained experimentally is indeed the real MWD of the polymer sample. There are two ways of answering this question, at least partially:

(1) The fractionation method can be tested by fractionation of a polymer possessing a theoretically predictable MWD, using for example a polycondensate with a "most probable" distribution according to Flory.

(2) Postulating all deviations from idealized behavior and all difficulties and shortcomings of the experimental procedure to result in a narrower "apparent MWD", different procedures inclusive of different solvent–nonsolvent systems can be compared and the broadest MWD obtained in this manner may then be assumed to be the best approximation of the real MWD. This time-consuming procedure proved to be suitable if nothing is known about the MWD-function of the polymer in question.

4. Further Methods derived from Phase Separation

The fractional precipitation is the theoretically best known method for characterizing the MWD of polymers. Its disadvantage is the time-consuming procedure. Numerous attempts were made to improve the fractional precipitation and to develop less time-consuming methods on the same theoretical basis.

(a) *Fractional Solution.* The purpose of fractional solution is to extract a thin polymer film by successive mixtures of solvent and nonsolvent, with increasing fractions of solvent. The unsatisfactory accessibility of the polymer molecules in the solid phase for dissolving is the principal shortcoming of the method. The thinner the film, the more quickly the solution equilibrium will be attained and the extraction process will be finished. The difficulty of this procedure is to fix the thin polymer films on the support, for example aluminium foil. Besides many polymers and copolymers, Fuchs (1956) fractionated acrylic polymers by this method also. Berger *et al.* (1966) used a woven material (cotton kalmuk) as a support and obtained a quick method with sufficient reproducibility.

(b) *Column Extraction.* When the polymer is precipitated on a spherical support (for example glass beads) and then the extraction process is performed in a column filled with non-loaded glass beads by a nonsolvent–solvent gradient, very sharp fractions can be obtained. In some cases a temperature gradient with an opposite effect increases the efficiency of the procedure (Baker–Williams method). For determining the MWD of fibre-forming polymers the method was surpassed by GPC. A review was given by Porter and Johnson (1967).

(c) *Turbidimetric Titration.* In turbidimetric titration the amount of polymer precipitated with increasing content of nonsolvent is measured by evaluating the turbidity of the system. The intensity of scattered light is a function of number and size of the particles precipitated. The shape and size of the particles are influenced by experimental conditions very strongly.

Thus, attaining good reproducibility of the method is troublesome work. Nevertheless, the method is often used to characterize the MWD of polymer samples because it is a very quick procedure, but it is almost completely unsuccessful in evaluating true MWD's. A review was given by Giesekus (1967).

5. Some Remarks about Copolymer Fractionation

Besides molecular weight distribution, copolymers have other distributions also. Firstly, the chemical composition is not the same in every macromolecule. The distribution of chemical composition caused by statistical fluctuations is very narrow, but those caused by concentration differences of monomers in the polymerization process are often rather broad and need to be taken into account. Other distributions are the distribution of sequence length, which was now often studied, and the distributions of branching or cross-linking degree, in special cases limited not only to copolymers. Here only the distribution of chemical composition may be considered. In this case the frequency distribution of polymer molecules is a function of the MW *and* the chemical composition. That means that copolymers possess a two-dimensional distribution function. Such a distribution area cannot be estimated by a single fractionation alone. A network of points must be determined in order to construct this distribution.

The theoretical basis of fractionation of copolymers was developed by Topčiev *et al.* (1962) using the Flory–Huggins theory of polymer solutions. Taking the symbols of Section III, C, 1, the quotient of volume fractions of a polymer of the polymerization degree P and chemical composition α is:

$$\ln \frac{\varphi_{II,P,\alpha}}{\varphi_{I,P,\alpha}} = (\sigma + K\alpha)P$$

where

$$K = (\chi_A - \chi_B)(\varphi_{II} - \varphi_I); \qquad \chi_A > \chi_B$$

and

$$\sigma = \varphi_I\left(1 - \frac{1}{P_{n,I}}\right) - \varphi_{II}\left(1 - \frac{1}{P_{n,II}}\right) - \varphi_I(1 - \varphi_I)\bar{\chi}_{n,I}$$
$$+ \varphi_{II}(1 - \varphi_{II})\bar{\chi}_{n,II} + (\varphi_{II} - \varphi_I)\chi_B$$

χ_A and χ_B are the interaction parameters of the homopolymers and $\bar{\chi}_n$ is the number average of the interaction parameters as a function of MW and chemical composition. The cross fractionation recommended by the authors with the conditions $\chi_A > \chi_B$ and $\chi_A < \chi_B$, respectively, is not always possible. If $\chi_A = \chi_B$, the fractionation takes place with respect to MW only. Therefore $\chi_A - \chi_B$ has to be as large as possible in order to get a fractionation with respect to the chemical composition, too. A theoretical expression for fractionating a polymer with respect to the chemical composition only, has not been known until now. A few contributions covered this fractionation path with cellulose derivatives (see Section IV, A) but other fibre-forming polymers have not been studied. However, Glöckner et al. (1971) found two solvent–nonsolvent systems suitable for fractionation of acrylonitrile–styrene copolymers. Methylenechloride-hexane fractionates this copolymer with respect to the chemical composition only and methylenechloride–methanol fractionates with respect to MW and chemical composition, but within a limited range the fractionation does not depend upon chemical composition. Here the fractionation takes place with respect to MW only.

Another possibility for obtaining data about the two-dimensional distribution of copolymers is to combine a fractional precipitation with respect to chemical composition and MW with gel permeation chromatography. In a limited range of composition GPC fractionates copolymers with respect to MW only.

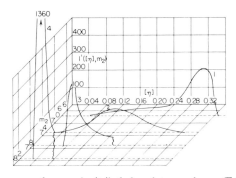

FIG. 4.   Distribution area of an acrylonitrilethylacrylate copolymer (Fritzsche et al., 1971).

Thus the distribution area is cut into two directions by the different methods. But to determine the whole of the distribution area it is not sufficient to carry out two sections only. Fritzsche et al. (1971) prepared some fractions of an acrylonitrile–ethylacrylate copolymer by fractional precipitation with DMF-decaline. The value $\chi_A - \chi_B$ should be as high as possible, so that the fractionation takes place with respect to MW and chemical composition. From each fraction possessing different chemical compositions the MWD was determined by GPC. Thus some MWD curves had been received from one copolymer sample. These MWD curves can be used to construct a curved area representing the distribution area of the copolymer in a three dimensional space (Fig. 4).

IV. Application of MW and MWD Measurements to Fibreforming Polymers and Some Correlations to Fibre Properties

In surveying the application of methods treated in Sections II and III, to special polymers relevant to fibre manufacture, four topics will be considered for each class of polymers: (1) necessary adaptation of analytical methods including especially the choice of solvent; (2) application of MW- and MWD-determination to polymerization and degradation processes; (3) correlation of MW and MWD to product properties and product quality; (4) special applications to copolymer problems.

A. CELLULOSE AND CELLULOSE DERIVATIVES

The most important methodological problem still offering some open questions is how to get cellulose into a rather molecular dispersed solution without degradation and without remaining gels or fibre residues (Gruber and Schurz, 1973), a special problem arising here from cross-linked samples (Agster, 1963). The classical way to achieve this dissolution is either: (a) by esterification to a triacetate or, preferably a trinitrate, these derivatives being stable and easily soluble in common, inexpensive and non-aggressive solvents to solutions of rather well-known physico-chemical properties and suitable for the application of MW- and MWD-methods covered here, or (b), especially for viscometric measurements, a reaction combined with dissolution of cellulose in an excess of some transition metal complexes ("Cuam", "Cuene") in aqueous solution. Recent work along these lines covers: (1) the broadening of the theoretical basis, for example by correlating viscometric data with solvent properties (Howard and Parikh, 1968) and coil dimensions (Schulz and Penzel, 1968); (2) optimization and rationalization of laboratory methods for obtaining MW and MWD of cellulose including the systematic comparison of different procedures (for example: Poller, 1969; Philipp and Linow, 1965, 1968); (3) the adaptation of GPC to the processing of cellulose trinitrate solutions for obtaining MW and MWD (Meyerhoff, 1970; Wadsworth et al. 1971; Segal, 1968); (4) the problem of chain degradation prior to measurement, especially in preparing the nitrate. The latter problem was tackled by modification of the experimental procedure for nitration, for example by use of nitric acid dissolved in acetic acid (Harland, 1954) and fractionation (for example Čöcieva et al., 1962), as well as by more fundamental discussions on the so-called "chain length difference" between nitrate- and Cuoxam-DP (Philipp and Linow, 1966, 1970). This problem of chain degradation obviously stimulated the search for new ways of dissolving cellulose, avoiding this degradation as far as possible. Along the route via stable derivatives soluble in common organic solvents, the reaction of cellulose with phenylisocyanate in pyridine to a tricarbanilate of cellulose conforms to this requirement rather well (Shaubhag, 1968), although the preparative procedure is still somewhat tedious and in using these tricarbanilated samples for fractionation somewhat long dissolving times have to be accepted to avoid gel residues (Linow and Philipp, 1970). A very useful method for nearly non-degradative cellulose dissolution was opened up by Jayme (1961) with the discovery of several new aqueous cellulose solvents based on metal complexes, especially the Cadmium–ethylenediamine complex ("Cadoxen") and the ferric–tartaric–acid complex in aqueous sodium hydroxide ("EWNN", "FeTNa"). These discoveries stimulated much research work on application of these new solvents in physical chemistry and analytics of cellulose. Edelmann and Horn (1958) studied the optimal preparation of these solvents, and Samsonova et al (1970) published work on the physico-chemical properties of the solutions and the shape of the cellulose molecule in these solvents. Measurement of viscosity in these solvents was a routine method for determining the DP of cellulosic fibres and textiles (Achwal and Vadza, 1970). MWD's were determined by phase-separating fractionation from FeTNa-solutions (Jayme and Kleppe, 1961), and turbidimetric studies have been performed on cellulose solutions in cadoxene (Geczy and Revai, 1971). Some problems arose from the high alkalinity of these solvents in making MW-determination by certain physical methods. In osmometry, with cadoxene solutions of cellulose, the membrane problem obviously could be solved by using a cellulose membrane cross-linked by formaldehyde (Usmanov et al., 1970); the use of a cadoxene solution for MWD-determination of hydroxyethyl cellulose by a sedimentation method was reported some years ago (Golubev et al., 1969). Fractionation by successive extraction or dissolution is prone to a superposition of separation according to MW and to accessibility. With cellulose, methods of this kind can be modified to give a preferential separation with regard to accessibility,   e.g. an emulsion xanthation with increasing concentration of NaOH (Philipp, 1962), or a fractional extraction of regenerated cellulose fibres by a series of FeTNa-solvents (Paulusma and Vermaas, 1963).

Of course, it is not possible to give a complete review of the literature published on applications of MW and MWD methods to chain-forming and chain-degradation processes. Only a few points can be stressed here and elucidated by examples. The question of MW and MWD of native cellulose fibres and especially that of "weak links" in the chains of these fibres still offers some problems (Schulz, 1974), although due to the development of rather nondegradative analytical methods the continuous enhancement of DP values reported for native cellulose has come to a stop at $DP_w \sim$ 10,000–15,000; considerable progress has been made in the understanding of the biochemical chain-forming process, especially by Marx-Figini

(1969). GPC and other fractionation methods are now being applied in combination with a controlled degradation for an experimental proof which one of the two models of cellulose structure now in discussion— folded chain model or fringe micelle model with extended chains (Muggli *et al.*, 1969; Muggli, 1968)—comes closer to reality. The numerous publications using MW and MWD determinations to elucidate degradation mechanism of cellulose by heat (Philipp *et al.*, 1969), by enzymes (Rinaudo *et al.*, 1969), and by acid hydrolysis can only be briefly mentioned here, emphasizing the importance of hydrolytic degradation kinetics and of the LODP-concept (Battista *et al.*, 1956) for gaining a deeper insight into cellulose structure. The monitoring role played by the determination of MWD in the technology of cellulose processing is self-evident, i.e. to avoid undue damage, but to carry the degradation sufficiently far to facilitate such processes as pulping and viscose manufacture. As shown theoretically by Brestkin and Frenkel (1969, 1970), in heterogeneous cellulose reactions a few occasions arise where chain rupture does not have a considerable effect on MW and MWD, e.g. a quasiperiodic distribution of unordered regions, and these are barely detected by these methods. Much data has been gathered in the last two decades on MWD of dissolving pulps for the viscose process, emphasizing the difference between sulphite and prehydrolysis sulphate pulps—the latter having a narrower distribution—and discussing a possible detrimental effect of high-molecular fractions on viscose filterability. GPC of cellulose trinitrate dissolved in tetrahydrofurane has been of great help in these systematic investigations in recent years (Huang and Jenkins, 1969), and has been found suitable, too, for following the degradation reactions in the viscose process (Phifer and Dyer, 1971), where research activities, of course, were centred upon kinetics and mechanism of the oxidative degradation during alkali cellulose ripening.

Concerning the use of MW and MWD determination in connection with evaluation of product quality of chemical fibres, three topics are to be considered with regard to cellulose itself: (1) the requirements of pulp for production of high quality regenerated cellulose fibres; (2) the correlation between MW and MWD and textile properties of regenerated cellulose fibres; and (3) MW and MWD in connection with after-treatment and ageing of cellulose fibres.

The numerous publications covering pulp data in relation to fibre performance generally come to the conclusion that in spite of the cellulose chain being degraded in alkali cellulose ripening, a high amount of accessible short chains in the starting material is detrimental to the tensile strength and fatigue resistance of tyre cord rayon (see for example Bachlott *et al.*, 1955) as well as to the textile quality of polynosic fibres (Drisch and Soep, 1953; Mikami and Ellefsen, 1962). For production of tyre cord rayon and of polynosic fibres (Siclari, 1967) generally a high α pulp of rather high DP is required, while in the production of HWM-type fibres demands on pulp parameters are not so high, less expensive pulp qualities being suitable. Also with ordinary rayon fibre too high an amount of short chains was found to decrease the bending resistance of the product.

Though regenerated cellulose fibres of similar textile quality may differ rather much in MW and MWD (Petterson and Treiber, 1969), and MW was shown, in a fundamental treatise by Perepelkin (1974), to be just *one* parameter in determining textile properties of chemical fibres, some definite correlations have been found between MW, MWD and textile properties, especially tensile strength, of regenerated cellulose fibres. As stated by Klare (1957), for the first time in such a definite and clearcut manner, an increase in fibre DP, at least up to DP ~ 570 leads to better tensile and elongation properties and out-weighs the influence of the cellulose concentration in the spinning solution, the maximum in tensile strength arrived at, 6–8% cellulose in the viscose, being more pronounced at higher DP. Important systematic contributions by Cumberbirch (1959) elucidated the interconnection between MW and spinning stretch in determining tensile properties of regenerated cellulose filaments, thus showing (1) that the DP is of much more influence to tensile properties on stretched filaments than hitherto assumed from results on isotropic model films, (2) that with fractions of secondary cellulose acetate spun under high stretch and subsequently saponified, tensile strength increased with MW, even up to a DP of ca. 800, (3) that the tenacity of filaments spun from blends of these fractions can be calculated from tenacity values of the fractions and MWD of the blend. With normal viscose rayon, this increase in tenacity with DP was traced back mainly to a shifting of the whole distribution curve to a higher average MW, and not to the presence of chains of an especially high DP (Kudrjavceva *et al.* 1972) While the MW has little effect only on "crystallinity" and "crystallite size" of viscose rayon, the primary swelling of the cellulose gel immediately after regeneration decreases remarkably with increase in DP (Tanzawa, 1960). In discussing correlations between structure and properties of various kinds of chemical fibres on the basis of definite morphological units, Krässig (1969) states that there is a linear relation between tensile strength and reciprocal DP for a series of "Meryl"-fibres. In ageing different textile fibres (including regenerated cellulose and cellulose acetate) by UV-radiation and by heat, tensile strength ("TS") was found by Sippel (1959) to depend on DP according to a generally applicable expression: ln TS = $k_1 - k_2/DP$. The effect of different chemical and physical treatments on MW and MWD of rayon print-cloth was studied by Dyer and Phifer (1971), who emphasized the considerable degradation especially of the longest chain molecules in mechanical abrasion and flexing experiments.

Turning now from cellulose itself to cellulose derivatives relevant to fibre-making, the first to be mentioned is, of course, cellulose acetate. Besides the necessary adaptation of methods to the special problems of cellulose acetate, concerning for example GPC (Tanghe *et al.*, 1970) or turbidimetric MWD-determination (Minjajlo and Zakuvdaeva, 1973), there arises, especially for products of lower DS ("$2\frac{1}{2}$-acetate"), the problem of a superposition of MWD and DS-distribution in fractionation experiments, typical for all copolymers. By proper choice of solvent and precipitant, a fractional precipitation separating predominantly either according to MW (heptane as a precipitant) or according to DS (water as a precipitant) has been achieved (Bischoff and Philipp, 1966). Recently it was shown that in thin-layer chromatography separation a binary mixture of methylenechloride/methanol is suitable for evaluation of acetyl content distribution, while a ternary mixture of methylenechloride/tetrahydrofurane/methanol preferentially separates according to MW (Kamide *et al.*, 1973a). Regarding correlations between data of cellulose material used and acetylation behavior, resp. acetate properties, a narrowing of MWD by partial hydrolytic degradation leads to a higher acetylation rate (Ivanov and Yablochko, 1969). The shape of the MWD of the starting materials is maintained to some extent in acetylation, while the final MW of the acetate mainly depends on acetylation procedure and not on MW of the starting material (Sushkevich *et al.*, 1966). The differences in MW and MWD obtained in comparing heterogeneous and homogeneous acetylation of the same starting material are stressed by Lalewa *et al.* (1970), while Kostrov *et al.* (1973) correlates MW and MWD of cellulose acetate with some properties of fibre obtained therefrom.

A rapidly expanding field still requiring new ideas on the combination of MW- and MWD-determination with other analytical methods is the preparation and evaluation of graft polymers of cellulose. GPC subsequent to a nitration procedure was sucessfully applied to radiation on degraded cellulose samples (Imamura *et al.*, 1972). Total hydrolysis of the cellulose backbone chain makes possible a MW- and MWD-determination of vinylic sidechains (Mileo and Nicolas, 1967), but there are still many questions as to the distribution of the side-chains. Length and distribution of grafted vinylic side-chains can be influenced by changing the structure of the starting material, for example by hydrolytic degradation (Ouchi *et al.*, 1973), as well as by the grafting process itself (Tanaka, 1971), for example by addition of emulsifiers. Of course, the properties of the grafts depend to a large extent on the number and MW of the side-chains, as shown for example by Druzhinina *et al.* (1968) for grafts of polymethacrylic acid on cellulose acetate, and by Wellons *et al.* (1967) for the permeation properties of films from polystyrene grafted cellulose acetate.

B. Fibre-forming Acrylic Polymers

Most of the acrylic fibres manufactured on a technical scale are based on copolymers of acrylonitrile. But, as already stated, the fractionation of copolymers is a problem open for further studies (see Section III, C). Therefore, and for other reasons, most of the methods developed for process control have nearly nothing to do with molecular parameters or properties. In the literature we mainly find that the results on product characterization determining the MWD of polyacrylonitrile and copolymers of acrylonitrile with rather small amounts of comonomers are obtained by fractional precipitation or fractional solution and only some results are obtained by other methods like GPC.

Selection of a suitable solvent/nonsolvent system is important to achieve good fractionation results. Taking dimethylformamid as a solvent, hexane, decaline and turpentine are the nonsolvents usually applied. Reviews are given by Fritzsche (1965) and Berger *et al.* (1966). In most solvent/nonsolvent systems polyacrylonitrile precipitates as a solid phase, which is more or less swollen. In this case it is difficult to reach the thermodynamical equilibrium between the phases. As a result the reproducibility is poor. With decaline as a precipitant polyacrylonitrile separates as a liquid phase, so that the fractional precipitation of poly-acrylonitrile in the dimethylformamide/decaline-system, in connection with utilizing the mixing gap between dimethylformamid and decaline, is a really suitable method of good reproducibility (Fritzsche, 1971a) which requires only a moderate amount of time.

Another problem with the adaptation of methods for estimating MW and MWD of acrylonitrile copolymers is the influence of ionic groups. A few ionized sulphonate groups per average polymer chain affect the dilute solution properties of the polymer to a large extent. Cha (1969) has carried out the characterization of copolymers containing sulphonate groups in $0 \cdot 1$ M LiBr-solution in DMF. Polyacrylonitrile in DMF gave unimodal GPC curves and the polydispersity index is somewhat larger than light scattering and osmometric results. However, the GPC chromatograms of the sulphur-containing polymers showed multiple peaks. Although the molecular weights of the polyacrylonitrile homopolymer sample and the sulphur-containing (S) polymer are not much different, the elution of S-polymer starts much earlier. This indicates that the apparent molecular sizes of S-polymers in DMF are much larger than those of the homo-polymer. It was suggested that the multiple peaks in the GPC curve are due largely to different distributions of sulphonate groups in the polymer chain, leading to different molecular configurations in solution. In other words, the higher GPC peaks could be due to less-charged molecules, i.e. molecules with smaller sulphonate concentration. On the other hand Coppola *et al.* (1972) supposed association between the macromolecules carrying sulphonate groups. Using 0·1 M LiBr-solution in DMF as a solvent system the elution volume of the S-polymer increases and the multiple peaks observed in DMF disappear.

Viscometric and sedimentation studies by Linow and Philipp (1971a) showed, that the α-value in the Mark–Kuhn–Sakurada equation for an acrylonitrile/methylacrylate–copolymer containing sulphonate-groups in DMF is $\alpha = 0.94$ and in 0·1 M LiCl-solution in DMF is $\alpha = 0.89$. That means, the thermodynamic state of solution is somewhat poorer in the salt solution than in DMF alone but still better than that of polyacrylonitrile in DMF ($\alpha = 0.75$). The sedimentation constants decrease with increasing sulphonate content and are larger in the LiCl/DMF system than in DMF.

Recently Kenyon and Mottus (1973) recommended the study of acrylonitrile copolymers with comonomers of somewhat larger polarity (vinylether) by GPC with LiBr-solution in DMF. The presence of LiBr often results in normal (unimodal) distribution curves. The authors suppose that the presence of polar comonomers in the chain favours an association of polymer and solvent molecules, but they gave no theoretical explanation for this assumption.

Another problem has been the determination of the "real MWD" of the samples. As shown in Section III, C, generally the rule is valid that using different methods or solvent/nonsolvent-systems the broadest MWD found experimentally can be considered as the best approximation to the real existing MWD. But time-consuming work is needed to determine this best approximation. To find a time-saving method of solving this problem, it can be divided into two steps of experimental work.

In the first step the experimental distribution curves obtained by one method only are compared for the different samples. For example, Takahashi and Watanabe (1961) compared the MWD's determined by fractional precipitation with the DMF/heptane : ether (1 : 1)-system of Orlon and Dralon samples and redox polymerizates. Using the Tung-evaluation they found inhomogeneity coefficients for the Orlon samples $P_w/P_n = 1.8 \ldots 2.0$, for the Dralon sample 2·2 and for the redox polymerization 2·4.

Kiuti *et al.* (1964) studied acrylonitrile/methylmethacrylate copolymers polymerized in aqueous media with the redox systems persulphate/sulphite or persulphate/triethanolamine using different procedures of polymerization. The MWD is narrower using the redox system first mentioned and applying a continuous polymerization process in comparison to a discontinuous procedure.

Fritzsche (1971a) found that the distribution curve of a polyacrylonitrile sample polymerized in DMF with azoisobutyronitrile is narrower than that of samples obtained in aqueous suspension with redox systems.   The sample with the broadest distribution curve shown in Fig. 5 was not a "normal" one with regard to the polymerization process and led to a disturbance in the subsequent spinning process.

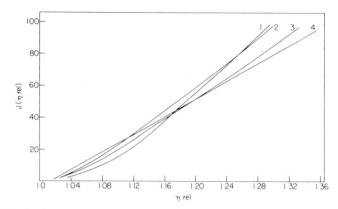

FIG. 5. Distribution curves of polyacrylonitrile samples. 1—solution polymerized in DMF; 2—aqueous laboratory polymerized with redox system; 3, 4—technical suspension polymerized (Fritzsche, 1971a).

Frenkel (1965) has published sedimentation diagrams of polyacrylonitrile samples prepared in solution, suspension, and emulsion. Samples polymerized in solution had always an unimodal MWD. But the samples polymerized in a heterophase system possessed in nearly all cases bi- or tri-modal MWD's. Therefore a superposition of two or three mechanisms òf polymerization should be taken into account here.

Ovsyannikova *et al.* (1973) observed a definite relationship between the MWD and the particle-size distribution of suspension-polymerized polyacrylonitrile. The statistical dispersion of the sedimentation coefficients decreased linearly with increasing percentage of particles with radii lower than 10 μm.

The second step in tackling the whole problem is then to look for the real MWD. Fritzsche (1967) compared results of preparative gel chromatographic experiments with results of fractional precipitations and concluded, that the fractional precipitation with dimethylformamid/decaline gives experimental MWD's which were probably the best approximation of the real MWD available at that time. Underlying a Tung function the polydispersities were calculated for a suspension polymerizate prepared with a redox system and thioglycolic acid as a regulator to $M_w/M_n = 2$. In technical polyacrylonitrile samples manufactured by suspension polymerization the value of M_w/M_n was 3 to 5. In the late 1950's, early 1960's there were some important contributions published on the influence of MWD on polyacrylonitrile fibre properties (e.g. Stefani *et al.*, 1959; Marzolph, 1962;

Beder and Pakshver, 1963; Kobayashi *et al.*, 1964). As a general rule a mechanical property, especially tensile strength, increases with the MW in a definite range of MW and then reaches a limit according to the function (Mark, 1956):

$$Z = Z_\infty - \frac{b}{P_n}$$

where Z_∞ is the maximum value of the mechanical property possible. Polyacrylonitrile fibres of acceptable tensile strength should have a M_w exceeding 50,000. An upper limit of MW exists only in connection with preparation and processing of the spinning solution. Interesting effects were reported recently by Dobrecov *et al.* (1969, 1970). In laboratory spinning tests they arrived at a super strength fibre of 120 kp/mm² tensile strength in the high MW range of about $3 . 10^6$, showing that there is an increase of tensile strength with MW up to higher MW-values than hitherto assumed. Nevertheless, an influence of the MWD on structure and properties of the fibre is probable. Kamalov *et al.* (1966, 1967) found no effect of the MW and MWD on fibre properties in the isotropic state. A significant effect is obtained as soon as orientation is high enough, and it increases with increasing orientation (stretching). The limit of orientation attainable depends on MW and MWD. In comparison with this the parameters under discussion do not greatly affect the relative loop strength and the elongation at break. Clearly, MW and MWD influenced the flex life of the fibre.

C. FIBRE-FORMING POLYCONDENSATES

The most important fibre-forming polycondensates, like polyamide 6.6, polyamide 6 and polyethyleneterephthalate, are produced in and spun out of melts of the polycondensates at temperatures high enough to reach chemical equilibrium between the polymer molecules of different MW. Therefore, in most cases the polycondensates will obey a definite MWD-function called Schulz–Flory or "most-probable distribution" function (Flory, 1953). The correlation between the MW and the properties of polymer and fibres is more important than the correlation between MWD and the properties because there is only a very limited possibility of changing the MWD. To arrive at this conclusion (of expecting similar MWD's in all these polycondensates) there were essential problems concerning the adaptation of the methods to the special polymers to be solved. Mertel and Heilmann (1973a) reviewed the relevant literature. Duveau and Piguet (1962) developed a continuous procedure for fractionation of polycondensates, especially of polyamide 6.6.

Reinisch *et al.* (1969) fractionated polyethyleneterephthalate with the solvent/nonsolvent system phenol-tetrachlorethane (1:1)/heptane. Starting with the concentration 1 g/100 ml of the polymer in the solvent mixture they obtained, by taking 25 fractions, a MWD which is a good approximation of a Schulz–Flory distribution. Lowering the concentration to 0·25 g/100 ml resulted in an essential decrease of the number of fractions needed to obtain a Schulz–Flory distribution. Almost quantitative separation was achieved with various model blends of polyethyleneterephthalate samples having different MW's. The MWD experimentally found corresponded rather well to the MWD calculated from a superposition of the MWD of the components of the blends.

These results cannot be simply transferred to the fractional precipitation of polyamide 6 in the same solvent/nonsolvent–system. Starting from initial concentrations of 0·25 g/100 ml and taking 10 fractions, distribution curves were obtained which deviated significantly from the Schulz–Flory distribution (Rafler and Reinisch, 1971). These deviations were traced back to a linear association of the polycaprolactam molecules. By acetylation the number of endgroups was diminished and thus the degree of association was minimized. The fractionation of these acetylated samples of polycaprolactam at an initial concentration of 0·125 g/100 ml proved to be an effective procedure which provided sufficient selectivity.

Sippel and Albien (1968) determined the MWD of polyamide 6.6 samples possessing different MW which were prepared by various amounts of chain length regulator or by different procedures of post-condensation. In nearly all cases the theoretical distribution was approximated by the experimental data rather well. That means that there is no necessity in regular cases to determine the MWD of polyamide 6.6 samples. Only by raising the temperature up to the beginning of decomposition is the MWD changed.

The MWD of polyamide 6 samples was mostly found to be a Schulz–Flory distribution (Mertel and Heilmann, 1973a). Generally the MWD of the samples prepared with chain length stabilizors like acetic acid was narrower than the MWD of polyamide 6 polymerized in the absence of stabilizors (Kudrjavcev *et al.* 1960; Mattiussi and Gechele, 1965; Mertel and Heilmann, 1973b). Mattiussi and Gechele (1965) also found that the temperature of polymerization had an influence on the MWD.

The MWD's of polyethyleneterephthalate samples, obtained by condensation with various metal ion catalysts in an open system and by degradation in a nitrogen stream under stirring at 280°C, have been investigated by Gehrke and Reinisch (1966). With good approximation, Flory–Schulz distributions have been found in all cases.

The few contributions concerning the influence of MW on processibility and properties of polyamide 6 was reviewed by Poludennaja *et al.* (1968).

Increasing the MW from 17,000 up to 30,000 an essential increase of tensile strength of the fibres was found. The stretchability and fibre properties of polyamide 6 are normally rather independent from MWD, as under all the usual conditions of processing nearly the same Flory–Schulz distribution is found. A narrower MWD does not affect the properties of the fibre, but an abnormally broad MWD causes a lower stretchability (Mertel and Heilmann, 1973b). Mal'ceva and Ajsenstein (1974) prepared two polyethylene terephthalate samples containing a different MWD. They found a broadening of the MWD to lower tensile strength and flex life of the fibres, but an increase in elongation at break.

Besides the linear polymer molecules possessing a most probable MWD, cyclic oligomers arise in the polycondensation process. Their amount in the polycondensate sample was predicted theoretically by Jacobson and Stockmayer (1950). In the early nineteen sixties the synthesis, separation and analysis of oligonomers was often studied (see for example Rothe and Rothe, 1963). Schwenke (1964) fractionated cyclo-oligomers of polyamide 6 by multiplicative distribution in the heptane-methanol system. In a more recent contribution, Zahn and Kusch (1967) separated, estimated and prepared a large number of oligomers and cyclo-oligomers of different types of polyamides and polyethyleneterephthalate. The content of oligomers in polyethyleneglycolterephthalate samples seems to cause two kinds of difficulties in the manufacturing of fibres and textiles, i.e. deposits in the dyeing equipment and redeposition on fibre material (Senner, 1972), while the influence of these oligomers on textile fibre properties is generally considered to be rather small.

References

Achwal, W. B. and Vadza, A. H. (1970). *Colourage* **17**, 29–36
Agster, A. (1963). *Textilpraxis* **18**, 577–583.
Bachlott, D. D., Miller, J. K. and White, W. D. (1955). *Tappi* **38**, 503–507.
Battista, O. A., Coppick, S., Howsmon, J. A., Morehead, F. F. and Sisson, W. A. (1956). *Ind. Eng. Chem.* **48**, 333–335.
Beder, N. M. and Pakshver, A. B. (1963). *Khim. Volokna* (5), 9–12.
Benoit, H. and Jacob, M. (1967). *Anal. chim. Acta* **39**, 245–257.
Benoit, H., Grubisic, Z., Rempp, P. and Zilliox, J. (1966). *J. Chim. Phys.* **63**, 1507; (*see also* Grubisic *et al.*, 1967).
Berger, R. (1969). *Plaste Kautschuk* **16**, 326–331, 572–576, 725–731, 822–824.
Berger, W., Hartig, S. and Sachmann, E. (1966). *Faserforsch. Textiltech.* **17**, 380–381.
Berry, G. C. and Cassassa, E. F. (1970). Macromolecular Reviews **4**, 1–66.
Bischoff, K. H. and Philipp, B. (1966). *Faserforsch. Textiltech.* **17**, 395–400.
Brestkin, Ju.V. and Frenkel', S. Ja. (1969). *Vysokomolekul. Soedin A* **11**, 2437–2443.

Brestkin, Ju.V. and Frenkel', S.Ja. (1970). *Sowj. Beiträge Faserforsch. Textilitech.* **7**, 122–126.

Brzezinski, J., Glowala, H. and Kornas–Catka, A. (1973), *European Polymer J.* **9**, 1251–1253.

Cantow, H. -J., Siefert, E. and Kuhn, R. (1966).*Chemie-Ing.-Techn.* **38**, 1032–1038.

Cha, C. Y. (1969). *J. Polymer Sci. B* **7**, 343–348.

Čočieva, M. M., Brestkin, Ju.V. and Nikitin, N. J. (1962). *Zh. prikl. Khim.* **35**, 2025–2035.

Coppola, G., Fabbri, P., Pallesi, B. and Bianchi, U. (1972). *J. Appl. Polymer Sci.* **16**, 2829–2834.

Cumberbirch, R. J. E. (1959). *J. Textile Inst.* **50**, 528–547.

Cumberbirch, R. J. E. and Harland, W. G. (1959). *J. Textile Inst.* **50**, 311–334.

Danilov, S. N., Samsonova, T. I. and Bolotnikova, L. S. (1970). *Russian Chem. Rev.* **39**, 156–168.

Dobrecov, S. L., Lomonosova, N. V. and Stel'mach, V. P. (1969). *Vysokomolekul. Soedin. B* **11**, 782;

Dobrecov, S. L., Lomonosova, N. V. and Stel'mach, V. P. (1970), *Sowj. Beiträge Faserforsch. Textiltech.* **7**, 118.

Drisch, N. and Soep, L. (1953). *Textile Res. J.* **23**, 513–521.

Druzhinina, N. N., Pen'kova, M. P., Livshits, R. M. and Rogovin, Z. A. (1968). *Vysokomolekul. Soedin. A* **10**, 2743–2752.

Duveau, N. and Piguet, A. (1962). *J. Polymer Sci.* **57**, 357–372.

Dyer, J. and Phifer, L. H. (1971). *J. Polymer Sci. C* **36**, 103–119.

Edelman, K. and Horn, E. (1958). *Faserforsch. Textiltech.* **9**, 493–500.

Flory, P. J. (1953). "Principles of Polymer Chemistry", Cornell Univ. Press. Ithaka, New York.

Flory, P. J. (1970). *Discussions Faraday Soc.***49**, 7–29.

Frampton, V. L. (1972). *Anal. Chem.* **44**, 866–869.

Frenkel', S.Ja. (1965). "Introduction into the statistical theory of polymerization", Publication "Nauka", Moskau, Leningrad, pp. 216–218.

Fritzsche, P. (1965). *Faserforsch. Textiltech.* **16**, 466–467.

Fritzsche, P. (1967). *Faserforsch. Textiltech.* **18**, 21–24.

Fritzsche, P. (1968). *Faserforsch. Textiltech.* **19**, 505–508.

Fritzsche, P. (1971a). *Faserforsch. Textiltech.* **22**, 535–540.

Fritzsche, P. (1971b). *Faserforsch. Textiltech.* **22**, 599–603.

Fritzsche, P., Klug, P. and Gröbe, V. (1971). *Faserforsch. Textiltech.* **22**, 250–254.

Fuchs, O. (1956). *Z. Elektrochem.* **60**, 229–236.

Gehrke, K. L. and Reinisch, G. (1966). *Faserforsch. Textiltech.* **17**, 201–207.

Géczy, I. and Révai, Gy. (1971). *Koloriszt. Ertesitó* **13**, 209–221.

Giesekus, H. (1967). In "Polymer Fractionation", (M.J.R. Cantow, eds.), pp. 191–249. Academic Press, New York and London.

Glöckner, G. (1966). *Plaste Kautschuk* **13**, 340–341.

Glöckner, G., Francuskiewicz, F. and Müller, K. -D. (1971). *Plaste Kautschuk* **18**, 654–656.

Golubev, V. M., Bilimova, E. S., Chin, N. N. and Prokof'eva, M. V. (1969). *Vysokomolekul. Soedin.* **11** B, 752–755.

Gruber, E. and Schurz, J. (1973). *Angew. Makromol. Chem.* **29/30**, 121–136.

Grubisic, Z., Rempp, P. and Benoit, H. (1967). *J. Polymer Sci. B* **5**, 753–759.

Harland, W. G. (1954). *J. Text. Inst.* **45**, 678–691.

Heitz, W. (1970). *Angew. Chem.* **82**, 675–689.

Heitz, W., Platt, K. L., Ullner, H. and Winau, H. (1967). *Makromol. Chem.* **102**, 63–72.

Howard, P. and Parikh, R. S. (1968). *J. Polymer Sci. A-1*, **16**, 537–546.

Huang, R. Y. M. and Jenkins, R. G. (1969). *Tappi* **52**, 1503–1507.

Huggins, M. L. (1942). *J. Amer. Chem. Soc.* **64**, 2716–2718.

Huggins, M. L. and Okamoto, H. (1967). In "Polymer Fractionation", (M.J.R. Cantow, eds.), pp. 1–42. Academic Press, New York and London.

Imamura, R., Ueno, T. and Murakami, K. (1972). *Bull. Inst. Chem. Res.* **50**, 51–63.

Ivanov, V. I. and Yablochko, L. N. (1969). *Izv. Akad. Nauk. Kirg. SSR*, 90–92.

Jacobson, H. and Stockmayer, W. H. (1950). *J. Chem. Phys.* **18**, 1600.

Jayme, G. (1961). *Tappi* **44**, 299–304.

Jayme, G. and Kleppe, P. (1961). *Papier* **15**, 272–278:

Johnson, J. F. and Porter, R. S. (1970). In "Progress in Polymer Science", (A. D. Jenkins, eds.), Vol. 2, pp. 203–256. Pergamon Press, Oxford/New York/Toronto/Sydney/Braunschweig.

Kamalov, S., Korotkov, A. A., Krasulina, V. N. and Frenkel', S.Ja. (1966). *Khim Volokna* (6), 9–14.

Kamalov, S., Korotkov, A. A., Krasulina, V. N. and Frenkel', S.Ja. (1967). *Sowj. Beiträge Faserforsch. Textiltech.* **4**, 123–128.

Kamide, K. and Co-workers (1968). *Chem. High Polymers Tokyo* **24**, 440–454, (*see also* Kamide *et al.* 1973b).

Kamide, K., Manabe, S. and Osafune, E. (1973a). *Makromol. Chem.* **168**, 173–193.

Kamide, K., Miyazaki, Y. and Yamaguchi, K. (1973b). *Makromol. Chem.* **173**, 157–174, 175–193.

Kennedy, J. W., Gordon, M. and Koningsveld, R. (1972). *J. Polymer Sci. C* **39**, 43–69.

Kenyon, A. S. and Mottus, E. H. (1973). *Amer. Chem. Soc., Polymer Preprints* **14**, 652–657.

Kiuchi, H. and Nishio, M. (1964). *Kogyo Kagaku Zasshi* **67**, 1476–1478; (ref. C. A. **62** (1965), 5338 h.)

Kiuti, X., Juguti, S. and Watanabe, M. (1964). *Kogyo Kagaku Zasshi* **67**, 1063–1068.

Klare, H. (1957). *Faserforsch. Textiltech.* **8**, 262–267.

Klein, J. (1970). *Angew. Makromol. Chem.* **10**, 21–48.

Kobayashi, H., Sasaguri, K., Fujisaki, Y. and Amano, T. (1964). *J. Polymer Sci. A* **2**, 313–331.

Koningsveld, R. (1967). *J. Polymer Sci. C* **16**, (3), 1775–1786.

Koningsveld, R. (1969). *Pure App. Chem.* **20**, 271–307.

Kostrov, Y. A., Iva, M. Ja., Bajabakova, Z. V., Golubev, V. M. and Perepečkin, L. P. (1973). *Khim. Volokna* (1), 44–46.

Kotera, A. (1967). In "Polymer Fractionation", (M. J. R. Cantow, eds.), pp. 44–60. Academic Press, New York and London.

462   B. PHILIPP AND P. FRITZSCHE

Kraemer, E. O. (1938). *Ind. Eng. Chem.* **30**, 1200–1203.

Krässig, H. (1969). *Textilveredlung* **4**, 26–37.

Kubin, M. (1969). *Collection Czech. Chem. Commun.* **34**, 703–707.

Kudrjavcev, G. I., Katoržnov, N. D. and Kratikova, A. G. (1960). *Khim. Volokna* (2), 30–33.

Kudrjavceva, A. G., Mogilevskij, E. M. and Papkov, S. P. (1972). *Khim. Volokna* (2), 50–52.

Lalewa, W., Georgiev, J. and Awramova, R. (1970). *Cell. Chem. Technol.* **4**, 511–523.

Linow, K.-J. (1978). *Faserforsch. Textiltech.* **29**, 452–458.

Linow, K.-J. and Philipp, B. (1968). *Faserforsch. Textiltech.* **19**, 509–513.

Linow, K.-J. and Philipp, B. (1970). *Faserforsch. Textiltech.* **21**, 255–258.

Linow, K.-J. and Philipp, B. (1971a). *Faserforsch. Textiltech.* **22**, 444–447.

Linow, K.-J. and Philipp, B. (1971b). *Plaste Kautschuk* **18**, 721–724.

Mal'ceva, N. G. and Ajzenstein, E. M. (1974). *Khim. Volokna* (1), 35–37.

Mandelkern, L. and Flory, P. J. (1952). *J. Chem. Phys.* **20**, 212–214.

Mark, H. (1956). In "Physik der Hochpolymeren", (Stuart, eds.), Vol. IV, p. 86. Springer, Berlin, Göttingen, Heidelberg.

Martin, A. F. (1962). *Amer. Chem. Soc.* Meeting in Memphis.

Marx-Figini, M. (1969). *J. Polymer Sci.* C **28**, 57–67.

Marzolph, H. (1962). *Angew. Chem.* **74**, 628–632.

Mattiussi, A. and Gechele, G. B. (1965). *Eur. Polymer J.* **1**, 147–157.

Mertel, H. and Heilmann, G. (1973a). *Faserforsch. Textiltech.* **24**, 333–336.

Mertel, H. and Heilmann, G. (1973b). *Faserforsch. Textiltech.* **24**, 456–464.

Meyerhoff, G. (1970). *Makromol. Chem.* **134**, 129–138.

Mikami, T. and Ellefsen, O. (1962). *Norsk Skogind.* **16**, 563–568.

Mileo, J. C. and Nicolas, L. (1967). *J. Appl. Polymer Sci.* **11**, 425–437.

Minjajlø, S. H. and Zakuvdeeva, N. P. (1973). *Khim. Volokna* (3), 44–46.

Molyneux, P. (1968). *Kolloid-Z.* **225**, 97–106.

Muggli, R. (1968). *Cell. Chem. Technol.* **6**, 549–567.

Muggli, R., Mühlethaler, K. and Elias, H.-G. (1969). *Angew. Chem.* **81**, 334.

Ouano, A. C., Horne, D. L. and Cregges, A. R. (1974). *J. Polymer Sci., Polymer. Chem. Ed.* **12**, 307–322.

Ouchi, T., Otsuka, S. and Imoto, M. (1973). *Chem. High Polymers (Tokyo)* **30**, (333), 46–50.

Ovsyannikova, S. A., Beder, N. M. and Nekrasov, I. K. (1973). *Khim. Volokna* (3), 75.

Paulusma, P. and Vermaas, D. (1963). *J. Polymer Sci.* C **2**, 487–497.

Pavlova, S. A., Timofeeva, G. I., Korshak, V. V., Vinogradova, C. V., Tur. D. R. and Sharoshi, D. (1971). *Vysokomolekul. Soedin.* **13**, 2643–2652.

Perepelkin, K. E. (1974). *Faserforsch. Textiltech.* **25**, 251–267.

Petterson, B. A. and Treiber, E. (1969). *Papier* **23**, 139–143.

Phifer, L. H. and Dyer, J. (1971). *Separ. Sci.* **6**, 73–88.

Philipp, B. (1962). *Svensk. Papperstid.* **65**, 197–208.

Philipp, B. and Linow, K.-J. (1965). *Zellstoff Papier* **14**, 321–326.

Philipp, B. and Linow, K.-J. (1966). *Papier* **20**, 649–657.

Philipp, B. and Linow, K.-J. (1968). *Zellstoff Papier* **17**, 195–199.

Philipp. B. and Linow, K.-J. (1970). *Faserforsch. Textiltech.* **21**, 13–20.

Philipp, B., Reichert, H., Tryonadt, A. and Gröbe, V. (1964). *Faserforsch. Textiltech.* **15**, 304–315.

Philipp, B., Baudisch, J. and Ruscher, Ch. (1969). *Tappi* **52**, 693–698.

Poller, S. (1969). *Faserforsch. Textiltech.* **20**, 71–76.

Poludennaja, V. M., Volkova, N. S., Archangel'skij, D. N., Žigockij, A. G. and Konkin, A. A. (1968). *Khim. Volokna* (5), 6–7.

Porter, R. S. and Johnson, J. F. (1967). In "Polymer Fractionation" (M. J. R. Cantow, eds.), pp. 95–121. Academic Press, New York and London.

Rafler, G. and Reinisch, G. (1971). *Angew. Makromol. Chem.* **20**, 57–69.

Reinisch, G., Rafler, G. and Timofejewa, G. I. (1969). *Angew. Makromol. Chem.* **7**, 110–120.

Rinaudo, M., Barwud, F. and Cherle, J. P. (1969). *J. Polymer Sci.* C **28**, 197–207.

Rothe, I. and Rothe, M. (1963). *Makromol. Chem.* **68**, 206–210.

Samsonova, T. I., Bolotnikova, L. S. and Frenkel', S. Ja. (1970). *Sintez, Struktura, Svoistva Polimerov*, 303–307; (see also Danilov *et al.*, 1970).

Schulz, G. V. (1940). *Z. Phys. Chem.* B **47**, 155–193.

Schulz, G. V. (1974). *Angew. Chem.* **86**, 454.

Schulz, G. V. and Blaschke, F. (1941). *J. prakt. Chemie* **158**, 130–135 and **159**, 146–154.

Schulz, G. V. and Penzel, E. (1968). *Makromol. Chem.* **112**, 260–280.

Schwenke, K. D. (1964). *Faserforsch. Textiltech.* **15**, 266–272.

Segal, L. (1968). *J. Polymer Sci.* C **21**, 267–282.

Senner, P. (1972). *Lenzinger Berichte* **34**, 44–51.

Shaubhag, V. B. (1968). *Arkiv Kemi* **29**, 1–22, 33–45, 139–161, 163–177.

Shibukawa, T., Sone, M., Uchida, A. and Iwahori, K. (1967). *J. Polymer Sci.* A–1 **5**, 2857–2865.

Siclari, F. (1967). *Pure Appl. Chem.* **14**, 423–433.

Sippel, A. and Albien, K. (1968). *Faserforsch. Textiltech.* **19**, 145–148.

Sippel, A. (1959). *Kunststoffe* **49**, 626–631.

Stefani, R., Chevreton, M., Terriev, J. and Eyrand, C. (1959). *Compt. Rend.* **248**, 2006–2008.

Sushkevich, T. I., Konotkova, T. F. and Usmanov, Kh. U. (1966). *Strukt. Modif. Khlop. Tsellyul.*, 41–45.

Takahashi, M. and Watanabe, M. (1961). *Sen-i Gakkaishi* **17**, 111–124.

Takeuchi, T. and Amanuma, K. (1966). *Kogyo Kagaku Zasshi* **69**, 1764–1767.

Tanaka, Z. (1971). *Sen-i Gakkaishi* **27**, 427–432.

Tanghe, L. J., Rebek, W. J. and Brewer, R. J. (1970). *J. Polymer Sci.* A–1, **8**, 2935–2947.

Tanzawa, H. (1960). *Kogyo Kagaku Zasshi* **63**, 2191–2194.

Teichgräber, M. (1968). *Faserforsch. Textiltech.* **19**, 249–252.

Topčiev, A. V., Litmanovič, A. D. and Stern, V. Ja. (1962). *Dokl. Akad. Nauk SSSR*, **147**, 1389–1391.

Turska, E. and Wolfram, L. (1958). *Zeszyty Nauk. Politech. Lodz., Chem.* **22**, 79–82.

Usmanov, Ch. K., Usmanova, M. I. and Machmudova, S. S. (1970). *Uzbeksk. Khim. Zh.* 43–44.

Wadsworth, J.I., Segal, L. and Timpa, J. D. (1971). (Pub. 1973). *Advan. Chem. Ser. No. 125*, 178–186.

Van der Want, G. M. and Kruissink, Ch. (1959). *J. Polymer Sci.* **35**, 119–138.

Wellons, J. D., Williams, J. C. and Stannett, V. (1967). *J. Polymer Sci.* A–1 **5**, 1341–1357.

Zahn, H. and Kusch, P. (1967). *Z. Ges. Textil-Ind.* **69**, 880–884.

Zimmermann, H. and Kolbig, Ch. (1967). *Fasserforsch. Textiltech.* **18**, 536–537.

Zimmermann, H. and Tryonadt, A. (1967). *Faserforsch. Textiltech.* **18**, 487–490.

12. The Thermal and Flammability Behavior of Textile Materials

LUDWIG REBENFELD, BERNARD MILLER,
and
J. RONALD MARTIN

Textile Research Institute, Princeton, New Jersey, USA

I. Introduction

One of the most important and far-reaching developments that has taken place in many countries recently is an increasing concern with consumer protection and product safety. In the case of textile materials, this concern has manifested itself in the form of regulations limiting the allowable flammability characteristics of certain textile products. In the United States stringent performance standards have been set on the flammability behavior of many products, for example, children's sleepwear, carpets, mattresses, and upholstery materials. It is reasonable to expect that even more stringent and more widespread standards will be promulgated in the future, and that such standards will be established in an increasing number of countries. In response to this critical challenge, a massive effort has been undertaken in many academic and industrial research laboratories, particularly in the United States, to develop flame-retardant textile materials. The task is by no means a simple one. Not only are the technical problems of flame retardancy quite complex, but the development of flame-retardant textile materials must be achieved with careful consideration of other textile properties and also with special attention to economics. It would be useless to develop flame-retardant textiles that did not satisfy consumer demands for aesthetics and performance and that did not fall within established price structures.

This chapter will deal with some of the important technical problems associated with the flammability behavior of textile materials. Special attention will be given to the measurement of various physical properties that are related to textile flammability and to a discussion of the controlling factors that define the flammability of multicomponent systems.

II. The Special Nature of Textile Flammability

The flammability of a textile material cannot in any way be regarded as a material constant that can be rigorously defined in thermodynamic or otherwise fundamental terms (Rebenfeld and Miller, 1974). Rather, flammability must be considered as a composite characteristic of a system that reflects not only the material itself, but also its geometric configuration and its physical environment.

The combustibility of textiles originates from the fact that the common textile fibers are composed of organic polymers. These polymers, which contain considerable amounts of carbon and hydrogen, are susceptible to thermal decomposition and in the presence of air or oxygen, eventually can ignite. Once ignition has taken place, a variety of endo and exothermic transformations and chemical reactions are available to propagate the flame.

In the case of textiles, the flammability problem is complicated by the fact that textile materials are very often composed of several different polymer types. Not only have engineered or intimate fiber blends become of increasing importance, but it is normal practice for textile fabrics to be combined in various ways at the consumer level. For example, the total clothing for an average person at any time is composed of several fabric types, and in such typical household items as curtains, drapes, and upholstery, we frequently find at least two layers of different fabrics. Rarely do we find a situation where the potential hazard arising from the inherent flammability of textile materials is due to any one particular fiber type. Thus, it is important that we understand the flammability behavior of mixed fiber systems. As will be discussed later, these blended and otherwise mixed textile materials pose unique problems.

There are a number of additional factors that make the textile flammability problem a particularly complex one. For example, there are always numerous finishing agents that have been added to fibers by the time the final textile product is produced. Normally anywhere from 2 to 10% of the weight of a textile material is non-fibrous in character. A typical textile product may contain two or three structurally complex dyestuffs, usually with a high degree of aromaticity, amounting to about 2% of the weight of the textile product, and about another 5% in the form of nitrogenous high molecular weight resins and additives to provide light stability, antistatic properties, soil resistance, durable-press properties, and a number of other aesthetic and performance characteristics. As a result, the exact chemical composition and structure of a textile product is not really known either by the producer or by the ultimate consumer.

Another factor that causes special problems is the extremely high surface-to-volume ratio which characterizes all textile-like substances. This provides a physical state most conducive to flaming combustion and makes it extremely difficult to predict and eventually control this material characteristic.

It is also important to bear in mind that, though the product manufactured by the textile industry is a two-dimensional planar sheet-like material, yet, in most cases, this is not what causes the potential hazard to the ultimate consumer. The two-dimensional fabric is transformed usually to a complex three-dimensional product, which may be an article of clothing, a curtain, a drapery, or a piece of furniture. The flammability behavior of any material is critically dependent on its shape and on its attitude and position with respect to the ignition source and the direction of the propagating flame front. In textiles, the unpredictable three-dimensionality of the material in the hands of the ultimate consumer poses special and unique problems in terms of predicting performance criteria.

III. Measurement of Flammability

One of the first problems confronting an investigator of fabric flammability is the lack of suitable methods for studying this complex phenomenon. The standards imposed upon certain classes of textile goods, considered to be sources of potential hazard, prescribe certain test procedures which generally provide only pass–fail data with no quantitative means of evaluating flammability. For example, the standard for children's sleepwear, set forth by the United States Department of Commerce, requires that bone dry samples of fabric (3·5 in. × 10 in) from which sleepwear is constructed must not burn upward to form a char length greater than 7 in. or with a residual flame time greater than 10 sec (U.S. Department of Commerce, 1971, 1973). Some workers have attempted to obtain quantitative data from this test by using the residual char length as an index of flammability. However, very little useful information has been gleaned from this approach. In the case of self-extinguishing fabrics, this measurement is highly unreproducible, and the interpretation of its meaning is not completely clear. For fabrics that have burned their entire length, the residual char length would simply be the length of the sample. This test, and most others like it, are simply methods of distinguishing those materials that carry a high flammability risk and do not provide a means for observing systematic differences in the flammability behavior of fabrics.

At the present time, the most prevalent form of quantitative flammability data is expressed in terms of oxygen index. This value, sometimes also referred to as the limiting oxygen index, is defined as the minimum oxygen concentration which will sustain downward, candlelike burning of a supported fabric specimen (Fenimore and Martin, 1966). This technique was first developed for plastic materials, but its application to textiles has become widespread (Tesoro and Meiser, 1970). Although the oxygen index is a reproducible material characteristic, proper interpretation of data leads to a realization that this quantity is related to only one facet of flammability behavior, the ease of extinguishability.

The most recent efforts in the development of test procedures have been concerned with assessing flammability hazard in terms of the heat transferred to a specific target during burning (Krasny, 1975). This technique, based on the NBS Mushroom Apparel Flammability Test apparatus, has been designed to simulate the situation that might occur when an apparel fabric is ignited while worn. Evolution of this procedure into a standard test method will involve some sort of classification of materials on the basis of ignitability and heat transmission.

Since the standard test methods provide very few suitable methods of quantifying flammability, research workers have been faced with the task of developing for themselves methods of measuring many of the various characteristics and consequences of fabric burning. A large part of any flammability research effort must be devoted to the development of such techniques which, once they are fully understood, can be applied to the study of specific problems which present themselves in this field of research.

Miller and Martin (1975) have suggested that flammability research be centered around a consideration of burning in terms of a number of independent steps and consequences which are considered individually. This approach acknowledges that the potential for and the consequences of burning cannot be evaluated by any one laboratory test, but must be studied in terms of a number of phenomena which may or may not be related. As a general practice, the following preignition and postignition characteristics should be included in a complete study of textile flammability:

Preignition:
   Thermal decomposition: temperatures, extent and rate of weight loss;
   Enthalpy changes before and during thermal decomposition;
   Products of thermal decomposition;
   Kinetics of the ignition process.

Postignition:
   Products of combustion;
   Flame temperatures;
   Heat emission during burning;
   Flame propagation rates;
   Mass burning rates;
   Extinguishability.

Methods for quantitatively evaluating these characteristics are available, and most have been refined to the point where they are adequate for systematic comparisons between materials.

IV. Flammability of Multicomponent Systems

The need for a thorough understanding of the thermal and flammability characteristics of multicomponent systems arises from numerous observations that the behavior of a multicomponent system cannot always be predicted from the behaviors of the individual components. Kruse (1969) observed visually noticeable changes in burning behavior when different fabrics were combined in multiple layers and burned. Bitterli (1971) reported that the addition of polyester or polyamide fibers to cotton or wool caused an increase in the flame propagation rate. Tesoro and Rivlin (1971) studied the behavior of a variety of woven blended fabrics and observed that the oxygen index value for blends could not be predicted from corresponding values of the polymer components. Abbott (1973) investigated fabrics coated with nonfibrous polymers and found that a number of combinations were surprisingly flammable. Miller *et al.* (1973a, b) in studies of the autoignition of polymers showed that in certain systems either chemical or physical interactions are taking place between the polymers themselves or between their thermal decomposition products. In view of the complex structures of the common textile polymers, it is not surprising that such interactions take place, particularly at the high temperatures that are involved in burning. In order to quantify studies of textile flammability, Miller and Martin (1975) have proposed criteria for noninteractive behavior of multicomponent systems in terms of the aforementioned flammability characteristics. These criteria are shown in Table I. On the basis of these it is possible to identify those mixtures whose flammability cannot be predicted from the parent components. In the remainder of this chapter we shall examine several typical multicomponent textile systems in terms of these measurable thermal and combustion characteristics, placing emphasis on those systems that appear to exhibit unusual interactive behavior.

V. Thermal Transitions and Enthalpy Changes

Thermal transition (i.e., decomposition) rates, temperatures, and associated enthalpy changes can be determined using standard thermal analysis techniques such as dynamic thermogravimetry (TG), differential thermal analysis (DTA), and differential scanning calorimetry (DSC). As indicated in Table I the response for mixed systems that do not interact should show decomposition and transition temperatures that are the same as those

TABLE I

Summary of criteria for noninteractive behavior of multicomponent systems

Preignition

Thermal decomposition temperatures	Unchanged
Thermal decomposition weight losses	Equal to the sum of individual losses
Enthalpy changes	Equal to the net enthalpic changes for the two components
Products of thermal decomposition or combustion	Equal to the sum of the contributions from the individual species
Kinetics of ignition	Identical to that of the faster igniting component

Postignition

Flame temperatures	Equal to that of the hotter burning component
Heat evolution during burning	Equal to the weighted average amount of heat produced by the components of the mixture
Flame propagation or mass transformation rates	Equal to that of the faster burning component
Mass burning rate (Unrestrained Vertical Burning)	Equal to the weighted sum of the individual rates

found for the individual components, weight losses should be the cumulative losses for the polymers involved, and enthalpy changes should be the net result of the endothermic and exothermic processes. Conversely, a shift in a decomposition or a transition temperature, or changes in weight loss or enthalpy should be indications of some form of interaction.

As an example, Table II summarizes TG data obtained at a heating rate of 20°C/min in a nitrogen atmosphere for rayon, acrylic, and a 50:50 rayon–acrylic mixture. The weight losses exhibited by the mixture over various temperature intervals are compared with those which would be predicted from the weight losses of the individual components. The total weight loss between 200 and 600°C is nearly identical to the expected weight loss. However, in the lower temperature interval (200–335°C) the weight loss is greater than expected, while in the upper temperature interval (335–600°C) the weight loss is less than expected. Such differences between observed and expected weight losses are indicative of chemical interactions that increase the thermal decomposition rate of one or both polymers. For comparison, Table II also shows similar data for a noninteracting acrylic–nylon mixture for which the weight losses within each temperature interval are the same as the expected values.

In studies of the nylon–cotton system, Miller and Turner (1976) reported DTA results using programmed heating (10°C/min) from room tempera-

TABLE II

TG weight losses (%) for multicomponent systems

	200–600°C	200–335°C	335–600°C
Acrylic	52·0	20·2	31·8
Rayon	80·6	12·6	68·0
50:50 Acrylic–Rayon (Observed)	67·3	28·9	38·4
50:50 Acrylic–Rayon (Expected)	66·3	16·4	49·9
Obs./Exp.	1·02	1·76	0·77
	200–600°C	200–380°C	380–600°C
Acrylic	52·0	40·9	11·1
Nylon	94·0	1·0	93·0
50:50 Acrylic–Nylon (Observed)	73·3	21·8	51·5
50:50 Acrylic–Nylon (Expected)	73·0	21·0	52·0
Obs./Exp.	1·00	1·04	0·99

ture to 500°C in nitrogen. Samples were placed in the aluminum pans commonly used for differential scanning calorimetry with and without covers. Figure 1 shows the thermal behavior of cotton fabric alone in a sealed pan provided with a pinhole leak in its cover. The result is typical for pure cellulose: a wide endothermic peak below 100°C resulting from moisture loss, and an endotherm peaking at 367°C corresponding to decomposition of the polymer. The minor exothermic peak at 325°C is an

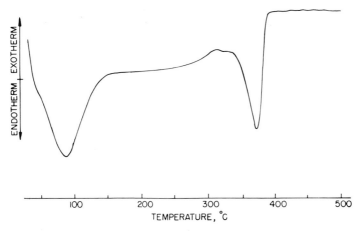

FIG. 1.   Differential thermal analysis of cotton fabric in a closed pan with a pinhole; heating rate = 10°C/min; N$_2$ atmosphere.

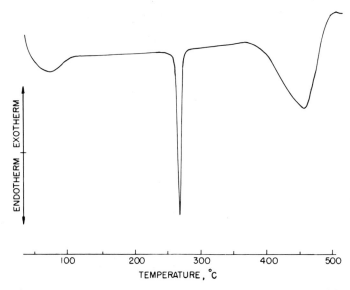

FIG. 2. Differential thermal analysis of nylon 6.6 fabric in a closed pan with a pinhole; heating rate = 10°C/min; N_2 atmosphere.

artifact resulting from a small amount of air that is trapped with the sample. The thermal behavior of nylon 6.6 in a closed pan with a pinhole (Fig. 2) shows moisture loss, melting at 260°C, and an endothermic decomposition starting at 375°C and peaking at about 450°C. Figure 3

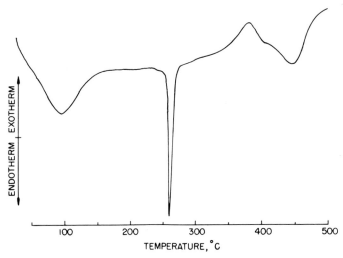

FIG. 3. Differential thermal analysis of a 50:50 nylon–cotton combination in a closed pan with a pinhole; heating rate = 10°C/min; N_2 atmosphere.

presents the comparable thermal behavior in a closed pan of a 50:50 combination of the nylon and cotton. The characteristic peaks for nylon 6.6 are still observed, but in place of the expected cotton endothermic decomposition peak we now see an exothermic response. In contrast, Fig. 4

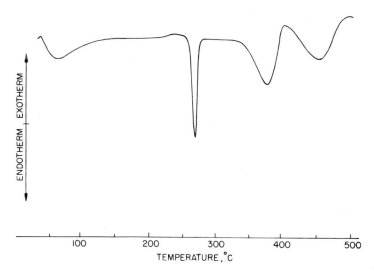

FIG. 4. Differential thermal analysis of a 50:50 nylon–cotton combination in an open pan; heating rate = 10°C/min; N_2 atmosphere.

shows the behavior of the same mixture when run in an open pan. The endotherm for cellulose decomposition has reappeared, and the combined thermogram is a simple composite of the nylon and cellulose responses. These data suggest that an interaction occurs upon heating and that a result of this interaction is the alteration of the thermal decomposition behavior of the cellulose component. Since the interaction does not occur when the mixture is left open to the nitrogen flow, it is considered that the interacting agent is volatile and can be swept away before it has the opportunity to modify the cellulose. The conclusion is that the active agent is a volatile decomposition product of nylon which is formed at a temperature lower than that at which cellulose decomposition takes place. Miller and Turner (1976) have provided strong evidence to suggest that this volatile nylon decomposition product is either ammonia or a low molecular weight amine. For example, Fig. 5 shows the gradual transformation of the decomposition endotherm of cellulose into an exotherm as the result of the treatment of cellulose with ammonia at increasing temperatures.

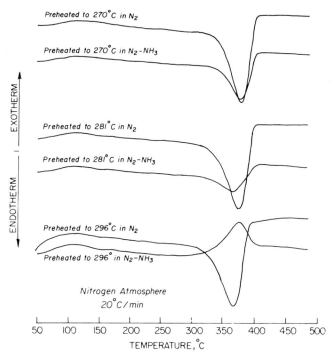

FIG. 5.   Differential thermal analysis of cotton preheated 30 min in nitrogen or nitrogen–ammonia atmospheres at 270, 281, and 296°C.

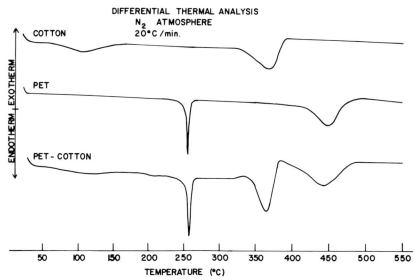

FIG. 6.   Differential thermal analysis of cotton, polyester, and a 50:50 polyester–cotton combination.

Figure 6 shows DTA curves for heating in nitrogen from room temperature to 550°C for cotton, polyester (PET), and a 50:50 combination of polyester and cotton. The cotton thermogram exhibits the normal endotherm below 100°C, indicating loss of water, and a decomposition endotherm peaking at 360°C. The polyester exhibits melting and decomposition endotherms. The polyester–cotton system shows all the parent endotherms with a slight perturbation in the character of the polyester decomposition (i.e., in the presence of cotton the polyester decomposition seems to begin at a lower temperature). Dynamic thermogravimetry (TG) data have also been obtained for polyester–cotton systems (Fig. 7). By obtaining the TG curves and their corresponding derivative curves (DTG) for each component, one can predict theoretical TG and DTG curves for a 50:50 polyester–cotton combination. The TG curve shows a slightly more rapid weight loss and less residual char than predicted; the DTG curve shows

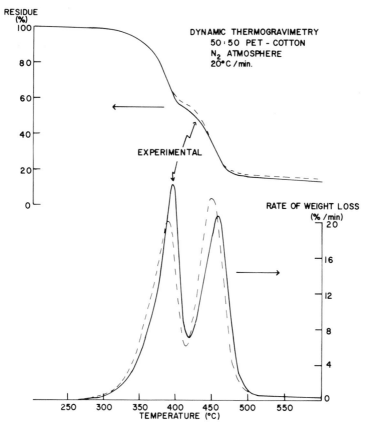

FIG. 7. Observed and expected dynamic thermogravimetry curves for 50:50 polyester–cotton.

faster weight loss for the cotton component and slower weight loss for polyester. Various flame-retardant polyester–cotton systems have also been investigated leading to the conclusion that early (low temperature) decomposition products of untreated or flame-retardant polyester affect the decomposition of untreated cotton (Miller *et al.*, 1975).

The thermal decomposition of polyester–wool blends provides another interesting example of interactive behavior (Martin and Miller, 1976). Figure 8 shows expected and observed TG and DTG curves for a 55:45

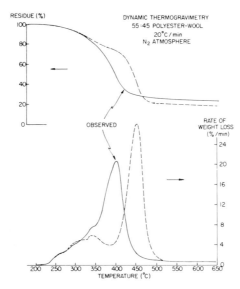

FIG. 8.   Observed and expected dynamic thermogravimetry curves for a 55:45 polyester–wool fabric.

polyester–wool blend (actually 42% wool when the composition is corrected for moisture content). The TG curve shows the major weight loss occurring sooner than expected (i.e., at a lower temperature) and a final char (at 650°C) greater than the theoretical value. The DTG curve shows that the polyester decomposition peak, normally at 450°C, is shifted to a lower temperature. The wool does not appear to be affected. Figures 9 and 10 summarize the thermogravimetry results in terms of the polyester peak temperature (from the DTG curves) and the residual char at 650°C (from the TG curve) as a function of wool content. These plots demonstrate that the presence of wool systematically decreases the decomposition temperature of the polyester and increases residual char.

The nylon–wool system also seems to exhibit interactive behavior during thermal decomposition (Martin and Miller, 1976). Figure 11 shows expected and observed TG and DTG curves for a 55:45 nylon–wool blend. The

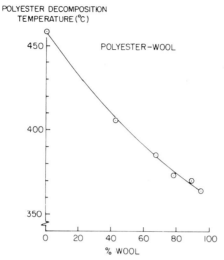

FIG. 9.   Polyester decomposition temperatures (from peaks of DTG curves) for polyester–wool blends.

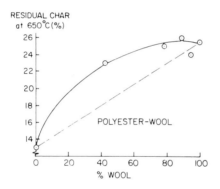

FIG. 10.   Residual char after inert atmosphere pyrolysis of polyester–wool blends.

TG curve has the major weight loss occurring at a slightly lower temperature than predicted and final char slightly greater than expected. The DTG
curve shows the nylon peak shifted to a slightly lower temperature and
decreased somewhat in intensity. However, the magnitude of these effects
is not as dramatic as those seen for polyester–wool.

VI. Thermal Decomposition Products

The products of the thermal decomposition of any given fiber type are
strongly dependent upon the rate of heating, temperature of decomposition, gaseous environment, and time. These products are of great

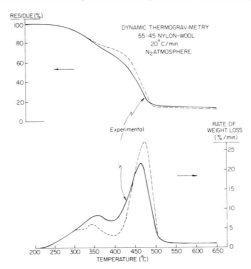

FIG. 11.   Observed and expected TG and DTG curves for a 55:45 nylon–wool blend.

importance, most particularly the combustible volatile components, since it is these species that actually provide the fuel that effects ignition and propagates the flame. Many studies have been conducted to identify the decomposition products of various fibers under specified laboratory conditions. Most of these studies have concerned themselves with pyrolytic conditions either in the presence or absence of air, while only limited studies have been conducted under combustion or flaming conditions. The importance of these studies in terms of generating a mechanistic understanding of flammability and flame retardancy cannot be overestimated.

Analysis of thermal decomposition products also provides a useful means of identifying interactive behavior in multicomponent systems. For example, Miller *et al.* (1976) have studied the pyrolysis of polyester–cotton systems brought to 350°C by heating at ~35°C/min in an inert helium atmosphere. The volatile decomposition products were collected in a liquid nitrogen cold trap and subsequently analyzed qualitatively by infrared spectroscopy and quantitatively by gas chromatography. At this pyrolysis temperature, polyester produced no detectable hydrocarbons and cotton produced only methane. However, the polyester–cotton mixed system produced a larger amount of methane, some acetaldehyde, and a significant quantity of ethylene which is not seen as a decomposition product of either component alone at this temperature. Figure 12 shows the ethylene generation as a function of pyrolysis temperature for polyester, cotton, and a 50:50 polyester–cotton combination. These results

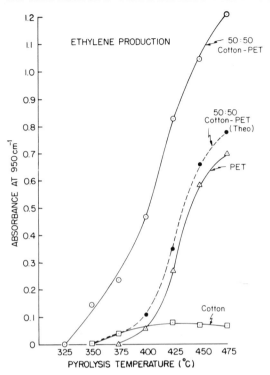

FIG. 12. Temperature dependence of ethylene evolution from pyrolysis of cotton, polyester, and 50:50 polyester–cotton.

show that ethylene is evolved from the mixture at a lower temperature and in greater amounts than predicted, indicating an interactive system.

Similar studies were conducted on the nylon–cotton system with attention again focused on volatile decomposition products (Miller and Turner, 1976). Mass spectrometry was used to monitor the products from these polymers, which were wrapped around a metal mesh cylinder serving as a sample holder during pyrolysis in an inert atmosphere. Three sets of combinations were studied: (1) nylon 6.6 and cotton yarns, (2) a needle-punched felt containing a 50:50 mixture of nylon 6.6 and cotton, and (3) a mixture of nylon 6 and cotton yarns. The results in Table III show a consistently greater CO_2 evolution than expected (between 31 and 38% greater) for all three combinations. It is interesting to note that, even though nylon 6 by itself gives off much less CO_2 than nylon 6.6, the effect of combining it with cotton yarns is to produce almost as much excess CO_2 as with nylon 6.6 (another indication that thermal behavior of the nylons probably influences the decomposition of the cellulose). Parallel oxidative decomposition studies were carried out using an environment of 21%

oxygen in argon. Once again there appeared to be a significant increase in CO_2 production when the polymers were intimately combined, indicating interactive behavior.

TABLE III

CO_2 production from the thermal decomposition of nylon and cotton in argon[a]

	Relative Partial Pressure of CO_2/gram sample		
Cotton (fabric)	40·2		
Nylon 6.6 (fabric)	31·5		
50 : 50 Nylon 6.6-Cotton (fabric)	49·5	$\dfrac{\text{Obs.}}{\text{Exp.}} = 1\cdot38$	
	35·8		
Nylon 6.6 (felt)	34·2		
50 : 50 Nylon 6.6-Cotton (felt)	48·8	$\dfrac{\text{Obs.}}{\text{Exp.}} = 1\cdot31$	
	37·8		
Nylon 6 (fabric)	19·8		
50 : 50 Nylon 6–Cotton (fabric)	39·4	$\dfrac{\text{Obs.}}{\text{Exp.}} = 1\cdot31$	
	30·0		

[a] Heated to 570°C at 10°C/min.

The results of interactions occurring during the decomposition of poly-ester–wool systems can be observed by study of the thermal decomposition products during pyrolysis in an inert helium atmosphere at ~35°C/min. The volatile decomposition products, collected in a liquid nitrogen cold trap, were analysed by infrared spectroscopy. The pyrolysis of wool at 450°C, a temperature at which both polyester and wool undergo significant decomposition, produces CO, CH_4, C_2H_6, HCN, and NH_3, while the pyrolysis products of polyester at this same temperature include CO, CO_2, CH_4, CH_3CHO, C_2H_2, and C_2H_4. At a lower pyrolysis temperature of 375°C wool undergoes nearly complete decomposition while polyester just starts to decompose with the generation of acetaldehyde. The pyrolysis products at 375°C of the blend, however, include all the decomposition products from wool, no acetaldehyde, but significant amounts of ethylene and acetylene, which are normally obtained only when polyester is heated above 375°C. Figures 13 and 14 show the cumulative evolution of ethylene and acetylene as a function of pyrolysis temperature for polyester alone and for a 70:30 polyester–wool mixture containing the same weight of polyester. These data show that ethylene and acetylene, both highly combustible, are evolved from polyester at a lower temperature when wool is present.

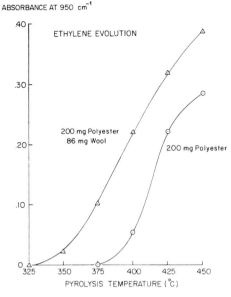

FIG. 13. Temperature dependence of ethylene evolution from pyrolysis of polyester and a 70:30 polyester–wool blend.

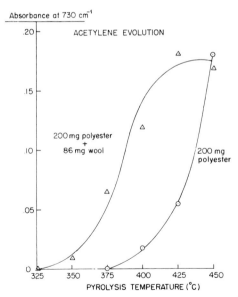

FIG. 14. Temperature dependence of acetylene evolution from pyrolysis of polyester and a 70:30 polyester–wool blend.

VII. Ignition Behavior

The generation of flaming combustion (ignition) is of obvious critical importance in attempts to understand the flammability problem. Ignition will occur when an adequate heat flux is supplied to a material, resulting in thermal decomposition and the generation of combustible volatile gases. These gases will ignite if and when their concentration reaches a critical value. The source of heat for the thermal decomposition may be in the form of a spark, a flame, radiant heat, convective heat, or conductive heat. Autoignition is defined as the generation of flaming combustion when a material is placed in contact with heated air in the absence of any other specific ignition source. The process depends on convective heat transfer and is a highly reproducible phenomenon, in contrast to ignition by radiant heating, flame impingement, or conductive heat transfer methods.

Several attempts have been made to describe autoignition in terms of ignition temperatures of various materials (Welker, 1970; Smith and King, 1970). More recently Miller *et al.* (1973a, b) have described autoignition in kinetic terms. These workers developed a technique for measuring and interpreting the time required for ignition at selected elevated air temperatures, which offers a simple and highly reproducible method for studying the kinetics of unpiloted ignition. In this technique, the temperature of the air in a furnace is suitably controlled and measured by a stationary thermocouple mounted in the furnace near the sample position. A rectangular sample is impaled on two prongs at the end of a removable holder. Positioned between the sample prongs but not touching the sample is a second thermocouple monitored by a strip chart recorder. With the furnace at the desired temperature, the sample holder is injected quickly into the oven. The time of injection is of the order of a fraction of a second, short enough to give a sharp inflection point on the recorder chart measuring the output of the thermocouple on the sample holder as a function of time. A second inflection point in the recorder output indicates the point of ignition, and the time between these two inflection points is called the ignition time, θ.

The observed ignition time, θ, has been shown to depend on sample mass as well as on the thermal and chemical degradative properties of the polymer. Sample mass can be conveniently varied by using multilayered samples. Over the mass range typical of textile materials the relationship between θ and sample mass is linear, and the extrapolated intercept of the line with the θ axis is termed the intrinsic ignition time, θ_0. This linear extrapolation of the relationship between ignition time and sample mass to zero mass results in an ignition time for a sample assumed to be instantaneously heated to its decomposition temperature. The extrapolation to zero mass is also equivalent to eliminating any effect from the furnace wall,

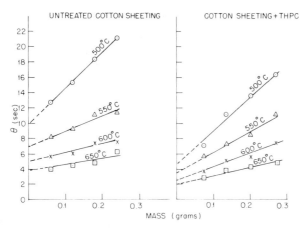

FIG. 15.   Autoignition of untreated cotton sheeting (left) and cotton sheeting + THPC (right).

so that the extrapolated intrinsic ignition time also represents the ignition time for a finite sample in an infinitely large furnace. Typical data for untreated and flame-retardant-treated cotton are shown in Fig. 15.

In Table IV are shown intrinsic ignition times at several furnace temperatures for a number of textile fibers. As expected, intrinsic ignition times for any given fiber decrease with increasing temperature, and the intrinsic ignition time for different fibers at any given temperature varies over a fairly wide range. It must be emphasized that the values shown in Table IV should not be considered as absolute material constants for any given fiber type. The specific value depends on a number of factors, including origin of the fiber, previous history, and fiber morphology. It is of some significance, however, that intrinsic ignition times are independent of the surface area of the material as evidenced by the observation that rayon and cellophane, different physical forms of regenerated cellulose, show identical ignition behavior (Miller *et al.*, 1973a).

TABLE IV

Intrinsic ignition times (sec)

	550°C	600°C	650°C	700°C	750°C
Cellulose acetate	8·1	7·0	4·9	—	—
Cotton	10·3	6·7	4·8	3·7	—
Polyacrylonitrile	14·6	11·2	6·8	5·0	—
Polypropylene	—	13·7	9·2	7·2	5·4
Polyester	—	12·9	9·5	7·4	6·0
Nylon 6.6	—	9·8	7·6	6·0	5·0
Nomex®	—	—	12·5	8·0	4·8

The temperature dependence of the intrinsic ignition time may be expressed in the form of an Arrhenius plot (log $1/\theta_0$ versus $1/T$). As can be seen in Fig. 16, the relationship is consistently linear with an apparent activation energy (obtained from the slopes) corresponding to 8–10 kcal/mole for all materials investigated, except Nomex®, which shows

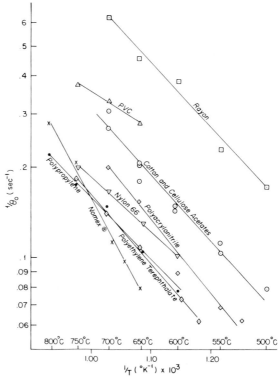

FIG. 16. Arrhenius plots of intrinsic ignition time, θ_0, as a function of furnace air temperature, for single-component systems.

an activation energy of 17 kcal/mole. These low activation energies, coupled with the fact that the same activation energy is obtained for nearly all chemical species investigated, suggest that the ignition phenomenon is controlled by physical processes (heat or mass transfer). The activation energy is derived using the temperature of the air in the oven and therefore probably represents the kinetics in the gas phase where diffusion of combustible gases from the fiber substrate would be important.

Turning to the behavior of multicomponent fiber systems, autoignition behavior can be used to identify those mixed fiber systems in which interactions between the components are taking place before the appearance of a flame. For a noninteracting multicomponent system, the ignition time (in particular, the intrinsic ignition time) should be that of the faster igniting component. Such is the case when rayon and acrylic yarns are combined. As can be seen in Table V, the intrinsic ignition time, θ_0, for the mixed system is the same as that for the rayon, the faster igniting of the two. In contrast, the behavior of a mixture of nylon and polyester yarns is such that the intrinsic ignition time for this mixture is shorter than that for either of the two parent components. This type of behavior identifies a combination that is undergoing some form of chemical interaction even before the appearance of a flame. Similarly, in the case of the polyester–cotton system, it appears that at 650°C, the mixed fiber system ignites faster than either one of the two parent fibers. This autoignition behavior of polyester–cotton blends confirms the conclusions regarding preignition interaction, drawn on the basis of other thermal decomposition studies (e.g., Fig. 12).

TABLE V

Intrinsic ignition times for several multicomponent fiber systems

	Temp.	θ_0 (sec)
Acrylic	600°C	6·0
Rayon		4·2
Acrylic–Rayon (50:50)		4·2
Nylon	650°C	8·1
Polyester		9·2
Nylon–Polyester (50:50)		6·0
Polyester	650°C	9·5
Cotton		4·9
Polyester–Cotton (50:50)		3·7

VIII. Flame Propagation Rate

Once ignition has occurred, it is evident that the rate of burning is of importance in any attempt to determine the general flammability behavior of textile materials. In most studies, fabric burning rate is measured by a direct determination of the linear velocity of the flame. Usually, the time for some designated part of the flame to move a given distance along a fabric sample is determined either visually or by means of some type of automatic device. Several research workers have generated useful and interesting information about textile flammability using this approach to the measurement of burning rate (Nielson and Richards, 1969; Feikema, 1973). However, as the flame front moves along a specimen in the normal burning process, a variable situation is created with respect to the gaseous environment, boundary conditions, and the heat transfer process in the vicinity of the flame front. As a result, burning rate studies that depend on a moving flame front generate data of poor reproducibility. A more reliable set of experimental conditions would be achieved if the flame front were stationary in space. To accomplish this, the specimen would have to be fed into the flame front at just the rate required to maintain the flame in a fixed location. This rate would be equal to the flame propagation rate. Furthermore, it would seem that other important flammability-related physical properties, such as flame temperature, heat emission, and smoke generation, could be measured more precisely if the flame were stationary in space. Miller and his associates (Miller and Meiser, 1971) at the Textile Research Institute have designed an apparatus to achieve such a steady-state burning condition. In the TRI Flammability Analyzer (Fig. 17), fabric is moved into its flame at a variable and known rate by mounting the specimen on the rims of a pair of linked wheels rotating on a horizontal axis, and arranging for the continuous adjustment of rotational speed. After ignition, the speed of the fabric holder is adjusted until the flame is observed to remain stationary at some predetermined point in the circular path. The rate of fabric movement is then equivalent to the rate of flame propagation. The wheels operate within an enclosure fed at the bottom

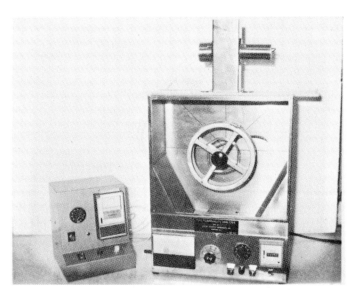

FIG. 17. The TRI Flammability Analyzer: flammability evaluation via moving fabric-stationary flame. (TRI Flammability Analyzer of U.S. Patent 3,667,277; inventors B. Miller, H. Lambert, and C. H. Meiser. Custom Scientific Instruments, Inc., Whippany, N.J., manufacture and market the Analyzer.)

from a gas-mixing system so that the gaseous environment, particularly the oxygen volume fraction and the total gas flow rate, can be controlled and varied systematically.

An interesting and particularly useful feature of the TRI Flammability Analyzer is that any direction of flame progress can be measured. With the flame kept at the top of the fabric path (i.e., at 12 o'clock), burning is in effect horizontal. However, the same type of experiment can be performed with the flame remaining at 3 o'clock (vertical upward burning), at 9 o'clock (candle-like downward burning), or at any predetermined angle of inclination. As is well known, entirely different flaming and combustion processes take place according to the orientation of the fabric specimen during burning.

The value obtained for the flame propagation rate (FPR) from the Flammability Analyzer can be multiplied by the fabric areal density (fabric weight) to give a measure of the rate at which fabric mass is transformed by the flame (i.e., the mass transformation rate, MTR). The TRI Flammability Analyzer can be readily used to study multilayers of fabrics and identify those systems in which either chemical or physical interactions are taking place. Not only are the results pertinent to the behavior of textile materials commonly used in multilayer configurations, such as carpet, lined draperies, furniture upholstery, and the total clothing assembly, but it has been found that multilayers can be used to simulate intimate fiber blends. Some typical data are shown in Table VI for a number of double-layered systems.

In terms of mass transformation rate, evidence has been accumulated that a noninteracting multicomponent system yields MTR values for a mixed system equal to that of its faster burning component. On the basis of the transformation rate data in Table VI, it would appear that the polyester–acrylic and the cotton–acrylic systems are noninteracting, while cotton–nylon, which burns faster than either component alone, and FR (flame retardant) cotton–acrylic, which burns at a rate between those of its two components, do interact.

Wool-containing systems are generally strongly interactive because of the high chemical reactivity of keratin. Interesting data for polyester–wool blended fabrics are shown in Fig. 18 in terms of MTR as a function of composition for both horizontal and 45° upward burning modes. In the 45° upward burning mode, the mass transformation rates for the blends are consistently higher than for either component; in the horizontal burning mode the mass transformation rates for the blends become greater than for either component near the 50:50 composition level. Both burning modes show maxima in MTR at 45 and 95% wool. Similar data for the nylon–wool system are shown in Fig. 19. Again one sees a complex burning behavior. MTR values for the blends are somewhat greater than the faster

TABLE VI

Horizontal mass transformation rates for fabric double layers

	MTR at 21% O_2 (oz/min-in.)
Noninteracting:	
Polyester-Acrylic	0·025
Polyester–Polyester	0·023
Acrylic–Acrylic	0·025
Cotton–Acrylic	0·027
Cotton–Cotton	0·014
Acrylic–Acrylic	0·025
Interacting:	
Cotton–Nylon	0·023
Cotton–Cotton	0·014
Nylon–Nylon	0·017
FR Cotton[a]–Acrylic	0·019
FR Cotton–FR Cotton	0·015
Acrylic–Acrylic	0·025

[a] Cotton + 1·2% Diammonium Phosphate.

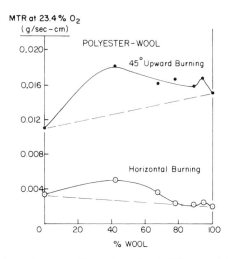

FIG. 18.　Mass transformation rates for horizontal and 45° upward burning of polyester–wool fabrics.

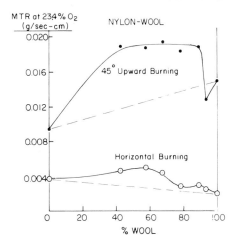

FIG. 19. Mass transformation rates for horizontal and 45° upward burning of nylon–wool fabrics.

burning component over a much wider composition range. The one exception is the blend containing 95% wool, which burns upward more slowly than expected.

Particularly meaningful data pertaining to fabric flammability behavior can be obtained from the TRI Flammability Analyzer by making measurements at various oxygen concentrations. It is of some interest to know the burning properties of fabrics at oxygen concentrations other than the nominal ambient value of 21% (by volume). It is well known that, without considerable ventilation, the oxygen concentration in the immediate vicinity of a flame front rapidly decreases after ignition. In such confined burning, normal convective currents do not seem to be adequate to keep the oxygen concentration at the 21% level. Also, in many environments textile fabrics are used at oxygen concentrations above the normal 21%, for example, in hospitals and aerospace applications. As might be expected, the burning rate in terms of MTR increases with oxygen concentration. Furthermore, the linear relationship between MTR and oxygen concentration, shown schematically in Fig. 20, allows the determination of two important indices of extinguishability: extrapolated oxygen index, $(O.I.)_0$, and oxygen sensitivity, $\Delta MTR/\Delta\% O_2$. The extrapolated oxygen index is the oxygen concentration corresponding to zero burning rate, and the oxygen sensitivity is a measure of the propensity of the burning to slow down as environmental oxygen is consumed (Miller *et al.*, 1973c). From a standpoint of hazard reduction, one would want a material with a low mass transformation rate (at 21% O_2), high extrapolated oxygen index, and high oxygen sensitivity.

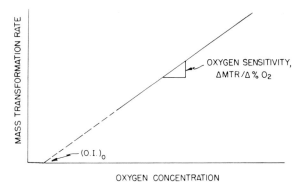

FIG. 20. Schematic diagram of mass transformation rate as a function of oxygen concentration showing extrapolated oxygen index, $(O.T.)_0$, and oxygen sensitivity, $\Delta MTR/\Delta\% O_2$.

Typical data for fabrics of several fiber types showing the dependence of burning rate (MTR) in the downward burning mode on oxygen concentration are shown in Fig. 21. The ease of extinction in terms of the extrapolated oxygen index $(O.I.)_0$ and the dependence of this quantity on the direction of burning is illustrated for these same materials by the data in Table VII.

Figure 22 shows MTR values as a function of oxygen concentration for untreated and flame-retardant-treated (FR) polyester–cotton double-

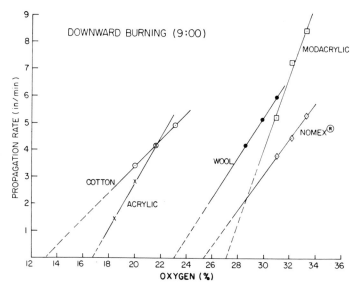

FIG. 21. Flame propagation rates for downward burning as a function of environmental oxygen concentration.

TABLE VII

Ease of extinction values

Fabric	$(O.I.)_0 \downarrow$ [a]	$(O.I.)_0 \leftarrow$ [b]	$(O.I.)_0 \uparrow$ [c]
Cotton	0·13	0·08	0·07
Acrylic	0·17	0·06	0·07
Wool	0·23	0·10	0·12
Nomex®	0·25	—	—
Modacrylic	0·27	0·14	0·13

[a] $\downarrow$ Indicates vertical downward burning. [b] $\leftarrow$ Indicates 45° upward burning. [c] $\uparrow$ Indicates vertical upward burning.

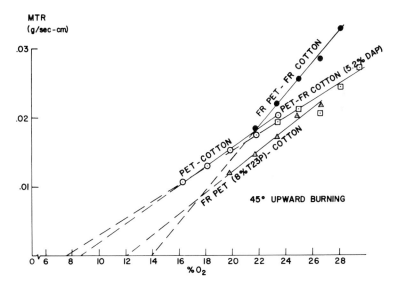

FIG. 22. 45° upward mass transformation rates as a function of ambient oxygen concentration for double layered polyester–cotton systems with and without added flame retardants.

layered combinations. The extrapolated oxygen index and oxygen sensitivity are essentially the same for both polyester– cotton and polyester–FR cotton systems. The FR polyester–cotton system shows a slightly higher extrapolated oxygen index, but no change in oxygen sensitivity. However, when both components are treated, the extrapolated oxygen index is even higher, and the oxygen sensitivity is significantly increased, both changes reflecting increased ease of extinguishability.

IX. Flame Temperatures

Interesting insight about flammability behavior, and particularly about multicomponent systems, can also be obtained from flame temperature measurements. Flame temperatures are sensitive indicators of interactions that may be occurring once a material commences burning. For flexible, sheet-like materials under ordinary flaming conditions, the measurement of flame temperature can be analytically difficult because the flame is transient in position and possibly in size. In contrast, the TRI Flammability Analyzer produces a steady flame which can easily be probed with a fine thermocouple to obtain its maximum temperature with considerable precision (Miller and Meiser, 1969). It must be emphasized that flame temperature values are strongly dependent on the burning conditions and the method of measurement. Nevertheless, determined under controlled conditions and considered as relative values, flame temperatures are revealing physical quantities. This is again particularly true in attempts to identify interactive behavior in multicomponent systems. Thus, if the observed flame temperature of a mixed system is the same as that for the component that burns at the higher temperature, it may be considered that no chemical interactions are occurring. Conversely, interactions are indicated when the flame temperature of a mixture varies in either direction from this value. Both these possibilities are illustrated in Table VIII.

TABLE VIII

Flame temperatures of fabric double layers
(Horizontal Burning)

	Maximum Flame Temperature at 21% O_2 (°C)
Noninteracting:	
Cotton–Acrylic	976
Cotton–Cotton	974
Acrylic–Acrylic	910
Cotton–Polyester	950
Cotton–Cotton	974
Polyester–Polyester	649
Interacting:	
Acrylic–Nylon	942
Acrylic–Acrylic	910
Nylon–Nylon	861
Cotton–Nylon	902
Cotton–Cotton	974
Nylon–Nylon	861

Cotton–acrylic and cotton–polyester mixtures both show about the same flame temperature in horizontal burning as that for cotton alone. On the other hand, a mixture of acrylic and nylon fabrics produces a flame temperature higher than that for either of the two homogeneous systems. The cotton–nylon combination burns with a flame temperature roughly between the high value for cotton and the lower value for nylon. The latter might seem to be what one would expect from a noninteracting combination; however, in general, noninteracting systems have been found to burn with a flame temperature close to that of the hotter burning component. On examining the effect of nylon–cotton blend composition (Fig. 23), it is seen that flame temperature increases systematically with the proportion of nylon in the blend. In fact, an 85 : 15 nylon–cotton combination has a flame temperature essentially equal to that of pure cotton, in spite of the fact that pure nylon burns with a lower flame temperature than cotton. It appears that when the two polymers are combined, the net combustion process is not what might be expected.

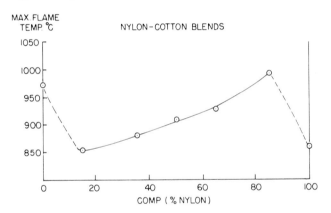

FIG. 23. Flame temperatures at 21% O_2, for horizontal burning of a series of nylon–cotton blends.

It is of interest to examine further the flammability behavior in terms of flame temperature of some common textile blends. As discussed above, it is frequently useful to measure flammability behavior as a function of the oxygen content of the immediate environment during burning. Normally, the flame temperature increases linearly with the oxygen concentration. This dependence (in terms of the slope of the generated lines) can be a sensitive indicator of any changes in flammability behavior. Data for polyester–cotton mixtures are shown in Fig. 24. At any given oxygen concentration, the flame temperatures of all mixtures of polyester and cotton are essentially the same as the flame temperature of pure cotton. During normal burning at 21% oxygen, the polyester material in the mixed system

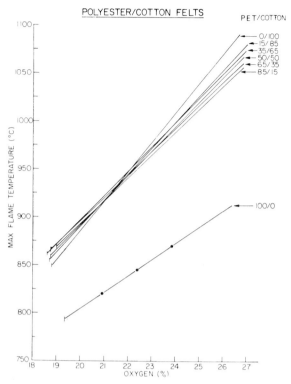

FIG. 24.   Flame temperature as a function of O_2 concentration for downward burning of polyester–cotton felts.

experiences a temperature at least 100 degrees higher than it would if it were burning by itself. Even if this increased temperature did not produce new chemical reactions, the effect on the kinetics of the combustion processes would obviously be significant. Thus, we would predict from these data that polyester–cotton blends should show somewhat enhanced flammability over what might be expected from them on the basis of the two component materials. As previously mentioned, this has been confirmed by a variety of evaluation techniques.

Other systems have been studied with regard to the oxygen sensitivity of flame temperature. For example, data for combinations of polyester and wool are shown in Fig. 25. Here there is a clear indication that there is some chemical interaction when these two fibers are burned together. The oxygen sensitivity of the flame temperatures for the blends is considerably different from that of the two parent species, which coincidentally have just about the same oxygen sensitivities. In Fig. 26 are shown data for polyacrylonitrile and wool. Mixtures of these two fibers have the same oxygen sensitivity as that found for the pure components, and the only indication of

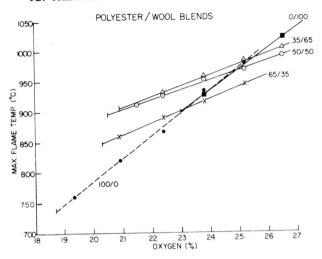

FIG. 25.   Flame temperature as a function of O_2 concentration for downward burning of polyester–wool blends.

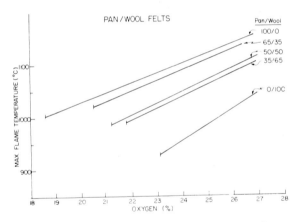

FIG. 26.   Flame temperature as a function of O_2 concentration for downward burning of polyacrylonitrile–wool blends.

interaction is the reduction in flame temperature below that of the pure acrylic.

X. Burning Rate

As indicated previously, the burning rate is of importance in any attempt to describe textile flammability behavior. Miller *et al.* (1975) have pointed out the limitations inherent in the measurement of flame propagation rates

particularly when the specimens must be restrained in some fashion so that flame movement can be followed. First, a flame propagation rate calculated from the distance between index points divided by the corresponding measured time interval is an average rate and cannot reveal any acceleration or deceleration that may occur. Indeed, the general use of this method has fostered the impression that flame propagation rates are unchanging during the burning process. Yet it is obvious that many fabrics do not burn at a constant rate, especially when burning upward. Secondly, it is customary to derive a "burning rate" by multiplying the observed linear propagation rate by fabric areal density. The result is the mass transformation rate, an apparent burning rate in terms of the rate at which a unit width of fabric is transformed by the advancing flame. This derived quantity gives no indication as to what fraction of the fabric is actually converted to volatile products, or what fraction is burned. Finally, restraining fabric samples during flammability measurements in a manner unrelated to their use in actual practice is quite unrealistic. Not only does the restraining frame serve as a heat leak path which may reduce the vigor of the burning process, but it also prevents the sample from moving in response to the heat of the flame. In real situations, most textiles exposed to a heat source undergo a variety of movements—curling, twisting, shrinking, etc.—depending on the chemical composition of the fibers, the construction of the fabric, and other factors associated with fiber and fabric structure.

Miller and his associates (1975) developed an apparatus that deals with all the aforementioned problems. In the TRI Vertical Burning Monitor, sample weight is measured during the upward burning of unrestrained fabric samples. It is thus possible to measure the instantaneous rate at which solid is changed into gaseous products directly and continuously. In the apparatus a fabric strip is suspended from a transducing system connected to a time-base chart recorder so as to provide a record of the weight of the fabric as it burns. To help achieve a smooth flame start, the strip is ignited at the tip of a "V" cut at its lower end. The transducer is protected from heat and gases by metal and asbestos shielding, and it has been demonstrated that flame-generated convective currents do not appreciably affect the weight reading. The data that are obtained from the TRI Vertical Burning Monitor are in terms of weight loss (mass) as a function of time; the slope of such a curve normalized by fabric width is the mass burning rate (MBR). It has been found that the most effective way of data presentation is to plot MBR as a function of sample weight loss (Δw) normalized by the fabric areal density. The slope of such a curve is a measure of the burning acceleration. Data for several fiber types are shown in Fig. 27. It is of interest to note that the several fibers fall into three categories of behavior. Cotton and rayon are the most rapidly burning

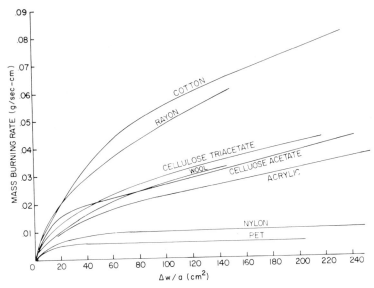

FIG. 27. Mass burning rate (upward) as a function of weight loss/fabric areal density ($\Delta w/a$) for unrestrained single fabric layers; samples 24 in $\times$ 2 in, all stitched with glass yarn.

fibers and also exhibit the greatest acceleration. The two thermoplastic fibers, nylon and polyester, exhibit only slight acceleration at the beginning of the burning process and then level off at a relatively slow rate of burning. The other fibers appear to be grouped together with respect to rate and acceleration, and form an intermediate category.

Multicomponent systems are again of particular interest in terms of mass burning rate behavior. Data in Fig. 28 for the polyester-cotton system show that burning rate and acceleration of the mixed system are controlled by the rapidly burning and accelerating cotton component. Examining the burning behavior of various flame-retardant systems, Fig. 29 shows the burning rate behavior of polyester–cotton combination in which the polyester component was made flame retardant by addition of 8% tris-2,3-dibromopropyl phosphate (T23P). Here again the cotton appears to be controlling the burning behavior of the blend. However, when flame-retardant cotton (treated with 5·2% DAP) is combined with either untreated or flame-retardant polyester, the polyester becomes the rate-controlling component as shown by the data in Figs 30 and 31.

Heat emission measurements were also made on these polyester–cotton combinations during unrestrained upward burning under a forced air flow. These measurements were made by continuously monitoring the temperature of the exiting gas stream. Table IX presents total heat emission (expressed as calories per gram of material volatilized), and also maximum

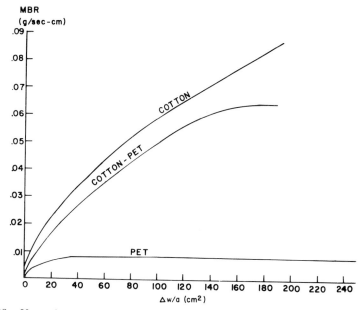

FIG. 28.   Upward mass burning rates for unrestrained double layers of polyester (PET) and cotton and a double layered combination of polyester and cotton.

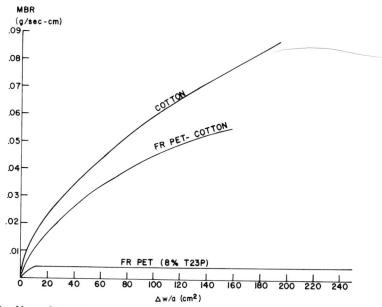

FIG. 29.   Upward mass burning rates for unrestrained double layers of cotton and FR polyester and a double layered combination of FR polyester and cotton.

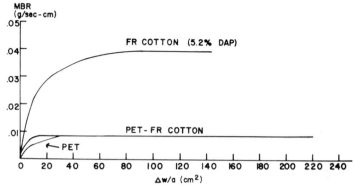

FIG. 30.   Upward mass burning rates for unrestrained double layers of FR cotton and polyester and a double layered combination of polyester and FR cotton.

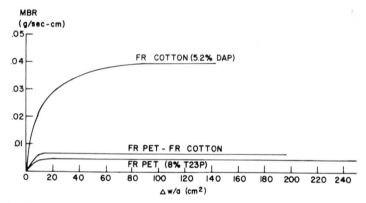

FIG. 31.   Upward mass burning rates for unrestrained double layers of FR cotton and FR polyester and a double layered combination of FR polyester and FR cotton.

TABLE IX

Heat emission from unrestrained double-layered polyester–cotton systems burning upward under an air flow rate of $6\cdot8$ cm/sec

	Total heat emission (cal/g)	Maximum rate of heat emission (cal/sec-cm)
PET–Cotton	3670	230
FR PET[a]–Cotton	3030	160
PET–FR Cotton[b]	3360	130
FR PET–FR Cotton	2340	40

[a] PET + 7% T23P; [b] Cotton + $2\cdot8$% DAP.

rates of heat emission (expressed as calories per second per unit width of fabric) for the four polyester–cotton systems where neither, one, or both components were treated with flame retardant. The data clearly show that the most significant decrease in both total heat emission and rate of heat emission occurs when both the polyester and cotton components have been treated with a flame retardant. It is of interest that these results correlate with the extinguishability data which show that ease of extinguishability is increased only when both components are treated with flame retardants (Fig. 22). This correspondence between heat emission and self-extinguishability is not unexpected, since feedback of heat to the fabric is one of the critical factors involved in maintaining fabric burning.

A particularly interesting example of a multicomponent system where interactive behavior has been observed is the polyester–acrylic system. As shown in Fig. 32, in this case the mixed system burns faster and accelerates more than either of the two components. This interactive behavior is probably due to the strong exothermic decomposition of polyacrylonitrile (Schwenker *et al.*, 1964).

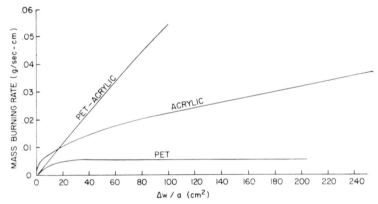

FIG. 32.   Upward mass burning rates for unrestrained double layers of acrylic and polyester (PET) and a double layered configuration of acrylic and polyester.

XI. Conclusion

It is evident that the flammability behavior of textile materials is quite complex, and that an important source of this complexity is the multifiber character of many textile products. In many mixed fiber systems either chemical or physical interactions take place which make it difficult, if not impossible, to predict the flammability of the mixed system from the behavior of the component fibers.

Significant progress has been made in our efforts to understand textile flammability behavior. By considering a variety of preignition and postignition events, it has been possible to define criteria for noninteractive behavior, and to identify those multicomponent systems where interactions seem to occur. On the basis of systematic studies of thermal decomposition and ignition phenomena, flame temperatures, burning rates, heat emission and extinguishability, it should be possible to develop new textile products and processes to satisfy both government regulations and consumer demands. Numerous important developments have already taken place, and we can expect many more exciting and useful developments in the years ahead.

References

Abbot, N. J. (1973). *J. Coated Fabrics* **3**, 135–141.

Bitterli, W. (1971). *Textilver.* **6**, 660–664.

Feikema, J. G. (1973). *Melliand Textilber.* **54**, 179–184.

Fenimore, C. P. and Martin, F. J. (1966). *Modern Plastics* **45**, 141–192.

Krasny, J. F. (1975). "NBS Activities in Apparel Flammability", Proceedings of Ninth Annual Meeting of the Information Council on Fabric Flammability, New York.

Kruse, W. (1969). *Melliand Textilber.* **50**, 460–469.

Martin, J. R. and Miller, B. (1976). "The Flammability Behavior of Mixtures Containing Wool and Synthetic Fibers", Schriftenreihe Deutsches Wollfor-schungsinstitut an der Technischen Hochschule Aachen.

Miller, B. and Martin, J. R. (1975). *J. Fire Flam.* **6**, 105–118.

Miller, B. and Meiser, C. H., Jr. (1969). "Flame Temperature Measurements on Textile Systems", Information Council on Fabric Flammability, Proceedings of the Third Annual Meeting, pp. 116–125.

Miller, B. and Meiser, C. H., Jr. (1971). *Textile Chem. Col.* **3**, 118–122.

Miller, B. and Turner, R. (1976). *Appl. Polym. Symp.* **28**, 855–868.

Miller, B., Martin, J. R. and Meiser, C. H., Jr. (1973a). *In* "Polymers and Ecological Problems" (James Guillet, ed), pp. 93–107. Plenum Press, New York and London.

Miller, B., Martin, J. R. and Meiser, C. H., Jr. (1973b). *J. Appl. Polymer Sci.* **17**, 629–642.

Miller, B., Goswami, B. C. and Turner, R. (1973c). *Textile Res. J.* **43**, 61–67.

Miller, B., Martin, J. R. and Meiser, C. H., Jr. (1975). *Textile Chem. Col.* **7**, 68–72.

Miller, B., Martin, J. R., Meiser, C. H., Jr. and Gargiullo, M. (1976). *Textile Res. J.* **46**, 530–538.

Nielson, E. B. and Richards, H. R. (1969). *Textile Chem. Col.* **1**, 270–277.

Rebenfeld, L. and Miller, B. (1974). *J. Fire Flam., Consumer Prod. Flam.* **1**, 225–239.

Schwenker, R. F., Beck, L. R. and Zuccarello, R. K. (1964). *American Dyestuff Reporter* **53**, No. 19, 30–40.

Smith, W. K. and King, J. B. (1970). *J. Fire Flam.* **1**, 272–288.

Tesoro, G. C. and Meiser, C. H., Jr. (1970). *Textile Res. J.* **40**, 430–436.

Tesoro, G. C. and Rivlin, J. (1971). *Textile Chem. Col.* **3**, 156–160.

U.S. Department of Commerce Standards for the Flammability of Children's Sleepwear, DOC FF 3–71 (as amended) and DOC FF 5–73. See: *Textile Chem. Col.* **4**, 71–76 (September 1972) and **6**, 130–131 (June 1974).

Welker, J. R. (1970). *J. Fire Flam.* **1**, 12–29.

Subject Index